8051
MICROCONTROLLERS
MCS 51 FAMILY AND ITS VARIANTS

SATISH SHAH

Professor
Department of Electrical Engineering
The Maharaja Sayajirao University of Baroda
Vadodara
Gujarat

OXFORD
UNIVERSITY PRESS

8051 MICROCONTROLLERS

MCS 51 FAMILY AND ITS VARIANTS

SATISH SHAH

Professor
Department of Electrical Engineering,
The Maharaja Sayajirao University of Baroda,
Vadodara
Gujarat

OXFORD
UNIVERSITY PRESS

Dedicated to my students, colleagues, and family members

OXFORD
UNIVERSITY PRESS

Oxford University Press is a department of the University of Oxford.
It furthers the University's objective of excellence in research, scholarship,
and education by publishing worldwide. Oxford is a registered trade mark of
Oxford University Press in the UK and in certain other countries.

Published in India by
Oxford University Press
22 Workspace, 2nd Floor, 1/22 Asaf Ali Road, New Delhi 110002

ISBN-13: 978-0-19-806357-5

Printed in India by Repro India Limited

Cover image: © XXXXXXXX/Shutterstock

For product information and current price, please visit www.india.oup.com

Third-party website addresses mentioned in this book are provided
by Oxford University Press in good faith and for information only.
Oxford University Press disclaims any responsibility for the material contained therein.

Preface

A microcontroller is a small computer on a single integrated circuit optimized to control electronic devices. In contrast to a general-purpose microprocessor, which is used in a personal computer, microcontrollers are designed for controlling specific tasks.

Microcontrollers are employed in process control systems in order to monitor and control the process parameters or variables in industrial controllers. Both analog and digital controllers are used in process industries. The extensive use of microcontrollers in different industrial applications has motivated manufacturers to develop different varieties of microcontrollers. The varieties being different in terms of the number of bits of operation, internal and external memory capacity, the number of timers present in the architecture, the number of interrupt facilities and incorporation of some special features such as timers, event counters, and watchdog.

Different types of microcontrollers, their architecture, and special features are taught at the undergraduate level to engineering students. Additionally, most projects to be done by students at the undergraduate and postgraduate levels are based on the design and applications of microcontrollers. Given this, an understanding of the basics of this important subject is essential for future engineers.

About the Book

This text has been designed keeping in mind the requirements of students taking a course on microcontrollers as part of their B Tech or BE curriculum. This textbook covers the fundamentals of microcontrollers such as the architecture of the microcontroller chip, instruction set, commonly used I/O, and interfacing concepts.

The first seven chapters of the book focus on the theoretical concepts of the subject. Software development tools and concepts of assembly language programming are explained with examples. All chapters contain review questions and programming exercises (wherever applicable) for practice. The last seven chapters of the book discuss the applications of microcontrollers which will help the students in their year-end projects. A wide range of design applications are used to introduce most of the concepts required in digital system design with microcontrollers. The most popular 8-bit microcontrollers used in the Indian industries for designing intelligent digital instruments as well as controllers for a set of varied applications have been clearly explained and illustrated in the book.

The highlight of the text is the detailed exposure to product development using 8-bit microcontrollers MCS-51 family and their variants. The necessary background for application is also described to help in the development of hardware system and software design. The projects are designed with the help of components available in the Indian market and the associated software and hardware are also detailed.

Contents and Coverage

The book is organized in the form of 15 chapters and 2 appendices.

Chapter 1 discusses the basic architecture of a digital system employing three-bus architecture. Data transfer schemes and their implementation are also discussed.

Chapter 2 provides information regarding microcontroller families and derivatives such as pin details, hardware details, and enhancements. The PIC and flash controllers have also been discussed.

Chapter 3 introduces assembly language programming, software support tools, instruction set, and use of high-level language such as C for software development. A large number of examples are used to introduce basic techniques and concepts.

Chapter 4 presents advance programming techniques such as code converters, floating point arithmetic, software delay generation, techniques of processing arrays, handling subroutines and interrupts with the help of examples. Algorithms have been explained and codes have been depicted with appropriate comments. Users can develop and test their own programs by downloading a software development environment from the Online Resource Centre.

Chapter 5 deals with commonly used input-output devices used with microcontrollers. Memory elements such as RAM, ROM, and EPROM and their interfacing are discussed. Typical prototype boards with associated software are also developed in the chapter which may be used for digital or analog applications.

Chapter 6 focusses on programmable peripheral ICs such as Intel 8255, 8279, 8254, RTC, and 8254 with chip architectures details, programming and interfacing with microcontroller for I/O devices such as keyboards, and 7-segment LED and LCD using parallel data transfer schemes. Small program segments have also been developed for operation of prototype.

Chapter 7 presents common standards for serial communication. Architecture, programming, and interfacing of USART are explained with illustrative examples and small program codes. Interfacing of commonly used serial devices such as MODEMs and telephones using DTMF decoders is discussed. A simple application of using telephone for remote control operation is developed.

Chapter 8 discusses modes of operations of peripherals such as ports and timers on microcontrollers for serial as well as parallel communication. Single chip solutions are developed using on-chip ports and timers. It also includes IC simulator, IC tester, and the timers. Implementing software USART and handling the multiprocessor systems are also described.

Chapter 9 describes implementation of a three-phase energymeter using ASIC and ATMEGA128 microcontroller. It also describes another software development tool AVR Studio. Basics of prepaid energymeters employing IT are also discussed. Additionally, the design and development of a prototype hardware with an associated software is also explained.

Chapter 10 introduces basic concepts of digital communication. A new technique used for communication called power line carrier communication (PLCC) technology and home networking using power line are discussed. Protocols used for PLCC and their implementation using microcontroller are explained. A prototype hardware and operating software is also designed.

Chapter 11 describes the design of a microcontroller-based welding machine controller. A small animation is available in the Online Resource Centre to understand the operation of the system. Starting with a basic block diagram, the development of a prototype hardware setup is explained. The necessary software for operation is developed. Special feature of this chapter is the program module which explains how to handle operations associated with EEPROM.

Chapter 12 considers enhancement of telephone services. It describes building of a multi-phonemeter and auto-conferencing facilities with the help of a microcontroller.

Chapter 13 explains the basics of IR communication system and protocols. Design of a generalized multifunction digital controller using IR transmitter receiver is explained. Prototype microcontroller-based hardware setup and testing and troubleshooting procedure are discussed. Application software using function modules is described.

Chapter 14 presents the setup for RF communication system and its terminology. The fundamental concept of an RF system and antenna design are explained. Implementation of an RF-based intercontroller communication using microcontroller is discussed with the help of a hardware setup and algorithms for operation of a base and remote stations.

Chapter 15 explains the concepts of SMS and AT commands used to change the mode of mobile operation. An application explains how the keypad of the mobile can be used for monitoring temperature at the remote terminal.

The book also contains appendices on functional grouping of instructions and details of bytes and oscillator periods for different mnemonics.

Online Resources

The Online Resource Centre of the book contains a local web-tutor providing details of architecture, pin configuration, block diagram, and instruction set through a specially designed graphical user interface (GUI). It includes examples of various useful peripheral IC chips for complex system design such as 8255, 8253, 8279, RTC, and LCD. It also includes a specially developed soft-laboratory comprising software development and testing tools such as assemblers, C-cross compiler, downloader, simulator/debugger, and I/O drivers in the form of IDE f ready reference. Programming techniques are also explained via examples with detailed algorithms and codes. Presentation slides are also provided for better understanding of microcontroller hardware, software, and interfacing.

Acknowledgements

An effort to develop a text of this kind is a group activity and calls for generous help and support from a lot of people. I take this opportunity to express my gratitude to all without whom this text would have been a distant dream.

My students encouraged me to develop this book and it is my pleasure to acknowledge the help I received from my students leading to the successful completion of the projects, a few of which are used in Chapters 9 to 15 of the book. I would also like to acknowledge the enormous help extended to me by Prof. Tejas Chanpura (Government Polytechnic, Gandhinagar), Mr Urmil Shah (Bloomberg, NY, USA) and Prof. K R Bhatt (SVIT, Vasad).

I thank my students Mandar, Nisarg, and Dewang Sheth and his group for their help in developing and organizing material for the local web tutor included in the Online Resource Centre of the book. I am grateful to Mr Nitin Paranjape (MD, Edutech Systems, Baroda) for sparing the software tools for this project and providing me permission to use the same.

I express my sincere gratitude and thanks to the entire editorial team of Oxford University Press for publishing the book and helping me overcome all hurdles and difficulties faced during the implementation of the project.

I thank my wife for her patience and constant encouragement which made the completion of this project a reality.

Finally, I thank all who have helped me directly or indirectly during the project.

Satish Shah

Contents

Chapter 1

Evolution of Microcontrollers

Microcontrollers are used in process control systems in order to monitor and control the process parameters or variables. It is necessary to maintain parameter values through a scheduled procedure. The parameters are required to be tuned to achieve optimal control by matching responses such as proportional, integral, and derivative responses of the process dynamics. In process industries both analog as well as digital controllers are used.

Analog controllers use linear circuit components such as operational amplifiers. The controllers which employ digital or logical circuit components are called *digital controllers*.

Digital controllers are configurable. They can carry out control functions and have diagnostic features. There is a built-in data security. The digital controllers introduce additional capabilities, so as to increase the overall efficiency of plant at a reduced cost. The advantages of digital controllers over analog controllers are:

1. In analog controllers, considerable time is wasted in terms of man-hours, prior to purchase, to ensure that the controllers are complete and accurate for application.
2. Digital controllers have different algorithms implemented for different applications.
3. Maintenance procedures are more rigorous in analog controllers, while they are simple in digital controllers.
4. Digital controllers have built-in signal conditioning circuits. Hence, additional hardware is not required.
5. In case of controllers, the display can be in a variety of forms, e.g., bar charts, alphanumeric, etc. The keys on the panel can be used to select the mode of operation, while the display unit may be used for current mode display.

Some of the additional features provided by digital controllers are:

1. COMP: To select if a computer is to be connected or not.
2. AUTO: To switch between different modes of operations such as: Feed Forward, Track, Manual, and Cascaded.
3. REMOTE_CALL: To select the set point.
4. VAR SELECT: To select which variable has to be displayed.
5. ALARM: For error condition.
6. SCALE_LIMIT: Change scales.

Since digital controller is a digital system, we will study the basic components of a typical digital system, its architecture and organization. The development of microcontroller-based system along with different data transfer schemes is described in this chapter. The basic building blocks of X51 family of microcontrollers, their characteristics and their enhancements in case of high speed microcontroller (HSM) will be discussed.

1.1 DIGITAL SYSTEM ORGANIZATION

A digital system can be constructed using integrated chips or integrated circuits (ICs). ICs can perform arithmetic, logical, or control operations necessary for a processing unit in digital systems. A distinguishing characteristic of the processing unit is that it processes several bits grouped together as a word. This data word might contain 4, 8, 16 or more bits.

The system blocks associated with data flow and processing fall into three general categories:

Processor Processors manipulate and move data. They often require a small amount of memory to hold intermediate results of their manipulations. Control operations are considered to be part of the processing unit. The controller has more interaction with the processor than with other subsystems.

Memory Data memory holds operands and intermediate and final results. Operands are variables used by the processor to calculate the results stored in the memory.

Input/Output (I/O) A digital system must input operands from outside sources and output the results of its calculations. I/O subsystems provide connections or interfaces with people or machines that are not part of the digital system.

The above three subsystems are combined together in a generalized architecture. They can be interconnected in a large variety of ways.

The processor controls the operation of the entire system via control program stored in its read only memory abbreviated as ROM. It is essential to be able to change the ROM program to perform different functions for different applications. Each line of ROM contents specifies the control signals of the system in coded form. A hierarchical approach called stored-program control has been developed.

Stored-program control   In a stored-program system, the ROM within the processor contains a standardized program used for all applications. The standard program in ROM sequences all the control signals to perform the specified operation. Data movement, arithmetic, Boolean functions, and data shifting are commonly implemented operations. The programmer's instructions stored in the memory subsystem are responsible for a complex sequence of actions to be performed. The system can be applied to solve many problems by changing the instruction sequence. Microcomputers are advantageous as they reduce the development cost of complex digital systems.

Microcomputer operation   A microcomputer is a computer that uses a microprocessor as its CPU. A microprocessor requires external memory to execute programs and it cannot interact directly with the I/O devices. To add memory components and I/O devices to a microprocessor-based design, a *bus* is employed. *A bus is defined as a collection of electronic signals and signal paths that are grouped together according to the function performed by them.* It can be unidirectional or bidirectional depending on whether the information flows in one direction or both directions. Hence, it can be said that the signals in a specified bus have a common function.

A memory subsystem is connected with a single processor. The addresses are sent to the memory from the processor via address bus. So the address bus is unidirectional.

Data may be sent from the processor to the memory or from the memory to the processor. The data signal thus flows in both directions on data bus, which requires that it must be bidirectional.

Instructions are also transferred on the data bus from the memory to the processor. This simple processor system consists of two major parts: the accumulator and the arithmetic logic unit (ALU), which manipulate data transferred between processor and memory via the data bus. The rest of the processor sequences the operation of the instructions stored in the memory system's ROM.

The program counter register (PC) provides the address of the instruction to be performed. The IR or instruction register holds the instruction after it is extracted from memory. The control unit sequences all the control signals needed to perform these operations and those signals needed to perform the data operations specified by the instructions in the IR. The control unit also provides memory control signals. An I/O subsystem may be connected with I/O control signals from the processor's control unit.

1.1.1 Microprocessor-based System

A digital computer has a control unit consisting of logic units such as gates, counters, flip-flops interconnected to obtain combinational or sequential functions. It is possible to have a control unit which has functional elements and memory on a single chip. Such a unit generates instructions to execute user program instructions. This makes the above unit a flexible one in the sense that by changing few instructions, it is possible to change operation of the unit, i.e., same unit can carry out other sets of operation without changing the hardware. Such a unit is called a *processor*. The processor on a microchip is generally termed as microprocessor. Hence, microprocessor can be defined as, *a functional unit, which can locate and fetch instruction from memory and execute it*. By sending a series of such instructions or operational commands to microprocessor, it is possible to carry out any activity. The sequence of instructions is called a *program*. Thus, microprocessor is a control element which makes a system behave in the way the instruction sequence in the stored program dictates.

1.1.2 Three-bus Structure

Normally, a digital action in a microprocessor-based system can be described by a three-bus structure. A microprocessor-based system divides any instruction to be executed into certain basic activities. These basic activities are accomplished by generating one or more hardware actions, such as:

1. Data transfer between memory and processor
2. Data transfer between I/O and processor
3. Manipulation of internal registers
4. Direct memory access

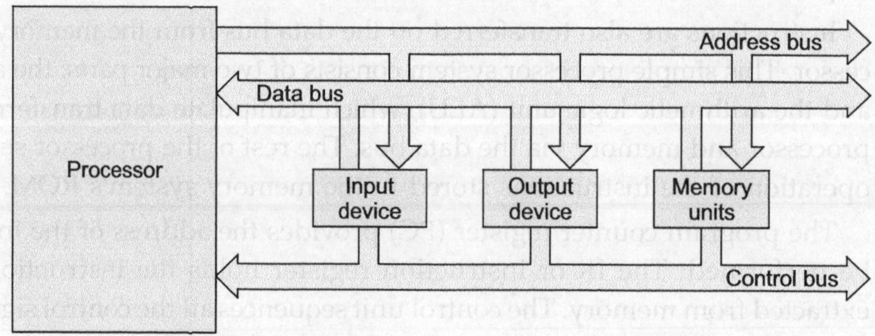

Fig. 1.1   Three-bus structure

The three-bus structure (Fig. 1.1) contains three buses: data bus, address bus, and the control bus. They are named based on the signals they carry.

Data bus   It provides a physical path for communication of data to be transferred. The signal is required to flow in both directions, hence it is bidirectional.

Address bus   It outputs the address to enable correct path for communication. The information is given by the processor, hence address bus is unidirectional.

Control bus   It provides signals to start and terminate the specified operation.

Bus Loading Effects

Since all devices are connected to a common bus, the selected device will put an external load on the bus. The loading from each interfacing device must be considered for proper operation of the system. Hence, it is necessary to calculate the load added by the hardware to the existing system. In fact, to establish reliable electrical path for transfer, problem of loading of address bus as well as data bus must be considered.

Cables, edge connectors, and additional peripherals connected to the processor may cause loading effect. This can be removed by installing buffers or current amplifiers.

In addition to loading, it is necessary for data bus that it must be capable of transferring data signals in both directions. This can be implemented by using bidirectional data buffers.

1.2 | INTERFACING OTHER LOGIC FAMILIES

The output of the processor and the peripherals may have a mismatch due to difference in the logic families. Different logic families have different electrical characteristics; they cannot be interfaced directly.

1.2.1 Interfacing TTL, CMOS Families

The problem with CMOS logic family is that there is a disparity between TTL and CMOS regarding output drive current and input current requirement. The response of CMOS family is slower as compared to TTL. The speed is considered for the simple reason that a processor output is TTL compatible. A strobe signal applied to CMOS interface may be delayed by few nanoseconds causing invalid data to be transferred due to strobe extension by delay. The problem can be easily solved by using the following:

1. A 74LS (low-power Schottky) buffer can be used. CMOS is capable of driving LS load, while LS family can drive std TTL load.

2. A special CMOS device CO4049/CO4050 can be used, which is capable of driving TTL loads.

1.2.2 Interfacing TTL, ECL Families

It is known that TTL and ECL use different power supplies and have different output voltages, it is necessary to have a translation for interfacing.

1.3 | MICROCONTROLLER ARCHITECTURES

In the microprocessor-based systems, external components such as I/O ports, timers, and memory elements are to be connected to implement a digital system. Microcontrollers aim at a single-chip solution for the digital system implementation. This section gives a brief overview of microcontroller architectures.

1.3.1 Von Neumann Architecture

Von Neumann architecture has common memory for storing data, programs, and auxiliary or final results.

The memory interface unit is responsible for arbitrating access to the memory space between reading instructions (based upon the current program counter) and passing data back and forth from the processor and its internal registers. The time required to execute a given instruction is used to fetch the next instruction (this is known as *pre-fetching*), so as to speed up the process.

It was better suited to the technology of the time. Using one memory was preferable because of the unreliability of current electronics, as it requires a single interface for memory.

The biggest advantage of Von Neumann architecture is that it simplifies the microcontroller chip design because only one memory is accessed. Hence, contents of random access memory (RAM) can be used for both variable data storage as well as program instruction storage. An advantage for some applications is the availability of program counter and stack contents. This helps for development of programs for real-time operating systems.

1.3.2 Harvard Architecture

Harvard's architecture uses separate memory banks for program storage, the processor stack, and the variable RAM. Harvard architecture was largely ignored until microcontroller manufacturers realized that the architecture had advantages for the devices they were then designing.

The Harvard architecture supports parallel operations. Because of parallel processing, execution of instructions is faster. Parallel operations means fetching of the next instruction and execution of the current instruction are done simultaneously. Executing this instruction in the Harvard architecture also takes place over two instructions, but the instruction read takes place while the previous instruction is being carried out. This allows the instruction to be executed in one instruction cycle.

1.3.3 CISC and RISC Architectures

CISC stands for complex instruction set computers. In CISC processors; there are large numbers of instructions, each carrying out different permutation of the same operation with instructions perceived to be useful by the processors designer.

RISC stands for reduced instruction set computers. In RISC processors the instructions are as bare a minimum as possible to allow the users to design their operations, rather than let the designer do it for them. Orthogonality or symmetry of the processor allows some operations to be unexpectedly powerful and flexible, e.g., conditional jumping.

In CISC system, a conditional jump is usually based on status register bits, whereas in RISC system, a conditional jump can be based on a bit anywhere in memory. By careful design of the processor's architecture, the flexibility can be increased so that a very small instruction set may be executed in very few instruction cycles. This provides implementation of extremely complex functions in the most efficient manner.

The 8051 microcontroller combines many of the features of RISC and the large instruction set of a CISC processor. This combination makes the creation of complex program functions surprisingly easy in many cases. This helps in development of a program which is fast and requires very little programming effort and memory space.

1.4 | DATA TRANSFER SCHEMES

The most important operation in any system is to transfer data between the peripherals and the controller or CPU. This can be carried out in different ways depending upon the way in which the I/O devices such as keyboard, displays, relays, etc. and controllers are connected with each other.

Each method has its own advantages and limitations. We will study three popular configurations used in designing systems with three-bus architecture such as:

1. Programmed I/O
2. Interrupt driven I/O
3. Direct memory access

1.4.1 Programmed I/O

It is known that the data transfer between the device and the microcontroller can be done, by using either parallel or serial data transfer.

Since the microcontroller is a parallel device, operations on all the 8-bits are done simultaneously. Programmed I/O is a scheme of data transfer in which each device is assigned a unique address. This is called a *port address*.

Port *A place from which actual data transfer takes place is called a port. Normally in a system there are two types of ports, input and output.*

The port from which data transfer direction is such that data is moved from device to the controller is called an *input port*.

The port from which data transfer direction is such that data is moved from controller to the device is called an *output port*.

The operational steps that are carried out in order to establish data transfer are as follows:

The microcontroller places the address on the address bus, sends a control signal, enables the device, and then the data transfer takes place.

Address Decoding

In order to impart or generate an address for the port, two most popular techniques are:

1. Memory-mapped I/O
2. I/O-mapped I/O

In the former method all the 16 address lines are used to generate unique port address, while only lower order address bus is used in case of I/O-mapped I/O scheme. In case of microprocessors, to reduce the hardware of the address decoding circuits, I/O-mapped I/O scheme is used. In the instruction set for microprocessors, there are special instructions called I/O group of instructions to support I/O-mapped I/O scheme. However, microcontrollers do not have specific I/O instruction, the former scheme is used.

In memory-mapped I/O, an I/O device is considered as a memory location in the address space. Hence, it can be accessed by normal memory reference instructions. This eliminates the need of special instructions for data transfer from I/O devices.

The advantages of memory-mapped I/O scheme are as follows:

1. Since all the address lines are used, there is no chance of multiple addresses for the device. This relieves system programmer from the fear of multiple device selection.
2. Even though the decoding logic is complex, due to a large number of address lines being used more I/O devices can be connected. Theoretically, it can be as large as (2^{16}).

3. Data transfer can take place between memory and any register or internal memory location.

The task of the address decoder is to decode the address bits and generate address select signals for each device in the system. In effect, it breaks the address space of processor into the blocks. Each address select signal indicates that the address on the address bus is within a predetermined range.

The decoding can be done using any one of the following methods:

1. *Linear decoding*   The method is very simple. It uses one address line for one device for selecting port address. The drawback of this scheme is that there may be multiple addresses for the same device.

2. *Absolute decoding*   It uses all the address lines. But in order to generate a unique port, address for the device gates/decoders/programmable ROM (PROM)/programmable logic arrays (PLA)/programmable array logics (PAL), etc. are used.

Using Address Decoders

The address lines being decoded, decides the range of addresses for which an address select signal is asserted. Figure 1.2 depicts the base diagram of the most commonly used address decoder, the TTL IC74LS138.

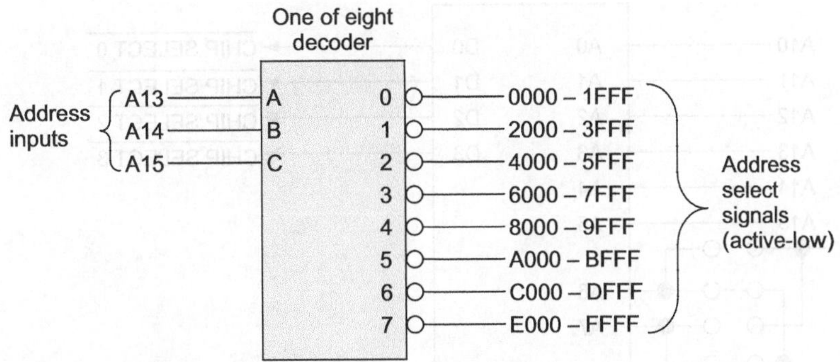

Fig. 1.2   Address decoder

It is a low power schottky version of TTL logic family. It has three enable inputs denoted as:

1. Two active-low (G2A and G2B)
2. One active-high (G1)

In order to function as an address decoder, the IC must be enabled first by applying appropriate signals. This can be done by pulling inputs (G2A and G2B) logically low and (G1) made high. Otherwise all the outputs of decoder will be high.

All the outputs are active-low, so the selected output is low and all other are high. The address decoder has three inputs, called address inputs. The combination of the address inputs will select any one of eight outputs low, while others are kept high.

Using PROM

This technique uses a specially programmed bipolar PROM as the address decoder. PROM is available with either open-collector or three-state outputs. The three-state versions are preferable, since their low-to-high transitions are much faster and output pull-up resistors are not required. Since the PROMs outputs are in the high-impedance state when chip select is negated, pull-up resistors are required even on the output. Commonly used bipolar PROMs are MMI 63S141, AMD 27S21, and 74S287 from the National Semiconductors.

Figure 1.3 shows a typical (256 × 4) PROM used as an address decoder. Address bits A15...A10 select one of 64 locations in the PROM. Address bits [A0 ... A9] from the microcontroller are ignored by the PROM, so each PROM location is accessed for a range of 1 K addresses. Table 1.1 shows an example of PROM programming

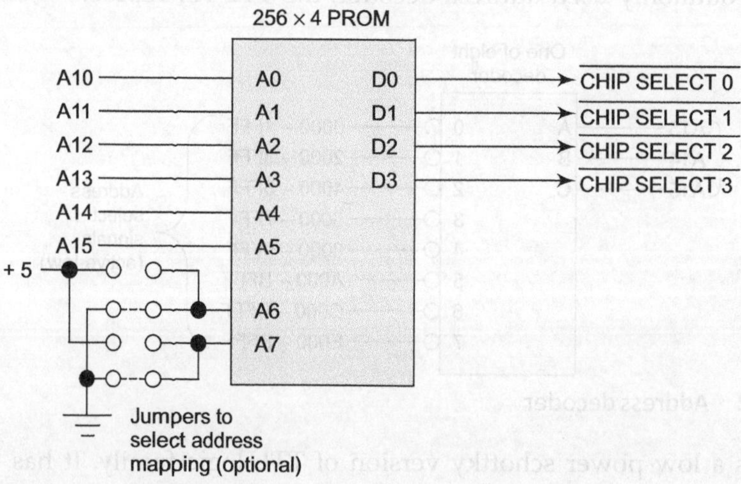

Fig. 1.3 PROM decoder

The first memory location is programmed with 1110; this makes only D_0 output low, when this location is accessed. Since this is the only memory location with a zero in the D_0 bit position, the D_0 output is asserted for a range of 1 K addresses. If some range of addresses are unused the contents of PROM locations may be made all HIGH, so it will not select any device. The range of address can be easily extended by programming more PROM locations.

Table 1.1   PROM decoder with contents

Address	Contents	Address range	Block size
000000	1110	0000-03FF	IK
000001	1111		Unused
000010	1111		Unused

1.4.2 Interrupt Driven I/O

Programmed data transfer is processor initiated, but this requires that processor should remain in a wait loop. In the interrupt driven data transfer, program sequences are altered.

These modes are device initiated, i.e., I/O device forces processor to suspend the normal activity. On being interrupted the processor jumps to an execution of different program segment called *interrupt service subroutine* (ISS). The starting address of ISS is called interrupt branch address (IBA). The processor begins execution of a subroutine stored from the IBA onwards. During this, the processor status is stored on stack. On completion of ISS, processor restores the status. Mostly this is hardware generated. In performing a simple interrupt transfer the following tasks are performed:

1. Interrupt request is generated by the peripheral device.
2. Acknowledge signal (ACK) is issued by the processor.
3. Contents of program counter (PC) are saved and program branches to the address containing first instruction of ISS.
4. Contents of registers are saved and data transfer is executed under software control.
5. Program execution returns to the pre-interrupt program sequence.

If more than one devices request for the interrupt service, it is known as multiple interrupt. Since it is difficult to identify source of interrupt in multiple interrupt, processor follows following steps:

1. Processor does not notice any new (interrupt) request.
2. It serves new routine.
3. It puts the request in a queue and takes only when routine is over.
4. It inquires whether the request is to be served or can wait.

The different sources are allotted priorities based on the speed with which they can handle data in the following order:

1. Power failure
2. Console, keyboard
3. Printer
4. Programming errors

The multiple interrupts can be handled in two different ways:

1. Vectored interrupts
2. Polled interrupts

Vectored Interrupts

This is a simple method used for resolving the multiple interrupts. Each of the interrupt pin on the processor corresponds to a fixed IBA in the processor's memory address space. This technique is called *vectored interrupts*. In this, the processor does not require any identification procedure for the interrupting device and the service required by the device. On getting a request from the device, the processor jumps to the specified IBA and executes the appropriate interrupt service routine, which specifies the operation to be performed by the processor to handle the device request.

In most of the processors the problem of handling multiple interrupts is solved by allocating priorities to the interrupt pins. Normally a priority is fixed for each pin on the processor. In case of multiple requests, the device connected to the interrupt pin, having the highest priority, will be entertained first. This is due to the inbuilt hardware support in the processor interrupt structure.

Polled Interrupts

This method can even be used to expand the interrupt handling capacity of the interrupt pins. This allows more than one device to be connected to the same interrupt pin.

This method makes use of *device polling*. The interrupt locating subroutine, (ILS) resides in the memory and asks each device whether it has interrupted or not. If the device has interrupted, processor will invoke appropriate ISS, execute and return to normal operation, otherwise the processor polls another device.

In the case of more than one device interrupt, priority will be the sequence in which, ILS scans the devices for the interrupt identification. There are two methods of polling the interrupts depending upon, whether the interrupt pin to which more than one device are connected, is a vectored or non-vectored one. They are software polling and hardware polling.

Software polling   The interrupt locating subroutine ILS is implemented using a program code/software. Typical feature of this interrupt is that all the interrupts are combined using an OR gate and output of the gate is connected to the vectored pin of the processor. Polling process begins on getting the device request, with the status flag checking according to the routine written.

Figure 1.4 depicts the polling scheme. The device interrupt status is connected to a data bus in tristate. The enable line of the buffer is connected to one of the output of a decoder and it is ANDed with device select lines. When the processor gets an interrupt request, it sends an appropriate address on bus and one of the bits is transferred generating a control signal for decoder. The output of the decoder is ANDed with strobe input, which READS the signal sent by the processor and is used for enabling output buffer.

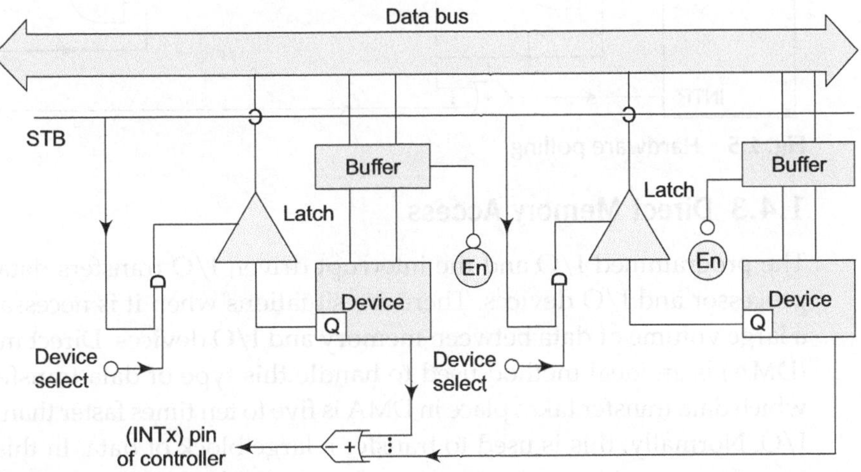

Fig. 1.4   Software polling

After reading the status, it checks whether status is at logic level '1' or not. If it is '1', it jumps to interrupt service routine (ISR) otherwise continues polling. Polling stops after each device has been scanned. For proper operation the interrupt status of the device, which has transferred the data needs to be reset. The reset output may be at the (logic HIGH/LOW) depending upon the software convenience. In later chapters, we will study how to implement this scheme for microcontrollers.

Hardware polling   Figure 1.5 depicts configuration of setup employing hardware polling. It is also known as a *daisy chain*. The priority of the devices connected is determined by the location of the device in the hardware layout. The device nearest to the CPU will have the highest priority.

The basic assumptions in this scheme are as follows:

1. All the devices are connected to same non-vectored interrupt pin.

2. Acknowledge signal is issued in the beginning of next machine cycle.

3. During this interval the processor will receive instructions from the device and not from the memory.

It is possible to use an acknowledge pin, in case of processors, which generates INTA signal. Microcontrollers do not have provision for INTA pin, this technique cannot be used directly. However, it is possible to generate INTA signal and make it available on any one of the port pin of the microcontroller by writing a program.

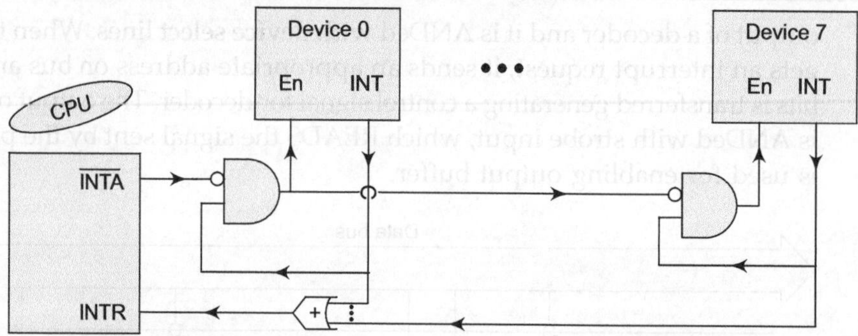

Fig. 1.5   Hardware polling

1.4.3 Direct Memory Access

The programmed I/O and the interrupt driven I/O transfers data between the processor and I/O devices. There are situations when it is necessary to transfer a large volume of data between memory and I/O devices. Direct memory access (DMA) is an ideal method used to handle this type of data transfer. The rate at which data transfer takes place in DMA is five to ten times faster than programmed I/O. Normally, this is used to transfer a large block of data. In this type of data transfer operation, processor transfers control of buses to the DMA so that operation is done without intervention of the processor.

Most of the I/O devices are not capable of handling DMA operation themselves; hence the data transfer using DMA is controlled through an external hardware set up called *DMA controller*. A typical DMA setup is depicted in Fig. 1.6.

The device initiates the data transfer process operation. Device puts a request for DMA operation by setting the request line of controller at logic level HIGH. The controller puts the request to the processor through HOLD pin. The processor acknowledges this by generating, the acknowledgement signal (HLDA). The address bus is tri-stated. This signal is used to disable the buses and they are isolated from the processor. The external controller takes over the control of processor buses.

Control information provided to DMA controller specifies whether it is the memory read or writes operations. This helps controller to set direction of data transfer. The address register is loaded with starting address in memory from which data transfer has to take place. The address register is incremented by one every time a data byte is transferred. When transfer is over, controller will issue command to reset the flip-flops, which puts READY pin of the processor at logic HIGH. During DMA operation processor polls this READY pin, processor regains the control of buses, when the pin is at logic level '1', the peripheral may accept or transmit the data in a serial or a parallel form.

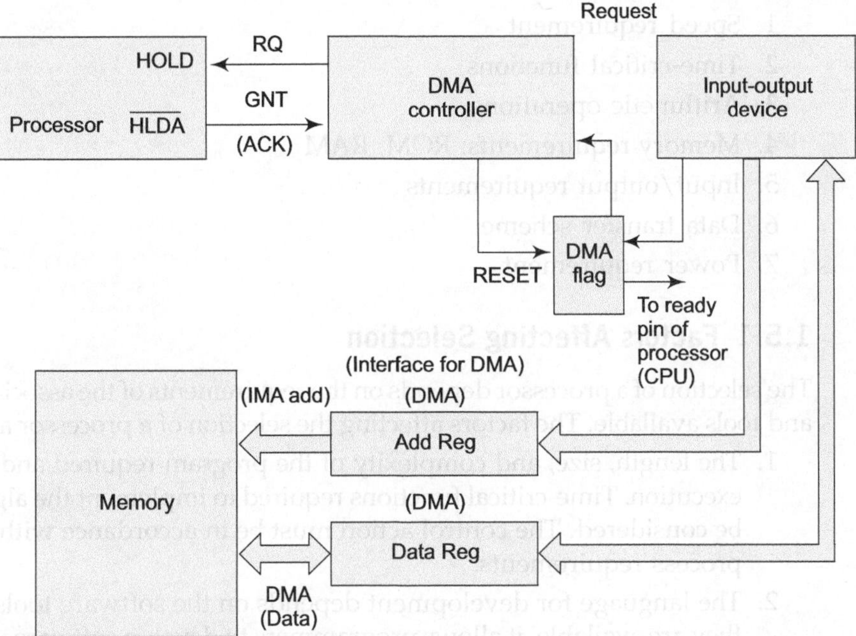

Fig. 1.6   Direct memory access

DMA is not supported by microcontroller X51 family. The technique is described for the completeness.

1.5 | MICROCONTROLLER-BASED SYSTEMS

A microprocessor-based digital controller system requires interfacing of additional external components such as I/O ports, timers, and memory elements to implement a digital system. A single chip, which has most of the required additional components on it leads to a simple digital controller for many applications. Such a chip is termed as a microcontroller. Single-chip microcontrollers are commonly used in control applications. However, for more complex control applications requiring large amounts of I/O, memory or high-speed processing additional peripheral chips can be added.

The selection of a particular microcontroller must be done with the application in mind. It is also important to determine which characteristics of the application are the most important. In order to design a microcontroller-based system various aspects of the application influences the selection.

It is necessary to properly define the project, before deciding the microcontroller to be used. The product should be thoroughly thought out and defined. The user interface should be fully defined, considering following characteristics:

1. Speed requirement
2. Time-critical functions
3. Arithmetic operations
4. Memory requirements: ROM, RAM
5. Input/output requirements
6. Data transfer scheme
7. Power requirement

1.5.1 Factors Affecting Selection

The selection of a processor depends on the requirements of the associated software and tools available. The factors affecting the selection of a processor are as follows:

1. The length, size, and complexity of the program required and the speed of execution. Time-critical functions required to implement the algorithm must be considered. The control action must be in accordance with the physical process requirements.

2. The language for development depends on the software tools available. If they are available, it allows programmers to develop software more quickly. However, programs requiring frequent changes, e.g., where peripheral chips are changed, the software must be modified accordingly.

3. The stack, interrupt vectors, and other special memory areas may be modified accordingly.

4. The amount of memory needed has two aspects: program memory requirements and data memory requirements. A system that stores large amounts of temporary data, such as characters, writing to be printed or transmitted, may require much more data memory. If the device is to be used in its single-chip mode, the program must fit in the on-chip ROM. The on-chip RAM must be adequate for data storage.

5. The expected lifetime of a product, the degree of change and the enhancement anticipated is important.

6. Portable applications require the system to have very low power consumption. Most CMOS microcontrollers include power saving halt modes. These modes are initiated under program control and terminated by an interrupt or by an internal timer reaching its count limit.

7. Power consumption of CMOS devices is also a function of clock speed.

8. The execution rate is a critical factor.

What Next?

Having gone through the preliminaries of microcontrollers and their features in general, we will consider a specific microcontroller family in the subsequent chapters and study the design trends and techniques of designing digital controllers for different industrial applications.

EXERCISES

1.1 What are the three blocks of a processing unit? Explain functions of each block.

1.2 What is a stored program control? What are the advantages of it?

1.3 What is the need of control signals?

1.4 Explain the operation of a microcontroller system.

1.5 What is another name for an architecture?

1.6 State merits and demerits of a three-bus architecture.

1.7 What do you understand by bus loading? How they can be eliminated?

1.8 Why is a combination of CISC and RISC preferred in microcontrollers?

1.9 Explain different data transfer schemes used in a digital system.

1.10 Why is DMA not used in microcontroller-based systems?

1.11 What is a programmed I/O? State the advantages of this scheme over interrupt driven I/O?

1.12 Define the terms: PORT and PORT address.

1.13 What does address decoding mean? What is its need?

1.14 Explain the address decoding schemes used in design of microcontroller-based system?

1.15 What do you mean by an interrupt?

1.16 State the advantages of interrupt driven data transfer scheme.

1.17 Define multiple interrupts. Explain various methods used to handle multiple interrupts.

1.18 Define polled interrupts.

1.19 State the characteristics to be considered for designing a microcontroller-based system.

1.20 How can the software for the microcontroller-based system be developed? State important factors affecting the choice.

Chapter 2

Introduction to Microcontroller Families

Microprocessors and microcontrollers are making a great impact on various activities in the daily functioning of both the developed and the developing industries. The microcontroller is the outcome of a trend towards smaller systems, which are faster, cheaper, accurate, and more reliable.

With the advancement in the semiconductor technology, it became possible to fabricate the whole CPU of a digital computer on a single chip, which is called the *microprocessor*.

A CPU with built-in I/O ports, timers, serial communication control and even ADC is called a *microcontroller*. The most popular array of first microcontrollers was INTEL 4048. Later versions of microcontrollers are 8051, 8751, and 8096. With on-chip RAM, ROM/EPROM, timer, serial port, I/O port, and powerful Boolean processing capabilities, microcontrollers are now more suitable for industrial control applications. This chapter provides a brief overview of the members of microcontroller families.

MICROPROCESSORS

Microprocessors can be defined as the CPU on one/more chips having capabilities to fetch the instruction code from the memory, decode, and execute it. The microprocessors have:

1. Many operational codes for money data from external memory to CPU.
2. One/two types of bit handling instructions.

The aim is to have rapid movement of data or code from memory to chip.

MICROCONTROLLERS

Microcontrollers do not have many external components connected to them. The aim is to give more processing power at relatively low assembly time requirement or provide a single-chip solution. They have

1. One/two types of data transfer instructions.
2. Many types of bit handling instructions.

Microcontrollers can function as a complete computer, with no additional digital parts.

2.1 | PACKAGING INFORMATION

Microcontrollers are connected to other digital integrated circuits in the system. In order to impart connection feasibility and standardization, they are made available in standard packages. The selection of the package depends on the other components in the system. The standard packages can be classified as

Through Hole Mounted Packages

DIP	:	Plastic dual-inline pack
SDIP	:	Plastic shrink dual-inline pack
HDIP	:	Plastic heat dissipating dual-inline pack
DBS	:	Plastic dual-inline bent from a single-inline package
SIL	:	Plastic single-inline pack

Surface Mount Packages

SO	:	Plastic small-outline package
SSOP	:	Plastic shrinks small-outline package
TSSOP	:	Plastic thin shrinks small-outline package
VSO	:	Plastic very small-outline package
QFP	:	Plastic quad flat package
LQFP	:	Plastic low profile quad flat package
SQFP	:	Plastic shrink quad flat package
TQFP	:	Plastic thin quad flat package
PLCC	:	Plastic leaded chip carrier

This information helps to select the type of soldering method to be used. Wave soldering is used when both types of components are mixed on printed circuit board (PCB), however, for thickly populated PCB the wire flow soldering is used. Microcontrollers are available in the 20, 40, 42, 44, 48, 52, 68, 84, and 100 pin packages.

2.2 | MICROCONTROLLER FAMILIES

There are many families of microcontrollers developed by Intel and others.

8048 Family

8035 It is a 40-pin DIP package with 8-bit microcontroller + ROM less + 64-byte data memory + 27 I/O lines + timer + 2 interrupts.

8048 It is a 40-pin DIP package with 8-bit microcontroller + 1 K masked ROM + 64-byte data memory + 27 I/O lines + timer + 2 interrupts.

8748 It is a 40-pin DIP package with 8-bit microcontroller + 1 K EPROM + 64-byte data memory + 27 I/O lines + timer + 2 interrupts.

8039 It is a 40-pin DIP package with 8-bit microcontroller + ROM less + 128-byte data memory + 27 I/O lines + timer + 2 interrupts.

8048 It is a 40-pin DIP package with 8-bit microcontroller + 2 K masked ROM + 128-byte data memory + 27 I/O lines + timer + 2 interrupts.

8749 It is a 40-pin DIP package with 8-bit microcontroller + 2 K masked EPROM + 64-byte data memory + 27 I/O lines + timer, 2 interrupts.

8022 It is a 40-pin DIP package with 8-bit microcontroller + 2 K masked ROM + 64-byte data memory + 27 I/O lines + timer + interrupt + 2 channel ADC.

8021 It is a 28-pin DIP package with 8-bit microcontroller + 1 K masked ROM + 64-byte data memory + 27 I/O lines + timer + no interrupt.

Intel 8051 Family

8031 It is a 40-pin DIP package with 8-bit microcontroller + ROM less + 128-byte data memory + 32 I/O lines + Two 16-bit timers + 5 interrupts + Universal asynchronous receiver transmitter (UART) + supports 64 K program memory + 64 K data memory.

8051 It is 40-pin DIP package with 8-bit microcontroller + 4 K masked ROM + 128-byte data memory + 32 I/O lines + Two 16-bit timers + 5 interrupts + UART + supports 64 K program memory + 64 K data memory.

8751 It is a 40-pin DIP package with 8-bit microcontroller + 4 K EPROM + 128-byte data memory + 32 I/O lines + Two 16-bit timers + 5 interrupts + UART + supports 64 K program memory + 64 K data memory.

8032 It is a 40-pin DIP package with 8-bit microcontroller + ROM less + 128-byte data memory + 32 I/O lines + three 16-bit timers + 6 interrupts + UART + supports 64 K program memory + 64 K data memory.

8052 It is a 40-pin DIP package with 8-bit microcontroller + 8 K masked ROM + 128-byte data memory + 32 I/O lines + Two 16-bit timers + 5 interrupts + UART + Supports external 56 K program memory and 64 K data memory.

8752   It is a 40-pin DIP package with 8-bit microcontroller + 8 K masked EPROM + 128-byte data memory + 32 I/O lines + Two 16-bit timers + 5 interrupts + UART + Supports external 56 K program memory and 64 K data memory.

80552   It is a 68-pin PLCC package with 8-bit microcontroller + 8 K masked ROM + 128-byte data memory + 40 I/O lines + Two 16-bit timers + 5 interrupts + UART + Supports external 56 K program memory and 64 K data memory + 10-bit ADC.

It is manufactured by M/s SIEMENS.

Intel 8096 Family

8095   It is a 48-pin DIP package with 16-bit microcontroller + ROM less + 232-byte data memory + 40 I/O lines + Two 16-bit timers + HSIO + 21 interrupts + UART + Supports external 56 K program memory + ADC.

8096   It is a 48-pin DIP package with 16-bit microcontroller + ROM less + 232-byte data memory + 40 I/O lines + Two 16-bit timers + HSIO + 21 interrupts + UART + Supports external 56 K program memory + no ADC.

8097   It is a 48-pin DIP package with 16-bit microcontroller + ROM less + 232-byte data memory + 40 I/O lines + Two 16-bit timers + HSIO + interrupts + UART + Supports external 56 K program memory + ADC.

8395   It is a 48-pin DIP package with 16-bit microcontroller + 8 K masked ROM + 232-byte data memory + 40 I/O lines + Two 16-bit timers + HSIO + 21 interrupts + UART + Supports an external 56 K program memory + ADC.

8396   It is a 48-pin DIP package with 16-bit microcontroller + ROM less + 232-byte data memory + 40 I/O lines + Two 16-bit timers + High speed I/O (HSIO) + 21 interrupts + UART + Supports external 56 K program memory + no ADC.

8397   It is a 48-pin DIP package with 16-bit microcontroller + ROM less + 232-byte data memory + 40 I/O lines + Two 16-bit timers + HSIO + 21 interrupts + UART + Supports external 56 K program memory + no ADC.

8795   It is a 48-pin DIP package with 16-bit microcontroller + 8 K EPROM + 232-byte data memory + 40 I/O lines + Two 16-bit timers + HSIO + 21 interrupts + UART + Supports external 56 K program memory + ADC.

8796   It is a 48-pin DIP package with 16-bit microcontroller + 8 K EPROM + 232-byte data memory + 40 I/O lines + Two 16-bit timers + HSIO + 21 interrupts + UART + Supports external 56 K program memory + no ADC.

8797   It is a 68-pin PLCC package with 16-bit microcontroller + 8 K EPROM + 232-byte data memory + 40 I/O lines + Two 16-bit timers + HSIO + 21 interrupts + UART + Supports external 56 K program memory + no ADC.

8098   It is similar to 8096 family, except that it has 8-bit external data path.

2.3 | THE PRECEDERS

The aim of the development of microprocessor-based system was to reduce the component count so that the size of the digital instrument using it can be reduced.

MCS-48 and UPI-41 were the first families of microcontrollers, which were developed by Intel. Each device comprised of CPU, I/O ports, and timer. Typically, the former family was used as primary controllers for stand alone systems while the latter one found applications in the field of designing programmable and intelligent peripherals. The MCS-48 is available in six functionally similar versions, the 8048 and 8049 microcomputers with read-only (ROM) program memory, the 8748 microcomputer with erasable and programmable ROM (EPROM), the 8035 and 8039 microcomputers which use no resident program memory, and the 8021 microcomputer, the lowest cost component in the MCS-48 family.

UPI-41 is based on the 8041/80414A microcomputers (with ROM) or the 8748/8041A microcomputer (with EPROM). The 8048, 8748, and 8035 are equivalent except for their program memories (ROM/EPROM). The 8035 is used with external program memories in prototype and reproduction system. The 8049 and 8039 are also equivalent, except for program memory and have the same instruction set as that for the 8048 group. Because of the different usage of the external bus, the 8041, 8741, and 8021 have a slightly different instruction set.

2.4 | X51 FAMILY

Microcontrollers are devices used in products that require sophisticated and flexible control. They have RAM, ROM, and I/O, as well as the CPU, integrated onto the same chip with smaller instruction set. The use of microcontrollers can be very cost-effective, in such mass-produced items as smart modems and video cassette recorders.

A good example of an 8-bit microcontroller is the *Intel MCS-51 Family* of devices. This family, typified by the 8051, has been designed mainly for sequential control application; in this chapter we will examine the hardware and software features. However, for a complete and detailed description, the students are advised to read the Intel manuals.

When microcontrollers are used for design of a digital controller for any application, it is desired to have one chip solution. This means that the microcontroller selected has all the features and components on it so that, as far as possible, there is no need to connect the external digital circuits or peripheral chips.

Generally, the processor of a digital system requires additional features such as I/O ports, timers, memory elements, etc. to build up a system. The manufacturers of microcontrollers include these features on the single IC chip, e.g., basic X51 family of microcontrollers.

2.4.1 Basic Building Blocks of X51

Clock

The microcontrollers clock is used to sequence the execution of the instructions. For each instruction, a specific number of *instruction cycles* are required for it to be executed. Providing operating resources to a microcontroller is generally easy.

The 8051 can be connected to a variety of power-clocking and reset sources still run. Most microcontrollers are capable of running from a clock, which has an in-built circuitry to allow a simple connection of a crystal or other hardware, such as ceramic resonators or an external clock source.

The microcontroller must have an internal *oscillator* built in, which allows it to run without any external components other than power.

Each instruction cycle is made up of a number of clock cycles. As we will study later, in the basic architecture of the 8051, each instruction cycle takes 12 clock cycles to execute. Requiring 12 cycles for each instruction is somewhat unusual in the microcontroller world, with some of them capable of running at one clock cycle per instruction.

8051 manufacturers have redesigned the basic 8051-processor core, so that the instruction will run faster (i.e., each instruction cycle run in few clock cycles). The Dallas Semiconductor HSM (high-speed microcontroller) devices, MCS-151 and MCS-251 from Intel and "XA" architectures from Philips Semiconductor, are examples of modifications. These are called 8051-compatible microcontrollers or derivatives. They require less clock cycles per instruction cycle. The instruction execution speed is many times that of a basic 8051. Each of the compatibles runs the 8051 instructions in fewer clock cycles.

Generally, the application is to run on the slowest microcontroller clock speed possible to minimize the power consumed by the microcontroller as the system is powered by a battery. This tends to minimize the power consumed as well as the amount of electrical noise generated by the application.

I/O Pins

A simple I/O capability on the microcontroller is desirable, so that the needs of additional peripheral chips are eliminated. The parallel I/O is implemented in microcontrollers; such that each bit's output driver-enabled operation is controlled by a single bit in the control register.

I/O pins are dedicated to specialized functions that further complicate them. These specialized functions include serial ports, analog I/O, external device buses, etc. Many microcontrollers provide pins with wire AND characteristic output to allow them to be used on buses. Many bus protocols require open collector circuits for correct operation.

The 8051 family is unusual in that all I/O pins are designed as open-drain outputs when used as parallel I/O pins. As we will study, most of the 8051's I/O pins have a weak built-in pull up. Using open drain I/O pins exclusively in the 8051 means that you have to add external pull ups for some applications with sinking current and never source it.

Interrupts

Interrupts may/may not be entertained. They are also called external requests because on acceptance, an interrupt request is acknowledged. Then a special set of events or routines is followed to handle the interrupt. It is also shown in Section 1.4 that the interrupt-driven data transfer is faster in comparison with the programmed data transfer scheme.

These special routines are known as *interrupt handlers* or *interrupt service routines* and are located at a special location in the memory. There are three different ways to respond:

1. *The first method of handling the interrupt is to ignore it.*
2. *Second method is to accept it but tell it to wait.*

The 8051 uses this method. This feature includes a separate vector for each type of interrupt along with space (8-bytes) devoted to each vector to allow a small interrupt handler to be located without having to jump to another location.

3. *The third method is to accept the interrupt and process it as it comes in.*

In 8051 as well as most other controllers vectored techniques are used. Different interrupt sources are given different priorities. A higher priority interrupt is serviced before a lower priority one.

Nesting of interrupts means that interrupts are re-enabled inside an interrupt handler. If another interrupt request codes in while the first interrupt handler is executing, processor acknowledge the new interrupt and jump to its vector. To avoid this, you might want to disable (mask) the interrupts during the handlers that you think should have the highest priority and should never be interrupted. Hardware support is made available on the chip and there are pins on the chip for the application of interrupt signals.

Timers

Timing information is one of the most critical functions required by any digital system. Along with providing real-time information, timers are used in microcontrollers to provide a number of other tasks helpful in the operation of the application. Generally, the information is provided by the manufacturer in the form of a counter or timer on the processor chip. The counter itself is usually 8- or 16-bits wide. Typically, the value in the counter can be read from or written to during operation. When the counter overflows, it generates an interrupt request.

Enhancements are made which include circuitry to measure pulse width, provide repeating interrupts at a specific interval or output a pulse width modulated (PWM) signal. This circuit is often used for driving peripherals that need a constant clock. In the 8051, this circuit is used to drive the asynchronous serial ports data rates.

The basic X51 timers do not support the pulse input measuring or pulse width modulated outputs, but they support a number of operating modes that will allow to implement these features in the applications very easily.

Peripherals

Instead of providing only I/O ports, sometimes peripherals are included on the chip. These peripherals include interfaces to microcontroller network devices to simplify the connections. These interfaces allow a microcontroller to be simply wired to digital devices. Many a times, peripherals are built directly onto the chip.

Microcontrollers with enhanced features might appear to be costlier, in comparison with basic microcontrollers. They are advantageous in terms of the performance.

2.4.2 8051 Hardware Overview

There are three basic members of the MCS-51. They are 8031, 8051, and 8751. The 8031 has no on-chip ROM and uses external memory for program storage. The 8051 contains 4 K of ROM, 128-bytes of RAM, 32 I/O lines, two 16-bit counter/timers, five interrupt sources (two external), a duplex serial port and a bit level Boolean processor. The 8751 has on-chip EPROM and is relatively expensive.

The three newer devices 8052, 8032, and 8752 are expanded versions with 8 K of ROM, 256-bytes of RAM, and three timers. The low-power CMOS versions designated 80C51, 80C31, and 87C51. They are available in a 40-pin DIP package.

2.4.3 Architecture

Figures 2.1(a) and (b) depict the block diagram of the X51 chip and pin configuration, respectively. It contains special function registers, accumulator, timing, and control units.

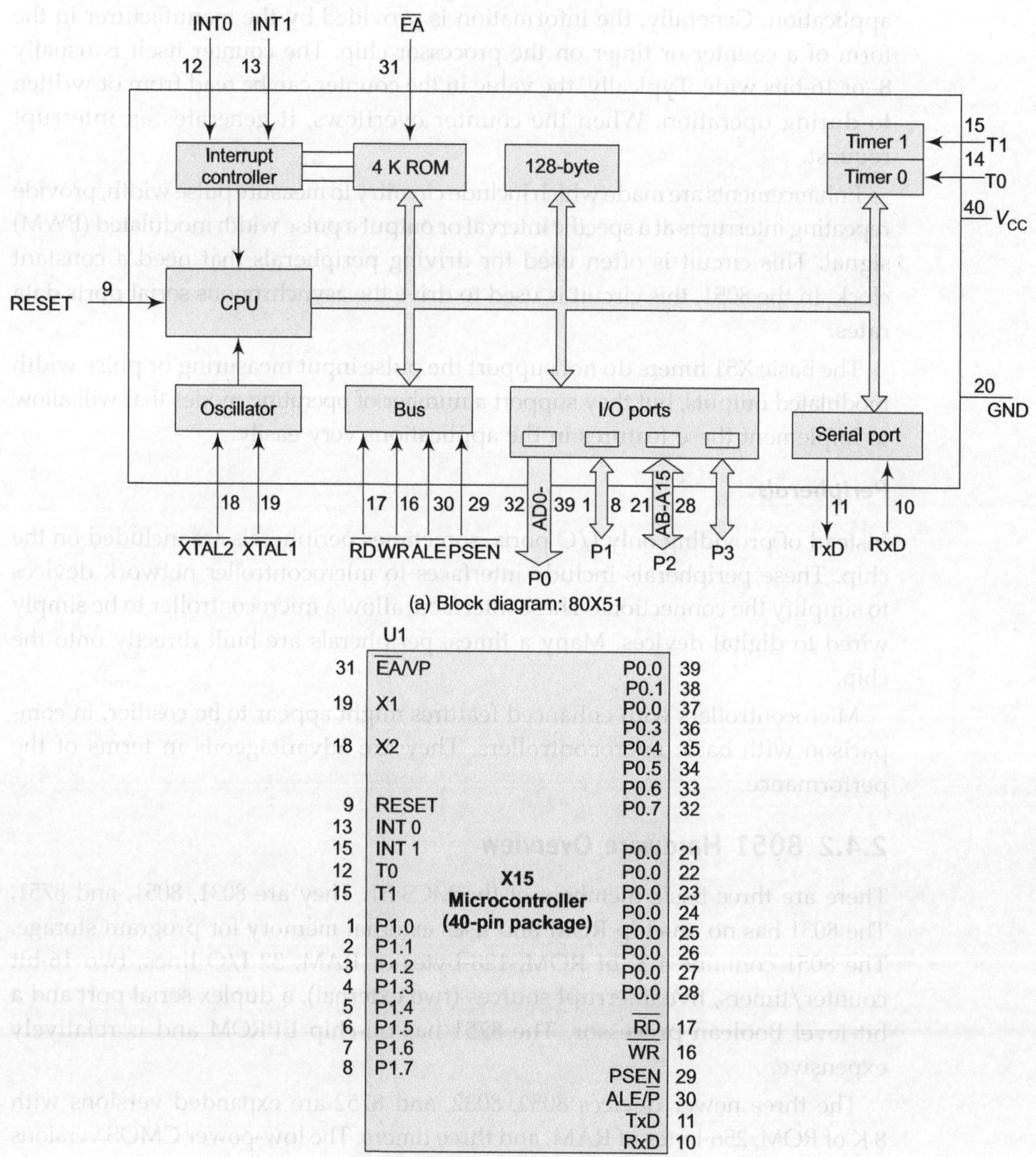

(a) Block diagram: 80X51

(b) Pin configuration (diagram): 80X51

Fig. 2.1

System clock Since, microcontroller is a sequential device, it needs a signal for operation (Fig. 2.2). The smallest unit of CPU timing is the oscillator period. It is called the *Clock*. 8051 contains an on-chip oscillator, which is driven by a crystal to be connected between the pins designated XTAL1 and XTAL2.

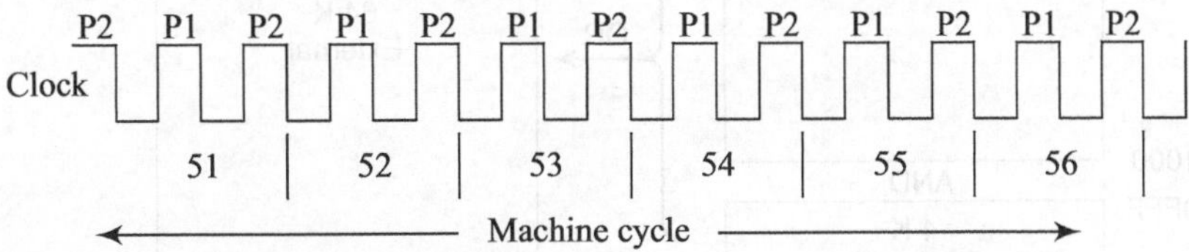

Fig. 2.2 Clock pulses for 8051

8051 can be driven by an external clock signal using any of the following methods:

1. For standard (HMOS) device, the external clock is applied to XTAL2, while XTAL1 is connected to ground (V_{ss}).
2. For CMOS device, external clock is applied to XTAL1 and XTAL2 is left unconnected.

   The maximum clock frequency is currently 12 MHz.

Reset A high level applied to the RST pin for at least two machine cycles will reset 8051. This is done with the help of an *RC* time constant or by using a digital counter.

For power-on reset, RST pin is connected to ground through a resistor (8.2 K) and connecting capacitor 10 µF from RST to the V_{CC}. It is assumed that the rise time is small and clock starts within msec.

2.4.4 Memory Organization

The microcontroller has separate address for program memory and data memory. The program memory can be up to 64 K long. The lower 4 K may reside on-chip. Figure 2.3 shows a map of program memory.

It has 128-byte of on-chip RAM plus a number of special function register (SFR). The lower 128-bytes of RAM can be accessed either by direct addressing (mov data addr) or by indirect addressing (mov @ri). Figure 2.5 shows the organization of internal RAM area.

The microcontroller can address up to 64 K of the data memory to the chip (Fig. 2.4). The movx instruction is used to access the external data memory.

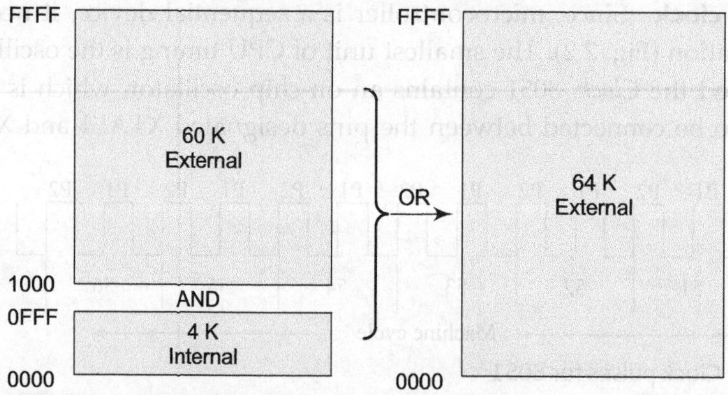

Fig. 2.3   Program memory

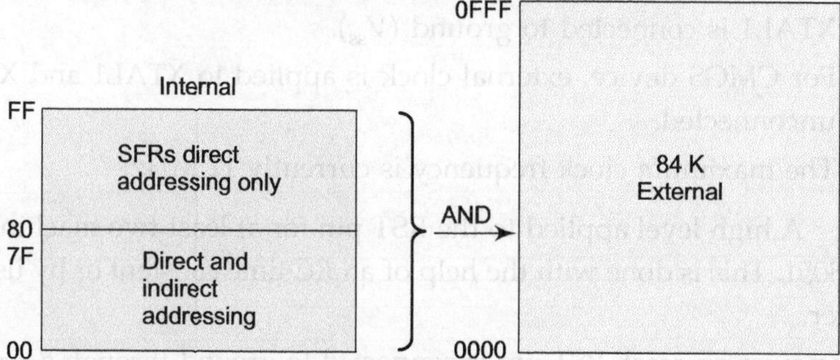

Fig. 2.4   Data memory organization

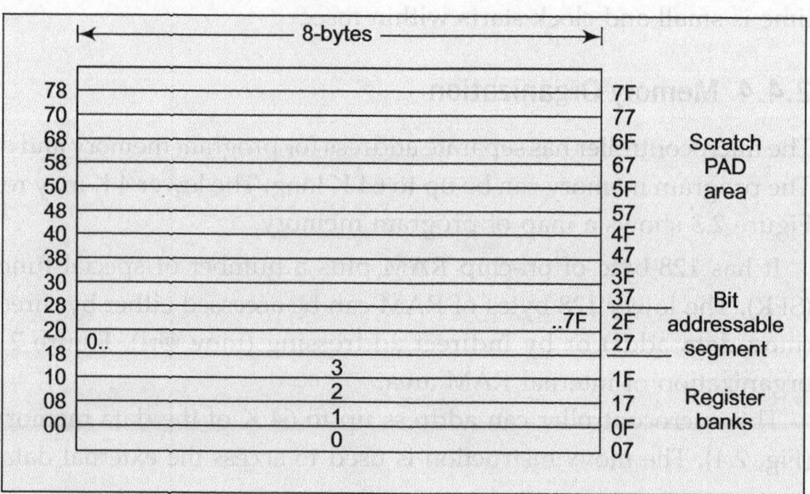

Fig. 2.5   Internal RAM organization

Figure 2.6 depicts internal RAM organization with bit addressable.

The lowest 32-bytes (00h–1Fh) of the lower 128 are grouped into four banks of eight registers. Since only one bank is active at a time, it can be selected by means of 2-bits in the processor status word (PSW) register through software. The eight registers in bank are designated $(R_0, R_1, ..., R_7)$. They are accessible to the programmer. On reset, stack pointer (SP) points to the top register of the lowest bank (R7). It can be modified by the instruction.

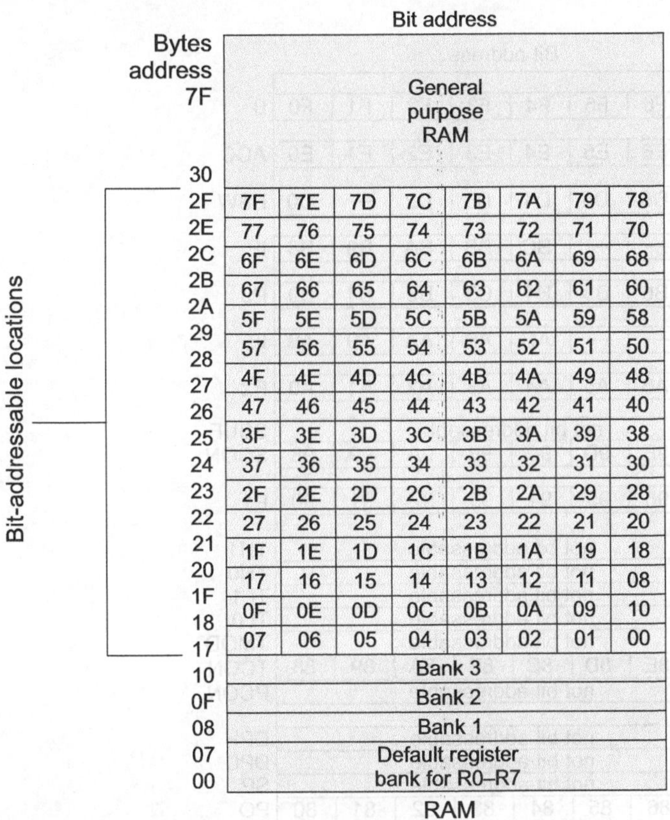

Fig. 2.6 Internal RAM organization

The next 16-bytes (20h–2Fh) form a block that can be addressed as either bytes or as 128 individual bits. The bit addresses range from 00h to 7Fh. The instruction that uses the address easily determines whether a byte or bit is being referenced without confusion.

2.4.5 Special Function Registers

Figure 2.7 depicts the organization and hardware address of special function registers (SFRs).

The registers, which are associated with important functions of the 8051, are assigned memory locations in the internal memory, so that they can be used by the programmer. The SFR can be either byte or bit addressable (Fig. 2.7). They are initialized on reset.

The unoccupied addresses are not implemented, and so the result of reading or writing to them is indeterminate. Unoccupied addresses in SFR space are reserved for use in future expansion.

Bytes address	Bit address								
FF									
F0	F7	F6	F5	F4	F3	F2	F1	F0	B
E0	E7	E6	E5	E4	E3	E2	E1	E0	ACC
D0	D7	D6	D5	D4	D3	D2	—	D0	PSW
B8	—	—	—	BC	BB	BA	B9	B8	IP
B0	B7	B6	B5	B4	B3	B2	B1	B0	P3
A8	AF	—	—	AC	AB	AA	A9	A8	IE
A0	A7	A6	A5	A4	A3	A2	A1	A0	P2
99	not bit addressable								SBUF
98	9F	9E	9D	9C	9B	9A	99	98	SCON
90	97	96	95	94	93	92	91	90	P1
8D	not bit addressable								TH1
8C	not bit addressable								TH0
8B	not bit addressable								TL1
8A	not bit addressable								TL0
89	not bit addressable								TMOD
88	8F	8E	8D	8C	8B	8A	89	88	TCON
87	not bit addressable								PCON
83	not bit addressable								DPH
82	not bit addressable								DPL
81	not bit addressable								SP
80	87	86	85	84	83	82	81	80	PO

Special Function Registers

Fig. 2.7 SFR: Internal organization

Accumulator (ACC) and B register When referring to the accumulator as a location in the SFR, the mnemonic ACC is used. It is also used as an index register. B register performs a specific function in the multiply and divide operations.

Program status word The PSW contains flag bits. A dot convention is used to address particular bit in PSW. A specific bit can be designated by defining bit position after the name of the register followed by a decimal point. The most significant bit (MSB) is position 7 and the least significant bit (LSB) is position 0. For example, the symbol AC (auxiliary carry), can also be designated as PSW.6.

The PSW does not have a zero flag, but the 8051 have specific instructions to test the accumulator for zero. The PSW also contains two programmable bits (RS0 and RS1) to make the resister bank active.

Stack pointer (SP)   The SP is 8-bits wide. It allows a maximum stack size of 256-bytes. The stack in the 8051 grows upward in the memory, therefore the SP is incremented before data is stored as result of a PUSH or CALL instruction. The stack may reside anywhere in the internal RAM. It can be initialized, by loading the appropriate address into the SP.

After reset, the SP contains the address 07h, causing the stack to start at location 08h in a register bank. The stack is usually moved higher in the RAM by loading a new address into SP before performing any PUSH or CALL instructions.

Data pointer (DPTR)   The DPTR is a 16-bit register, which can also be accessed as two 8-bit registers. The quantity may be held in two 8-bit parts: They are: high byte in register portion called DPH and low byte in the DPL portion.

The main purpose of the DPTR is to hold a 16-bit address for certain instructions, which try to access external memory. It can be used as a single 16-bit register.

Port latches (P0, P1, P2, P3)   The 32 I/O pins are organized into four 8-bit ports designated as P0, P1, P2, and P3. Each port has an associated 8-bit latch, the output of which, drive the matching I/O pins. The contents of the latch can be read from or written in the SFR.

Program status word (PSW)   This contains the information about the flags and active register bank. The format of PSW is shown below:

D7	D6	D5	D4	D3	D2	D1	D0
CY	AC	FO	RS1	RS0	OV	---	P

CY	: Carry flag, also used as a Boolean accumulator.
AC	: Auxiliary Carry or borrow out of bit 3 of accumulator.
OV	: This indicates whether an overflow has occurred.
P	: Parity flag.
FO	: This is the user defined flag.
RS1 & RS0	: Register Bank Select [00' indicates bank 0, '01' indicates bank 1, '10'

Serial data buffer (SBUF)   The SBUF is actually two separate registers sharing a common address:

1. Read-only; for reading the received data.
2. Write only; for writing the data to be transmitted.

Timer registers   Registers TH0 and TL0 are the high and low bytes, respectively, of the 16-bit counting register for Timer/Counter 0. Likewise, TH1 and TL1 are the high and low bytes for Timer/Counter 1.

Control registers   The SFR contains registers used for the control and status of the interrupt system, timer/counters, and serial port. They are as follows:

IP (Interrupt priority)   If the bit is 0, the corresponding interrupt has a lower priority and if the bit is 1, the corresponding interrupt has the highest priority.

D7	D6	D5	D4	D3	D2	D1	D0
---	---	---	PS	PT1	PX1	PT0	PX0

---	IP.7	Reserved for future use.
---	IP.6	Reserved for future use.
---	IP.5	Reserved for future use.
PS	IP.4	Define the serial port interrupt priority level.
PT1	IP.3	Define the Timer 1 interrupt priority level.
PX1	IP.2	Define the external interrupt 1 priority level.
PT0	IP.1	Define the Timer 0 interrupt priority level.
PX0	IP.0	Define the external interrupt 0 priority level.

Priority within level   Priority within the level is only to resolve simultaneous request of the priority level. From high to low, interrupt sources are also shown in Fig. 4.6: IE0, TF0, IE1, TF1, RI, and TI.

IE (Interrupt enable) (Bit addressable)

D7	D6	D5	D4	D3	D2	D1	D0
EA	---	---	ES	ET1	EX1	ET0	EX0

EA	IE.7	Disables all interrupts. If EA = 0, no interrupt will be acknowledged.
---	IE.6	Reserved for future use.
---	IE.5	Use for 8052.
ES	IE.4	Enable or disable serial port interrupt.
ET1	IE.3	Enable or disable Timer 1 overflow interrupt.
EX1	IE.2	Enable or disable external interrupt 1.
ET0	IE.1	Enable or disable Timer 0 overflow interrupt.
EX0	IE.0	Enable or disable external interrupt 0.

TMOD *(Timer mode)*

D7	D6	D5	D4	D3	D2	D1	D0
GT1	C/T	M1	M0	GT1	C/T	M1	M0
Timer-1				Timer-0			

GATE - When TR in TCON is set and GATE=1, timer/counter will run only while INT pin is high
(hardware control). When GATE=0, timer/counter will run while TR =1 (software
control)

C/T - Timer or counter selector.

M1 - Mode selector bit.

M0 - Mode select

TCON: *Timer/Counter control register (Bit addressable)*

D7	D6	D5	D4	D3	D2	D1	D0
TF1	TR1	TF0	TR0	IE1	IT1	IE0	IT0

TF1- Timer 1 overflow flag (set by h/w when o/f occurs)

TR1- Timer 1 Run control bit (set or cleared by s/w)

TF0- Timer 0 overflow flag (set by h/w when o/f occurs)

TR0- Timer 0 Run control bit (set or cleared by s/w)

IE1- Interrupt 1 edge flag (set by h/w when o/f occurs)

IT1- Interrupt 1 type control (set or cleared by s/w)

IE0- Interrupt 0 edge flag (set by h/w when o/f occurs)

IT0- Interrupt 0 type control (set or cleared by s/w)

SCON *(Serial port control)*

D7	D6	D5	D4	D3	D2	D1	D0
SM0	SM1	SM2	REN	TB8	RB8	T1	R1

SM0	SCON.7	Serial mode specifier
SM1	SCON.6	Serial mode specifier
SM2	SCON.5	Enable for multiprocessor Mode 2 and Mode 3
REN	SCON.4	Set/clear to enable/disable reception
TB8	SCON.3	Set/clear in Mode 3 if the 9th data bit is to be transmitted
RB8	SCON.2	1, if SM2 =0, RB8 is a stop bit. Not used in Mode 0
TI	SCON.1	Transmit interrupt flag
RI	SCON.0	Receive interrupt flag

PCON *(Power control)*

D7	D6	D5	D4	D3	D2	D1	D0
SMOD	--	---	---	GF1	GF0	IDL	PD

SMOD	:	Double baud rate bit
--	:	Reserved for future use
GF1	:	General purpose flag bit
GF0	:	General purpose flag bit
PD	:	Power down bit
IDL	:	Idle mode bit

2.4.6 I/O Ports

One of the most useful features of 8051 is the I/O consisting of four bidirectional ports. Each port has an 8-bit latch in the SFR space, an output driver, and an input buffer. The ports can be used for following functions:

1. General I/O
2. Address and data lines
3. Certain special functions

Input, loading, and output drive   Ports 1, 2, and 3 have the equivalent of internal pull-up resistors.

When used as inputs: The pins of P1, P2, and P3 will be high (logic 1) when open, but will source current when pulled low by an external device. Since, Port 0 does not have the pull up feature, it is in high-impedance state, when used as an input. The reading of the pin and the latch is different and it is governed by the instruction used.

When used as outputs: Ports 1, 2, and 3 each can drive the equivalent of four LS TTL inputs, while the Port 0 can drive eight such equivalent inputs.

Alternate port functions   All the pins of Port 3 have an alternate function. To enable the alternate function, '1' must be written to the corresponding bit in the port latch.

2.4.7 Accessing External Memory

The 8051 family has separate program memory and data memory and it uses different hardware signals to access the corresponding external storage devices.

The PSEN (program store enable) signal is used as the read strobe for program memory RD and WR are used as the read and write strobes to access data memory. Accesses to external program store always use a 16-bit address while accesses to external data store may use either an 8-bit or a 16-bit address.

1. *In the case of a 16-bit address:* The high order 8-bits of the address are output on Port 2, where they are held constant during the entire memory access cycle. The previous contents of the Port 2 latches in the SFR are not lost but are restored after the memory access cycle.

2. *In the case of 8-bit address:* The contents of Port 2 are unchanged, which allows some of the pins of Port 2 to be used to select 256-byte pages for the lower 8-bits of the address.

The low order 8-bits of the address are multiplexed with the data byte on Port 0, when used in this mode. The address latch enable (ALE) signal must be used to latch the low order address.

2.4.8 Timers/Counters

The 8051 family has two 16-bit registers that can be used as either timers or counters. They are designated as Timer 0 and Timer 1. These registers are in the SFR as a pair of 8-bit registers.

When used as a timer: The register is incremented once per 12-clock period.

When used as a counter: The register is incremented on a 1 to 0 transition (a negative edge) applied to the appropriate input pin T0 or T1. Since, T0 and T1 are alternate functions of the port pins, it takes two complete machine cycles for the 8051 to see the transition. So the input must be held high for at least one cycle and low for at least one cycle.

The mode of operation of the timer is determined by a byte written to the TMOD register. The bits M0 M1 are used to select one of four operation modes: Mode 0, Mode 1, Mode 2, or Mode 3. Both counters operation is similar in Modes 0, 1, and 2. However there is a change in Mode 3.

Mode 0: 13-bit counter In Mode 0 (Fig. 2.8), the timer is configured as a 13-bit counter that can be considered as an 8-bit counter preceded by a 5-bit divide-by-32 pre-scalar. The 8-bit count is in the THX register (TH0 or TH1), depending on counter to be used.

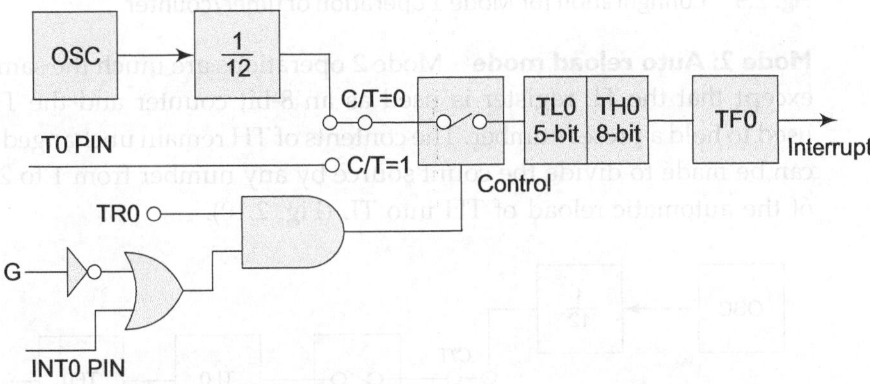

Fig. 2.8 Configuration for Mode 0 operation of timer/counter 0

5-bits of the TL register are used for count while the upper 3-bits of the TL register are ignored. As the 13-bits count in TL and TH goes from all 1s to all 0s, the timer interrupt flag (TF0 or TF1) is set. The timer interrupt is the connection between the counter hardware and the program software.

Counter/timer (C/T) bit in TMOD is used to select the source of input to the counter. The counting process can be turned ON and OFF (enabled or disabled), independent of the input.

In order for the counting process to be turned, the TR bit (TR0 or TR1) in the TCON register has to be at logic level 1. Also, either the appropriate GATE bit in the TMOD register is a 0, or the appropriate INTX pin has to be held low.

The use of GATE or TR allows software controlled counting while the use of INT allows external hardware controlled counting.

Mode 1: 16-bit counter It is the same as Mode 0, except that the timer register is 16-bits long with all 8-bit of TL being used (Fig. 2.9).

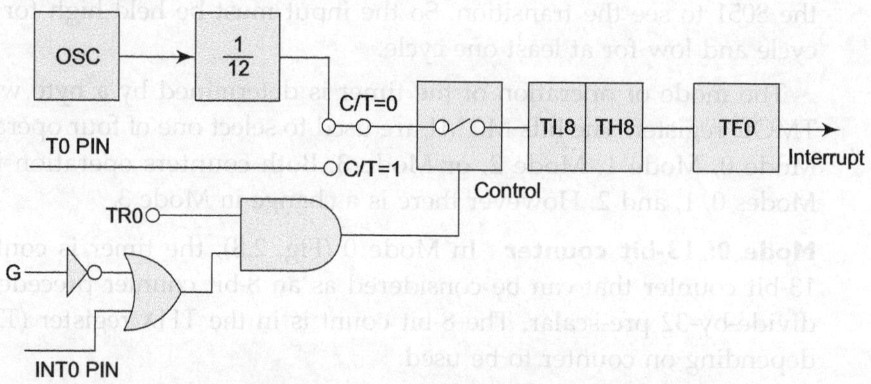

Fig. 2.9 Configuration for Mode 1 operation of timer/counter

Mode 2: Auto reload mode Mode 2 operations are much the same as Mode 0, except that the TL register is used as an 8-bit counter and the TH register is used to hold a preset number. The contents of TH remain unchanged. The counter can be made to divide the count source by any number from 1 to 255 by means of the automatic reload of TH into TL (Fig. 2.10).

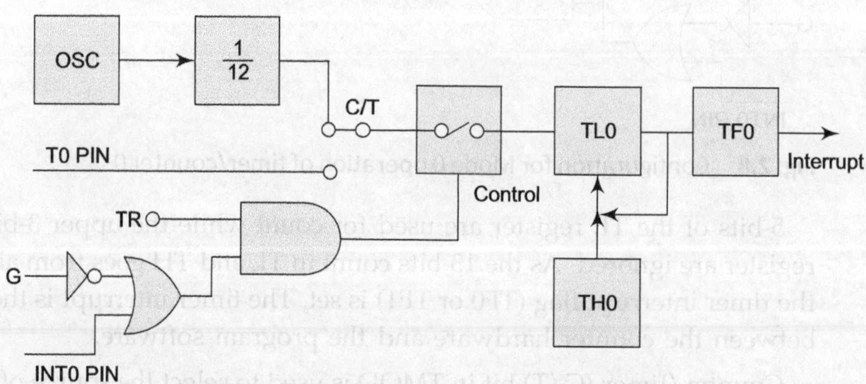

Fig. 2.10 Configuration for Mode 2 operation of Timer/Counter 0

Mode 3: Mode 3 operation differs from the operations of the other modes. Here, Timer 1 is disabled but holds its count; thus it is essentially frozen (Fig. 2.11).

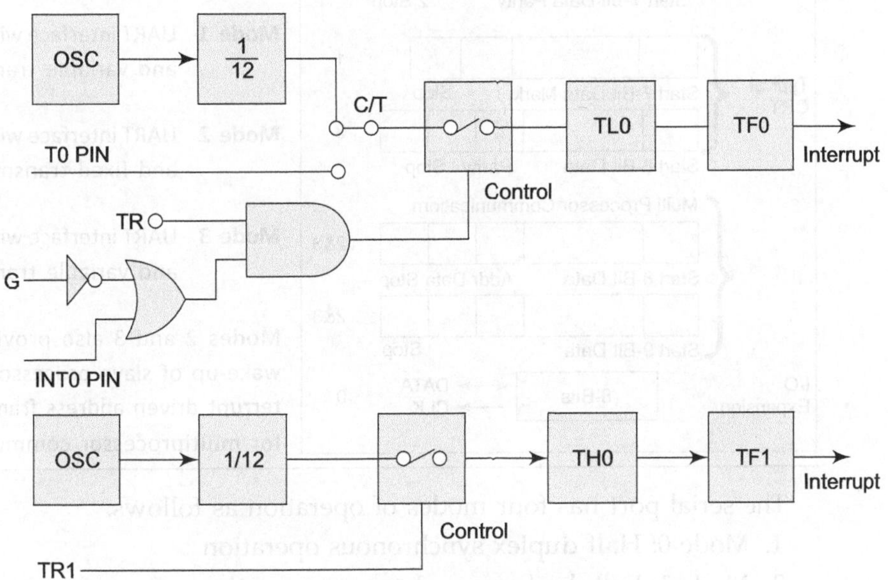

Fig. 2.11   Configuration for Mode 3 operation of timer/counter

Timer 0 is split into two separate counters. The first counter is the same as Mode 0. TL0 is used as an 8-bit counter without a pre-scalar.

The second counter uses TH0 as an 8-bit counter. Enable control is given by TR1 bit in the TCON register. The first counter sets the TF0 interrupt flag while second counter sets TF1.

2.4.9 Serial Port Interface

The 8051 family has a full duplex serial port that allows data to be transmitted and received simultaneously, while controller is busy doing other activities.

A serial port interrupt is generated by the hardware to get the attention of the program in order to read or write serial port data.

The receiver hardware is double buffered, meaning that a received frame of data can be held for reading, while a second frame is being received. Double buffering allows the receiver interrupt service routine to be less time critical, but the stored frame must be read before reception of the second frame is complete or the stored frame will be overwritten and lost.

Table 2.1 Formats of the serial communication frames for serial port on X51

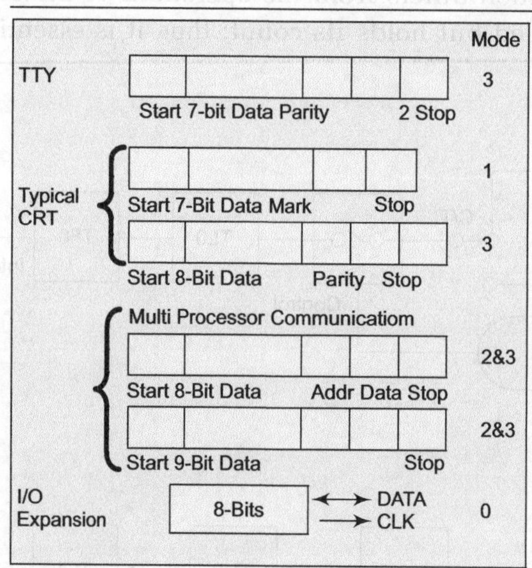

Mode 0 Synchronous I/O expansion using TTL or CMOS shift registers

Mode 1 UART interface with 10-bit frame and variable transmission rate

Mode 2 UART interface with 11-bit frame and fixed transmission rate

Mode 3 UART interface with 11-bit frame and variable transmission rate

Modes 2 and 3 also provide automatic wake-up of slave processors through interrupt driven address frame recognition for multiprocessor communications.

The serial port has four modes of operation as follows:

1. Mode 0: Half duplex synchronous operation
2. Mode 1: Full duplex asynchronous operation
3. Mode 2: Full duplex asynchronous operation with variable baud rate
4. Mode 3: Full duplex asynchronous operation with programmable baud rate

Table 2.1 show the frame formats used in each mode.

Serial Port Control Register

SCON is the serial port control and status register. Bits SM0 and SM1 are used to select the operating mode, while bit SM2 are used in a multiprocessor system.

Mode 0: Half duplex synchronous operation   Data is sent and received through the RxD pin in the 8-bit frames with LSB first. The bit rate is fixed at one-twelfth of the oscillator frequency. The shift clock is sent out to the TxD pin during both transmission and reception and is used to synchronize the receiver with the sender. A shift clock edge occurs during the valid state of each data bit. Note that the TxD and the RxD are alternate functions of Port 3 pins. This is also called shift register mode of operation.

Mode 1: Full duplex asynchronous operation   Data is sent out the TxD and received through the RxD by using a 10-bit code called *frame*. A complete frame consists of a start bit always 0, 8 data bits (LSB first) and a stop bit which is always 1. The start and stop bits are added by the hardware. Data transfer takes place through SBUF.

The baud rate is variable and can be generated by Timer 1, which is used as a baud rate generator in this mode of operation. Due to operation being similar to an 8-bit USART, this mode is also called standard 8-bit USART mode.

Mode 2: Full duplex asynchronous operation with variable baud rate   This mode is used to handle communication in multiprocessor environment. It uses an 11-bit frame instead of a 10-bit frame as in Mode 1. A ninth data bit is inserted before the stop bit. Hence it is also called 9-bit USART mode.

While transmitting, the ninth bit is obtained from the TB8 in SCON, which must be initialized. While receiving, the ninth bit can be read from the RB8 in SCON. Ninth bit acts as a parity bit for 8-bit data.

The baud rate is variable and can be set as 1/32 or 1/64 of the oscillator frequency by the SMOD bit in the PCON register. When the SMOD bit is 1, 1/32 is used as a multiplier, 1/64 is used if SMOD bit is 0.

Mode 3: Full duplex asynchronous operation with programmable baud rate   Like Mode 2, this mode is used to handle communication in multiprocessor environment. It uses an 11-bit frame instead of a 10-bit frame as in Mode 1. A ninth data bit is inserted before the stop bit. Hence it is also called 9-bit USART mode. It differs from Mode 2 in selection of baud rate. In this mode, baud rate is programmable and selection is similar to one in the Mode 1 operation, which uses Timer 1.

Reception for Modes 1, 2, and 3 is enabled when the remote enable (REN) bit in SCON is set to 1 but the actual reception is initiated by a high to low transition at RxD pin. Transmission is initiated by writing to SBUF in any serial mode.

Baud Rate Generator

It is very common to use a timer in auto-reload mode (Mode 2). The baud rate is then given by the expression:

$$\text{Baud rate} = \text{Oscillator frequency}/[N\,\{256 - (\text{TH1})\}]$$

where N depends on the SMOD bit in the PCON register

$$\text{SMOD} = 0, \rightarrow N = 384$$

$$\text{SMOD} = 1, \rightarrow N = 192$$

The TH1 represents the contents of the register TH1.

By definition, value stored in TH1 must be an integer. Oscillator frequency must be an integral multiple of both baud rate and N, in order to obtain an exact baud rate using above expression.

PCON Register

Only bit 7, SMOD, is implemented in the standard 8051. Bits 0, 1, 2, and 3 are implemented in the CMOS version. Bits 0 and 1 are used in power saving modes and bits 2 and 3 are used as general-purpose flags.

2.4.10 Interrupt Structure

The 8051 family has five sources of interrupts: external pins (INT0 and INT1) timer/counters (TF0 and TF1) and serial port (T1 or S1). A useful feature of the 8051 is that the interrupt sources are associated with bit location in registers. These bits can be set/reset by instruction, similar to an external hardware signal.

All the interrupts or each individual interrupt can be enabled or disabled by setting or clearing the appropriate bit in the IE register. If enabled, an interrupt will cause a call to one of the predefined locations in the RAM. If the interrupt occurs while it is disabled, or while a higher priority interrupt is running, it becomes pending. Pending interrupt enabled, will cause a call unless it was cancelled by the software while it was still pending or a higher priority interrupt was simultaneously made active.

The timing of events in external devices usually has no relationship to the timing of the CPU; these events are asynchronous to the processor. In order to monitor and control the external devices, a microcomputer must respond to I/O requests and other external events in a timely manner.

I/O devices often require immediate service while the processor is in the middle of doing something else. The interrupt is a software-controlled hardware feature that forces the processor to suspend its current activity in order to service the I/O device. When it is finished with the I/O, the processor resumes where it left off, much the same as a subroutine. Interrupts that can be blocked are called *maskable*, while others are called *non-maskable*.

An I/O device requests service by activating an interrupt pin on the CPU. If the CPU has enabled its interrupts in software, it initiates its response, often with an acknowledgment signal to the I/O device. The CPU push the return address onto the stack and branches to a predefined part of the memory, where it expects to find the interrupt service routine which will handle the interrupting device. The last instruction in the routine is RETURN, which pop the return address off the stack and into the PC register the CPU then resumes program execution from the point where it was interrupted.

It is possible to have more than one interrupt source. When an interrupt occurs, the processor must first determine which device has caused. One way is to poll the devices. But polling can be time-consuming. A faster way is to have each interrupting device to point to the place in memory where its service routine is stored. The CPU can then go there directly. Such a system is called a *vectored interrupt*. The 8051 have such a priority scheme in addition to a priority structure. The 8051 have a two-tier priority structure.

The top tier has two priority levels: These are *high* and *low*. Each interrupt source can be assigned to either high-level or low-level status by setting the appropriate bits in the IP register. When two interrupts of different levels are received simultaneously, the high-level interrupt is serviced first.

The second tier priority is used to resolve simultaneous interrupts within the same level. The priority within level ordering from highest to lowest is fixed as: IE0, TF0, IE1, TF1, R1, or T1.

Interrupt flags are assessed by the 8051 hardware at the end of a typical machine cycle. During the following machine cycle if the flag is found to be set, then a call to the appropriate interrupt vector starts with the fifth cycle and continues for as long as is required by the routine.

The hardware will not generate the interrupt call in the following situations:

1. If an interrupt of equal or higher priority is already in the program. A lower priority interrupt can itself be interrupted by a higher priority interrupt assuming that the higher one is enabled.

2. If the current machine cycle is not the final cycle of the instruction being executed. The instruction in progress must be completed before jumping to the interrupt service instructions.

3. If the instruction in progress is RET1 or any instruction that write to the IE or IP register. At least one additional instruction must be executed following those before jumping to the interrupt vector address. The reason for this condition is to prevent an interrupt from occurring in the middle of a routine that is in the process of reconfiguring the interrupts.

4. The time between the activation of an interrupt and the start of execution of the service routine is the response time. The shortest response time is three machines, the worst-case condition must be assumed to happen.

In general, interrupts can be either level triggered or edge triggered. A transition-activated event is, by definition, when the flag bit associated with the interrupt is cleared. Clearing the flag bit of a level activated interrupt will have no effect if the external level causing the interrupt stays active. In the 8051, the external interrupts INT0 and INT1 can be configured to be either transition activated or level activated, depending on the value of the bits IT0 or IT1 in the TCON register.

The flag bit will be set when the interrupt occurs and cleared automatically when the call is made to the interrupt vector. The TF0 and TF1 flags are set when the count in the corresponding counter register rolls over. The counter interrupts are in effect transition activated. TF0 and TF1 are also cleared automatically during interrupt service. An 8051 assembler will typically recognize the predefined code addresses, given in table below.

Symbol	Address	Interrupt source
RESET	00 h	Power up or reset
EXTI0	03 h	External interrupt 0
TIMER 0	0B h	Timer 0 interrupt
EXTI1	13 h	External interrupt 1
TIMER 1	1B h	Timer 1 interrupt
SINT	23 h	Serial port interrupt

2.4.11 Enhanced Architecture

There are several new processor cores in the market that significantly reduce this cycle of 12-clock cycles per instruction and provide devices that offer significantly more efficiency than the original 8051 while running at the same clock speed.

INTEL MCS-151/251 Intel MCS-151 and MCS-251 microcontrollers are an enhancement of the original 8051 with a number of changes and improvements as follows:

1. An enhanced instruction set for 16/32-bit data transfer,

2. Up to 12 M of SRAM addressing ability

3. A stack capable of being 64 K

4. The instruction can be executed in as few as two or four clock cycles

The MCS-151 and MCS-251 are the versions that are pin and software (binary code) compatible with the 8051.

The MCS-151 is basically a replacement for the 8051 with the same instruction set and features. The MCS-251 is an enhancement to the 8051. The MCS-251 can run up to 15 times faster than a stock 8051. Both the MCS-151 and MCS-251 use a hardwired processor capable of pipelining (reading ahead) instructions, which is a primary reason for the faster program execution.

High speed microcontrollers (HSM) Dallas Semiconductor was the first to come up with the idea of changing the 8051 microcoded processor code with a hardwired one. This change resulted in instructions taking 4, 8, 12, or 16 clock cycles per instruction cycle resulting in an improvement of 1.5 to 3 times over a true 8051 running at the same clock speed. This change in the instruction timing meant that the Dallas Semiconductor HSM parts cannot just be dropped into an application and expected to run with just a slower clock.

Control store and external memory   These enhancements give some options in adding external memory (both control store and RAM) that make the design easier and give flexibility not present in basic 8051 microcontrollers.

This makes it possible to resize the control store ROM. This feature allows to the test code destined for other devices or mix-and-match the amount of erasable PROM (EPROM) and/or RAM that is available for the control store.

The built-in control store is controlled by the register at address 0C2h known as ROMSIZE register. The three RSX bits in register are written to by a timed-access instruction sequence for the modification. On power-up, ROMSIZE register is initialized to the maximum size of the microcontroller's EPROM control store. For example in the DS87C520, these bits will be set to 101 because the device has 16 K of on-board EPROM available.

Adding wait status   With the increase in speed of HSM, there are a number of potential problems. An application that earlier used an 8051 and was running without any problems, requires that the wait states must be included to make it compatible with the HSM.

One of the most obvious concerns is the interface to the external memory. In both the 8051 and the HSM, the external memory access is timed as part of the instruction cycle. By changing to the HSM with a shorter instruction cycle, external memory access that used to work, might not work because of the faster memory access.

In 8051, two control store reads take place in a 12-clock cycle period, and one external memory access takes place in a single 12-clock cycle period. In HSM, one control store byte read requires 4-clock cycles, and one external memory access nominally requires 8-clock cycles. This means that, in an HSM running at the same clock speed as an 8051, both the control store reads and the external RAM accesses run 33% faster than in a regular 8051.

The most obvious solution to the problem of faster external memory accesses is run the HSM at a slower speed than the 8051 or replace the memory with faster devices. If the external control store reads are unreliable then one of these two solutions has to be implemented.

To time the actual access time, the number of clock cycles used is multiplied by the HSM clock period. For a 33-MHz HSM clock, accessing external memory will take 485 ns (30 ns clock period times 16).

Scratchpad RAM   It is assumed that SFR can be accessed directly, while additional scratchpad RAM can be accessed indirectly.

Using an index register (R0 or R1) to access anywhere in the first 256 addresses of the enhanced 8051 data space only access RAM will be accessed, regardless of what address is loaded into the index register.

With 256-bytes of scratchpad RAM, you can actually address 384 addresses in the first 256 addresses of the enhanced 8051 data space. It is important to remember that, for these devices, the upper 128 addresses are SFR when direct addressing is used and additional scratchpad RAM when indirect addressing is used.

Timers The basic 8051 timer design is quite easy to understand and use. Unfortunately, it is also quite limited in its capabilities to simplify interfacing with external devices and providing advanced functions. For many enhanced interfaces, specialized timers are available. In the 8052 version of the 8051, a third 16-bit-only timer has been added to provide some additional software timing options in the 8051.

In the normal 8051, a timer tick takes one instruction cycle (12 clock cycles). This creates a problem for HSM, because of the fewer clock cycles that the HSM executes.

The solution to this problem was given by providing a switch between the 4- and 12-cycle delay for the timers. The timer count switch is controlled by the clock control register located at the address 08Eh called *CKCON*.

This option of dividing the clock into timers by four means that an HSM should really work at one-third the speed of an 8051, not divided by 2.5. If an application is currently running at 12 MHz with a straight 8051, an HSM should run at 4 MHz to ensure that the timers will work exactly the same between the two applications.

Timer 2 *In 8052 the set of enhancements to the basic 8051 is the inclusion of a third 16-bit timer. This timer called Timer 2 also provides some additional I/O capabilities.* In serial data transmission the rate of data transfer is measured in terms of baud rate bits per sec bps. It represents the number of bauds (a packet of data) transferred per second. A clock generator is required to provide pulses for serial transfer. Timer 2 can also be used as a secondary baud rate generator for the first serial port. Timer 2 can be run in two different modes. Those are:

Capture mode. It uses Timer 2 as a free running clock, which saves the timer value on each high to low transition. This mode can be used for recording bit lengths when receiving Manchester-encoded data.

Auto-reload mode. When the timer has overflowed, a value is written into the TH2/TL2 registers from the RCAP2H/RCAP2L registers. In this mode, the T2 pin on the microcontroller can also be used to initiate a reload of the timer. This feature can be used to implement a system watchdog timer.

If the reset signal is not received in a timely manner, the Timer 2 interrupt can be used to initiate a system reset by the 8052. The T2CON register (at address 0C8h) is used to control the operation of the Timer 2 known as TMR2.

Like TMR0 and TMR1, TMR2 can use either the internal clock or an external one for the clock source.

Unlike TMR0 and TMR1, when an interrupt request is acknowledged and the TMR2 interrupt handler is executing, overflow flag (TF2) or reload/capture flag (EXF2) must be reset in software and not rely on the instruction RETI (return from interrupt) to reset them. For a typical 8052, the clock source is identical to TMR0 and TMR1.

In HMS Dallas, the clock divider is controlled by the control divider bits (CD bits) of the PMR register (at address 0C4h).

When TMR2 runs in auto-reload mode, the overflow can be used as a data rate generator for the serial port (or the first serial port in the Dallas Semiconductor HSM parts). By setting the RCLK and TCLK bits, the serial port will be driven from TMR2 and not TMR1. In the Dallas Semiconductor HSM parts, this means the two serial ports can be driven by different clocks and run at different speeds.

Watchdog timers   A watchdog timer (usually referred to as WDT) is used to protect an application in case the controlling microcontroller begins to run amok and execute randomly rather than the preprogrammed instructions written for the application. The purpose of a watchdog timer is to reset the microcontroller if execution is erroneous.

WDT is made ON and the application resets, timer before the WDT overflows and resets the application. If the application does not reset the WDT at appropriate time, it indicates malfunctioning of microcontroller.

The microcontroller clock is passed through a series of counters, and the overflow of each is passed either to another stage or used to trigger the watchdog timer reset.

The first divisor selects the first delay in clock cycles for the incoming signals. This delay is specified by CD0 and CD1 bits of the PMR an SFR. If the PMR register is not present in the HSM, then the clock is not divided at this stage (or can be thought of as a divisor of 1). The WDT divisor circuit cans time out after two to the power of 17, 20, 23, or 26. These values are selected from the WDT control register called WDCON at address 0D8h. A better way of describing the watchdog timer time-out intervals is after 131072, 1048576, 8388608, or 67108864 clock cycles. If the HSM was running at 12 MHz, with a divide by 256, this can be increased to almost 24 min.

To start the WDT, bit 1 of WDCON is set to start the WDT running. Before the watchdog reset occurs, an interrupt can be generated by the watchdog timer hardware. Using the WDT interrupts request circuitry, the application will be warned 512 clock instruction cycles before the reset, to allow the application time to either power down or reset the WDT.

2.5 | AT89C2051: FLASH CONTROLLERS

The AT89C2051 is a low-voltage CMOS 8-bit microcomputer, with 2 K flash programmable EEROM. The device is a versatile 8-bit CPU with flash on a monolithic chip. It is a powerful microcomputer, which provides a highly flexible and cost effective solution to control applications. AT89C2051 provides standard features like 2 K of flash, 128-bytes of RAM, 15 I/O lines and two 16-bit timers. Figure 2.12 shows internal block diagram and Fig. 2.13 depicts the pin diagram of a flash controller.

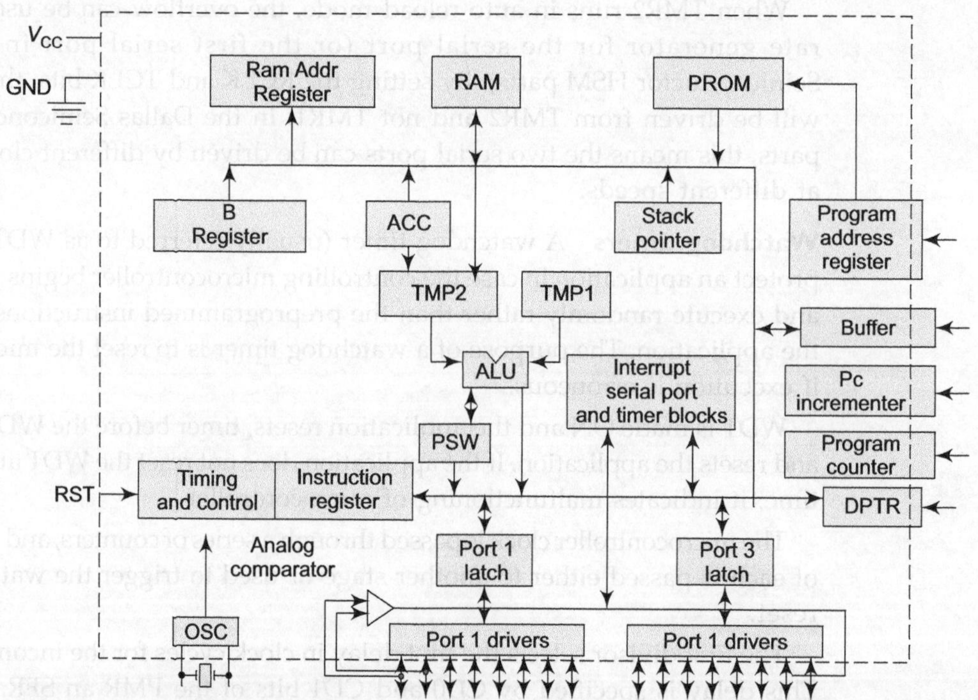

Fig. 2.12   Internal block diagram of a flash controller

2.5.1 Characteristics of AT89C2051 Flash Controllers

The flash controllers are compatible with MCS-51 products. The characteristics can be summarized as:

1. 2 K of reprogrammable flash memory
2. Endurance: 1,000 Write Erase Cycle
3. Data retention: 10 years
4. 2.7 V to 6 V operating range
5. Fully static operation: 0 Hz to 24 Hz
6. Two level program memory lock
7. 120×8-bit internal RAM

8. 10 programmable I/O line
9. Two 16-bit timer/counter
10. Five interrupt source
11. Programmable serial UART channel
12. Direct LED drive output
13. Low power idle and power down mode

They are available in a package size varying from 20 to 64-pins. For the simple applications, controller with 20-pins can be used. The complex applications can use chip with additional features, such as additional timer or ports. Here we will describe the 20-pin controller, which can be considered as simplified X51 chip.

In AT89Cx051, voltage comparators are also included. This helps to compare two-voltage levels and can be used to provide voltage monitoring function between different analog voltage sources. The comparator is driven by voltage inputs at port pins P1.0 and P1.1 output of comparator is made available on port pin P3.6.

2.5.2 AT89C2051: Pin Layout and Description (20-Pin Controller)

Port I

Port 1 is an 8-bit bidirectional I/O port. P1.2 to P1.7 are provided with internal pull-ups, while P1.0 and P1.1 require external pull-ups, and also serve as the positive input (AIN0) and the negative input (AIN1), respectively, of the on-chip precision analog comparator. The Port 1 output buffer can sink 20 mA current and can drive LED display directly. They can be used as inputs. When pins P1.2, P1.7 are used as inputs and are externally pulled low, they source the current (IIL) because of the internal pull-ups.

Port 1 also receive code data during flash programming and program verification.

```
                Pin configuration
                   PDIP/SOIC

         RST ——— 1        20 ——— V_CC
   (RXD) P3.0 ——— 2       19 ——— P1.7
   (TXD) P3.1 ——— 3       18 ——— P1.6
       XTAL2 ——— 4        17 ——— P1.5
       XTAL1 ——— 5        16 ——— P1.4
   (INT0) P3.2 ——— 6      15 ——— P1.3
   (INT1) P3.3 ——— 7      14 ——— P1.2
    (T0) P3.4 ——— 8       13 ——— P1.1 (AIN 1)
    (T1) P3.5 ——— 9       12 ——— P1.0 (AIN 0)
         GND ——— 10       11 ——— P3.7
```

Fig. 2.13   Pin diagram of a flash controller

Table 2.2   Alternate functions of Port 3 pins

Port pin	Alternate functions
P3.0	RXD (serial input port)
P3.1	TXD (serial output port)
P3.2	INT0 (external interrupt 0)
P3.3	INT1 (external interrupt)
P3.4	T0 (Timer 0 external input)
P3.5	T1 (Timer 1 external input)

Port 3

Port 3 pins P3.0 …P3.5 and P3.7 are seven bidirectional I/O pins with internal pull-ups. P3.6 is hard-wired as an input to the output of the on-chip comparator and is not accessible as a general purpose I/O pin. The Port 3 output buffer can sink 20 mA current.

When 1s are written to Port 3 pins, they are pulled high by the internal pull-up and can be used as inputs. As inputs, Port 3 pins that are being pulled low externally, will source current (I) because of the pull-ups. Port 3 also performs functions given in Table 2.2.

Port 3 also receives some control signals for flash programming and programming verification.

RST

It is the reset input. All I/O pins are reset as soon as the RST goes high. Holding the RST pin high for machine cycles, while the oscillator is running, resets the device.

XTAL1

Input to the inverting oscillator amplifier and input to the internal clock operating circuit.

XTAL2

Output from the inverting oscillator amplifier.

Oscillator Characteristics

XTAL1 and XTAL2 are the input and output, respectively, of an inverting amplifier which can be configured for use as an on-chip oscillator (Fig. 2.14(a)), or driven from an external clock source. XTAL2 is left unconnected, while XTAL1 is driven as shown in Fig. 2.14(b).

There are no requirements on the duty cycle of the external clock signal since the input to the internal clocking circuit is through a divide by two flip-flop but the minimum and the maximum voltage and low-time specifications shown in Table 2.3 must be observed.

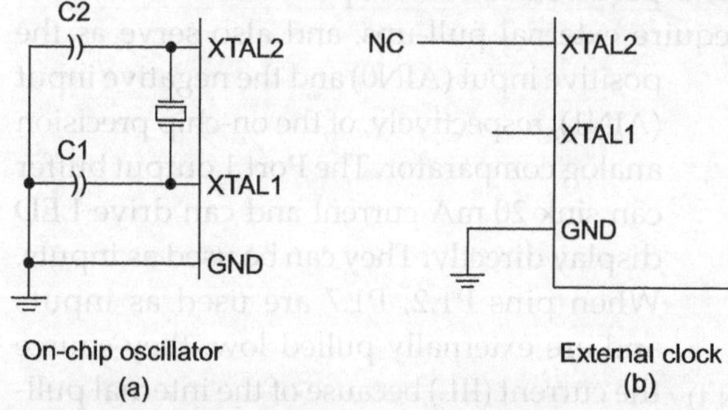

Fig. 2.14   XTL1 and XTL2 connections

Table 2.3   Timing specifications for external clock drive

Symbol	Parameter	Min	Max	Unit
$1/t_{CLCL}$	Oscillator frequency	0	24	MHz
t_{CLCL}	Clock period	41.6		ns
$t_{CHCX}P$	High time	15		ns
$t_{CLCX}P$	Low time	15		ns
t_{CLCH}	Prise time		20	ns
t_{CHCL}	Fall time		20	ns

2.5.3   Memory Organization

The memory organization of AT89C2051 is similar to X51 family with certain additional features that are described below.

Special-function Registers

A map of the on-chip memory area called the SFR space is similar to normal architecture described in the previous section.

User software should not write to these unlisted locations, since they may be used in future products to invoke new features. In that case, the reset or inactive values of the new bits will always be 0.

Program Memory Lock Bits

On the chip are 2-bits, which can be left unprogrammed (U) or can be programmed (P) to obtain the additional features listed in Table 2.4.

Table 2.4   Lock bit protection modes

Mode program lock bits		Protection type
	LB1	LB2	
1	U	U	No program lock features
2	P	U	Further programming of the flash is disabled
3	P	P	Same as Mode 2, also verify is disabled

The locks bits can only be erased with the chip erase operation.

2.5.4 Modes of Operation

The different modes of operation of a flash controller are as follows:

Idle Mode

In the idle mode, the CPU puts itself to sleep while all the on-chip peripherals remain active. The mode is invoked by software. The content of the on-chip RAM and all the SFRs remains unchanged during this mode. The idle mode can be terminated by any enable interrupt or by a hardware reset.

P1.0 and P1.1 should be set to '0' if no external pull-up is used and set to '1' if external pull-ups are used.

It should be noted that idle mode is terminated by a hardware reset. The device normally resumes program execution, from where it left off, up to two-machine cycle before the internal reset algorithm takes control.

On-chip hardware inhibits access to internal RAM in this event, but access to the port pins is not inhibited. To eliminate the possibility of an unexpected write to a port pin when idle mode is terminated by reset, the instruction following the one that invokes the idle mode should not be one that writes to a port pin or to external memory.

Power Down Mode

In the power down mode, the oscillator is stopped and the instruction that invokes the power down is the last instruction executed. The on-chip RAM and SFRs retain their values until the power down mode is terminated.

The only exit from power down mode is a hardware reset. Reset redefines the SFR, but does not change the on-chip RAM. The reset should not be activated before V_{CC} is restored to its normal operating level and must be held active long enough to allow the oscillator to restart and stabilize.

P1.0 and P1.1 should be set to '0' if no external pull-ups are used, and to '1' if external pull-ups are used.

2.5.5 Programming the Flash Controllers AT89C2051

The AT89C2051 is available with the 2 K bytes of on-chip EPROM code memory array in the erased state and ready to be programmed.

The code memory array is programmed 1-byte at a time. Once the array is programmed, then to reprogram any nonblank byte, the entire memory array needs to be erased electrically.

Internal Address Counter

AT89C2051 contains an internal EPROM address counter, which is always reset to 000h on the rising edge of RST and is advanced by applying a positive going pulse to pin XTAL1. The connections are shown in Fig. 2.15.

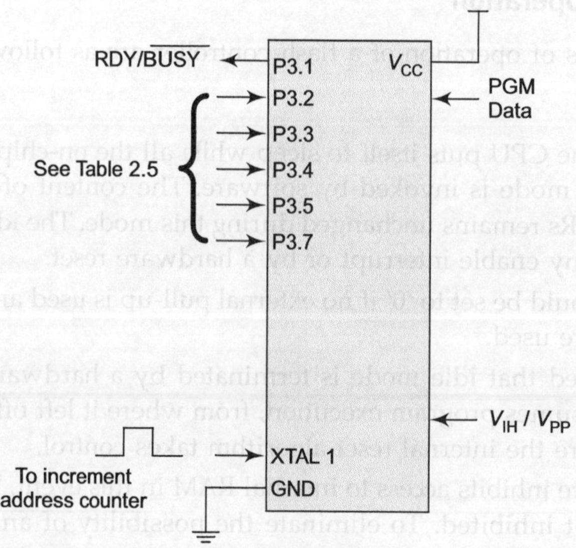

Fig. 2.15 Programming flash memory

Program Algorithm

To program the AT89C2051, the following sequence is recommended by the manufacturer:

1. Power-up sequence: Apply power between the V_{CC} and the GND pins. Set RST and XTAL1 to GND with all other pins floating and wait for at least more than 10 ms.
2. Set pin RST to 'H'.
   Set pin P3.2 to 'H'.
3. Apply the appropriate combination of 'H' or 'L' logic level to pins P3.3, P3.4, P3.5, P3.7 select one of the programming operations shown in Table 2.5.

Table 2.5 EPROM programming modes

MODE	RST	P3.2/PROG	P3.3	P3.4	P3.5	P3.7
Write code data [1.3]	12V	‾_/‾	L	H	H	H
Read code data [1]	H	H	L	L	H	H
Write lock	12V	‾_/‾	H	H	H	H
Bit -1	12V	‾_/‾	H	H	L	L
Chip erase	12V	‾_/‾	H	L	L	L
Read signature byte	H	H	L	L	L	L

To Program and Verify the Array

Apply the byte to be written at the starting location in memory on the Port 1 pins. Apply reset pulse raised to +12 V, which enables programming mode. Apply logic 'high' at port pin P3.2 which will write the byte in the EPROM array or the lock bits. The byte-write cycle is self-timed and typically takes 1.2 ms. To verify the programmed data, lower RST from 12 V to logic 'H' level and set pins P3.3 … P3.7 to the appropriate levels as shown in Table 2.5. Read the Port P1 pins for output data for verification.

Apply pulse at XTAL1 pins once which advances the internal address counter. Apply new data to the Port P1 pins. Repeat above procedure steps with new data and advancing the address counter for the 2 K array or until the end of the object file is reached.

Power-off Sequence

1. Set XTAL to 'L'
2. Set RST to 'L'
3. Float all other I/O pins
4. Turn V_{CC} power OFF

Data polling The AT89C2051 features data polling to indicate the end of a write cycle. During a write cycle an attempted read of the last byte written will result in the written data on P1.7 being complemented. Once the write has been completed, true data is valid on all outputs, and the next cycle may begin. Data polling may begin any time after a write cycle has been initiated.

Ready/Busy The progress of byte programming can also be monitored by the RDY/BSY output signal. Pin P3.1 is pulled low after P3.2 goes high, to indicate BUSY. P3.1 is pulled high again when programming is done to indicate READY.

Program Verification

If lock bits LB1 and LB2 have not been programmed, code data can be read back via the data lines for verification.

1. Reset the internal address counter to 000h by bringing RST from 'L' to 'H'.
2. Apply the appropriate control signals for read code data and read the output data at the Port P1 pins.
3. Pulse pin XTAL1 once to advance the internal address counter.
4. Read the next code data byte at the Port P1 pins.
5. Repeat Steps 3 and 4 until the entire array is read.

The lock bits cannot be verified directly. Verification of the lock bits is achieved by observing that their features are enabled.

Chip Erase

The entire EPROM array 2 K and the two lock bits are erased electrically by using the proper combination of control signals and by holding P3.2 low for 10 ms. The code array is written with all ls in the chip erase operation and must be executed before any nonblank memory byte can be reprogrammed.

Reading the signature bytes The signature bytes are read by the same procedure as normal verification of location 000h, 001h, and 002h, except that P3.5 and P3.7 must be pulled to logic low. The values returned are as follows:

1. (000h) = 1Eh indicates manufactured by Atmel
2. (001h) = 21h indicate 89C2051
3. (002h) = FFh indicates 12 V programming

The important points to be remembered are as follows

1. The internal EPROM address counter is reset to 000h on the rising edge of RST and is advanced by a positive pulse at XTAL1 pin.
2. Chip erase requires a 10 ms PROG pulse.
3. P3.1 is pulled low during programming to indicate RDY/BSY pulse at XTAL pin.

2.6 | HIGH SPEED DERIVATIVES

In previous sections, we have seen some common features of the MCS-51 family of microcontrollers. The extensive use of microcontrollers in different industrial applications has motivated many manufacturers to develop different varieties of microcontrollers. The difference may be in terms of number of bits of operation, internal and external memory capacity, number of timers present in the architecture, number of interrupt facility and incorporation of some other special features. In compatibles, special features are made available, so that depending on the requirements a suitable controller may be chosen. The special features include:

Watchdog timer   WDT is user programmable free running timer that can serve as a time-base generator, an event timer and system supervisor. Incorporation of WDT in controllers is a commonly used technique for verifying proper program execution. If program execution diverts from its proper way, the WDT cannot be restarted before its time-out, which causes a reset after a predetermined time interval. If there is no fault the WDT is restarted before its time-out occurs.

Pulse width modulation   PWM is a technique where the duty cycle of a signal is varied between 0% and 100% so as to vary the average voltage of the particular signal from 0 to maximum value. The PWM signals are used for different control actions like DC motor speed control, temperature control by on-off control, etc.

Analog comparator   AC compares an unknown analog voltage with a reference analog voltage. When analog voltage exceeds the reference voltage, the output of the comparator gives a logic change.

Inter integrated circuits I^2C serial bus is a twin-line serial communication bus. The system is unique because data transport, clock generation, address recognition and bus arbitration are controlled by hardware. It is designed primarily for efficient IC control. This system is composed of two bus lines SCL (serial clock), SDL (serial data) that carry information between the ICs connected to them.

The I^2C bus allows the designer to implement intelligent application oriented control circuits without encountering numerous interfacing problems. Proven I^2C applications are currently being implemented in digital control/signal processing circuits for audio and video systems, a lower cost alternative for RS-232 bus standard. Some I^2C peripherals are as follows: PCF8570 RAM chip, SAAI064 LED driver, etc.

Global serial channel GSC is a high-speed multiprotocol synchronous serial communication interface. It uses packeted data frames that consist of a beginning of frame (BOF), address byte(s), data byte(s), a cyclic redundancy check (CRC) bit and an end of frame (EOF).

Control area network CAN is industrial standard network protocol that permits any processor to be connected together for data interchange. It is a replacement for I^2C bus communication. This results in much reduced wiring harness and enhanced diagnostic and supervising capabilities.

Programmable counter array PCA provides more timing capability with less CPU intervention. The advantages of PCA are to reduce software over head, improved accuracy, and higher speed for different operations. It has five 16-bits captured/compare modules associated in it. Modules can be operated in either of the following modes. Rising edge and/or falling edge capture mode, 16-bits software timer mode, high-speed output mode, WDT mode and PWM mode. The clock input to the PCA timer can be selected from the following four modes:

Oscillator frequency	Mode 0
Oscillator frequency	Mode 1
Timer 0 overflow	Mode 2
External input	Mode 3

In the *capture mode* the counts of the counter can be captured at different transitions of input signals which allows the PCA flexibility to measure timer periods, pulse widths, duty cycle and phase difference on up to five separate inputs.

In the *compare mode* 16-bits value of the PCA timer is compared with a 16-bit value placed in the module's compare registers. When there is a match one of the three events can happen.

Interrupt: software timer mode

Toggle of a port pin: high-speed output mode

Reset: watchdog timer mode

In PWM mode the PCA generates 8-bit PWM outputs by comparing the low byte of PCA timer with the low byte of the compare register.

Analog to digital converters ADC is the feature, which converts one or more input analog signal(s) into 8/10/12/16-bits digital signals.

Direct memory access DMA is a special feature where the data transfer between memory and peripherals occurs without CPU's intervention. Hence, the CPU idle times decreases and speed of program execution increases.

Even though the additional features are useful in complex system design, we will restrict our study to the designing with basic MCS-51 family of micro-controllers.

2.7 │ AT89C52

The AT89C52 is a low-power, high-performance CMOS 8-bit microcomputer with 8 K bytes of flash programmable and erasable read-only memory (PEROM) and is compatible with the industry-standard 80C51 and 80C52 instruction set and pin out. The on-chip flash allows the program memory to be reprogrammed in-system or by a conventional nonvolatile memory programmer. The features includes: compatibility with MCS-51™, 8 K in-system reprogrammable flash memory; 256 × 8-bit internal RAM, 32 programmable I/O lines, three 16-bit timer/counters, eight interrupt sources, programmable serial channel, low-power idle and power-down modes. Figure 2.16 shows pin layout of 89C52, while Fig. 2.17 depicts internal block diagram.

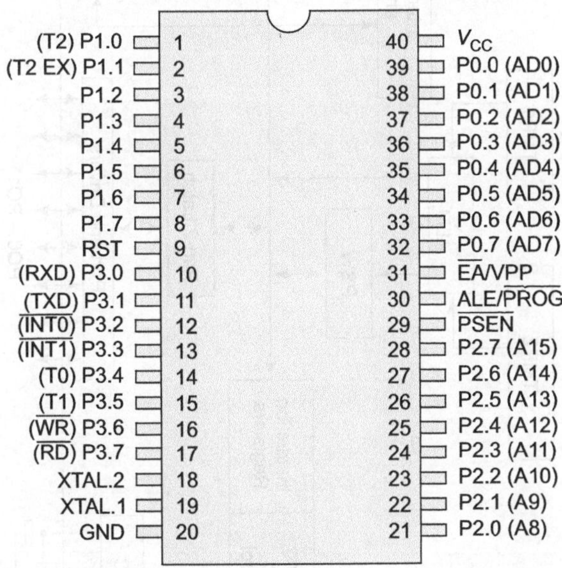

Fig. 2.16 89C52: Pin diagram

The special features of variations in 89C52 with reference to MCS-51 are described below:

Port Pins P1.0 and P1.1 P1.0 and P1.1 can be configured to be the Timer/Counter 2 external count input (P1.0/T2) and the Timer/Counter 2 trigger input (P1.1/T2EX), respectively.

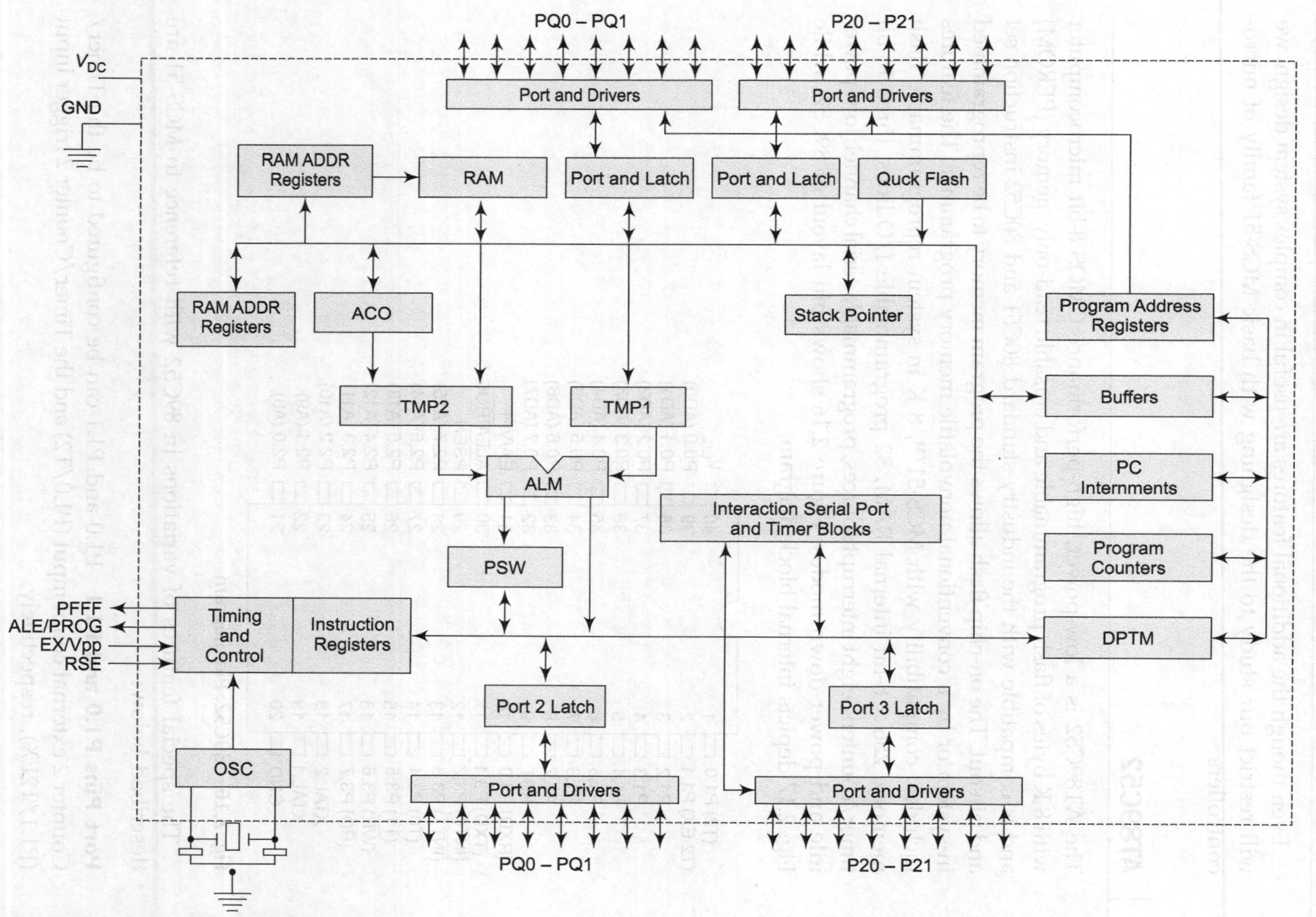

Fig. 2.17 89C52: Internal Diagram

ALE/PROG   Address latch enable is an output pulse for latching the low byte of the address during accesses to external memory. This pin is also the program pulse input (PROG) during flash programming. In normal operation, ALE is emitted at a constant rate of 1/6 the oscillator frequency and may be used for external timing or clocking purposes. Note, however, that one ALE pulse is skipped during each access to external data memory. If desired, ALE operation can be disabled by setting bit 0 of SFR location 8EH. With the bit set, ALE is active only during a MOVX or MOVC instruction. Otherwise, the pin is weakly pulled high. Setting the ALE-disable bit has no effect if the microcontroller is in external execution mode.

Table 2.6   89C52: SFR map and reset value

0F8H									0FFH
0F0H	B 00000000								0F7H
0E8H									0EFH
0E0H	ACC 00000000								0E7H
0D8H									0DFH
0D0H	PSW 00000000								0D7H
0C8H	T2CON 00000000	T2MOD xxxxxx00	RCAP2L 00000000	RCAP2H 00000000	TL2 00000000	TH2 00000000			0CFH
0C0H									0C7H
0B8H	IP xx000000								0BFH
0B0H	P3 11111111								0B7H
0A8H	IE 0x000000								0AFH
0A0H	P2 11111111								0A7H
98H	SCON 00000000	SBUF xxxxxxxx							9FH
90H	P1 11111111								97H
88H	TCON 00000000	TMOD 00000000	TL0 00000000	TL1 00000000	TH0 00000000	TH1 00000000			8FH
80H	P0 11111111	SP 00000111	DPL 00000000	DPH 00000000				PCON 0xxx0000	87H

EA/VPP   External access enable (EA) must be strapped to GND in order to enable the device to fetch code from external program memory locations starting at 0000H up to FFFFH. Note, however, that if lock bit 1 is programmed, EA will be internally latched on reset. EA should be strapped to V_{CC} for internal program executions. This pin also receives the 12 V programming enable voltage (V_{PP}) during flash programming when 12 V programming is selected.

Special function registers   A map of the on-chip memory area is called the special function register (SFR). The space is shown in Table 2.6. All of the addresses are occupied, and unoccupied addresses may not be implemented well in general return random data, and write accesses will have an indeterminate effect. User software should not write 1s to these unlisted locations, since they may be used in future products to invoke new features. In that case, the reset or inactive values of the new bits will always be 0.

Interrupt registers   The individual interrupt enable bits are in the IE register. Two priorities can be set for each of the six interrupt sources in the IP register.

Data memory   The AT89C52 implements 256-bytes of on-chip RAM. The upper 128-bytes occupy a parallel address space to the special function registers. That means the upper 128-bytes have the same addresses as the SFR space but are physically separate from SFR space. When an instruction accesses an internal location above address 7FH, the address mode used in the instruction specifies whether the CPU accesses the upper 128-bytes of RAM or the SFR space. Instructions that use direct addressing access SFR space. Instructions that use indirect addressing access the upper 128-bytes of RAM. Since the stack operations are examples of indirect addressing, so the upper 128-bytes of data RAM are available as stack space.

Timer 2   is a 16-bit timer/counter that can operate as either a timer or an event counter. The type of operation is selected by bit C/T2 in the SFR T2CON (Table 2.7). Timer 2 has three operating modes: capture, auto-reload (up or down counting), and baud rate generator. The modes are selected by bits in T2CON. Control and status bits of Timer 2 are contained in registers T2CON and T2MOD. The register pair (RCAP2H, RCAP2L) are capture/reload registers in 16-bit capture mode or 16-bit auto-reload mode.

Table 2.7   T2CON: Time/counter register

Hardware address: 0C8H				Reset value: 0000 0000B			
D7	D6	D5	D4	D3	D2	D1	D0
TF2	EXF2	RCLK	TCLK	EXEN2	TR2	C/T2	CP/RL2

TF2	Timer 2 overflow flag (set by h/w when o/f occurs). It will not be set if RCLK or TCLK IS '1'
EXF2	Timer 2 external flag run control bit (set or cleared by s/w)
RCLK	Receiver clock enable
TCLK	Transmitter clock enable
EXEN2	Timer 2 external enable; allows capture or reload
TR2	Start/Stop control for Timer 2
C/T2	Counter/timer mode for Timer 2
CP/RL2	Capture or reload select (capture if ='1' , reload otherwise)

2.8 | PIC18F4431 MICROCONTROLLER

Figure 2.18 depicts pin configuration of PIC18F4431 microcontroller. It is a flash microcontroller with nanowatt technology, high performance PWM, and A/D. The features of this variant are:

1. **14-bit Power Control PWM** Module which has four channels with complementary outputs with edge- or center-aligned operation, flexible deadband generator, and fault protection inputs. It supports simultaneous update of duty cycle and period.

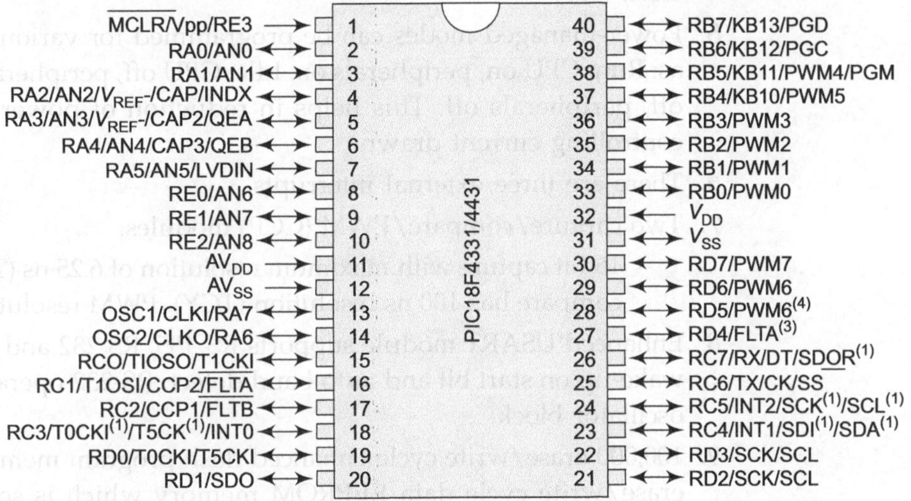

Fig. 2.18 PIC18F4431: Pin diagram

2. **Motion Feedback Module** It contains three independent input channels with special hall sensor interface. The channels can be programmed for period and pulse-width measurement. The module has special event trigger output to other modules as well as two phase inputs and one index input from encoder. It provides for high- and low-position tracking with direction status and change of direction interrupt as well as velocity measurement. This module features a quadrate encoder interface (QEI) and an input capture (IC) module. The QEI accepts two-phase inputs (QEA, QEB) and one index input (INDX) from an incremental encoder. The QEI supports high and low precision position tracking, direction status and change of direction interrupt, and velocity measurement. The input capture features 3 channels of independent input capture with Timer 5 as the time base, a special event trigger to other modules, and an adjustable noise filter on each IC input.

3. High-speed, 200 kbps 10-bit A/D converter up to nine channels with simultaneous sampling for two-channel sampling and sequential sampling for 1, 2 or 4 selected channels. It has auto-conversion capability and 4-word FIFO with selectable interrupt frequency. The acquisition time can also be programmed.

4. It has flexible oscillator structure having four crystal modes up to 40 MHz and two external clock modes up to 40 MHz. Internal oscillator block can be operated at any of the selectable frequencies between 31 kHz to 8 MHz. Fail-safe clock monitor: allows for safe shutdown of device if clock fails.

5. Power-managed modes can be programmed for various operations such as: Run: CPU on, peripherals on, Idle: CPU off, peripherals on; Sleep: CPU off, peripherals off. This helps in reduction of power consumption by controlling current drawn.

6. There are three external interrupts.

7. Two capture/compare/PWM (CCP) modules:
   – 16-bit capture with maximum resolution of 6.25 ns (TCY/16) while the compare has 100 ns resolution (TCY). PWM resolution is 1 to 10-bits.

8. Enhanced USART module supports RS-485, RS-232 and LIN 1.2 has auto-wake-up on start bit and auto-baud detect. RS-232 operation uses internal oscillator block.

9. 100,000 erase/write cycle enhanced flash program memory and 1,000,000 erase/write cycle data EEPROM memory which is self-programmable under software control.

10. Extended watchdog timer (WDT) is programmable period from 41 ms to 131 s.

11. Single-supply in-circuit serial programming™ (ICSP™) via two pins.

12. In-circuit debug (ICD) via two pins.

13. **Reset**   It can differentiate between various kinds of reset such as: power-on reset (POR), MCLR reset during normal operation or sleep, watchdog timer (WDT) Reset during execution, programmable brown-out reset (BOR), RESET instruction stack full or under flow reset.

2.8.1 Memory Organization

There are three memory types in enhanced MCU devices: program memory, data RAM, and data EEPROM. Data and program memory use separate buses, which allows for concurrent access of these types. A 21-bit program counter is capable of addressing 2 M program memory space (Fig. 2.19). Accessing a location between the physically implemented memory and the 2 M address will cause a read of all 0s (an NOP instruction). It has 8 K flash memory which can store up to 4 K single-word instructions and 16 K flash memory and can store up to 8192 single-word instructions. The reset vector address is at 000000h and the interrupt vector addresses are at 000008h and 000018h.

Flash program memory is readable, writable, and erasable during normal operation over the entire V_{DD} range. A read from program memory is executed on 1-byte at a time. A write to program memory is executed on blocks of 8-bytes at a time. Program memory is erased in blocks of 64-bytes at a time. A bulk erase operation may not be issued from user code. While writing or erasing program memory, instruction fetches cease until the operation is complete. The program memory cannot be accessed during the write or erase, therefore, code cannot execute. An internal programming timer terminates program memory writes and erases.

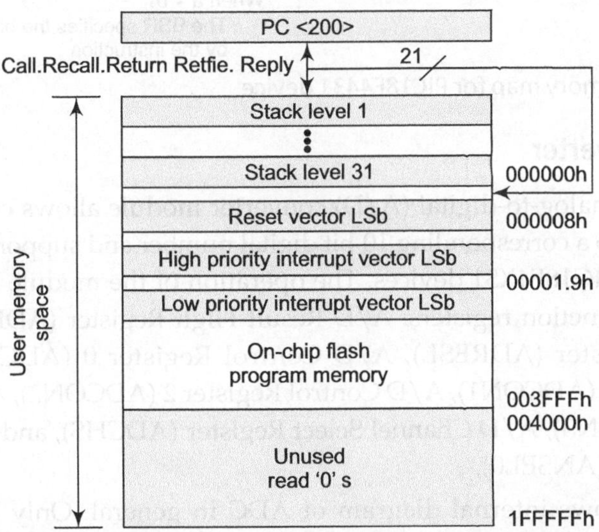

Fig. 2.19   Program memory map and stack: PIC18F4431

Data EEPROM memory is readable and writable during normal operation over the entire V_{DD} range. The data memory is indirectly addressed through the special function registers (SFR), EECON1, EECON2, EEDATA, and EEADR. It allows byte read and write and holds the 8-bit data for read/write and EEADR holds the address of the EEPROM location being accessed. These devices have 256-bytes of data EEPROM within address range from 00h to FFh. The write time is controlled by an on-chip timer. Figure 2.20 depicts the memory map.

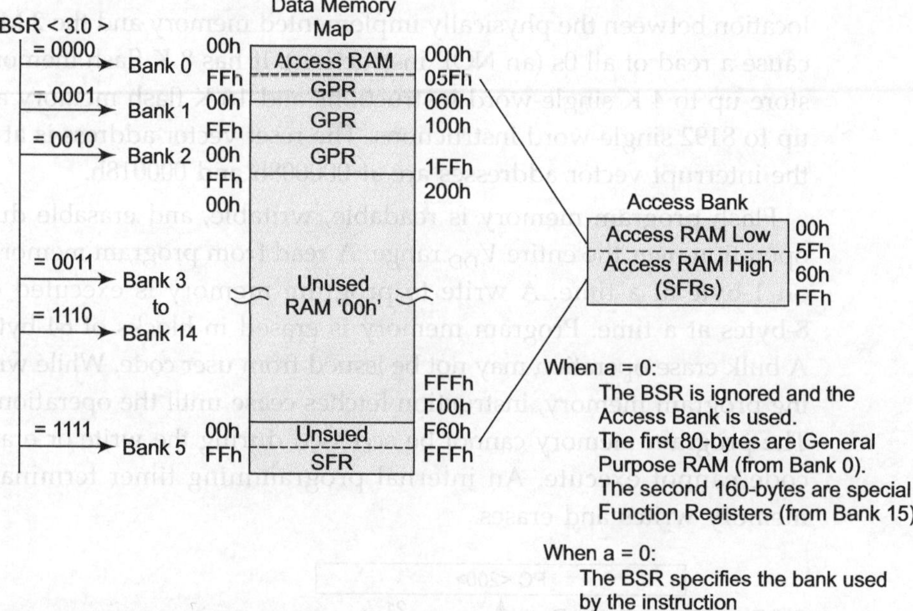

Fig. 2.20 Data memory map for PIC18F4431 device

2.8.2 A/D Converter

The high-speed analog-to-digital (A/D) converter module allows conversion of an analog signal to a corresponding 10-bit digital number and supports up to nine channels on the PIC18F4X31 devices. The operation of the module is controlled by nine special function registers: A/D Result High Register (ADRESH), A/D Result Low Register (ADRESL), A/D Control Register 0 (ADCON0), A/D Control Register 1 (ADCON1), A/D Control Register 2 (ADCON2), A/D Control Register 3 (ADCON3), A/D Channel Select Register (ADCHS), and Analog I/O Select Register 0 (ANSEL0).

Figure 2.11 shows internal diagram of ADC in general. Only PIC18F4X31 devices have AN5-AN9 channels available.

The A/D channels are grouped into four sets of two or three channels. For the PIC18F4X31 devices, AN0, AN4, and AN8 are in Group A, AN1 and AN5 are in Group B, AN2 and AN6 are in Group C and AN3 and AN7 are in Group D. The selected channel in each group is selected by configuring the A/D channel select register, ADCHS.

The analog voltage reference is software selectable to either the device's positive and negative analog supply voltage (AVDD and AVSS), or the voltage level on the RA3/AN3/VREF+/CAP2/QEA and RA2/AN2/VREF-/CAP1/INDX, or some combination of supply and external sources. Register ADCON1 controls the voltage reference settings. The A/D converter can operate while the device is in Sleep mode. To operate in Sleep, the A/D conversion clock is derived from the A/D's internal RC oscillator.

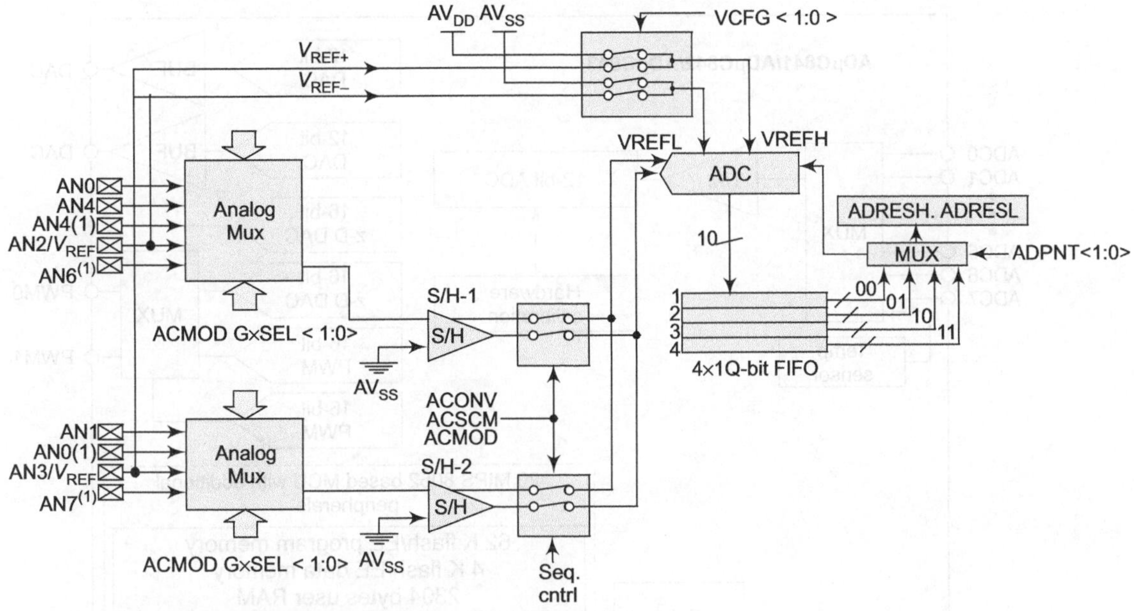

Fig. 2.21 Block diagram: AD section

2.9 | ADµC842

The ADµC842 is a complete smart transducer front end, that integrates a high-performance self-calibrating multichannel ADC, a dual DAC, and an optimized single cycle 20 MHz 8-bit MCU (8051 instruction set compatible) on a single chip. It is also called *microconverter*, both ADCs and DACs are in-built and hence no external chip interfacing is required which otherwise would have added to system unreliability. Both the ADCs and DACs are having 12-bit resolution and are highly accurate and have high-sampling rates. The controller provides in-circuit re-programmability (ICP) and hence makes it easy to perform desktop experiments with ease of debugging.

It has single cycle 20 MIPS 8052 core, 62 K chip flash program memory, 4 K chip flash data memory, two flexible PWM outputs/16-bit SD DACs, 32 kHz external crystal and on-chip programmable PLL, temperature monitor, precision voltage reference (20 ppm/°C), serial interface ports (UART, I²C and SPI), watchdog timer, time interval counter (TIC), power supply monitor, power-on-reset (POR), and embedded download/debug and emulation features. It is pin-compatible upgrade to ADµC832. Figure 2.22 depicts the functional block diagram of the chip.

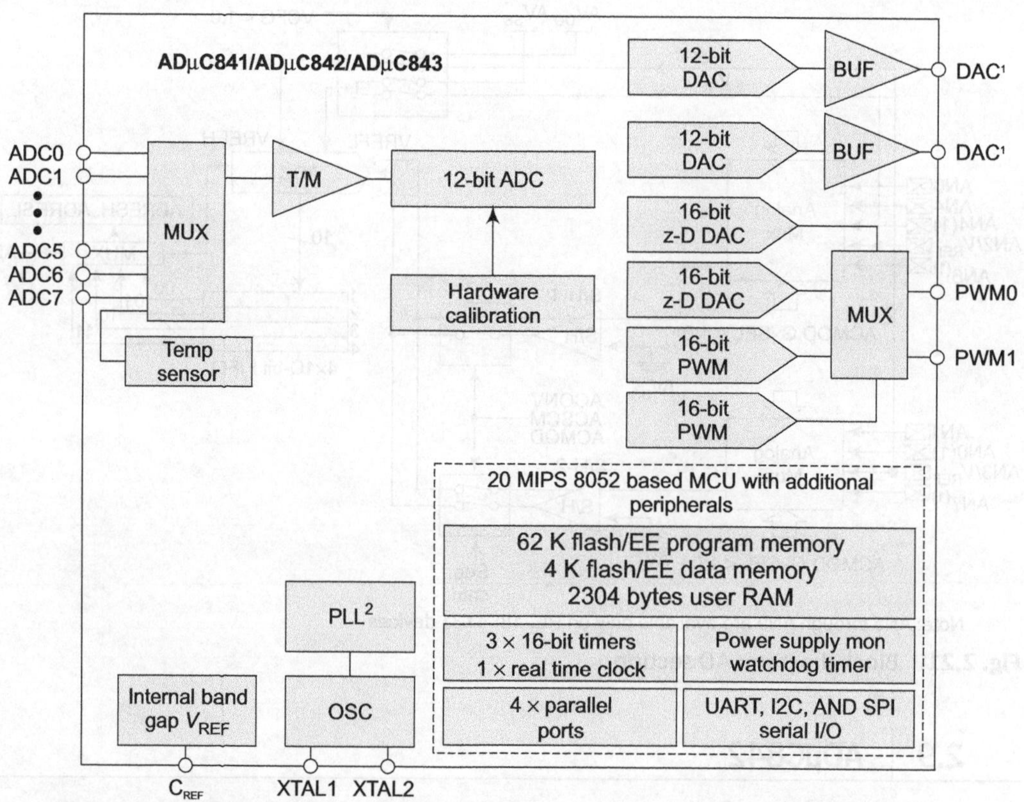

Fig. 2.22 ADµC842: Internal block diagram

It uses a 32 kHz crystal with an on-chip PLL generating a programmable core clock up to 16.78 MHz. The microcontroller is an optimized 8052 core offering up to 20 MIPS peak performance. Three different memory options are available offering up to 62 kB of non-volatile flash/EE program memory. The 4 K of non-volatile Flash/EE data memory, 256-bytes RAM are also integrated on-chip. The option allows SPI operate separately on P3.3, P3.4 and P3.5 while, I²C uses the standard pins. The I²C interface has also been enhanced to offer repeated start, general call, and quad addressing. Figure 2.23 depicts pin configuration of the chip, while Table 2.8 depicts pin definitions.

The ADC block provides an eight channel, 12-bit single supply ADC with track-and-hold on-chip reference, calibration features. The ADC block can be configured using ADCCON1-3 control registers. The 12-bit result word is stored in the ADCDATA (L/H) registers. The upper 4-bits of ADCDATAH contains the channel number. Table 2.9 shows the format of the result.

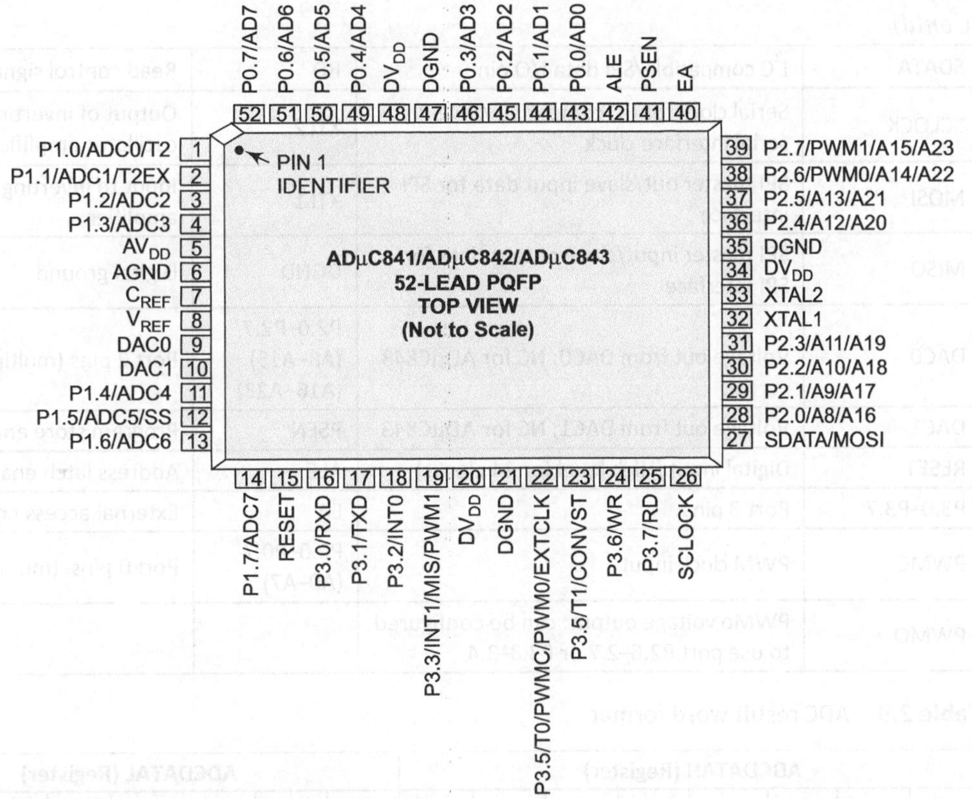

Fig. 2.23 ADμC842: Pin configuration

Table 2.8 Pin definitions of ADμC842

Mnemonic	Function	Mnemonic	Function
DV$_{DD}$	Digital positive supply (3V or 5V)	PWM1	PWM1 voltage output
AV$_{DD}$	Analog positive supply (3V or 5V)	RxD	Receiver data input
C$_{ref}$	Decoupling input connect. $C = 0.47\ \mu F$ between AGND and this pin	TxD	Transmitter output data
V$_{ref}$	NC	INT0	Interrupt 0
AGND	Analog ground	INT1	Interrupt 1
P1.0–P1.7	Port 1 pins	T0	Timer/Counter Input 0
ADC0–ADC7	Analog input. channel selection by SFR	T1	Timer/Counter Input 1
T2	Timer 2 digital input	CONVST	Start convert (active low)
T2EX	Digital input: capture/reload trigger for Timer 2	EXTCLK	External clock input
SS	Slave select input for SPI interface	WR	Write control signal

(Contd)

(Contd)

SDATA	I^{2}C compatible/SPI data I/O pin	RD	Read control signal
SCLOCK	Serial clock for I^{2}C compatible/SPI SPI serial interface clock	XTL2	Output of inverting oscillator amplifier
MOSI	SPI master out/slave input data for SPI interface	XTL1	Input to inverting oscillator amplifier
MISO	SPI master input/slave output data for SPI interface	DGND	Digital ground
DAC0	Voltage out from DAC0; NC for ADµC843	P2.0–P2.7 (A8–A15) (A16–A23)	Port 2 pins (multiplexed)
DAC1	Voltage out from DAC1; NC for ADµC843	PSEN	Program store enable
RESET	Digital input (High level for 24 clocks)	ALE	Address latch enable
P3.0–P3.7	Port 3 pins	EA	External access enable
PWMC	PWM clock input	P0.0–P0.7 (A0–A7)	Port 0 pins. (multiplexed)
PWMO	PWMo voltage output: can be configured to use port P2.6–2.7 or P3.3–3.4		

Table 2.9   ADC result word format

ADCDATAH (Register)								ADCDATAL (Register)							
D15	D14	D13	D12	D11	D10	D9	D8	D7	D6	D5	D4	D3	D2	D1	D0
ADC: Channel information				Upper 4-bits of result				Lower 8-bits of 12-bit Result							

ADC calibration modes can be easily initiated by user software. The ADCCON3 SFR is used to calibrate the ADC. Bit 1 (typical) and CS3 to CS0 (ADCCON2) setup the calibration modes.

- Device calibration is used to compensate for significant changes in operating frequency, analog input range, reference voltage, and supply voltages. Offset calibration uses internal AGND selected via ADCCON2 register Bits CS3 to CS0 (1011), and gain calibration uses internal V_{REF} selected by bits CS# to CS0 (1100). It must follow offset calibration.

- System calibration can be initiated to compensate for both internal and external system errors.

The DAC block incorporates two 12-bit voltage output DACs on-chip which can be operated in 12-bit/8-bit modes. The voltage output buffer is capable of driving 10 kW/100 pF. Each has two selectable ranges, 0 V to V_{REF} and 0 V to AVDD.

A PLL block locks onto a multiple of basic frequency to provide a stable 16.78 MHz clock for the system. The core can operate at this frequency or at binary submultiples of it to allow power saving in cases where maximum core performance is not required. The default core clocks are also derived from the PLL clock, with the modulator rate being the same as the crystal oscillator frequency. The PLL can be configured using the SFR termed as PLLCON. At 5 V the core clock can be set to a maximum of 16.78 MHz, while at 3 V the maximum core clock setting is 8.38 MHz.

The chip can be configured to serial download mode to download the code using the standard UART serial port. The user can download code to the full 62 K of Flash/EE program memory while the device is in circuit in its target application hardware. All registers are set to their default state and program execution starts at the reset vector once the RESET pin is deasserted.

2.10 | FLASH CONTROLLERS (SX FAMILY)

SX family is high-performance 8-bit microcontrollers fabricated in an advanced CMOS process technology, combined with a RISC-based architecture. They allow high-speed computation, flexible I/O control, and efficient data manipulation. Throughput is enhanced by operating the device at frequencies up to 50 MHz and by optimizing the instruction set to include mostly single-cycle instructions. On-chip functions include:

- Two 16-bit timers with 8-bit pre-scalar supporting different operating modes (PWM, simultaneous PWM/capture, and external event counter).

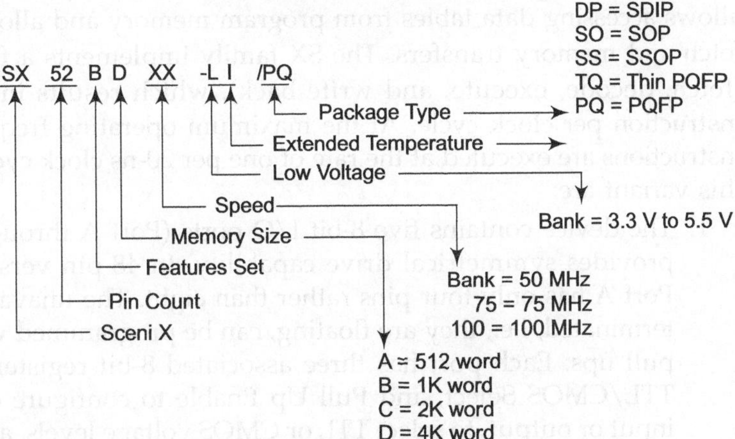

Fig. 2.24 SX Family: Part Number reference

- A general purpose 8-bit timer with pre-scalar, an analog comparator, a brown out detector, a watchdog timer, a power-save mode with multi-source wakeup capability, an internal R/C oscillator, user-selectable clock modes, and high-current outputs. Figure 2.24 describes the part number reference.

The SX48BBD and SX52BD are functionally the same, except for the package type and pinout (Fig. 2.25). The SX48BD has four fewer pins and has only four rather than eight I/O pins for Port A.

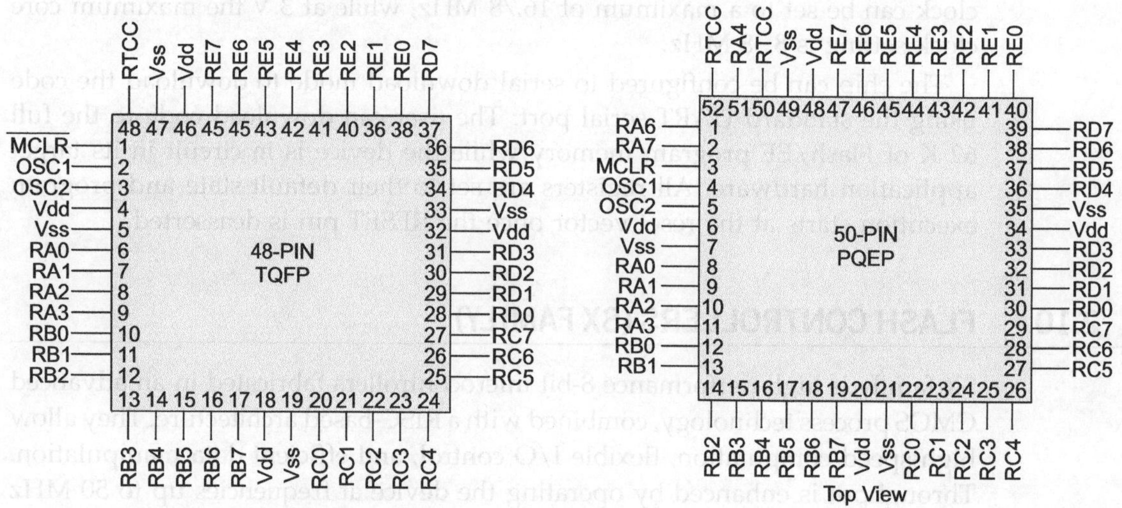

Fig. 2.25   SX Family: SX48BBD and SX52BD

The SX devices use a modified Harvard architecture which uses two separate memories with separate address buses, one for the program and one for data, while allowing transfer of data from program memory to SRAM. This ability allows accessing data tables from program memory and allows overlapping of fetch and memory transfers. The SX family implements a four-stage pipeline (fetch, decode, execute, and write back), which results in execution of one instruction per clock cycle. At the maximum operating frequency of 50 MHz, instructions are executed at the rate of one per 20-ns clock cycle. The features of this variant are:

1. The device contains five 8-bit I/O ports (Port A through Port E). Port A provides symmetrical drive capability. In 48-pin version of the device, Port A has only four pins rather than eight. The unavailable pins are not terminated, i.e., they are floating, can be programmed with internal weak pull ups. Each port has three associated 8-bit register: Direction, Data, TTL/CMOS Select, and Pull Up Enable to configure each port as Hi-Z input or output, to select TTL or CMOS voltage levels, and enable/disable the weak pull-up resistor. The least significant bit of the register corresponds to the least significant port pin. To access these configuration register, an appropriate value must be written into the Mode register. On power-up, all bits are initialized to 1.

2. Port B also supports the on-chip differential comparator. Port RB1 and RB2 are the comparator negative and positive inputs, respectively, while Port RB0 is the comparator output pin. Port B supports the multi-input wakeup

feature on all eight pins. Port B and Port C also support the multi-function timers T1 and T2.

3. RB4 and RB5 are the T1 capture inputs, RB6 is the T1 PWM output, and RB7 is the T1 external event counter input. Similarly, RC0 and RC1 are the T2 capture inputs, RC2 is the T2 PWM output, and RC3 is the T2 external event counter input.

4. The five ports are memory-mapped into the data memory address space as the RA, RB, RC, RD, and RE file registers at data memory address 05h through 09h, respectively. Writing to a port data register sets the voltage levels of the corresponding port pins that have been configured to operate as outputs. Reading from a data register reads either the voltage levels of the corresponding port pins or the data contained in the port data register depending on the status PORTRD bit contained in the T2CNTB register.

5. The mode register controls access to the port configuration registers and Timer T1/T2 control registers. Because the MODE register is not memory-mapped, it is accessed by the special-purpose instructions.

6. When power is made ON, all the port control register are initialized to FFh. Thus, each port pin configuration is to operate a high-impedance input that senses TTL voltage levels, with no internal pull-up resistor connected. The MODE register is initialized to 1Fh, which allows immediate write access to the data direction register using the "MOV !rx, W" instruction.

7. The device contains an on-chip differential comparator. Ports RB0–RB2 support the comparator. Ports RB1 and RB2 are the comparator negative and positive inputs, respectively, while Port RB0 serves as the comparator output pin. To use these pins in conjunction with the comparator, the user program must configure Ports RB1 and RB2 as inputs and Port RB0 as an output. The CMP_B register is used to enable the comparator, to read the output of the comparator internally, and to enable the output of the comparator to comparator output pin.

8. Power-on-reset, brown reset, watchdog reset, wakeup reset, or external reset initializes the device. Each one of these reset conditions causes the program counter to branch to the top of the program memory (FFFh).

2.10.1 Special Function Registers

Table 2.10 depicts the set of SFR to control operation of the device. The CPU register includes an 8-bit working register (W), which serves as a pseudo accumulator. It holds the second operand of an instruction, receives the literal in immediate type instructions.

A set of 31 file register serves as the primary accumulator. One of these registers holds the first operand of an instruction and another can be program-selected as the destination register. The first 10 file register include the RealTime Clock/Counter register (RTCC), the lower 8-bits of the 12-bits program counter (PC), the 8-bit STATUS register, five port control register for Ports A through E, the 8-bit File Select Register (FSR), and INDF. The five low order bits of the FSR register select one of the 31 file register in the indirect addressing mode.

Table 2.10   Special function registers

Addr	Name	Function
00h	INDF	Used for indirect addressing
01h	RTCC	Real time clock/counter
02h	PC	Program counter (low byte)
03h	STATUS	Holds status bits of ALU
04h	FSR	File select register
05h	RA	Port RA data register
06h	RB	Port RB data register
07h	RC	Port RC data register
08h	RD	Port RD data register
09h	RE	Port RE data register

Status Register

The STATUS register holds the arithmetic status of the ALU, the page select bits, and the reset state. The STATUS register is accessible during run time, except that bits PD and TO are read-only. It is recommended that only SETB and CLRB instruction be used on this register. Care should be exercised when writing to the STATUS register as the ALU status bits are update upon completion of the write operation, possibly leaving the STATUS register with a result that is different than intended.

D7	D6	D5	D4	D3	D2	D1	D0
PA2	PA1	PA0	TO	PD	Z	DC	C

PA2–PA0: Page Select Bit: 000-page0 (000-01ffh) … 111:page7 (0E00-0FFFh)
TO   : Time-out bit
PD   : Power down
Z   : Zero flag
DC   : Digital carry (similar to auxilliary carry)
C   : Carry flag

Option Register

When the OPTIONX bit the FUSE word is cleared, bits 6 and 7 of the OPTION register function as described below. When the OPTIONX bit is set, bits 7 and 6 of the OPTION register read as 1s.

D7	D6	D5	D4	D3	D2	D1	D0
RTW	RTE_IE	RTS	RTE_ES	PSA	PS2	PS1	PS0

RTW	:	RTCC/W; '0'=W and ' 1'= RTCC
RTE_IE	:	RTCC Edge Enable
		0 = RTCC roll-over interrupt is enabled
		1 = RTCC roll-over interrupt is disabled
RTS		RTCC increment select
RTE_ES		RTCC edge select
PSA		Pre scalar Assignment
PS2–PS0		Pre scalar divide ratios :

	Divide rate	→ RTCC	Watchdog
000		1:2	1:1
001		1:4	1:2
010		1:8	1:4
011		1:16	1:8
100		1:32	1:16
101		1:64	1:32
110		1:128	1:64
111		1:256	1:128

2.10.2 Device Configuration Registers

The SX device has three registers (FUSE, FUSEX, DEVICE) that control functions such as operating the device in Turbo mode, extended (8-level deep) stack operation, and speed selection for the internal RC oscillator. These register are not programmable "on the fly" during normal device operation. Instead, the FUSE and FUSEX register can only be accessed when the SX device is being programmed. The DEVICE register is a read-only, hard-wired register, programmed during the manufacturing process.

FUSE Word

D10	D9	D8	D7	D6	D5	D4	D3	D2	D1	D1
TURBO	SYNC	OPTIONX	STACKX	IRC	DIV2	DIV0	CP	WDTE	FOSC1	FOSC0

TURBO	Turbo mode enable
SYNC	Synchronous input enable (for turbo mode):
OPTIONX	Option register extension enable
STACKX	Stack extension enable
	0 = 8 level (stack extension enabled)
	1 = 2 level (stack extension disabled)
IRC	Internal RC oscillator divider
CP	Code protect enable
WDTE	Watchdog timer enable

(Contd)

Fosc1: FOSC0	External oscillator configuration (Valid when IRC = 1)
00b =	LP – low power crystal
10b =	HS – high speed crystal
01b =	XT – normal crystal
11b =	Rc network – OSC2 is pulled high by weak pullup (no CLKOUT output)

FUSEX Word

D10 - D9	D8 D7 D6	D5	D4 D3 D2	D1	D0
SLEEPCLK	WDRT2- WDRT0	CF	IRCTRIM2 ...IRCTRIM0	BOR1	BOR0

SLEEPCLK	Sleep Clock Disable
WDRT2	Delay Reset Timer (DRT) timeout period. [100-No Delay;0.06 – 1920 msec)
CF	Active low – makes the carry flag an input to ADD and SUB instructions
IRCTRIM2-0	Internal RC Oscillator Trim. Operate within the target frequency range of 4.0 MHz plus or minus 8%
BOR1 : BOR0	Brown-Out Reset

2.10.3 Memory Organization

The program memory is organized as 4 K, 12-bit wide words. The program memory words are addressed sequentially by a binary program counter. The program counter starts at zero.

A page is composed of 512 words. The lower 9-bits of the program counter are zeros at the first address of a page and ones at the last address of a page. This page structure has no effect on the program counter. The program counter will freely increment through the page boundaries. The program counter contains the 12-bit address of the instruction to be executed. The lower 8-bits of the program counter are contained in the PC register (02h). The three upper bits are specified by the STATUS register (PA0, PA1, PA2). Changing the status bits is necessary to cause jumps and subroutine calls *across* program memory page boundaries. Prior to the execution of a branch operation, the user program must initialize the upper bits of the STATUS register to cause a branch to the desired page. On reset, the program counter is initialized with 0FFFh.

The subroutine stack consists of eight 12-bit save register A physical transfer of register contents from the program counter to the stack or vice versa, and within the stack, occurs on all operations affecting the stack, primarily calls return. The stack is physically and logically separate from data RAM. The program cannot read or write the stack.

The data memory is a RAM-based register set consisting of 262 general-purpose register and 9 special-purpose register. All of those register are 8-bits wide. The data memory is organized into 16 banks. Designated Bank 0 through Bank F, each containing 16 registers, plus an additional bank of 15 *global register*. Because the register are organized into banks or "files" these memory-mapped are called *file register*.

2.10.4 Interrupt Support

The device supports both interrupt and external maskable interrupts. The internal is generated as a result of the RTCC rolling over from FFh to 00h.

This interrupt source has an associated enable bit in the located in the OPTION register and pending flag bit in the Timer T1 Control B register. In addition, Timer T1 and T2 each has three interrupt sources associated with counter overflow compare match, and input capture.

Port B provides the source for eight external software selectable, edge sensitive interrupt, when the device is not in the power down mode. These interrupt sources share logic with the multi-input wakeup circuitry. The WKEN_B register allows interrupt from Port B to BE individually enabled or disabled clearing a bit in the WKEN_B register enables the interrupt on the corresponding Port B pin. The WKED_B select the transition edge to be either positive or negative.

The EKPND_B register comes up a with random value upon reset. The user program must clear the WKPND_B register prior to enabling the interrupt. All interrupt are global in nature; that is, no interrupt has priority over another. Interrupt are handled sequentially.

Once an interrupt is acknowledged, all subsequent global interrupts are acknowledged, all subsequent global interrupts are disabled until return from servicing the current interrupt. The PC is pushed onto the single-level interrupt stack. And the contents of the FSR, STATUS, and W register are saved in their corresponding shadow register. The status bits PA2, PA1, and AP0 are cleared after STATUS has been saved in its shadow register. The interrupt logic has its own single-level stack and is not part of the CALL subroutine stack. The vector for the interrupt service routine is address 0.

Once in the interrupt service routine, the user program must poll all interrupt pending bits to determine the source of the interrupt. The interrupt service routine should clear the corresponding interrupt pending flag. Normally, it is a requirement for the user program to process every interrupt without missing any. To ensure this, the longest interrupt routine must take less time than the shortest delay between interrupts.

If an external interrupt occurs during the interrupt routine, the pending register will be updates but the trigger will be ignored unless interrupts are disabled at the beginning of the interrupt routine and enabled again at the end. This also requires that the new interrupt does not occur before interrupts are disabled in the interrupt routine. If there is a possibility of additional interrupt occurring before they can be disabled, the device will miss those interrupt triggers. In other words, using more than one interrupt, such as multiple external interrupts, can result in missed or, at best jittery interrupt handling should one occur during the processing of another.

When handling external interrupts, the interrupt routine should clear at least one pending register bit. The bit that is cleared should represent the interrupt being handled in order for the next interrupt to trigger.

Upon return from the interrupt service routine, the contents of PC, FSR, STATUS and W register are restored from their corresponding shadow register. The interrupt service routine should end with instructions such as RETI or RETIW. RETI pops the interrupt stack and the special shadow register used for storing W, STATUS, and FSR (preserved during interrupt handling). RETIW behaves like RETI but also adds W to RTCC. The interrupt return instruction enables the global interrupts.

If a MIWU interrupt occurs during a pre-existing interrupt service routine, the MIWU interrupt flag is set immediately, and the MIWU interrupt is serviced upon completion of the pre-existing interrupt service routine.

2.10.5 Real Time Clock (RTCC)/Watchdog Timer

The device contains an 8-bit real time clock/counter (RTCC) and 8-bit watchdog timer (WDT). An 8-bit programmable pre-scalar extends the RTCC to 16-bits. If the pre-scalar is not used for the RTCC, it can serve as post scalar for the watchdog timer.

Timer consists of a 16-bit counter register supported by a 16-bit capture register and a 16-bit comparison register. The timer I/O pins are alternate functions of Port B pins for Timer T1 and Port C pins for Timer T2. It can be configured to operate in one of the: pulse width modulation (PWM) mode, software timer mode, external event mode or capture/compare mode. Table 2.11 lists I/O port pins associated with the Timer T1 and Timer T2 I/O functions.

Table 2.11 Timer T1/T2 pin assignments

I/O Pin	Timer T1/T2 function
RB4	Timer T1 capture Input 1
RB5	Timer T1 capture Input 2
RB6	Timer T1 PWM/compare output
RB7	Timer T1 external event clock source
RC0	Timer T2 capture Input 1
RC1	Timer T2 capture Input 2
RC2	Timer T2 PWM/compare output
RC3	Timer T2 external event clock source

There are two 8-bit control registers associated with each timer, called the Control A and B register. The Control A register contains the interrupt enable bits and interrupt flag bits associated with the timer. The Control B register contains bits for setting the timer operating mode, the clock pre-scalar divide by factor, and the contains one device configuration bit not related to operation of the multi-function timers. Tables 2.11 and 2.12 depict formats of timer registers.

Table 2.12 Timer T1 counter A register (T1CNTA)

D7	D6	D5	D4	D3	D2	D1	D0
T1CPF2	T1CPF1	T1CPIE	T1CMF2	T1CMF1	T1CMIE	T1OVF	T1OVIE

T1CPF2	Timer T1 capture flag 2.
T1CPF1	Timer T1 capture flag 1.
T1CPIE	Timer T1 capture interrupt enable.
T1CMF2	Timer T1 comparison flag 2.
T1CMF1	Timer T1 comparison flag 1
T1CMIE	Timer T1 comparison interrupt enable.
T1OVF	Timer T1 overflow flag. This flag is automatically set to 1 when the timer counter overflows from FFFFh to 0000h. The flag stays set until is cleared by the software.
T0VIE	Timer T1 overflow interrupt enable.

Table 2.13 Timer T1 counter B register (T1CNTB)

D7	D6	D5	D4	D3	D2	D1	D0
RTCCOV	T1CPEDG	T1EXEDG	T1PS2-T1PS0			T1MC1-T1MC0	

RTCCOV	RTCC overflow flag.
T1CPEDG	Timer T1 capture edge.
EXEDG	Timer T1 external event clock edge.
T1PS2–T1PS0	Timer T1 prescaler divider field. This 3-bit field specifies the divide- by factor for generating the timer clock from the on-chip system clock. [000-1; 111-128].
T1MC1–T1MC0	Timer T1 mode control filed.

	00 →	Software timer mode
	01 →	PWM mode
	10 →	Capture/compare mode
	11 →	External event mode

Table 2.14   Timer T2 control B register (T2CNTB)

D7	D6	D5	D4	D3	D2	D1	D0
PO RT RD	T2C PED G	T2 EX ED G		T2PS2–T2PS0		T2MC1–T2MC0	

PORTRD	Port read mode.
T2CPEDG	Timer T2 capture edge.
T2EXEDG	Timer T2 external event clock edge.
T2PS2–T2PS0	Timer T2 prescaler divider field. This 3-bit field specifies the divide- by factor for generating the timer clock from the on-chip system clock. [000-1; 111-128]
T2MC1–T2MC0	Timer T2 mode control filed.

00 →	Software timer mode
01 →	PWM mode
10 →	Capture/compare mode
11 →	External event mode

EXERCISES

2.1 What are the special features of microcontrollers?

2.2 Define the terms: DIP, PLCC, CLCC, and QFP.

2.3 Make a comparative study of X48 family members.

2.4 Make a comparative study of X51 family members.

2.5 Make a comparative study of X96 family members.

2.6 Which are the basic members of X51 family?

2.7 Which are the new members?

2.8 What is a clock? How an external clock can drive X51?

2.9 How does a power on RESET can be implemented?

2.10 What is the minimum time for which reset signal must be applied?

2.11 Describe the organization of internal RAM of '51.

2.12 What is the function of B register?

2.13 What is a dot convention? Where is it used? Explain with the help of example.

2.14 What is stack?

2.15 How can you access a stack?

2.16 What is the maximum size of a stack in case of X51 microcontrollers? Why?

2.17 What is the value in the SP on reset?

2.18 How can you access external memory address space?

2.19 Enlist the control registers in X51.

2.20 Describe the function and format of following SFRs:

IP (interrupt priority)

TCON (timer control)

IE (interrupt enable)

SCON (serial port control)

TMOD (timer mode)

PCON (power control)

2.21 Why Port 0 pins are in high-impedance state when used as input?

2.22 What is the advantage of separate data and program memory?

2.23 What is the basic difference in data and store memory from accessing point of view?

2.24 Explain the difference in operation of internal Timer and Counter.

2.25 Explain different modes of operation of timers.

2.26 Why Mode 0 is not suitable for communication?

2.27 Write a program to compute the baud rates in Mode 1.

2.28 How overrun can be detected?

2.29 In which mode it is possible to detect a framing error?

2.30 What is the need for two stop bits?

2.31 Is it possible for a microcontroller operating in Mode 0 to communicate with another operating in Mode 3? Why?

2.32 What is the significance of TI flag when it is cleared to 0? When it is set to 1?

2.33 What is an interrupt?

2.34 What are sources of interrupt in '51 microcontrollers?

2.35 What is priority scheme in '51 to handle interrupt? How interrupts can be controlled?

2.36 When the hardware does not generate an interrupt call?

2.37 How the interrupts can be configured?

2.38 Explain the difference between the microprocessors and microcontrollers.

2.39 What are the popular architectures used in design of microcontroller-based design?

2.40 What are the three buses in microcontroller-based systems?

2.41 What is an address? How does it relate to memory or I/O?

2.42 How program is executed by the microcontroller? Explain which bus is used in each step and what information is carried on the bus.

2.43 Describe the machine cycle and instruction cycle.

2.44 State the advantage of separate data and code memory.

2.45 Name the alternate functions of Port 3 pins.

2.46 Describe SFR section of controller. Name 10 SFR and describe their uses.

2.47 Describe each step of the machine cycle necessary to transfer data from the memory to the controller.

2.48 Why two internal port pins be tied together even though one may output the Logic 1 when the Logic 0 is output at the other pin? What will be seen on these pins?

2.49 Describe importance of accumulator.

2.50 How are the flags used?

2.51 What is a stack overflow? What is the cause?

2.52 What do you mean by the term embedded?

2.53 How many clock pulses and phases are there in 8051 machine cycle?

2.54 Define bits in PSW.

2.55 Can jump-on-zero be done without a zero flag bit? How?

2.56 What are contents of SP on power ON?

2.57 Which register is associated with the SBUF?

2.58 Which register is associated with the timers?

2.59 Which port can take heaviest load?

2.60 How does PSEN differ from RD and WR?

2.61 When the contents of Port 0 latches may be lost?

2.62 When is the counter advanced from an external source?

2.63 How does Mode 2 differ from Mode 0 in a counter?

2.64 What does double buffering mean?

2.65 Which serial port mode allows full duplex asynchronous operation with variable baud rate?

2.66 What is pending interrupt?

2.67 Explain two-tier priority scheme.

2.68 Which interrupt has higher priority, RI or TF0?

2.69 What controls state of flag for external level activated internal?

2.70 How many instruction types are there in 8051 instruction set?

2.71 Explain significance of Boolean processor.

2.72 Write program that will read a byte from Port 1 and use it as an index into jump table starting at 03b5h.

2.73 Write program that will read a byte from Port 0 and use it as an index into any one of the eight locations in jump table depending on which bit is one. Use bit wise instructions.

2.74 Which flag is not cleared automatically, when the hardware generates the call to the service routine?

2.75 What is handshaking? Why and when is it required?

2.76 What are the advantages of flash controllers such as AT89C2051?

2.77 Explain modes of operation of flash controllers.

2.78 Differentiate between powers down and idle mode.

2.79 How microcontroller can come out of idle mode?

2.80 How microcontroller can come out of power down mode?

2.81 Explain the procedure steps to be carried out for programming flash controllers.

2.82 How does bit addressable registers are differentiated from byte addressable SFR?

2.83 What is meant by the term: enhanced architecture?

2.84 What are the additional features introduced in basic architecture by other chip manufacturers?

2.85 What is different in high speed microcontrollers HSM?

2.86 How waits status is controlled in HSM?

2.87 What is the requirement of third timer in 8052? What are its modes of operation?

2.88 What is a watchdog timer? How is it useful?

2.89 How can you control operation of a watchdog timer?

Chapter 3

Introduction to Assembly Language Programming

Microcontroller is a programmable device. The desired operation can be realized by giving a set of instructions called *program*. This program has to be loaded in the memory unit of the microcontroller. Since the MCS-51 family represents an 8-bit microcontroller, the instructions are also 8-bit wide. However, the operation is represented by a hexadecimal code.

The program development is made simpler by providing support tools such as assemblers, loaders, the cross-compilers, etc. Assemblers allow the user to write a program in the assembly language of the microcontroller that can be converted into hex code and loaded into the memory. Cross-compilers allow program development in higher languages, such as C, so that the user is not required to use instructions from the set directly.

This chapter describes in detail the stages in the program or software development process. The features of instruction sets are explained along with illustrative programming examples.

3.1 | SOFTWARE DEVELOPMENT

The tools used for the product development process are shown in Fig. 3.1. The user develops an application program on a computer either in an assembly language or in a higher language such as C. The source is converted into microcontroller code by an assembler or a cross-compiler. It is linked with other modules from library using linker. This creates an executable module, which is converted into Intel hex format.

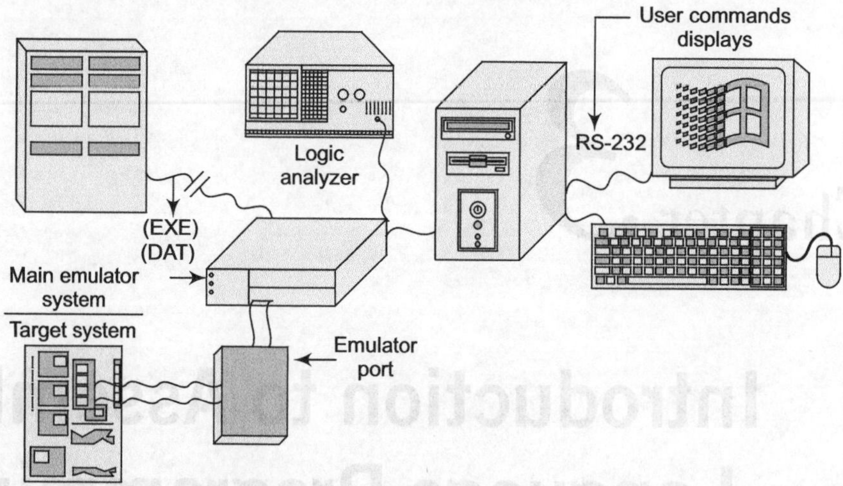

Fig. 3.1 A typical setup for development of microcontroller-based system

A *simulator* can be used to check the program execution and errors can be identified and corrected using a debugger. Using a downloader it is loaded in the memory of a microcontroller system via the RS-232 link. An *emulator* with hardware support tools such as logic analyzer can be used to test the system operation. Figure 3.1 shows a setup for the product development process.

Figure 3.2 depicts the sequence of operations performed for the product development.

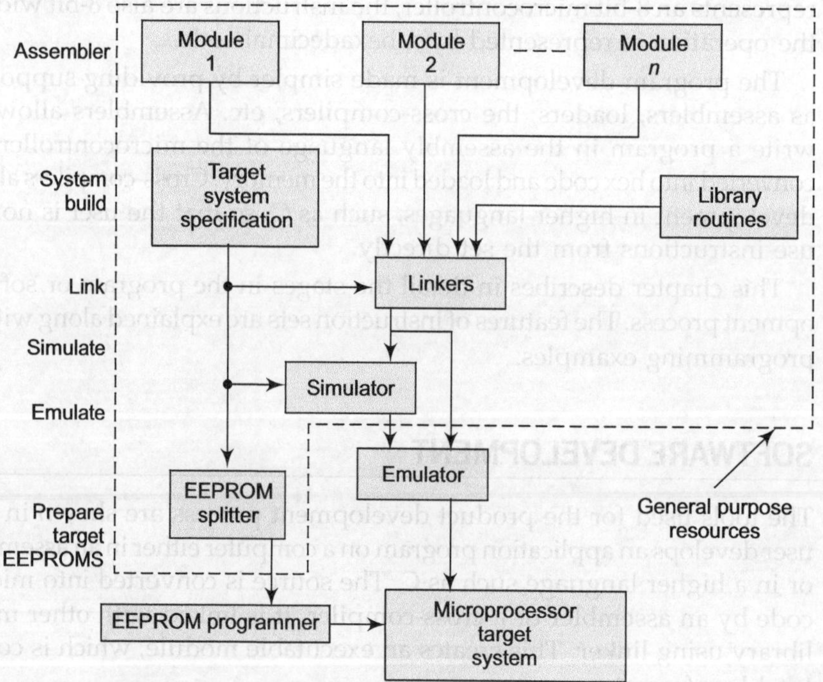

Fig. 3.2 Stages of software development

3.1.1 Types of Assemblers

Although all assemblers perform some task, their implementation varies greatly. Based on the task performed they can be classified as:

Cross-assembler   It is the assembler that runs on a computer other than the one for which it assembles the object program.

Resident/Self-assembler   It runs on the same computer for which it assembles the program.

Macro assembler   It allows the programmer to define sequences of instructions as macros.

Micro assembler   It is used to write the microprograms while defining the set of instructions of a computers. It has nothing to do with microprocessor.

Meta assembler   It is an assembler which can handle many different instruction sets.

One pass assembler   It goes through secure program only once.

Two pass assembler   It goes through the program twice. First time the assembler collects and defines all the symbols while second time it replaces references with the actual definitions. Most of the microprocessor assemblers are two-pass assemblers.

3.1.2 Assembly Process

The input to the assembler is a source module, which is a file created by the text editor. The output of the process is the object module and a listing. Most of the assemblers scan the source code twice and are called *two-pass assemblers*. Figure 3.3 depicts block schematic of an assembly process.

The purpose of the first pass is to provide the assembler with the location of labels, while that of the second pass is to generate codes. To locate the labels, the assembler contains a variable called *location counter* (LC). As the program is being scanned, LC is incremented by an amount equal to number of bytes required by the statement. The LC can be thought of as a pointer that dynamically indicates the relative position/offset within each segment as assembly progresses.

During the first pass, it uses the LC to construct the table called a *symbol table*. The table allows the second pass to use the offset of a label for generating end addresses. The initial value of the LC is called *origin*. It can be initialized by an assembler directive ORG.

If the origin is not specified, it is assumed to be zero. The ORG can be used more than once in the program. First ORG is used to set the origin and the next ORG is used to set the LC to the value specified by the operand.

It is convenient to refer to a location by noting its position relative to the current location of the LC. A special symbol is used to represent the current value of LC in the statement. In the assembly process the first pass builds the symbol table. The second pass scans the source code and converts it into machine code using the symbol table to insert addresses as needed.

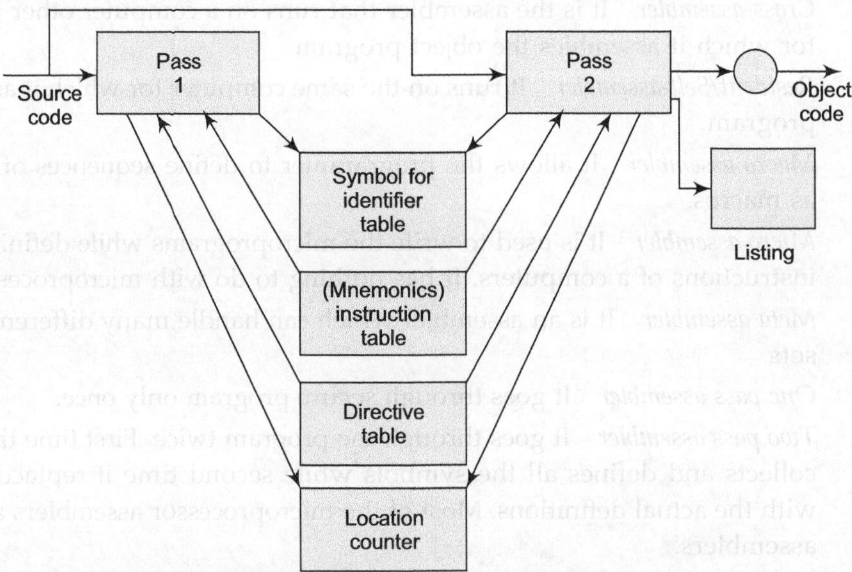

Fig. 3.3 Assembly process

When a label is found in the leftmost field of the statement, it is said to be defined. At this point the label is associated with an offset by placing the current location of LC in symbol table along with the label's symbolic representation with attributes. If the same label is defined in the module twice, an error is indicated. In order to identify the keywords there is a permanent symbol table that contains the instructions or the directives that might be needed by the assembler. This includes mnemonics and the length of instruction (number of bytes) information. END directive indicates it is the end of the assembly process. Then the assembler enters the second pass. During this pass, it inserts the preassigned constants that occur in the data definition directive and prepares other information that may be required to combine the output object module with other object modules. Since the instructions and the directives may include expression, the entry may involve the computation of offset rather than simply looking the entry in the table, it is done during second pass.

The listing contains the source along with the translation. If comments are given, it may include a list of errors detected during the process. A complete listing contains the mnemonics, listing, error messages, and a cross-reference symbol table.

3.1.3 Linking and Loading

The general process for creating and executing a program is shown in Fig. 3.4. The loading is done alone by the OS. The arrow indicates that corrections must be done after any one of the major steps. Linker/loader must perform following functions:

1. Find the object module to be linked.
2. Construct the load module by assigning positions of all of the object modules using linker.
3. Fill in all addresses that could not be determined by the assembler.
4. Load the program for execution.

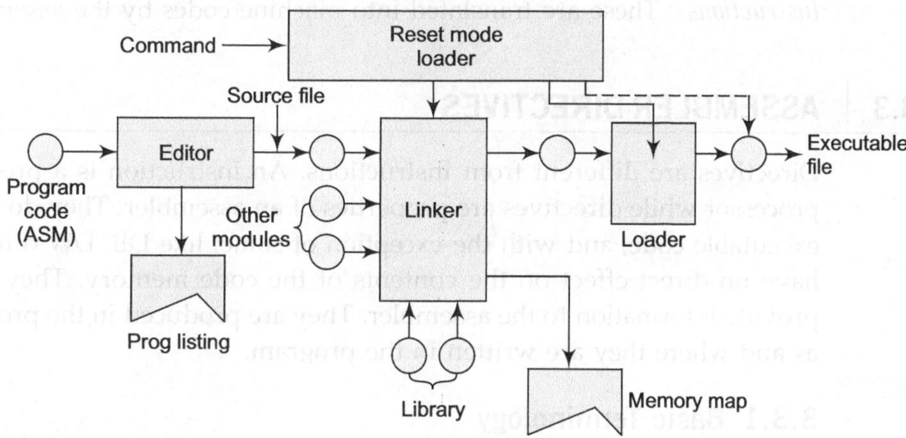

Fig. 3.4   Complete software development process

The object modules to be linked are determined by naming them in the command to the linker (which will be explained later) and searching through libraries. Because several modules may be used to produce a program, every module must have a name defined by:

Label Name   **Module Name**

and every module must have an END directive.

3.2 | ASSEMBLY LANGUAGE PROGRAMMING

There are various ways of programming a computer. It is possible to use assembly language instructions and an assembler to convert them into machine instruction codes. Since assembly language is dependent on the machine, it gives the programmer a control in using a machine.

Because of the acronyms called mnemonics, it is easier to write in assembly language. It is also easier to use the character strings, called *identifiers*, to represent addresses/numbers.

The features that are available in a high-level language must also be available in the assembly language. In addition to including ways of performing branches/

loops/I/O, arithmetic and assignment, the assembler needs to provide the preassignment storage allocation, available in high level language, naming of constants, comments and structuring data. Since an assembler is nothing but a translator program, the format and syntax of the instruction and directives depends on how assembler in written. There are two types of statements in an assembly language :

Directives   These give direction to the assembler during the assembly process, but are not converted into machine codes.

Instructions   These are translated into machine codes by the assembler.

3.3 | ASSEMBLER DIRECTIVES

Directives are different from instructions. An instruction is a property of the processor while directives are properties of an assembler. They do not produce executable code, and with the exception of some, like DB, DW directives, they have no direct effect on the contents of the code memory. They are used to provide information to the assembler. They are produced in the program listing as and where they are written in the program.

3.3.1 Basic Terminology

Symbols

They are used to give user defined names to the data or addresses so that it becomes easier to handle them in the program. They must begin with a letter or a special character (? or _) followed by a letter, digit or character. Up to 255 characters in a symbol name can be used but only the first 31 characters are significant. A symbol name can contain upper or lowercase characters, but the assembler converts to uppercase characters for internal representation.

The instruction mnemonics, assembly-time operators, predefined bit and data addresses, segment attributes, and assembler directives may not be used as user-defined symbol names as they have special meaning to the assembler.

Labels

A label is a symbol. All the rules for forming symbol names apply to labels. A statement label is the first field in a line, but it may be preceded by any number of tabs or spaces and must be terminated by a colon (:) to identify it as a label. Only one label is permitted per line. They are optional which means that every statement in a program need not carry a label.

When a label is defined, it receives a numeric value and a segment type. The numeric value will always be the current value of the location counter of the currently selected segment at the point of use. The value of the label will be relocatable or absolute depending on the relocatability of the current segment. A label, once defined, should not be redefined.

3.3.2 Classification of Assembler Directives

Based on the function they perform or the purpose for which they are used, directives are divided into the following categories:

1. Symbol definition
2. Storage initialization
3. Program linkage
4. Assembler control
5. Segment selection
6. Usage

Symbol Definition

The symbol definition directives are used to create symbols that can be used to represent segments, registers, numbers, and addresses. None of these directives may be preceded by a label. Symbols defined by these directives may not have been previously defined and may not be redefined by any means. The SET directive is the only exception to this. The various directives used for the purpose are:

Segment directive   It is used to define the type of segment. The segment type specifies the address space where the segment will reside. The allowable segment types are:

- CODE   The code space
- XDATA   The external data space
- DATA   Internal data space accessible by direct addressing (0 ...127)
- IDATA   Internal data space accessible by indirect addressing (0... 127)
- BIT   The bit space (overlapping locations 32 ...47 of the internal data space)

When used in expressions, the segment symbol stands for the base address of the combined segment. Any DATA or IDATA segments may be used as a stack (there is no explicit stack segment).

EQU directive   The format for the EQU directive is shown below. Note that a label is not permitted.

Symbol_name   EQU   Expression

The EQU directive assigns a numeric value or special assembler symbol to a specified symbol name. The symbol name must be a valid name based on the following rules:

1. The expression must be without any unknown reference or undefined systoles.

2. The special assembler symbols A, R0, R1, R2, R3, R4, R5, R6, and R7 can be represented by user symbols defined with the EQU directive. If a symbol is defined to be a register value, it will have a type 'REG'.

3. A symbol defined by the EQU directive cannot be defined anywhere else.

SET directive   The SET directive operates similar to EQU with a difference, that the defined symbol can be redefined later, using another SET directive.

BIT directive   The BIT directive assigns a bit address to the specified symbol name. The symbol gets the segment type BIT. A symbol defined as BIT may not be redefined elsewhere in the program.

DATA directive   The DATA directive assigns an on-chip data address to the specified symbol name. The expression must be an absolute or simple relocatable expression.

IDATA directive   The IDATA directive assigns an indirect internal data address to the specified symbol name. The expression must be an absolute or simple relocatable expression. Absolute expressions may not be larger than 127 for the 8051.

CODE directive   The CODE directive assigns a code address to the specified symbol name. If the expression dose not evaluate to a number, its segment type must be CODE. The symbol that a segment type gets defined by the CODE directive may not be redefined elsewhere in the program.

Storage Initialization and Reservation

The storage initialization and reservation directives are used to initialize and reserve space in either word, byte, or bit units. The space reserved starts at the point indicated by the current value of the location counter in the currently active segment. These directives may be preceded by a label.

DS directive   The DS directive reserve space in byte units. It can be used in any segment except a BIT type segment. The expression must be a valid assembly-time expression with no forward references. The location counter of the current segment is incremented by the value of the expression. The sum of the location counter and the specified expression should not exceed the limitations of the current address space, or those set by the current relocation type.

DBIT directive   The DBIT directive reserves a space in bit units. It can be used only in a BIT type segment. The expression must be a valid assembly-time expression with no forward references. When the DBIT statement is encountered in a program, the location counter of the current (BIT) segment is incremented by the value of the expression.

DB directive   The DB directive initializes code memory with byte values. Therefore, a CODE type segment must be active. The expression list is a series of one or more byte values or strings separated by commas (,). A byte value can be represented as an absolute or simple relocatable expression or as a character string. Each item in the list (expression or character string) is placed in memory in the same order as it appears in the list.

The DB directive permits character strings longer than two characters, but they must not be part of an expression. A null character string defined as an item in the list, generates no data. If the directive has a label, the value of the label will be the address of the byte in the list.

For example:

PRIMES   ; DB 1, 2, 3, 5, 7, 11, 13, 17, 19, 23, 29, 31, 37, 41, 43, 47, 53
            ; This DB lists the first 17 prime numbers.
            ; (PRIMES is the address of 1.)

DW directive   The DW directive initializes code memory with a list of words (16-bit) values. Therefore, a CODE type segment must be active. The expression list can be a series of one or more word values separated by commas (,). Word values can be absolute or simple relocatable expressions. If you use the location counter ($) in the list, it evaluates to the code address of the word being initialized. Here, more than two characters are not permitted in a character string. Each item in the list is placed in the memory in the same order as it appears in the list, with the high-order byte first followed by the low-order byte.

Program Linkage

Program linkage directives allow the separately assembled modules to communicate by permitting intermodule references and the naming of modules.

PUBLIC directive   The PUBLIC directive allows symbol to be known outside the currently assembled module. If more than one name is declared PUBLIC, the names must be separated by commas (,). Each symbol name may be declared PUBLIC only once in a module. Any symbol declared PUBLIC must have been defined somewhere else in the program.

EXTRN directive   The EXTRN directive lists symbols to be referenced in the current module that are defined in other modules. This directive can appear anywhere in the program.

The macro is a string replacement facility. It permits you to write repeatedly used sections of code and then insert that code. If several programmers are working on the same project, a library of macros can be developed and shared by the entire team. Conditional assembly sections of code help to achieve the most compact code possible.

All macro processing of the source file is performed before the code is assembled. Because of this independent processing of macros and assembly of code, we must differentiate between macro-time and assembly-time. At macro-time, assembly language symbol-labels, SET and EQU symbols, and the location counter are not known. Similarly, at assembly-time, no information about macros is known. The macro scans the source file looking for macro calls. A macro call is a request to the controller to replace the call pattern of a built-in or user-defined macro-controller with its return value

NAME directive The NAME directive is used to identify the current program module. All the rules for naming apply to the module name. The NAME directive should be placed before all directives and machine instruction in the module. Only comments and control lines can precede the NAME directive. The symbol used in the NAME directive is considered undefined for the rest of the program unless it is specifically defined later.

Assembler State Controls

END directive The END statement must not have a label, and only a comment may appear on the line with it. The END statement should be the last line in the program; otherwise this will produce an error.

ORG directive The ORG directive is used to alter the assembler's location counter to set a new program origin for statements that follow the directive. The syntax is: ORG expression.

The expression should be an absolute or simple relocatable expression referencing the current segment and containing no forward references. When the ORG directive is encountered in a program, the value of the expression is computed as the new value of the location counter specifying the address at which the next machine instruction or data item will be assembled in the current selected segment.

If the current segment is absolute, the value will be an absolute address in the current segment; if the segment is relocatable, the value will be offset from the base address of the instance of the segment in the current module.

The ORG directive modifies the location counter, it does not generate a new segment. That is, when location counter is incremented from the current value, the space between the previous and the current location becomes part of the current segment. In an absolute segment, the location counter must not be decremented to an address below the beginning of that segment.

For example: `org  100h`   ; set location counter to 100

Segment Selection Directives

The segment selection directives will divert the succeeding code or data into the selected segment until another segment is selected by a segment selection directive. The directives may select a previously defined relocatable segment, or optionally create and select absolute segments.

The format for relocatable segment selection directives will divert the succeeding code or data into the selected segment until another segment is selected by a segment selection directive. The directives may select a previously defined relocatable segment, or optionally create and select absolute segments.

CSEG, DSEG, ISEG, BSEG, and XSEG select an absolute segment within the code, internal data, indirect internal data bit, or external data address spaces, respectively. If you choose to specify an absolute address (by including 'At absolute address'), the assembler terminates the last absolute segment, if any, of the specified segment type, and creates a new absolute segment starting at that address. If you do not specify an absolute address, the last segment of the specified type is continued. If no absolute segment of this type was selected and the absolute address is omitted, a new segment is created starting at location 0. You cannot use any forward references and the start address must be an absolute expression.

Each segment has its own location counter; this location counter is always set to 0 in the initial state. The default segment is an absolute code segment; therefore, the initial state of the assembler is location 0 in the absolute code segment.

When another segment is chosen for the first time, the location counter of the former segment retains the last active value. When that former segment is reselected, the location counter picks up at the last active value. You can use the ORG directive to change the location counter within the currently selected segment.

Macros

Some assemblers are designed so that programmers need to write the code only once and refer to it as often as required. Every time it is called, the assembler will insert the code in the program module. A program segment written this way is called a *macro*. Each macro is given a name. The association of the name with the macro is called a *macro definition*.

Assemblers define macros. Macro assemblers are the assemblers capable of working with macros. They do not conserve memory or reduce the number of instructions. They only help programmers by allowing them to unite a set of code only once. The macros can also be stored in libraries.

A macro definition includes:

(a) Header: which identifies the code that follows as being a macro defined.

(b) Prototype code to be inserted when expanded.

(c) A terminating directive to mark the end of macro.

A macro is a string replacement facility. It permits you to write repeatedly used sections of code and then insert that code. If several programmers are working on the same project, a library of macros can be developed and shared by the entire team. Conditional assembly sections of code help to achieve the most compact code possible.

All macro processing of the source file is performed before the code is assembled. Because of this independent processing of macros and assembly of code, we must differentiate between macro-time and assembly-time. At macro-time, assembly language symbol-labels, SET and EQU symbols, and the location counter are not known. Similarly, at assembly-time, no information about macros is known. The macro scans the source file looking for macro calls. A macro call is a request to the controller to replace the call pattern of a built-in or user-defined macro-controller with its return value.

When a macro call is encountered, the macroprocessor expands the call to its return value. The return value of a macro is then place in a temporary work file, and the macroprocessor continues. All characters that are not part of a macro call are copied into the temporary work file. The return value of a macro is the text that replaces the macro call. The return value of some macro is the null string, a character string containing no characters.

3.3.3 Addressing Modes

The programming model for the microprocessors is essentially the accumulator, the flags, a few registers, and a relatively simple vectored interrupt system. The assembly language of 8051 contains features such as Boolean operations, not seen in similar processors, that give it flexibility and power as a control device.

The addressing mode indicates how the controller gets hold of an operand from any one of the memory spaces as per specification in the instruction. The 8051 supports five addressing modes.

Direct Addressing

Instructions using direct addressing are 2-bytes long: an 8-bit op-code followed by an 8-bit address. The address is a location in the internal RAM or SFR area. The address refers to either a byte location or a specific bit in a bit-addressable byte. Assemblers for 8051 usually have predefined mnemonic symbols corresponding to important bit and byte addresses.

For example

`mov a, 04`　　; move contents of RAM from byte location 04 to the accumulator.

`clr 07`　　　　; clear bit address 07 (it is also the MSB of byte address 20h).

(The symbol ; indicates the starting of comment in the instruction).

Indirect Addressing

An instruction using direct addressing specifies the fixed address of the operand. An instruction using indirect addressing specifies a register that contains the address of the operand. As the contents of the register can be changed in the program, indirect addressing is a powerful technique. Most of the instructions that use indirect addressing are only 1-byte long, making them efficient in memory space and execution time. The indirect addressing mode is indicated by symbol @ in the instruction.

Indirect addressing can access external as well as internal RAM area. The addresses can be 8-bits or 16-bits.

1. 8-bit addresses use the registers designated as R0...R7.
2. 16-bit addresses use the dptr register. dptr is actually two 8-bit registers DPH and DPL treated as one 16-bit register.
3. Different banks of registers can be selected for R0...R7, depending on bits in the PSW.

For example

```
mov a, @r3          ; load accumulator by the data byte from internal RAM
                      address.
                    ; specified by register r3.
movx a, @dptr       ; load accumulator by the data byte from external RAM
                      address.
                    ; specified by register dptr.
```

Register Addressing

Register instructions use the contents of one of the registers, typically R0...R7 as the operand. All register instructions have the efficiency of being 1-byte long. A few register instructions are register specific; they do not allow the programmer to specify the operand register.

For example

```
mov a, r5   ; move contents of the register r5 to the accumulator.
div a b     ; divide the contents of a accumulator by the contents of b register.
```

Immediate Addressing

Instructions using an immediate operand have an 8- or 16-bits data following the op-code. This mode is indicated by a symbol # in the instruction.

For example

```
mov a, #55h      ; move the 8-bit data 55h to the accumulator.
mov dptr, #2500h ; move the 16-bit data 2500h to the data pointer register.
```

Indexed Addressing

8051 uses indexed addressing for reading data tables from program memory space and implementing jump tables. In either case, a 16-bit register holds the base address and the accumulator holds an 8-bit displacement or index. The address of the data byte is the sum of the 16-bit base and the unsigned 8-bit displacement. It results in a forward reference from the base of 0 to 255-bytes. The base register is either DPTR or PC.

For example

```
mov a, @a+pc        ; move code byte relative to pc to the accumulator.
jmp @a+dptr         ; jump relative to dptr.
```

3.3.4 Operand Modifiers: @ and

The 8051 assembly language uses two special symbols to distinguish operand types: They are the symbols @ and the sign # .

The @ before an operand means that indirect addressing is being used, while the symbol # before an operand means it is an immediate operand.

For example

```
add a, r5       ; add to the accumulator the contents of register r5 of the selected
                  bank.
add a, @r5      ; add to accumulator the contents of the RAM location whose
                  address is specified by the register r5 of the selected bank.
add a, 25       ; add to the accumulator the contents of RAM address 25h.
add a, #25      ; load the accumulator with the number 25h.
```

3.4 | INSTRUCTION SET

As already explained, an instruction set can be defined as a set of instructions which can be understood by the microcontroller. The instruction set helps the user to carry out operations or activities related with control, arithmetic manipulation, and serial I/O. 8051 has Boolean (bit processing capability), which allows the user to carry bit manipulations. The purpose of this section is to make user familiar with the instruction set, along with the information regarding operations supported by 8051.

When an instruction is not available in the set to carry out an operation, a set of two or more instructions has to be used to realize the activity. Such a group of instructions, which when executed sequentially leads to the realization of the activity, is called a *program*.

For 8051, being an 8-bit microcontroller the instruction op-code consists of an 8-bit combination, represented by 2-hex digits. Instruction op-code has value 00h to FFh. Hence, there are 256 possible instruction codes. Each combination (code) represents a unique activity, which can be carried out by the microcontroller.

Instead of attempting to study them one by one in order, we can group them in terms of operations. Instructions which carry out similar operations are grouped into a single functional group. Such a classification leads to the following main groups:

- Data transfer group
- Arithmetic group
- Logical group
- Branch group

3.4.1 Data Transfer Group

Instructions in this group, are concerned with the movement of data. In fact this means copying data from one location, termed as *source* to another location, called *destination*.

Since the memory organization of 8051 generates five types of address spaces, it may be required to copy data from one space to another space. Five address spaces are:

1. Register   (rn : r0 to r7)   : (rn)
2. Accumulator   (a)   : (a)
3. Internal RAM   (00K to FFh)   : (direct)
4. External memory
5. Code memory

The operation is represented by different mnemonics.

For example

mov	; data transfer in internal memory.
movx	; data transfer from external memory.
movc	; data transfer from code memory.

To be more specific, there are 28 operations supported by the instruction set. In addition to the above-mentioned address spaces, one more place from where data can be obtained is from the instruction itself. As already explained in case of immediate (data) addressing mode, data is an integral part of the instruction. It is important to note that the data transfer group of instructions do not affect the flags and hence no flags can be used as a condition for branching.

The following chart (Fig. 3.5) classifies 28 instructions in terms of the required data transfer, or allowed transfer by the instruction set. In fact 8051 family is, one of the few processors which allow memory to memory data transfer directly, even though it is restricted only to internal *RAM*.

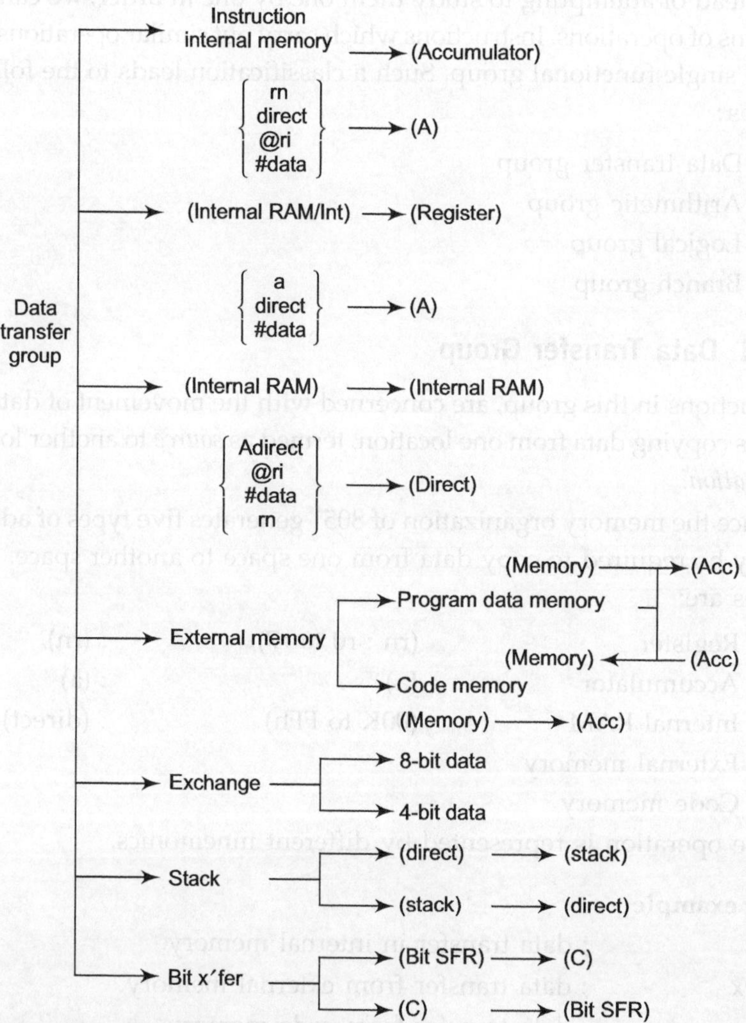

Fig. 3.5 Data transfer operations

Internal RAM→Accumulator

These instructions are used to transfer data in the internal RAM to the accumulator. Since the source can be specified using various addressing modes, instruction will specify the location of source, while the destination is the accumulator. The format of instructions in this subgroup is

$$\text{mov a,} \begin{Bmatrix} \text{rn} \\ \text{direct} \\ \text{@ri} \\ \text{\#data} \end{Bmatrix}$$

The symbol { } (curved bracket) means that at a time, any of the quantities listed within the braces can be used to write a valid instruction. The instructions can be written as:

1. `mov a, rn`   ; register addressing mode.

   Move contents of register specified in r0–r7 to the accumulator.

2. `mov a, direct`   ; direct addressing mode.

   Move contents of memory location specified to the accumulator.

3. `mov a, @ri`   ; indirect addressing mode.

   Move contents of memory location specified by ri to the accumulator.

4. `mov a, #data`   ; immediate addressing mode.

   Move data byte specified in the instruction to the accumulator.

For example

1. `mov a,  r3`   ; move contents of r3 to A.
2. `mov a,  35h`   ; load contents of 35h to accumulator.
3. `mov a,  @ r0`   ; load acc with contents of the location specified by r0.
4. `mov a,  #35h`   ; load acc with 35h.

Internal RAM→Register

$$\text{mov rn,} \begin{Bmatrix} \text{a} \\ \text{direct} \\ \text{\#data} \end{Bmatrix}$$

The specified data will be moved to rn, i.e., specified register r0–r7. The important thing to notice is that indirect addressing is not allowed. The valid instructions are:

1. `mov rn , a`   ; copy acc into specified register.
2. `mov rn , direct` ; copy contents of RAM location to the specified register.
3. `mov rn, #data`   ; load specified register with the byte in the instruction.

For example

1. `mov r0, a`   ; copy contents of the accumulator to register r0.
2. `mov r2, #35h`   ; load 35h into r2.
3. `mov r4, 35h`   ; copy contents of RAM location, 35h is loaded to r4.

Internal RAM→Internal RAM

These instructions can be used to move data within internal RAM to another location of RAM. All addressing modes are supported. The general syntax is

$$\text{mov direct,} \begin{Bmatrix} \text{a} \\ \text{rn} \\ \text{@ ri} \\ \text{direct} \\ \text{\# data} \end{Bmatrix}$$

The valid instructions are:

1. mov direct, a   ; register addressing
2. mov direct, rn   ; register addressing
3. mov direct, @ ri   ; indirect addressing
4. mov direct, direct   ; direct addressing
5. mov direct, #data   ; immediate addressing

For example

1. mov 3Fh, a   ; move contents of acc to RAM location 3Fh.
2. mov 32h, 02h   ; move contents of RAM location 02h to RAM location 32h.
3. mov 31h, @r1   ; move contents of RAM addressed by r1 to RAM location 31h.
4. mov 39h, #05h ; replace contents of RAM location 39h with 05h.
5. mov 38h, r2   ; move contents of r2 to RAM location 38h.

Transfer To / From External Memory

The external memory in a microcontroller-based system may be:

1. Data memory
2. Code memory

Two different groups of mnemonics are specified for the operation.

Data memory   In case of data memory transfer, the mnemonics used is movx. The letter x in the name indicates the external memory. A byte of data will be transferred between the accumulator and a memory location whose address is specified by either 8-bit in (r0) or (r1), or 16-bit in [dptr].

There are two operations which can be done: store accumulator and load accumulator.

Store Accumulator

$$\text{movx} \begin{Bmatrix} \text{@ dptr} \\ \text{@ ri} \end{Bmatrix} \text{, a}$$

The contents of the accumulator are stored in external memory, whose address is specified, either by data pointer (dptr) or registers (r0) or (r1).

Valid instructions are:

1. movx @ dptr, a
2. movx @ r1 , a

Load accumulator

$$\text{movx a, } \begin{Bmatrix} @ \text{ dptr} \\ @ \text{ ri} \end{Bmatrix}$$

The accumulator is loaded from the external memory location specified by dptr or ri. Since there is no specific I/O instruction in the set, these are used to access data from the input port or send data to the output port via accumulator.

Code memory   This is read-only type of data transfer. It is used to read a code byte or a constant from code memory. This is identified by a letter C in the mnemonics for move operation. In code memory, i.e., [movc] is used to access the data.

$$\text{movc a, } \begin{Bmatrix} @ \text{ a + dptr} \\ @ \text{ a + pc} \end{Bmatrix}$$

The effective address is the sum of either dptr or pc and the accumulator.

The contents of this effective address is moved to accumulator. This carries out 16-bit addition to carry propagation from low byte addition. PC is incremented, to point to the next instruction, before being added. The valid instruction is:

1. `movc a, @ a + dptr`

   Contents of dptr and acc are added to form an effective external code address and the accumulator is loaded with contents of this effective address.

2. `movc a, @ a + pc`

   Contents of pc and acc are added to form an effective external code address and the accumulator is loaded with the contents of this effective addresses.

For example

```
mov a, #05h
movc a, @ a + dptr
```

Let the contents of (dptr) be: 3005h

The effective address is calculated as sum of dptr and a ; 3005 h + 05 h = 300 A h.

Hence contents of memory location 300Ah will be loaded to the accumulator.

Data Exchange

These instructions are used to exchange data between internal RAM and accumulator. The exchange may be 4-bit or 8-bit wide. In case of 8-bit of data exchange, direct, indirect, and register modes are supported, while only indirect addressing mode is supported for 4-bit exchange. The exchange is supported on byte or bit basis.

Byte exchange This instruction loads the accumulator with the contents of the internal memory operand specified. At the same time the internal old value of the accumulator is written to the specified location. General syntax is

$$\text{xch a,} \begin{Bmatrix} \text{rn} \\ \text{@ri} \\ \text{direct} \end{Bmatrix}$$

The valid instructions are :

1. xch a, rn   ; register addressing
2. xch a, @ri   ; indirect addressing
3. xch a, direct ; immediate addressing

For example

1. xch a, r3   ; contents of a and register r3 are exchanged.

2. xch a, @r1   ; contents of a and location specified by r1 are exchanged.

Bit exchange This exchanges the nibble (low order) with that of the internal RAM location indirectly addressed by the specified register. The valid instructions are:

1. xchd a, @ ri
2. xchd a, @ r0
3. xchd a, @ r0

For example

Let the contents of accumulator be:

3F

and the contents of location 35h which is pointed by r0 be:

42

After execution of instruction:

xchd a, @ r0

The contents will be:

$$(A) = 32h$$
$$(35h) = 4Fh$$

This means that only the lower nibbles are exchanged, while the upper nibbles are unchanged.

Stack Operation

Stack is a special area defined in the internal RAM. The address of the stack is maintained by the special function register (SP). This uses indirect as well as implicit addressing mode. The difference between normal memory and stack memory operations is that in the former case, registers r0 or r1 are used to hold the address while latter uses SP.

The stack grows vertically upward, when data is written. The operations with stack are:

1. **Write a stack**   The mnemonic specified for the operation is push. The steps taken by microcontroller to carry out operation are as follows:
   1. The contents of SP are incremented by 1.
   2. The operand specified by the instruction is stored at the location specified by SP.

2. **Read a stack**   The mnemonic specified is POP. The steps are reverse of the write operation. They are listed below:
   1. The contents of the location specified by SP are moved to the address specified in the instruction.
   2. SP is decremented by one.

Whatever operation is performed with the stack, it is to be noted that SP is updated by the microcontroller so that it always points to the top of the stack. On increment if SP reaches FFh, the subsequent write operation causes it to roll over to (00h), which is the address of register r0. Since the RAM address ends at 7F, pushing beyond 7F, results in an error. The direct addresses are to be used, while register names cannot be used as operand for these operations.

The valid format of instruction is

$$\begin{Bmatrix} \text{push} \\ \text{pop} \end{Bmatrix} \text{direct}$$

1. `Push direct` ; push data into the address or the stack
2. `Pop direct` ; pop data from the top of the stack to the address specified in
                 ; instruction.

On power up, stack SP is initialized to (07) (it is the register r7 in Bank). Using stack beyond this may lead to destroying data in the SFR. As already mentioned, the names of SFR cannot be used to initialize the stack pointer.

For example
To initialize stack pointer ;
    `mov SP, #20`   ; is invalid since it uses the SFR name
The correct way to do is:
    `mov 81h, #21h` ; 81h is the byte address of SP

Bitwise Data Transfer

In case of bitwise operation, the Boolean accumulator is either the source or the destination. The operations are performed via Boolean accumulator. The flag C works as an accumulator. However, there is no direct bit to bit movement. It has to be done through the carry flag. In absence of special I/O instructions, these instructions are used to read/write at the I/O port pins. They are defined between the carry and bit addressing registers (SFRs). The general syntax of operations is

$$\{c\} \rightarrow \left\{ \begin{matrix} \text{Bit add} \\ \text{SFR} \end{matrix} \right\}$$

$$\left\{ \begin{matrix} \text{Bit add} \\ \text{SFR} \end{matrix} \right\} \leftarrow \{a\}$$

For example

1. movc, 35h   ; move contents of 35h to carry flag.
2. mov 90h, c   ; output status of carry hit on bit 0 of Port 1.

Sample Programs

Instead of attempting directly to write programs for the complete algorithms, we will develop simple programs, preferably using instructions from this group only, so that the use of instructions from data transfer group is well understood.

Example 3.1   Write a program to select register bank-1 and initialize r3 and r4 with FFh and 00h, respectively.

Solution

It is known that register banks can be selected by selecting bits RS0, RS1 in PSW. The address of PSW is (000h)

RS1	RS0	
0	0	Bank 0
0	1	Bank 1
1	0	Bank 2
1	1	Bank 3

In order to select bank-1 we will have to choose (01) combination. The PSW byte will be:

Cy	Ac	Fo	RSI	RS0	Ov	--	P
0	0	0	0	1	0	0	0

```
Banksel:
        mov 0D0, #08h      ; select bank 1
        mov r3, #FFh       ; initialize r3
        mov r4, #00h       ; initialize r4
```

Example 3.2   Write a program to save contents of A, r7, and Port 1 on the stack.

Solution

The first step is to initialize the stack pointer SP, the program can be written as:

```
Save_ prog:
        mov  sp, #30
        push r7
        push P1
        push A
```

All statements in the program have the same error, which is the restriction that names cannot be used with stack operation only, direct addresses must be used. The correct program is

```
Mode_save:
        mov  81h, #30
        push 07              ; 07 address of r7
        push 90              ; 90h address of P1
        push 0E0             ; 0E0 address of A
```

Example 3.3   Write a program to transfer the contents of location 2400h to 2500h in external memory.

Solution

Since there is no direct instruction available for such a movement, a program algorithm has to be developed. The operation can be divided into small sequential steps described below:

First method

Let us use address (40h, 41h) to hold the

First address:

24	00

i.e.,   (40) = => 00h

(41) = = > 24h

Also locations (42h, 43h) to hold the 2nd address

25	00

i.e.,   (42) = => 00h

(43) = =>25h

The dptr is used as a pointer which holds 16-bit memory address. The program to carry out the specified operation is listed below.

```
        org  100h            ; this is assembler directive, it initializes
                                  location counter

MEM_TRANS:
        mov  40, #00h
        mov  41, #24h
        mov  42, #00h        ; initialize the 2nd address
        mov  43, #25h
```

```
        mov  dpl, 40h        ; set data pointer to 1st address
        mov  dph, 41h
        movx a @dptr         ; save data to be moved in acc
        mov  dpl, 42h        ; set dptr to 2nd address
        mov  dph, 43h        ; move acc to the memory
        mov  @dptr, a
        end                  ; this will be used as a terminator for the
                             ; time being until we study corresponding
                             ; instruction.
```

Second method The method describes the approach to carry out an operation using direct addressing mode instructions.

```
        org  100h
MEM_TR:
        mov  dptr, #2400     ; load pointer with 1st address
        mov  a, @dptr        ; get first data
        mov  dptr, #2500     ; load pointer with 2nd address
        mov  @dptr, a        ; move data from acc to memory
```

3.4.2 Arithmetic Group

The 8051 family supports most common arithmetic operations viz.: addition, subtraction, multiplication, and division. The arithmetic operations modifies the flags C, AC, OV, which are called *arithmetic flags*. These flags are SET/RESET based on the result of arithmetic operation performed by the instruction. The status of flags are used as the test conditions for execution of various control or branch instructions. The accumulator is one of the source and also the destination for the result in most of this group of operations.

Addition

The instruction set of the 8051 supports half as well as full addition. The full addition includes carry flag in the addition operation. The mnemonics indicates whether carry is to be included or not.

> **For example**
> add: add without carry flag
> addc: add with carry flag

The instructions in the set support all the addressing mode: The most general format can be written as:

$$\begin{Bmatrix} \text{add} \\ \text{addc} \end{Bmatrix} a, \begin{Bmatrix} \text{rn} \\ \text{direct} \\ \text{\# data} \\ @ \text{ ri} \end{Bmatrix}$$

The valid instructions are:

1. add a, rn   ; register addressing mode
2. add a, rn   ; register addressing mode
3. add a, direct   ; direct addressing mode
4. addc a, direct   ; direct addressing mode
5. add a, # data   ; immediate addressing mode
6. addc a, #data   ; immediate addressing mode
7. add a, @ ri   ; indirect addressing mode
8. addc a, @ri   ; indirect addressing mode

The setting of flag based on the result of operation is as follows:

carry flag c is set to 1   ; if there is a carry out of addition of MSB of
                                ; operands.

auxiliary carry Ac is set to 1; if there is a carry out of addition of lower
                                ; nibbles of operands.

Overflow flag OV is set to 1   ; if there is carry out of addition of operand
                                ; bits [(b7) OR (b6)]

Usually the ADD instruction is used for low-byte addition in case of multibyte addition, while ADDC is used for other byte. ADDC can be used for all bytes provided carry flag Cy is properly initialized.

For example

1. add a, 25h   ; (a) + (25h)   (A)
                  ; 25h will be added to the contents of acc and result is
                  ; stored in the acc.

2. addc a, 25h   ;   (a) + (c) + (25h)   (A)
                  ; 25h as well as status of carry flag will be added to the
                  ; contents of acc and result is stored in the acc.

Subtraction

The 8051 instruction set has a support for performing signed and unsigned subtraction. In this operation also the accumulator is one of the source and the destination for the result. This supports full subtraction, which means that the status of carry flag is combined as a borrow, and is used in the operation. The flag setting based on result for C, AC, and OV are similar to the addition operation. It is necessary to initialize the carry flag properly in the subtraction operation. This instruction for this operation in the 8051 set has support for all addressing modes.

The syntax or the general instruction format is

$$\text{subb a,} \begin{Bmatrix} \text{rn} \\ \text{@ r0} \\ \text{direct} \\ \text{\# data} \end{Bmatrix}$$

The valid instructions are:

1. subb a, rn   ; register addressing
2. subb a, @ r0   ; indirect addressing
3. subb a, direct   ; direct addressing
4. subb a, # data   ; immediate addressing

Increment and Decrement

These are special instructions for addition and subtraction. They can be used to add 1 or subtract 1 from the specified operand. They are supported by instructions in the set for all the addressing modes. The most important point is that unlike other arithmetic operations, accumulator may not be a source or a destination. The operations can be done anywhere in the internal RAM. Even though these are arithmetic instruction, no flags are affected.

The difference between increment and decrement operation is that, increment operation is applicable to contents of dptr register, however decrement operation cannot be applied to the contents of dptr register. The overflow of dptr from FFFF is to 0000h.

$$1. \text{ inc} \begin{Bmatrix} \text{a} \\ \text{rn} \\ \text{@ r0} \\ \text{direct} \\ \text{dptr} \end{Bmatrix} \qquad 2. \text{ dec} \begin{Bmatrix} \text{a} \\ \text{rn} \\ \text{@ r0} \\ \text{direct} \end{Bmatrix}$$

Multiplication

A, B registers are used as source as well as destination in this operation. The operation performs the multiplication of unsigned 8-bit number by another unsigned 8-bit number. The specific algorithm has to be developed for multiplying two 8-bit signed number. The syntax of the instruction is

mul a b

Contents of the accumulator A are multiplied by the contents of B register. The result of the multiplication operation result may be 16-bit. LSB of the result is stored in the accumulator A and the MSB of the result is stored in B register. If the result is > FF, (OV) flag is set to indicate that B are the high byte of the result.

Division

Operation is exactly similar to multiply except that data operand will be divided so that operation A/B is carried out. The OV flag set to 1 indicates divide by zero. The instruction is destructive, hence the original contents of accumulator A and B register are lost. The general syntax is:

```
div ab
```

After execution of the instruction, contents of accumulator A represents the quotient of the division operation, while remainder is stored in B register.

BCD Arithmetic Support

The arithmetic operations on the operands are done assuming all data items to be in hexadecimals or binary. However for inputting the data or display of the result, it is convenient to have the operands in a BCD from the instruction DAA, in the 8051 set converts the hex result into equivalent BCD. The syntax is

```
daa:            ; convert the result to BCD after addition
```

The accumulator, carry flag C, and auxiliary carry (AC) take part in the operation. The instruction is effective only after addition.

The set of operations performed by the controller to execute the instruction are as follows:

Let us assume that the result of adding two BCD is in the accumulator and the flags Cy and AC are set in accordance with the result of the addition operation.

B7	B6	B5	B4	B3	B2	B1	B0
H1 (High nibble)				H0 (Low nibble)			

The controller checks:
1. If H0>9 or AC=1

    then add (06h) to the result.
2. If H1>9 or C=1

    then add (60h) to the result.   (H0: Low Nibble; H1: High Nibble)

After the operation, carry flag is set or reset as per the final result after correction is applied.

Sample Programs

Example 3.4   Write a program to add two BCD numbers stored in memory location 41h, 42h, store the result in the memory location 43h.

```
BCD_ADD:

        mov r0, #41          ; immediate addressing mode
        mov a, @r0           ; load acc with first data
        inc r0
        add a, @r0           ; indirect addressing mode
        daa                  ; convert to BCD
        mov 43h, a           ; store result-direct addressing mode
        end
```

Example 3.5 Write a program to subtract two BCD numbers.

Solution

Since, BCD subtraction is not directly supported by the microcontroller, a variation of complementary method may be used, i.e., the BCD subtraction can be carried out using:

1. **9's complement method.**

> **For example**
>
> Subtract $(29)_{10}$ from $(46)_{10}$ using 9's complements method.
>
> The steps are :

(a) Subtract the number from the largest number in the base. The largest number in the base. The largest number is (99).

(b) 9's complement of $(29)_{10}$ is:
$99 - 29 = 70$; 9's complement of no. $(29)_{10}$

(c) Add complement to the number
$46 + 70 = 116$

(d) Add carry to answer to obtain (BCD) result
$16 + 1 = 17 \rightarrow$ answer

```
comp_9:

        mov 41h, #46h        ; load 1st no.
        mov 42h, #29h        ; load 2nd no.
        mov 0D0, #00h        ; reset carry flag
        mov a, #99h          ; load Acc with 99
        subb a, 42h
        add a, 41h
        addc a, #00h         ; add carry bit
        mov 43h, a           ; (43) location will contain result
        end
```

2. **10's complement method.**

For example

Subtract $(29)_{10}$ from $(46)_{10}$; using 10's complement method.

The steps are:

(a) Find 9's complement method using process described above, which is $(70)_{10}$

(b) Add 1 to it to get 10's complement, which gives $(71)_{10}$

(c) Add 10's complement to the number and neglect carry:

$46 + 71 = 117 \rightarrow$ answer

Since BCD addition is supported, above algorithm steps can be used to carry out BCD, subtraction.

comp _ 10:

```
        mov  41, #46

        mov  42, #29h

        mov  0D0, #00h

        mov  a, #99h

        subb a, 42h          ; 9's comp of number

        inc  a               ; 10's comp of number

        add  a, 41h

        mov  43h, a

        end
```

It is to be noted that since we have not studied other instructions, this is a partial program. The program will not give correct answer if result is negative. We will modify the program to incorporate this in later sections.

3.4.3 Logical Group

The 8051 support logical operations such as AND, OR, EX-OR, NOT. The operations specified are done between corresponding bits of operands if operation is binary. The operations can be done on a byte or on a single bit. In case of the byte operation the accumulator or any location in the internal RAM is one of the source and destination, while bit operation makes use of a carry flag.

It is already known that for the Boolean processing, carry serves as a Boolean accumulator. The logical operation supported, can be classified as:

1. Byte operation
2. Bit operation

Byte Operation

The operation specified in the instruction is done on all the 8-bits in a byte in parallel. The effect of operation is same on every bit in a byte. Except operation with PSW, no flags are affected as a result of these operations. When port SFR addresses are used, destination of logical operation is the port address and not the actual port pin, however the source of data in the instruction is port pin. The formats of the instructions in this group are given below. Any one of the formats based on whether the expected destination is a direct address or accumulator is selected.

$$\left.\begin{matrix} rn \\ direct \\ @ri \\ \#data \end{matrix}\right\} \longrightarrow a$$

$$2nd\ operand$$

$$\left.\begin{matrix} \#data \\ a \end{matrix}\right\} \longrightarrow direct$$

The logical operations can be binary or unary. The above formats are valid for binary operations only which require two operands.

Binary Operations

As described earlier, the general syntax for logical operations such as AND, OR, EX_OR is given by

$$\left.\begin{matrix} anl \\ orl \\ xrl \end{matrix}\right\} \quad a, \quad \left\{\begin{matrix} rn \\ direct \\ @ri \\ \# data \end{matrix}\right.$$

$$\left.\begin{matrix} anl \\ orl \\ xrl \end{matrix}\right\} \quad direct, \quad \left\{\begin{matrix} \# data \\ a \end{matrix}\right.$$

Unary Operations

These operations are performed on a single operand. The source and destination for these operations is the accumulator. The operand must be in an accumulator before the instruction is issued for the operation.

The unary operations supported by the 8051 microcontroller instruction set are:

Clear The mnemonics used for clear operation is CLR.

   CLR A : clear accumulator

It is used to initialize accumulator to zero.

Complement (NOT) This is used to generate 1's complement of the data in the accumulator, i.e., 0 is replaced by 1 and vice versa in the data byte. The instruction syntax is,

   CPL A : Complement accumulator

Rotate This is a special logical shift operation. However, it differs from a normal shift operation in a sense that none of the data bit in a byte is lost. This operation helps to check a bit without doing a logical test. The relation can be with and without involving the carry flag. Generally, carry flag is used to perform rotate operation on a word or multi-byte data. The rotation can be performed in left or right direction on the original data in the accumulator. This leads to the four possible operations shown in Fig. 3.6:

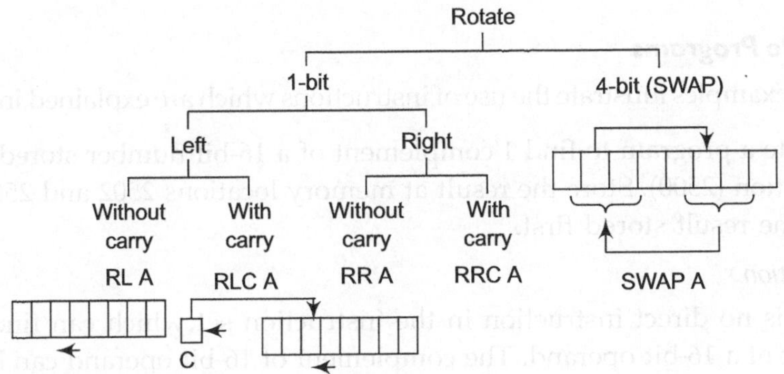

Fig. 3.6

Swap The mnemonic swap defines an operation used for exchanging nibbles in a data byte. The operation can be considered as a 4-bit rotation operation performed on a data byte either to the left or to the right.

Bit Operations

This can be done on any addressable RAM or SFR bit location. The carry flag is the accumulator for this Boolean operation. It is one of the source as well as the destination for most of the bit operations. The SFR register is affected for ports, however the port pins are not affected by the bit operation. The bit operations supported by 8051 instruction set are OR, AND, SET, RESET, and NOT operations.

The general format of these bit operations is shown below:

$$\begin{Bmatrix} anl \\ orl \end{Bmatrix} \quad c, \quad \begin{Bmatrix} bit \\ / bit \end{Bmatrix}$$

where bit represents any addressable bit in the RAM location or in the bit addressable SFR registers. It is important to note that the original data bit is not changed, however after operation specified by the instruction, carry flag will reflect the result. (/) indicates that the operation is to be done with the complement of the contents of specified bit address. The valid instructions are;

1. anl c, bit   ; and c with status of bit specified
2. anc c, /bit   ; and c with status of complement of the specified location
3. orl c, bit   ; or c with status of the specified bit location
4. orl c, /bit   ; or c with status of complement of the bit location

For other operations, the syntax is

$$\left\{\begin{array}{c} \texttt{clr} \\ \texttt{setb} \\ \texttt{cpl} \end{array}\right\} \quad \left\{\begin{array}{c} \texttt{bit} \\ \texttt{c} \end{array}\right\}$$

They are used for initialization of carry or bit addressable locations to the desired value.

Sample Programs

These examples illustrate the use of instructions which are explained in this section.

Example 3.6 Write a program to find 1'complement of a 16-bit number stored at memory location (2500). Store the result at memory locations 2502 and 2503 with LSB of the result stored first.

Solution

There is no direct instruction in the instruction set which can find 1's complement of a 16-bit operand. The complement of 16-bit operand can be found by considering 1-byte of the operand at a time. Another important point to be remembered is that the operand must be in an accumulator for the complement operation.

The program is attempted using two methods.

First method It is assumed that the operand is stored in the external data memory. Direct addressing mode is used for loading the external memory pointer in the dptr register.

```
Comp-16:
        mov  dptr, #2500h    ; load data pointer with address
        movx a, @dptr        ; load acc, from memory location pointed by
                             ; the dptr register
        cpl  a               ; find 1's complement of data
        mov  dptr, #2502h    ; load dptr with the destination address to
                             ; store low byte of the result
        movx @dptr, a        ; store LSB of the result
        mov  dptr, #2501h    ; load the upper byte address in dptr
        movx a, @dptr        ; load acc, from memory location pointed by
                             ; the dptr register
        cpl  a
        mov  dptr, #2503h    ; load dptr with the destination address to
                             ; store upper byte of the result
        movx @dptr, a        ; store LSB of the result
        end
```

Second method   The process can be simplified by using increment instruction. First the 16-bit operand can be moved to accumulator byte by byte and result can be stored temporarily at internal RAM memory location and after the complement operation is over, results can be moved to the desired external memory locations.

```
Mod_comp 16:
        mov dptr, #2500h        ; load data pointer
        movx a, @dptr           ; get LSB of 16-bit data
        cpl a                   ; find 1's complement
        mov 41h, a              ; save the complement of LSB at the address
                                ; 41h

        inc dptr
        movx a, @dptr           ; get MSB of 16-bit data
        cpl a                   ; comp MSB
        mov 42h, a              ; save the complement of MSB at the address
                                ; 42h

        inc dptr                ; dptr points to the address 2502h, which is the
                                ; location to store LSB of the result
        mov a, 41h              ; get LSB of the result in accumulator
        movx @dptr, a           ; save LSB at 2502h
        inc dptr                ; dptr points to the address 2503h, which is the
                                ; location to store MSB of the result
        mov a, 42h              ; get MSB of the result in accumulator
        movx @dptr, a           ; save MSB of the result
        end
```

Example 3.7   Write a program to shift the 16-bit data to the left by 1-bit.

Solution

We will develop programs using two different algorithms, using:

Shift logic

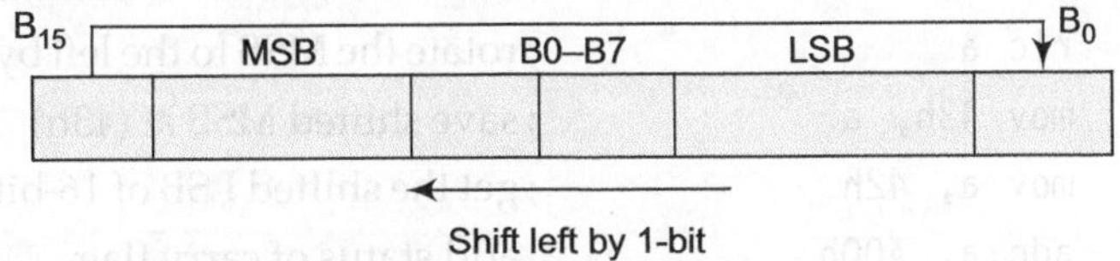

Let us assume that bit B_0 denotes the least significant bit of 16-bit operand, while the bit B_{15} represents MSB of the 16-bit operand.

When the 16-bit data is shifted by 1-bit to the left, all bits are shifted to the left by one position. For example, B_0 becomes B_1, B_1 becomes B_2 and so on. The blank is created at LSB position (B_0), which must be filed by MSB bit B_{15}, so that no data bit is lost.

We will carry out the operation on 1-byte at a time. A carry flag is used to hold the overflowing bit due to shift operation on LSB of 16-bit data.

Assume that data is stored at location:

40h: LSB of 16-bit data

41h: MSB of 16-bit data

The following steps are performed to realize operation:

1. Reset the carry flag.

2. Get LSB in the accumulator and rotate it with carry. The overflow bit B_7 will sit in to carry flag, while B_0 will reset, as the carry is reset to zero. Save the shifted data in RAM location (42h).

3. Get MSB in the accumulator and rotate with carry. The bit B_7 stored in C will become B_8, while the overflowing bit B_{15} is saved in carry flag. Save status of carry in the memory location (43h).

4. Get the shifted data from (42h).

The status of carry flag indicate the LSB (i.e., if the carry flag is set LSB was '1', otherwise LSB was '0'). This operation can be done by adding a carry bit to the accumulator. The program describes implementation of procedure steps:

```
SHL-16:
        mov  40h, #LSB          ; get the LSB of 16-bit
        mov  41h, # MSB         ; get the MSB of 16-bit data
        clr  c                  ; clear carry flag
        mov  a, 40h             ; load LSB in the accumulator
        rlc  a                  ; rotate accumulator data to the left with carry,
                                ; C flag will now contain status of bit B7 of
                                ; LSB
        mov  42h, a             ; save shifted LSB
        mov  a, 41h             ; load MSB in the accumulator
        rlc  a                  ; rotate the MSB to the left by 1-bit
        mov  43h, a             ; save shifted MSB at (43h)
        mov  a, 42h             ; get the shifted LSB of 16-bit data
        adc  a, #00h            ; add status of carry flag
        mov  42h, a             ; store result in memory at (42h)
        end
```

The internal RAM locations 43h, 42h will have 16-bit number shift to left by 1-bit.

Addition logic When a number is added to itself, it is similar to a shift left operation by 1-bit. The difference between the shift and rotate operation is that in shift operation overflowing bit is lost and has to be saved, while in rotate the operation no bit is lost. Hence, when we consider a 16-bit data, the overflowing bit can be saved in a carry flag which can be moved later to the LSB positions done in shift logic method.

For example

Let, a 16-bit data (9234) be represented in binary as:

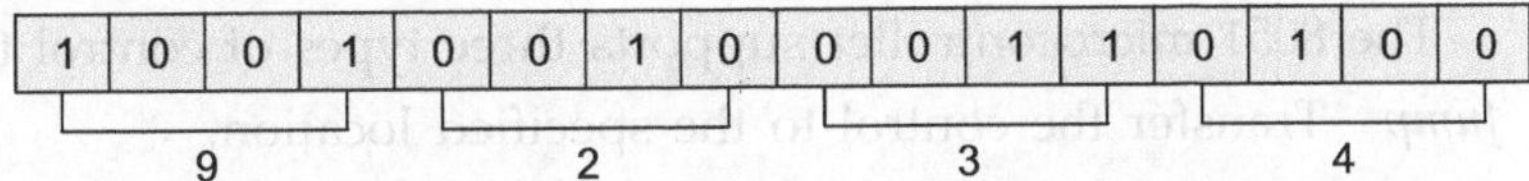

When it is shifted by 1-bit, new shifted data becomes

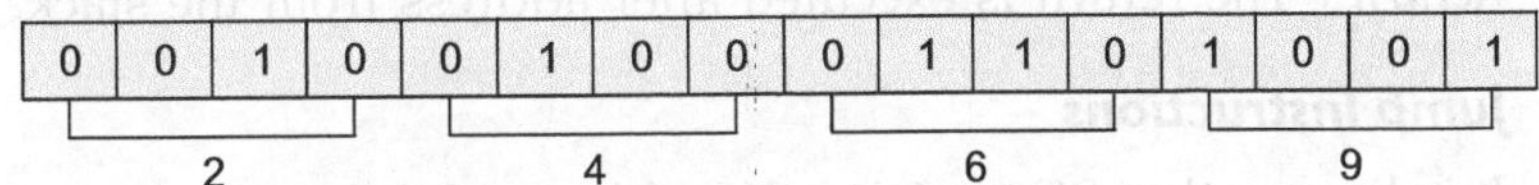

Let us add the number to itself.

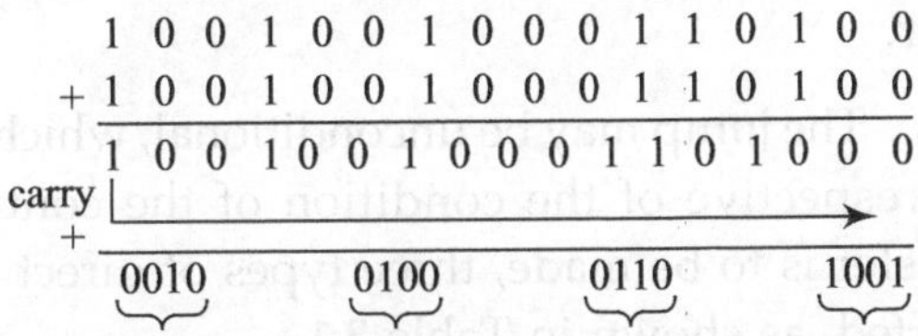

The algorithm is implemented in the following program:

```
mod_SHL_16:
        mov 40h, #LSB       ; get the LSB of 16-bit data in the RAM
        mov 41h, # MSB      ; get the MSB of 16-bit data in the RAM
        mov a, 40h          ; move LSB in the accumulator
        add a, 40h          ; add LSB to itself
        mov 42h, a          ; save shifted data in RAM
                            ; status of with B7 is stored in carry flag
        mov a, 41h          ; repeat for MSB
        addc a, 41h         ; shifted MSB in accumulator, B15 status
                            ; saved in carry flag
        mov 43h, a          ; save shifted MSB of 16-bit data
        mov a, 42h          ; get LSB of shifted 16-bit data
        addc a, #00         ; consider carry flag status
        mov 42h, a          ; save resulting LSB in RAM location 42h
        end                 ; (42h, 43h) have shifted 16-bit data
```

3.4.4 Branch Group

Normally, a program is executed sequentially from top to bottom. Whenever it is necessary to change this sequential flow of execution, the branch instruction can be used. The branch instructions are used preferably after arithmetic group/logical group of instruction. As a result of certain operations it is required to take a decision regarding the executing of program, based on flag condition. These instructions transfer the program control to any desired location rather than to the next instruction in a program.

The 8051 microcontroller supports three types of control transfer. They are:

Jump   Transfer the control to the specified location.

Call   Transfer the entry to the location to execute the program and remember.

Return   The return is executed after address from the stack.

Jump Instructions

It is known that PC contains the address of the next instruction to be executed. The control transfer can be implemented simply by loading the desired address in a program counter.

Unconditional jump   The jump may be unconditional, which loads PC with the specified address irrespective of the condition of the controller. Depending upon where the transfer is to be made, three types of direct as well as indirect addressing is supported, as shown in Table 3.1.

Table 3.1   Types of unconditional jumps supported by the 8051 microcontrollers

Type of control transfer	Addressing size	Instruction syntax
Signed (+127, 127)	8-bit	SJMP address$_8$
Within a page	11-bit	AJMP address$_{11}$
Anywhere in addressing	16-bit	LJMP address$_{16}$
Indirect addressing	Unsigned 8-bit	JMP @A + dptr
		JMP @A + PC

In indirect addressing mode, the jump address is computed by adding 8-bit unsigned displacement from the accumulator to (dptr), using 16-bit addition.

Conditional jump   These instructions load the PC with the new address provided the condition specified in the instructions are satisfied. The condition to be tested are derived based on the status of any one of the three flags or on bit condition at bit addressable locations in RAM or flag status after decrement/compare operations. Thus, conditional jump may be caused by

Flags Z, C or bit   The syntax, for the flag Z, C is

```
jz   addr₈   |
jnz  addr₈   |
jc   addr₈   |
jnc  addr₈   |                    ; 8-bit address in (– 127, 128) signed
                                  ; displacement
jb   bit_add, addr₈    ; the control is transferred to the address if
jnb  bit_add, addr₈    ; the bit addressed is set/reset
```

Decrement and Jump   The contents of Rn or RAM add is decremented by one and result (flag setting) is tested before making a decision for transfer. These instructions support direct as well as register addressing mode. The operation is valid for data in internal RAM only. The syntax is

```
djnz rn, addr₈
djnz direct, addr₈
```

Compare and Jump   CJNZ op1, op2, label(addr₈)

Compare can be used to compare two operands and a jump is executed if they are not equal. The comparison is not destructive, i.e., none of the operand is changed.

$$\text{If} \quad op1 < op2 \qquad : \quad C = \text{'1'} \; ; \text{ carry is SET}$$
$$op1 > op2 \qquad : \quad C = \text{'0'} \; ; \text{ carry is RESET}$$

The jump is relative 8-bit offset with reference to the next instruction address (Fig. 3.7). This supports register as well as indirect addressing mode. The syntax of operation is

```
CJNE A, direct addr, add8
CJNE rn, @ ri, # data, add8
```

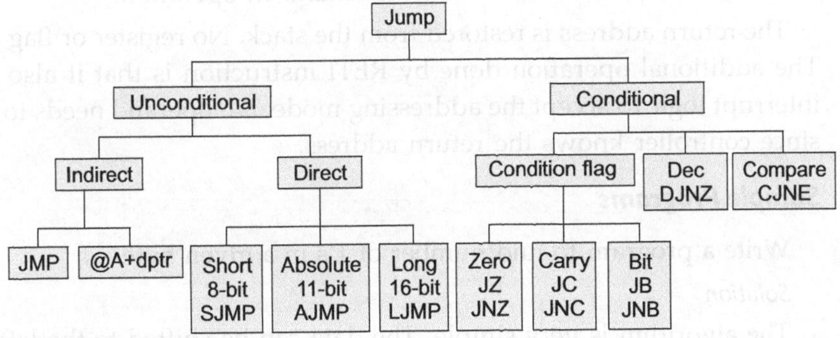

Fig. 3.7

Call Instructions

Since the length of program in assembly language is very large, modular programming is preferred. A program is divided into small functional modules.

The program modules are called *procedures* or *subprograms*. The program modules are coordinated by a main program. The call is used to transfer control to the starting address of the subprogram or a procedure. Depending upon where the module is located, the address specified in the instruction may be 11-bit or 16-bit.

i.e.,

Within a page (2 K) : 11-bit add ; ACALL add_{11}

Anywhere in (64 K) memory address space : 16-bit add ; LCALL add_{16}

Whenever this instruction is encountered, the microcontroller first saves the address of the next instruction on the stack, which is the return address after executing the program starting at the (11-bits/16-bits) address specified in the instruction.

The important thing is that only the return address is saved, hence any register contents which are to be saved must be taken care of by the programmer, if the register is used by the subprogram module.

Return Instructions

There are two types of program modules invoked by the main program:

1. Subprogram
2. Interrupt service program

On completion of the subprogram, control is transferred back to the next instruction in the main program from where it was called. The mnemonics are RET and RETI, both instructions are similar in operation.

The return address is restored from the stack. No register or flag are affected. The additional operation done by RETI instruction is that it also restores the interrupt logic to accept the addressing mode. No operand needs to be specified since controller knows the return address.

Sample Programs

Example 3.8 Write a program to find number of 1's in a given byte.

Solution

The algorithm is very simple. The data can be shifted to the left or right by 1-bit (to the left or right is not important). The status of the carry is checked. If it is 1, it indicates that the bit is one and the counter should be incremented. The process is repeated 8 times for the byte.

```
No_of_1:
        mov  41h, #08h          ; memory location 41h is a used as a counter.
                                ; it is loaded with count 08.
        mov  42h, #00h          ; (42h) is initialized to zero, it will contain the
                                ; answer
        mov  a, #data₈
        clr  c                  ; clear carry flag
loop:   rrc
        jnc  skip
        inc  42h                ; if bit is 1, increment count
skip:   djnz 41h, loop          ; repeat if all bits are not tested
        end
```

Example 3.9   Write a program to find the largest number in an array of numbers. The number of elements in an array (length) is stored at location (2500h) while the array starts from (2501h) onwards and store the result at (2600h).

Solution

Table 3.2 Data array

2500	06	← Length
2501	07	← Array starts
2502	32	
2503	04	
2504	42	
2505	72	
2506	35	

Table 3.2 depicts an array of length 06. The contents of the memory location 2500h represents length, while the contents of memory locations 2501h onwards are elements of an array. We wish to write a program which will find out the largest number in the array. The program should give 72 as the answer if this array is used as data.

The process of finding the largest number can be described as a sequence of the following steps:

1. Get the length of the array and store it in some location in the RAM location 41h.
2. Assume that the first element is maximum and store it in the RAM location 42h.
3. Read next data byte and compare it with data in (42h), if the new byte is larger, replace the old assumed value by the new one.
4. Repeat Step 3 for all data bytes in the array.
5. Store the result at location (2600h).

The algorithm is implemented in the following program:

```
max-8:
        mov  dptr, #2500h       ; initialize data pointer
        movx a, @dptr           ; get length
```

```
              mov 41h, a              ; save length in (41h)
              inc dptr
              movx a, @dptr           ; store it in (42h)
        ok:   inc dptr
              djnz 41h, cont          ; continue if are all bytes are not checked
              mov a, 42h              ; move largest number in accumulator
              mov dptr, #2600h
              mov @dptr, a            ; store the result at the location 2600h
              sjmp end1
        cont: movx a, @dptr          ; get new element of the array
              cjne a, 42h, chk        ; compare it with assumed largest value
              jmp ok
        chk:  jnc ok
              mov 42h, a              ; replace largest value with new element
              sjmp ok
        end:  end
```

Example 3.10   Write a program to reverse bits in a byte.

Solution

The desired operation can be explained with the help of Fig. 3.8.

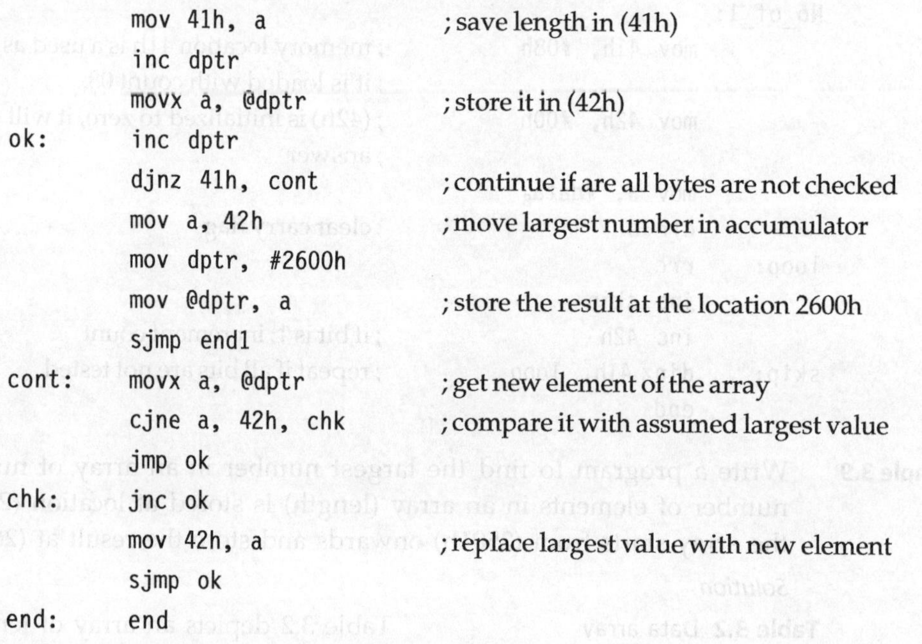

Fig. 3.8   A byte representing an 8-bit group

It is clear from Fig. 3.8 that bit B0 represents LSB of the byte while MSB is denoted by B7. When the bits are reversed, (B0, B7), (B1, B6), ..., (B3, B4) will exchange their positions in the byte. This requires the bit operations of the 8051 to be used. After execution of the program the position of bits are changed as following:

$$\text{bit } 7 \rightarrow \text{bit } 0$$

$$\text{bit } 6 \rightarrow \text{bit } 1$$

$$\vdots$$

$$\text{bit } 0 \rightarrow \text{bit } 7$$

The required operation can be realized by bit operation via Boolean accumulator, i.e., carry flag (C) will be used. The program listing is given below:

```
Bit_rev:
        mov  20h, #data input    ; starting add of bit add_RAM
        mov  c, 07
        mov  0e0, c
        mov  c, 06
        mov  0e1, c
        mov  c, 05
        mov  0e2, c
        mov  c, 04
        mov  0e3, c
        mov  c, 02
        mov  0e4, c
        mov  c, 02
        mov  0e5, c
        mov  c, 04
        mov  0e6, c
        mov  c, 00
        mov  0e7, c
        mov  r2, 20              ; r2 : original data byte
        mov  r3, A              ; r3 : reversed data byte
        end
```

Example 3.11   Write a program to reverse an array. The length of the array is stored in (2500h), while the data is stored from (2501h) onwards.

Solution

Table 3.3 Data array

	Original array		Array after reversal
2500	05		05
2501	11		55
2502	22		44
2503	33		33
2504	44		22
2505	55		11

Array on the left-hand side in Table 3.3 depicts an array of length 06. The contents of the memory location 2500h represents length, while the contents of memory locations 2501h onwards are elements. After the reversal operation, right-hand side of the table represents the new array.

This is an example of realisation of exchanging data in external memory when there is no direct instruction in the set. The process is to be repeated $L/2$ times, where L is the length of an array. Divison by 2 can be carried out by a shift right operation. Let us assume that length of the array is less than 255-bytes.

```
Array_Reverse:
            clr a                   ; clear accumulator
            clr c                   ; clear carry flag
            mov dptr, # 2500        ; load the starting address in dptr register
            movx a, @dptr           ; accumulator will contain length of an array
            mov 41h, a              ; get LSB of the address of last location of the
                                    ; array
            rrc a                   ; divide length by two to obtain loop count
            mov 42h, a              ; (42h) will contain the loop count
rept1:      inc dptr                ; get element of an array
            movx a, @dptr
            mov r3, a               ; r3 temp register for holding data from array
            push dpl                ; save lower byte of dptr on the stack- *
            mov dpl, 41h
            movx a, @dptr           ; get the data from bottom of the array
            mov r2, a               ; save it in register r2
            mov a, r3
            movx @dptr, a           ; bring data bottom location
            mov a, r0
            pop dpl                 ; set the top address
            movx @dptr, a
            dec 41h
            djnz 42h, rept1
            end
```

* With PUSH/POP instead of name of the register to be pushed, hardware address is to be used in the instruction.

Quick Charts

The purpose of these charts is to make quick summary of the instruction set in a pictorial form so that it becomes convenient to refer to the instruction. They help in selection of the operation for implementation of algorithm. The summary of the 8051 instruction set is represented in four charts, one each for the functional group.

Figure 3.9 represents allowed operations and corresponding instructions in a data transfer group. Similarly, Figs 3.10–3.12 represent arithmetic, logical, and branch group of operations and instructions.

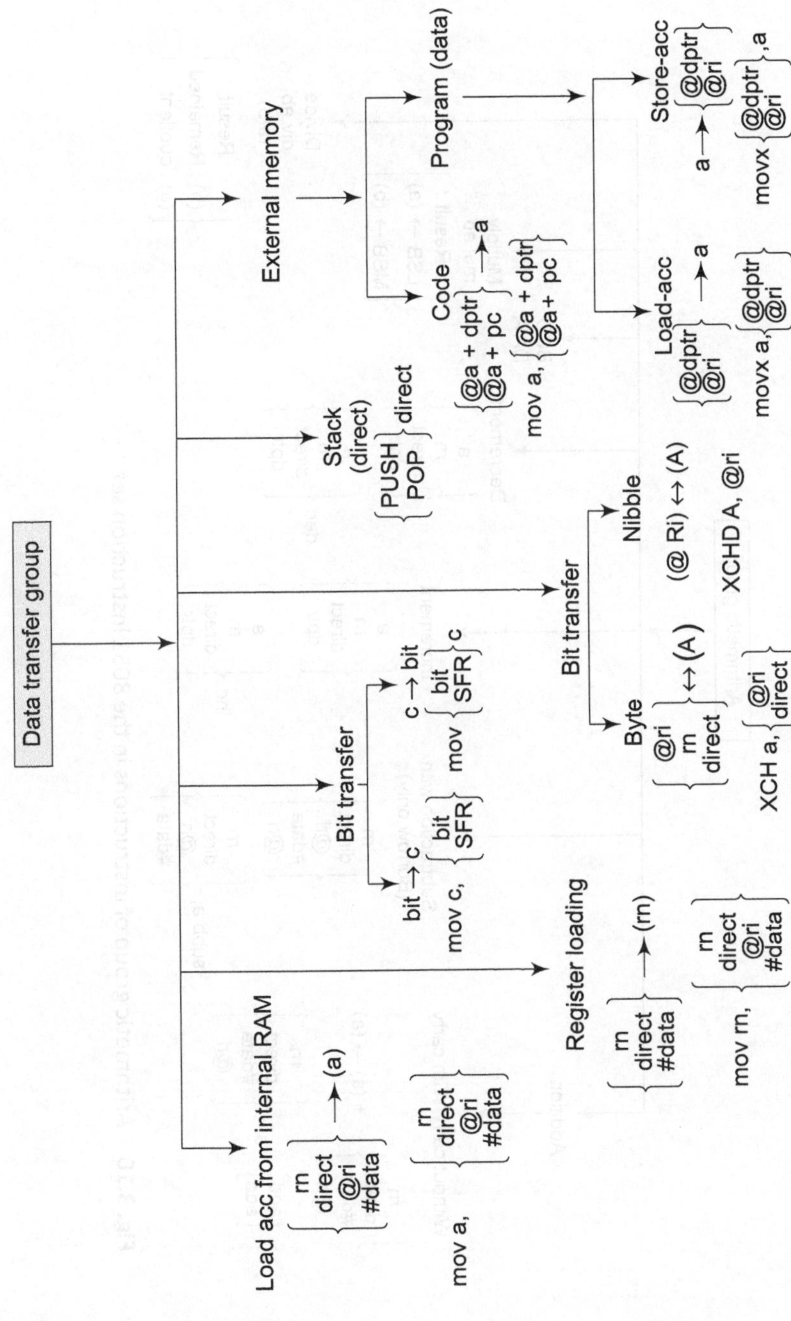

Fig. 3.9   Data transfer group of instructions in the 8051 instruction set

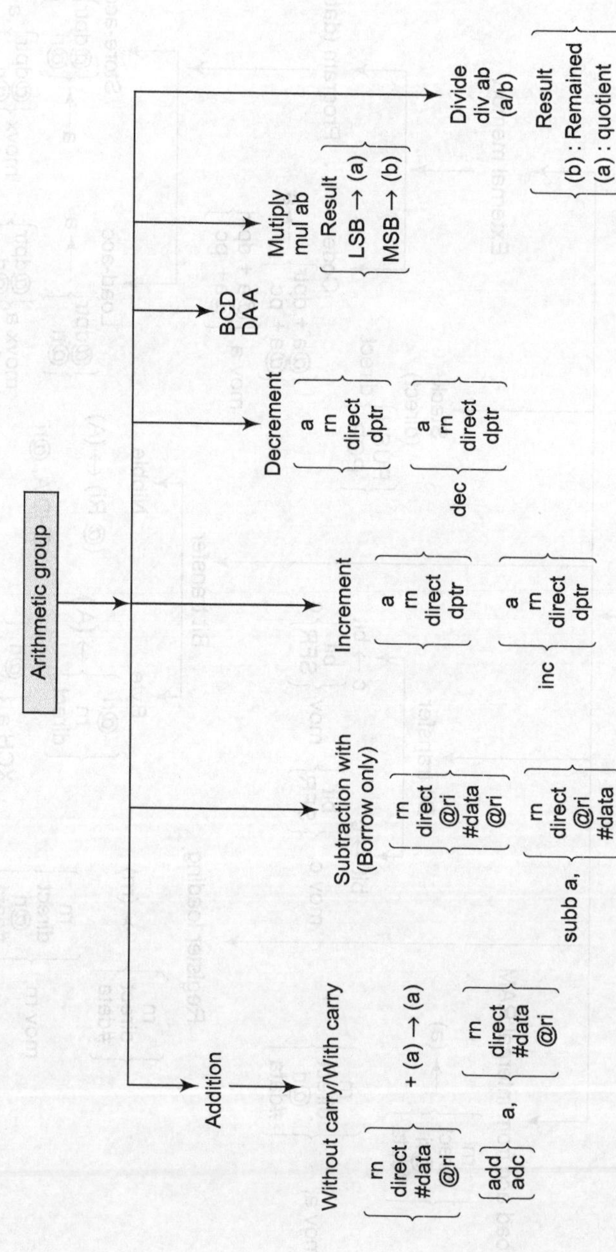

Fig. 3.10   Arithmetic group of instructions in the 8051 instruction set

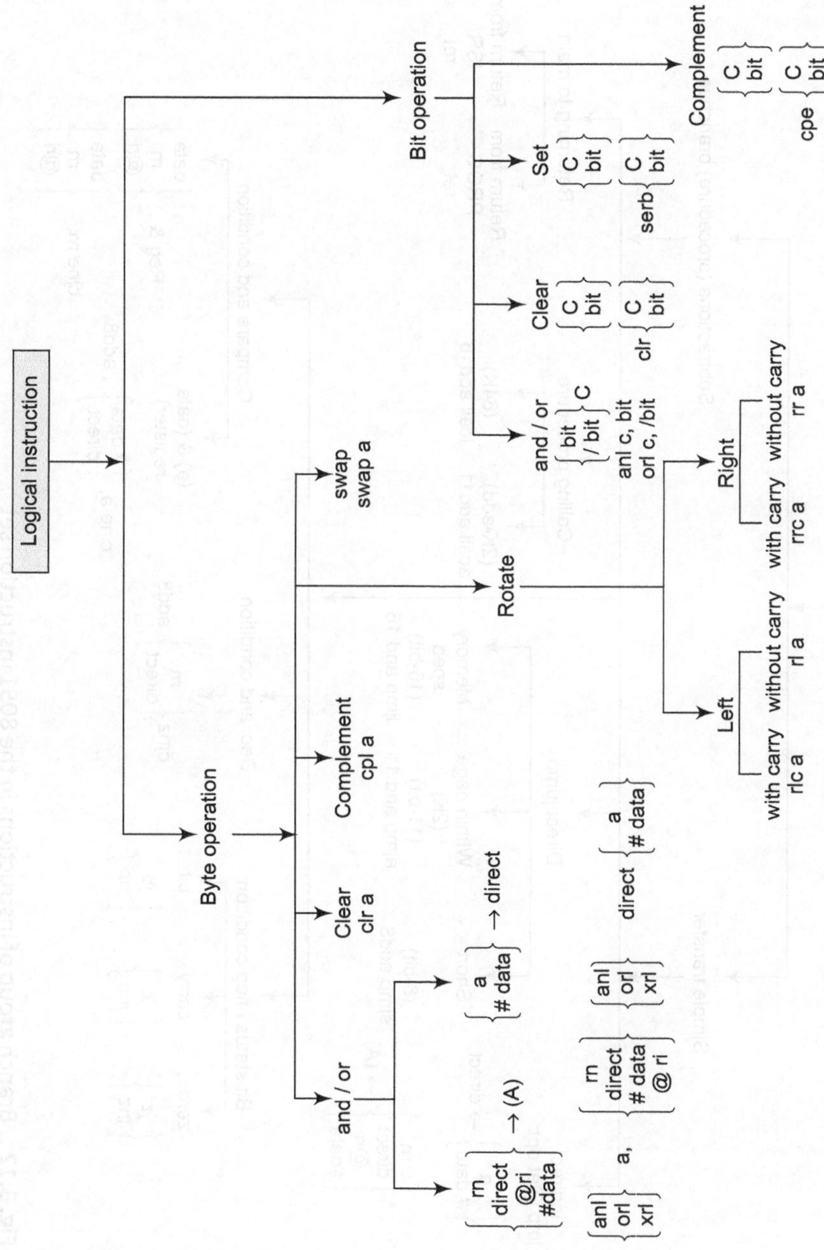

Fig. 3.11 Logical group of instructions in the 8051 instruction set

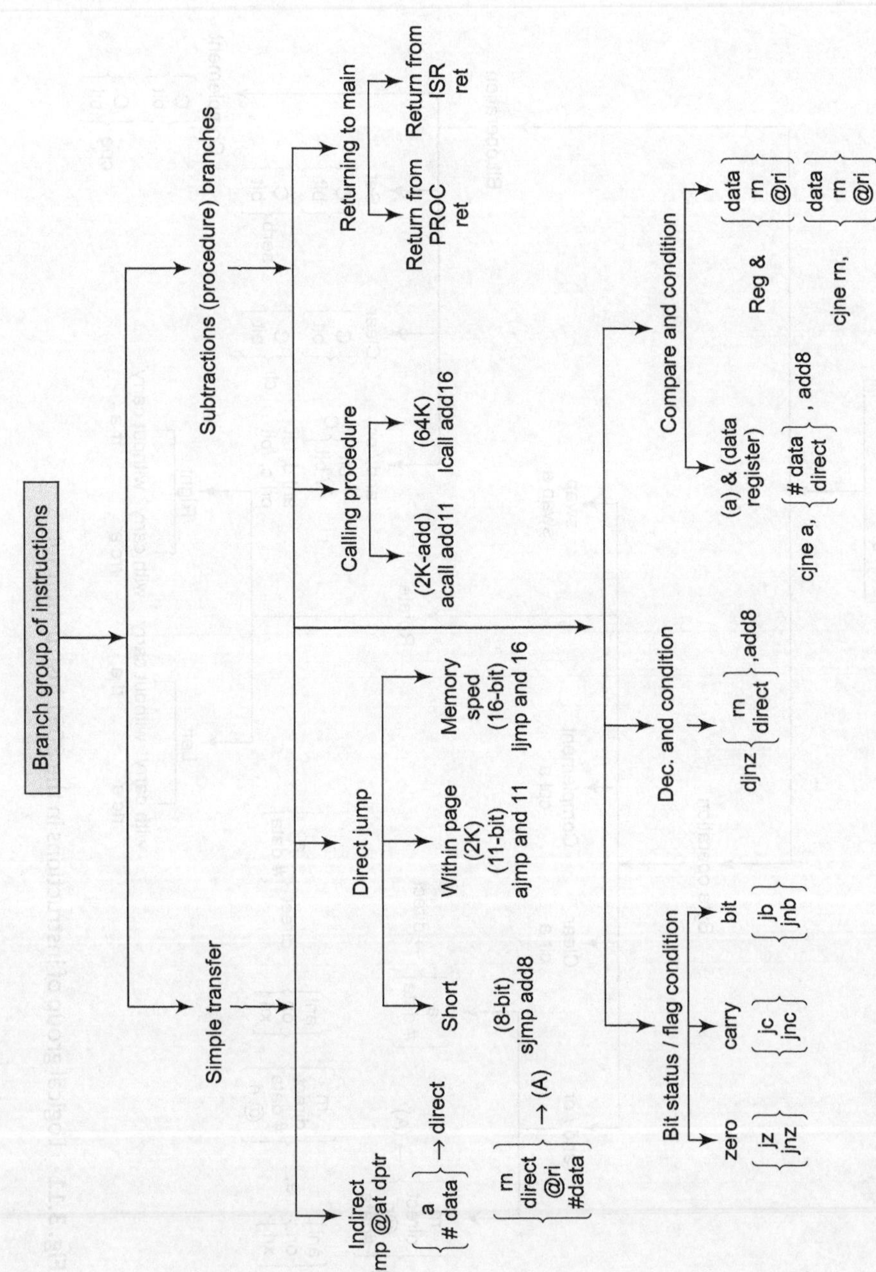

Fig. 3.12 Branch group of instructions in the 8051 instruction set

3.5 | PROGRAMMING EXAMPLES

In this section we will study the assembly language programming in detail with the help of simple examples. We will consider all types of programs which are needed in the development of a microcontroller-based design. Few of them may be used in the form of utilities or libraries so that they can be shared between applications.

3.5.1 Simple Programs

Example 3.12 Write a program to multiply a 16-bit number by an 8-bit number.

Solution

Let us assume that the registers used to store the values are:

Multiplicand 16-bit number	MSB LSB xx xx	MSB stored in register r4	LSB stored in register r3
Multiplier 8-bit number	LSB Yy	———	LSB (r5)
Answer 24-bit product	Byte-3 stored as prod + 2	Byte-2 stored as prod + 1	Byte-1 stored as prod

The following steps describes the multiplication algorithm:

Step 1

Multiply the LSB of the multiplicand with the multiplier and store result at prod and prod + 1, locations with the LSB first.

Step 2

Multiply the MSB of the multiplicand with the multiplier to get partial product.

Step 3

Add LSB of the partial product to the MSB of previous result and store the LSB at prod + 1.

Step 4

If there is a carry, increment the MSB of the partial product and store at prod + 2, else store the MSB without increment.

The program listing is given below:

```
MUL16_8:
        push 00h                ; save r0 on stack (some assemble accept
                                ; direct address instead of names of
                                ; register/SFR with PUSH/POP)
        mov r0, #prod           ; initialize the result pointer
        mov r3, #LSB_ multiplicand
```

```
        mov  r4, #MSB_ multiplicand
        mov  r5, #multiplier
        mov  a, r3
        mov  b, r5
        mul  ab              ; (a) are LSB and (b) are MSB of 16-bit result
        mov  @r0, a          ; store LSB ( byte-1) at the location prod
        inc  r0
        mov  @r0, b          ; store MSB (byte-2) at the location prod+1
        mov  a, r4
        mov  b, r5
        mul  ab              ; (a) are LSB and (b) are MSB of 16-bit result
        add  a, @r0          ; obtain final byte-2
        mov  @r0, a          ; store byte-2 of answer at prod+1
        mov  a, b
        addc a, #00h         ; consider carry and obtain byte-3 of answer
        inc  r0
        mov  @r0, a          ; store byte-3 at prod+2
        pop  00h             ; restore r0 (some assemble accept
                             ; direct address instead of names of
                             ; register/SFR with PUSH/POP)
        end
```

Example 3.13 Write a program to divide 16-bit number by another 16-bit number.

Solution

Dividend	16-bit	MSB	LSB
		(r4)	(r3)
Divisor	16-bit	MSB	LSB
		(r6)	(r5)
Result	8-bit		r7
Flag			r2

Let us assume that the registers used to store the values are as per above. Since a 16-bit operation is not supported directly, we will carry out subtraction taking 1-byte at a time.

```
Div16_16:
        mov  r7, #00h
        mov  r2, #00h
rept1:  mov  a, r4
        cjne a, r5, skip
        mov  a, r3
        cjne a, r5, skip
```

```
skip:       jc end1
            clr c
            mov a, r3
            subb a, r5
            mov 01h, C
            mov r3, a
            mov a, r4
            mov c, 01h
            subb a, r6
            mov r4, a
            inc r2
            cjne r2, #00h, rept1
            inc r7
            sjmp rept1
end1:       end
```

Example 3.14   Write a program to find factors of the given number *n*.

Solution

We will make use of the fact that the largest factor of the number can be one half of the number. To check if the number is divisible by some value, remainder can be checked. A zero remainder indicates that the value is a factor.

Let us assume that the factors of the number will be stored in memory locations, named as fact, onwards.

```
Factors:
            mov r4, #number
            mov r0, fact                 ; initialize pointer to memory
            mov a, r4
            mov b, #02
            div ab
            mov r5, a
            mov a, #01
            mov r3, a
            mov @r0, a                   ; save '1' as factor
            inc r0
rept1:      inc r3
            cjne r3, r5, next
            sjmp carry_on
next:       jnc end1
carry_on:   mov b, r3
            mov a, r4
            div ab
            mov r7, b
```

```
                 cjne r7, #00h, rept        ;not a factor, continue
                 mov a, r3
                 mov @r0, a
                 inc r0
                 sjmp rept1
end1:            mov a, r4
                 mov @r0, a                 ;number itself is last ;factor
                 end
```

Example 3.15   Write a program to find prime factors of the given number *n*.

Solution

We will make use of the fact that the largest factor of the number can be one half of the number. To check, if the number is divisible by some value, remainder can be checked. A zero remainder indicates that the value is a factor.

Let us assume that the factors of the number will be stored in memory locations, named as fact, onwards. This program differs from the last program in the sense that we are interested in prime factors only and not in all factors. This can be done by a minor modification in logic. Study the program listing carefully and point out the difference.

```
Prime_Factors:  mov r4, #09        ;#number
                mov r0, #0050h     ; fact-initialize pointer to memory
                mov a, r4
                mov b, #02
                div ab
                mov r5, a
                mov a, #01
                mov r3, a
                mov @r0, a          ; save '1' as factor
                inc r0
rept1:          inc r3
                mov a, r3
                cjne a, 05, next
                sjmp carry_on
next:           jnc end1
carry_on:      mov b, r3
                mov a, r4
                div ab
                mov r7, b
                cjne r7, #00h, rept1    ;not a factor, continue
                mov a, r3
                mov @r0, a
```

```
                        inc r0
                        mov a, r4              ; divide number by factor
                        mov b, r3
                        div ab
                        mov r4, a
                        clr c
                        rrc a
                        mov r5, a              ; largest factor
                        sjmp rept1
end1:                   mov a, r4
                        mov @r0 a
                        end
```

3.5.2 Arithmetic Operations

Example 3.16   Write a program to generate and store Fibonacci terms, which are less than (FFh)

Solution

The series is given as: 1, 1, 2, 3, 5, 8, 13, 21, 34...

Inspection of the series leads to the conclusion that next term of the series can be obtained by adding two immediately preceding terms. The steps of the procedure are:

1. Move the first number which is 1 in r3 as well as in the memory location.
2. Move the second number which is again 1 in r2 and store it in next memory location.
3. Increment the pointer to the memory so that it points to the next location.
4. Obtain the next term by adding r3 and r2.
5. If the term is <= FF
    (a) Store result in memory
    (b) Move 2nd number while it is in r2 to r3
    (c) Move current term SUM to r3 to prepare for next operation
    (d) Goto step 3

Else,

Stop the process.

The following program represents the implementation of the algorithm.

```
Factors:
Fibb:       mov dptr, #Fibb
            mov a, #01h            ; 1st number of series
            movx @ dptr, a         ; store in memory
            inc dptr
```

```
                movx @dptr, a            ; store 2nd number
                mov r2, a                ; 2nd number in r2
                mov r3, a                ;1st number in r3
        Next_term:
                inc dptr                 ; print to next memory location
                mov a, r3
                add a, r3                ; get new term
                jc end1                  ; no. is >FF, end the process
                movx @dptr, a
                mov r4, a
                mov a, r2
                mov r3, a
                mov a, r4
                mov r2, a                ; prepare for the next term
                sjmp Next_term
        end1:   end
```

Example 3.17   Write a program to find the sum of squares of 10 hex digits (data_in) stored in memory 2500 onwards. Store the LSB of the result at 2700h and MSB at 2701h. We will use a look-up table. The square of numbers is stored in memory in the form of look-up table.

Solution

Let the look-up table be stored at the memory locations square onwards.

square db 00, 01, 04, 09, 10, 19, 24, 31, 40, 51, 64, 79, 90, A9, C4, E1

Data_in

2500h	2501h	2502h	....	2509h
b	6	7		C

The sum may be 16-bit, hence we will use the register r2 to store the upper byte of the answer. The register used in the program are:

$$r3 = (LSB)\ of\ result$$
$$r2 = (MSB)\ of\ result$$
$$r4 = Counter\ (10)$$

```
        Square_add:
                mov dptr, data_in        ; initialize data pointer
                xrl a, a                 ; clear accumulator
                mov r3, a                ; initialize r2, r3
                mov r2, a
        rept1:  movx a, @dptr
                and a, #0Fh              ; mask upper nibble
```

```
                     push  dpl              ; some assemble nedd direct
                                            ; address instead of name of
                     push  dph              ; SFR/Register with PUSH
                     mov   dptr, square     ; base address of look-up table
                     movc  a, @a+dptr       ; look-up table is stored in code memory
                     add   a, r3            ; get LSB of the result
                     mov   r3, a            ; save it
                     jnc   skip
                     inc   r2               ; inc MSB of the result
   Skip:             pop   dph              ; some assemble need direct
                                            ; address instead of name of
                     pop   dpl              ; SFR/Register with POP
                     inc   dptr
                     djnz  r4, rept
                     mov   dptr, result     ; store LSB of the result
                     mov   a, r3
                     movx  @dptr, a
                     inc   dptr
                     mov   a, r2
                     movx  @dptr, a         ; store MSB of the result
                     end
```

Example 3.18   Write a program to find HCF or GCD (greatest common divisor) of two numbers N1 and N2 .

Solution

The method of division can be used for finding out GCD or HCF. The algorithm can be represented as:

1. Find remainder of the larger number divided by the smaller number
   dividend = larger number
   divisor = small number
2. If rem # 0
   Carry out division ; with: --> dividend = divisor and divisor = remainder
   goto Step 2
   Else ==> Stop the process, divisor is GCD or HCF.

In the following program listing, we have used ;
   r5 ; larger number (dividend)
   r4 ; Smaller number (divisor)
   r3 ; 1st Number (n1)
   r2 ; 2nd Number (n2)

```
   hcf:              mov   dptr, #number    ; memory pointer to data
                     movx  a, @dptr         ; get 1st number
```

```
                    mov   r5, a                    ; store as maximum
                    mov   r2, a                    ; save it also in r2
                    inc   dptr
                    movx  a, @dptr
                    mov   r3, a                    ; store 2nd number in r3
                    cjne  a, r5, carry_on
                    mov   r4, a                    ; two number are
                    sjmp  end1                     ; equal so any one is HCF
cary_on:            jnc   no-change
                    mov   r4, a                    ; larger number in r5
                    mov   a, r5                    ; smaller number in r4
                    sjmp  process
no_change:          mov   r4, a
process:            mov   a, r5
                    mov   b, r4
rept1:              div   ab                       ; get remainder check (B)
                    mov   r7, b
                    cjne  r7, #00h, next           ; rept, if the remainder is non-zero
                    sjmp  end1                     ; over, if rem = 0
next:               mov   a, r4                    ; prepare for the next division
                    mov   r4, b
                    sjmp  rept1
end1:               inc   dptr                     ; pointer in memory to store HCF
                    mov   a, r4
                    mov   @dptr, a
                    end
```

If this module is to be used as a subprogram, then registers used in the program, should be saved on stack using push instructions and popped before the return to main. The last instruction end has to be replaced by ret instruction.

Example 3.19   Write a program to find LCM (least common multiplier) of two numbers N1 and N2.

Solution

A simple formula used to find (LCM) of two numbers N1 and N2 is

$$LCM = (N1 \times N2) / HCF$$

which means that LCM can be found by dividing the product of two numbers by HCF

```
LCM:                mov   dptr, #number
                    movx  a, @dptr                 ; get 1st number
```

```
                mov r5, a              ; store as maximum
                mov r2, a              ; save it in r2
                inc dptr
                movx a, @dptr
                mov r3, a              ; store 2nd number in r3
                cjne a, r5, carry_on
                mov r4, a              ; two number are equal, so any
                                       ; one is HCF
                sjmp end1
carry_on:       jnc no_change
                mov r4, a              ; larger no. in r5
                mov a, r5              ; smaller in r4
                sjmp process
no_change:      mov r4, a
process:        mov a, r5
                mov b, r4
rept1:          div ab                 ; get remainder and check (B)
                mov r7, b
                cjne r7, #00h,next     ; rept, if rem is non-zero
                sjmp end1              ; over, if rem = 0
next:           mov a, r4              ; prepare for
                mov r4, b              ; next division
                sjmp rept1
end1:           inc dptr               ; pointer in the memory to store
                                       ; HCF
                mov a, r4
                mov @dptr,a
                inc dptr
                mov b, r4
                mov a, r2
                div ab                 ; (a) = (LSB)
                mov b, r3              ; (b) = remainder = 0
                mul ab                 ; (a) = (LSB) and (b) = (MSB)
                mov @dptr, a
                mov a, b
                inc dptr
                movx @dptr, a
end1:           end
```

3.6 | C-CROSS-COMPILER

C-programming language offers a truly unique and powerful alternative to assembly language programming. It has a wide variety of operations, extensive use of pointers, and an expandable base of run time functions. It gives the capability of a higher level as well as the flexibility of assembly language coding.

MICRO/c_51 is a representative of cross-compilers for the 8051 family of microcontrollers. It supports number of features that provide direct access to the architecture, e.g., assigned variable storage, pointer support for all memory mapping, direct and user expandable C source code, access to current and future SFR, bit map support as well as handling hardware interrupts.

The cross-compiler generates an assembly language source file compatible with relocatable assembler and linker or loader. The software development tools may include three components:

(i) Macro assembler

(ii) Relocatable loader

(iii) Object_Hex conversion program

A batch file can be developed to carry out the procedure steps for the generation of hex code using cross-compiler. A sample for c_51 is given below:

```
/* Let the bat file name be: c-51.bat  */
a:
mcc51 %1.c %2
b:
asm51 %1.src
rl51 a: %.obj, a:mc51.lib, ram size(255)
oh a: %1 to a:%. hex
a:
```

the command to be given is

c_51 FILE NAME

All the steps will be carried out and a hex file can be generated which can be executed on trainer kit or simulator.

3.6.1 Features of X51 Support

In addition to standard C, the cross-compiler has expanded support tools to support the architecture of X51 family.

Memory Maps

It supports all six memory maps, on default or variable basis. The variable declaration has a format:

[Storage Class] [map_spec.] Type_spec. List of locations.

The map_spec may be any one of the followings:

b_map	; On_chip bit address	→ {128-bits}
d_map	; On_chip byte address	→ {128-bits}
f_map	; On_chip indirect address	→ {128-bits}
p_map	; External page 8-bit address	→ {256-bits}
e_map	; External address	→ {128-bits}
c_map	; External address	→ {128-bits}

For example

b_map	bit	a,b,c
c_map	Char	buffer[100]

Special Function Registers

They are used to control the processor as well as on-chip peripheral devices.

For example

```
Int  pcon/* define a variable pcon        */
Main( )
{
ea     = 1              /* enable interrupt */
pcon = 0x54            /* define pcon       */
}
```

Register Bank

Since the architecture provides for a set of four 8-byte banks, compiler provides for the bank selection on a function by function basis.

For example

Function (———) using n ; where $n = 0, 1, 2, 3$ indicating bank.

Using it automatically restores the previous register bank upon exiting the function.

Interrupt Attributes

This is also on the function basis.

For example

Function (—— } interrupt : n ; where $n = [0, 15]$

The number n is called an attribute. More than one attribute can be assigned to a single function, if it is to be driven by more than one interrupt. The attribute puts the program code at the specified interrupt vector. The program may use all the ports, registers {a,b,PSW, dptr}, and executes the desired function. Declaring attributes does not affect IE or the priority. User must enable or disable as per processing requirement. The assignments are given in table:

Interrupt type assignments		
Interrupt number	Source	Vector address
0	EXT INT 0	0003
1	TIMER 0 OVERFLOW	000B
2	EXT INT 1	0013
3	TIMER 1 OVERFLOW	001B
4	SERIAL PORT	0023
5	TIMER 0 OVERFLOW	002B
6	SOFTWARE	0033
7	SOFTWARE	003B
8	SOFTWARE	0043
9	SOFTWARE	004B
10	SOFTWARE	0053
11	SOFTWARE	005B
12	SOFTWARE	0063
13	SOFTWARE	006B
14	SOFTWARE	0073
15	SOFTWARE	007B

Reset

The entry to all C programs is the function main. A reset vector at location 0 is generated by any module within the main. There should be only one main program.

Run Time Support

This is in the form of (*.LIB). It is for:

1. arithmetic → Support C operations
2. map   → Memory access and pointer operation
3. vector   → Reset and interrupt vector
4. function   → Standard and special C function

It is possible to modify them by adding library modules, individual modules can be computed and assembled separately. Once they are ready, the relocatable loader RL51 handles the job of assembling object modules into a single executable program.

For Example

Rl51 : xyz.obj, a.obj, a.lib, mc51.lib

Exception Processing

It provides the means to trap unexpected errors in the programs. There are built-in error handling modules for detection of errors. The errors may arise in

1. arithmetic,

2. memory operation, and

3. stack operation.

The error passes a code or a pointer argument to the [cexhl] run-time exception handler. Default returns to the calling program. However, the label can be used to develop the routine.

Expanding Pointer Access

A pointer support for additional memory spaces or peripherals can be added by redirecting the unsupported map access to certain routine. The pointer is

23 16	15 0
----- 8-bit map prefix -----	----- 16-bit map address -----

The map prefix is used for supporting the memory map of X51:

0	; On-chip bit address	→	{128-bits}
1	; On-chip byte address	→	{128-bits}
2	; On-chip indirect address	→	{128-bits}
3	; External page 8-bit address	→	{256-bits}
4	; External address	→	{128-bits}
5	; External address	→	{128-bits}

3.6.2 Procedure

1. MICRO/C51 accept as input a C source file created with a standard text editor. This file is required to have the extension (.c).

2. Assembly language source output is sent to filename .src.

3. Listing and error messages are full pathname support provided for both source and include files.

3.6.3 Command Line

MICRO/C51 has built-in command line processor which permits various options (switches) to control the compilation process.

The format of command line is

Mcc51 filename. c [/ switches..]

The switches are

1. c	:	include C source in the source file
2. dc	:	set default memory map
3. f	:	enable function trace
4. ln	:	C source listing control
5. p	:	define PDF [processor descriptor file] name
6. wn	:	set warning report level
7. t	:	generate statement labels

c-switch (Include C Source)

The c-switch causes C source statements to be included in the assembly source file as comments. This switch can be useful during the debugging phase of program development.

d-switch (Default Memory Map)

The d-switch is used to establish the default memory map for all variables not explicitly declared with a map specification :

Maps available include

b	-	bit addressable (128-bits)
d	-	direct addressable (128-bytes)
i	-	indirect addressable (128/256-bytes)
p	-	page addressable (256-bytes)
e	-	external data memory (64K bytes)

If the d-switch is not specified, the indirect map is automatically selected as the default map.

f-switch (Function Trace)

The f-switch enables function trace for all functions (excluding main) within the C source file. This switch causes a call to function _ f trace is passed one argument containing a pointer to a text string of the functions name.

l-switch (C Source Listing Control)

The l-switch is used the control C source listing generation. Formats available for this switch are

1. pretext substitution C source listing. C source is listed in its original form with line numbers and error/warning messages.
2. post-text substitution C source listing. C source is listed after text substitution occurs. Line numbers and error/warning messages are included.

If the l-switch is not specified, the C listing will include only error/warning messages with line numbers. The listing is directed to the standard output device, which is usually the console, but may be redirected to a disk file using MS-DOS output redirection control.

m-switch (Stack Monitor)

The m-switch enables stack monitor code generation.

p-switch (Define Processor Descriptor File)

The p-switch gives the capability to extend the processor-dependent bit and byte-special function registers (directly addressable in your C program) beyond the standard 8051 processor by defining a processor descriptor file. This file, available from your chip manufacturer, contains descriptions of all the special function registers for a specific 8051 family member, i.e., 8044, 8051.

In addition to providing special function register definition to the compiler, the p-switch also causes an include statement for the specified file to be inserted in the assembly language source output file to supply the MCS-51 assembler with the special function registers.

The format of this switch is

p "filename.pdf"

where "filename.pdf" would define the processor descriptor file available during compilation and assembly. If the p-switch is not specified, C-51 will recognize the standard 8051 processor special function registers.

w-switch (Warning Report Level)

The w-switch used to set the warning report level generated during the compilation process. The default warning level is 0.

t-switch (Statement Labels)

The t-switch additional statement labels to be generated for each statement and expression in the assembly source file. This switch can be useful during the debugging phase.

3.6.4 Compiler Messages

MICRO/C-51 generates three types of messages during the compilation process. These are:

1. Operator messages directed to the standard error device, typically the console. These messages are usually informative in nature and not related to the file being processed.

2. Warning messages directed to the standard output device, typically the console or redirected to a disk file. These messages are related to the file being processed and are reported according to the w-switch setting.

3. Error messages directed to the standard output device. These messages can be related to the file being processed or compiler operation.

All messages related to the file being processed indicates the file name and line number where the problem was detected.

3.6.5 Linking and Loading Format

The general process of creating and executing a program is shown in Fig. 3.5. The loading is done by (O.S.). The arrow indicates that corrections must be done after any one of the major steps. The RL51 program performs two functions, linking and loading for the programs, as follows:

1. The link function, combining a number of object modules specified in an input list.

2. The locate function, assigning absolute addresses to any relocatable addresses in the input modules.

The RL51 command has the overall format:

Directory device; RL51 input_list : TO output_file [control list]

where

Directory: directory where the RL51 is located

Input_list: List of file names to be included in the process. They must be separated by commas.

Output_list: Name of the file to receive the output.

Control list: They are used to select the options for linking as well as loading.

Linking Controls

The linking command allows the programmer to name the resultant output module and to specify which debug information is to be copied to the output module. Linking switches are, DEBUG SYMBOLS, DEBUG PUBLICS, and DEBUG LINES controls select, which specifies the debug information to be included in the output file. The default of any switch is always the positive form.

3.7 | DOWNLOADING PROGRAM FOR EXECUTION

It is known that file produced by the assembler is in

1. Intel-Hex format

2. S_ Records

The file is designed so that it can be loaded into the memory (EPROM) and executed by microcontroller. When program is to be developed, there is a need to download it into the system from the development system. As already described, the computer system memory contains such files. There is a need to transfer it through the program. This is normally done through a serial port.

The downloader can be made a part of monitor program and when executed, we get a hex-file from the system through the serial port and it is loaded at the appropriate memory location.

3.7.1 Intel-Hex Format

Figure 3.13 shows a sample line in the Intel-Hex file:

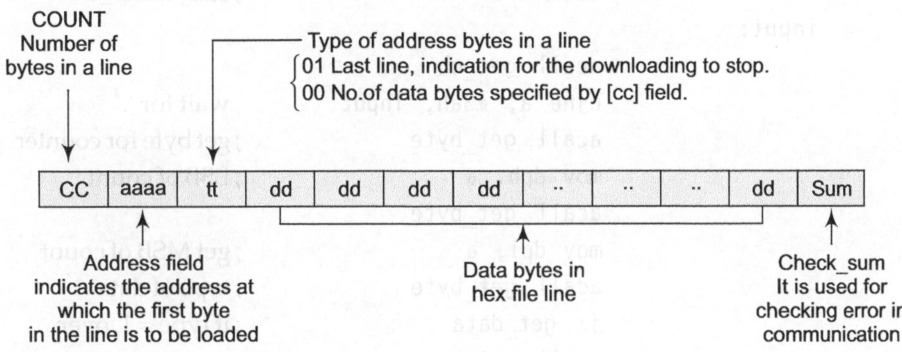

Fig. 3.13

The format of a line of a Intel-Hex.file contains four fields;

1. count
2. address
3. type
4. check sum

Check sum is used by the receiver to check the communications. The algorithm of computing check sum is

1. Add all bytes in a line, ignoring carry. bit.
2. Obtain 2's complement, the total of which is the check sum digit.

Steps of the algorithm for the downloader program are described below;

1. Initialize serial and linear ports
2. Wait for start character " : "
3. Read and assemble next 2-bytes to obtain the no. of data bytes.
4. Formulate the addition word, start in the data fares.
5. Formulate type byte, and terminate the process if it is zero.

The program implements the algorithm described above. The program is loaded in the memory at locations 2500h onwards. Once the program is loaded, on execution, it jumps to the address execute_module to start downloading program from computer to the memory address space of microcontroller.

The program listing is

```
            org  2500h
dn-loader:
            mov scon, #50          ; 8-bit UART enable rec.
            mov tcon, #20          ; set Timer 1 in mod 1
```

```
                mov tcon, #0
                mov th1, #0Fdh          ; reload value
                setb tr1                ; start timer_1
input:
                acall get_ch
                cjne a, #3ah, input     ; wait for '.'
                acall get_byte          ; get byte for counter
                mov dph, a              ; LSB of count
                acall get_byte
                mov dpl, a              ; get MSB of count
                acall get_byte          ; type of data
                jz get_data             ; if type = 1, over
                acall get_byte
                ljmp execute_module
get_data:
                acall get_byte          ; get and store data
                movx @dptr,a
                inc dptr
                djnz R4, get_data
                acall get_byte
                sjmp input              ; discard check_sum
get-byte:
                acall hex-digit         ; forms packed bcd_byte
                swap a
                mov R3, a
                acall hex-digit
                orl a, R3
                ret
hex-digit:
                acall rx_data           ; receives serial ASCII data
                jrb acc.6,skip
                clr acc.5
                add a, #9
skip:           anl a, #0Fh
                ret
rx-data:
                jnb ri, rx-data         ; waits for serial data
                clr ri
                mov a, sub t
                ret
```

3.7.2 S_Records

As shown in Fig. 3.14, lines in S file, contain four fields :

1. length
2. address
3. type
4. check sum

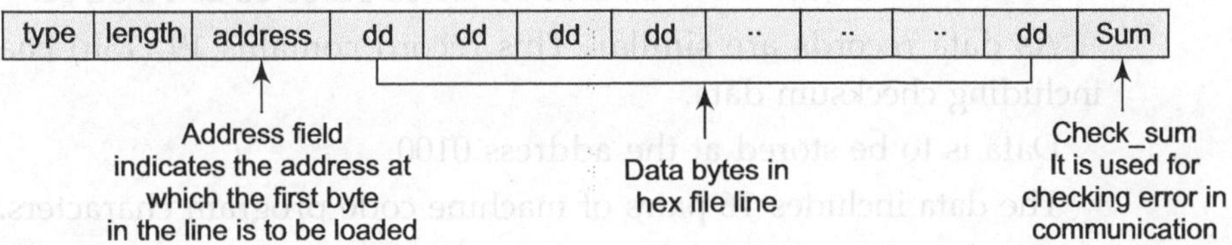

Fig. 3.14

Type:

S9 ; Last line, indication for the downloading to stop.

S1 ; Number of data bytes specified by the length field.

Address:

; It indicates the address of the first byte in the line which is to be

; loaded.

dd:

; Data Bytes

The program listing is

```
              org  2500h
dn-loader:
              mov  scon, #50        ; 8-bit UART enable rec.
              mov  tcon, #20        ; set Timer 1 in mod 1
```

Example 3.20   S-records are another form of hex file format. It is also similar to Intel's format. The following S-file contains S-records obtained by the assembler AS11, for MC68HC11.

; xyz..s19

S104000000FB

S1130100CE000186029B00A700088C000626F63E5E

S9030000FC

File Interpretation:

The First Line

S1 04 0000 00 FB

S1 : data record as indicated by its two left-most characters.

The next two characters (04) indicate record length, the number of the following S-record character pairs contain address information and data, including a pair of checksum characters.

The next four characters (0000) indicate the microcontroller [M68HC11's] memory address, where this S-record's data bytes will be stored.

The Second Line

S1 13 0100 CE 00 01 86 02 9B 00 A7 00 08 8C 00 06 26 F6 3E 5E

As data records are similar. This record contains 19 (13h) character pairs, including checksum data.

Data is to be stored at the address 0100.

The data includes 16 pairs of machine code program characters.

The checksum equals 5E

The Third Line

S9 03 00 00 FC

It contains a termination record S9. This record includes: (03) character pairs, including two pairs of address characters and a checksum. 0000 indicates the address where the first S-record's data will be stored. S9 record includes no data bytes.

EXERCISES

3.1 How many instruction types are there in 8051 instruction set?

3.2 Explain significance of Boolean processor.

3.3 Write a program that will read a byte from Port 1 and use it as an index into jump table stating at 03b5.

3.4 Write a program that will read a byte from Port 0 and use it as an index into any one of the eight locations in jump table depending on which bit is one. Use bitwise instructions.

Write program segments to carry out the following operations:

3.5 Set Port 2, bits 1, 2, 5, 1 to logic HIGH.

3.6 Clear bit 3 of RAM location 34h, without affecting other bits.

3.7 Complement the lower nibble of register r4.

3.8 Complement the upper nibble of location 2bh.

3.9 Rotate dptr 1-bit to the left.

3.10 Rotate dptr 1-bit to the right.

3.11 Multiply contents of memory locations 41h, 42h and put the result in 43h and 44h, respectively, with LSB first.

3.12 Add a byte in the location [02cdh] and internal RAM location [19h] and put the result in the dptr.

3.13 Decrement TL0, TH1, TH0, TL1.

3.14 Divide the contents of memory location [32cdh] by 12h and store the quotient in r0 and remainder in r1.

3.15 Count the number of 1's in a data in dptr and store the result in r5.

3.16 Move PC to dptr.

3.17 Move dptr to PC.

3.18 Initialize program counter PC.

3.19 Assuming a crystal frequency of 12MHz, and use Timer 0, write a program that will interrupt after time delay of 10 ms.

3.20 Write a look-up table: with dptr as base: to find square root [2-byte] of a number, where first byte is integer and second is a fraction.

3.21 Write a look-up table: with PC as base: to find, 1-byte square root [nearest integer value] of a number placed in the A.

3.22 Write a program to test operation of the serial port, the success is indicated by logic High on port PIN P1.0, and Low otherwise.

3.23 Write a program which displays contents of specified register as follows:

r0 = xx: where X ® Contents in hex.

3.24 Write a program segment to add contents of r0 to accumulator.

3.25 Which branch instruction is based on the contents of accumulator?

3.26 Write a program segment to clear memory locations 0×20 to 0×30. Use direct addressing and no loops.

3.27 Write a program segment to clear memory locations 0×20 to 0×30. Use indirect addressing and loops.

3.28 Write a program to read Port 1 and wait until bit 3 of the Port 1 is Low. When Low is detected, on bit 3, set all port pin High on Port 3.

3.29 Write a program to read Port 1 and wait until bit 3 of the Port 1 is Low. When Low is detected, on bit 3, set all port pin High or Port 3. Use Boolean instruction.

3.30 Write a code to initialize the Timer 1 to be externally clocked in Mode 0.

3.31 Write a program to edit ASCII characters so as to replace zeros by blanks. The string starts at address [string] specified by [dptr].

3.32 Write a program to multiply multi byte number.

3.33 Write a program to convert a gray code number into binary.

3.34 Write a program to convert 16-bit number into BCD number.

3.35 Write a program to convert four digits BCD number into hex.

3.36 Write a program to find maximum of an array of ten numbers.

3.37 Write a program to arrange of an array of ten numbers in ascending order.

3.38 Write a program to arrange of an array of ten numbers in descending order.

3.39 Write a program to check if an array numbers are in ascending order or not.

3.40 Write a program to add two multiple word binary numbers.

Whose length in bytes is in accumulator? The starting address of numbers is NUM1 and NUM2. Store the result in memory locations starting from result.

3.41 Write assembly language statements to simulate following C statements:

(a) if I > = 5

x = x + 7;

Else

x = x * x;

(b) If (I > = 5) || (j = 0)

x =I + 5;

Else

x = x /5 + I;

3.42 Write a program that will divide 16-bit number in [dptr] by an 8-bit number in r2. The quotient is in [r4], while remainder is in [Rz5]. If divisor is zero, branch to [ERRORZ].

If the magnitude of the quotient is >127, branch to [ERRORL].

The algorithm is as follows:

    (a) If remainder is zero branch to [ERROROZ].

    (b) Rotate dividend left by 1-bit.

    (c) Subtract divisor from [HB] of dividend, if negative branch to [ERRORZ]

    (d) Add divisor to high byte of dividend.

    (e) Set counter to [7] and clean carry.

(f) Rotate dividend left by 1-bit.

(g) Subtract divisor from HB of dividend.

(h) If result is negative, put 0 in carry and add divisor to HB of dividend otherwise put 1 in carry.

(i) decrement the count.

(j) if count is zero.

  stop

 else

   rotate LB of the divisor to left by 1-bit.

   goto step (f)

3.43 Controller is used to monitor six external relay settings and perform one of the three tasks depending on status of relays. The relays are numbered (0–5) and setting is brought to register r3. [0: Relay open; 1: Relay closed.]

(a) If relays 0,1,5 are closed, perform Task 1.

(b) If Task [t1] is not performed and relay (2 or 3) is closed and 1 is closed, perform Task 2.

(c) Task 3 is to be performed if Task 2 or 1 is not performed.

Write a program for the monitor and control operations.

3.44 Write program segments [ABSOL] that will take the absolute value of 8-bit binary number.

1. Assume that number is in [A] and absolute value is to be placed in [A].

2. Assume that number is pointed to by the dptr and absolute value is to be placed in [A].

3. Assume that number is pointed to by the dptr and absolute value is to be placed in [dptr + 2].

3.45 Write macros for the following operations:

(a) [ABS]: Takes absolutes value of first location and puts result in the second location.

(b) [APPM]: Performs 16-bit memory-to-memory addition.

(c) [SPPM]: Performs 16-bit memory-to-memory subtraction.

(d) [SHIFT_R]: Performs multi-byte shift right operation. The operand to be shifted is in memory pointed by dptr, length of data in bytes in [r5}.

(e) [SHIFT_L]: Performs multi-byte shift left operation. The operand to be shifted is in memory pointed by dptr, length of data in bytes in [r5}.

3.46 Use Intel's format hex codes to identify the number of data bytes and list them. Also identify the address at which each byte will be placed. Recalculate check sum to verify the data.

: 0730100006e3a1887f8a326

Chapter 4

Advanced Programming Techniques

The microcontroller is a programmable device having predefined instruction set. In case the desired operation is not available as part of the instruction set, it is necessary to develop an algorithm for realization, which can be coded using features available in the set of instructions. Advanced programming techniques used in program development in assembly language are described in this chapter. The techniques are illustrated through examples.

4.1 FLOATING POINT REPRESENTATION

The instructions in the set cannot handle floating point numbers. This section deals with representation of floating point numbers and programming algorithms for carrying out operations on floating point numbers.

In scientific and engineering calculations, real numbers are conventionally represented in terms of floating point numbers. In this representation, a number has four parts:

1. Sign of number (positive or negative)
2. Mantissa (fractional part)
3. Sign of exponent (positive or negative)
4. Power (exponent)

Real number	Floating point representation	
	Mantissa	Exponent
31421.1	+ 0.31421	+ 05
− 214.305	− 0.214305	+ 03
0.001999	+ 0.1999	− 02
− 0.000375	− 0.375	− 03

A conventional form to represent this number in memory is using a 7-byte code.

Sign of number	Packed BCD 1	Packed BCD 2	Packed BCD 3	Packed BCD 4	Packed BCD 5	Expont. MSB is a sign bit

(a) Byte for the sign of number is   00   : for positive number

                                               01   : for negative number

(b) MSB of the last byte is   0   : for positive exponent

                                               1   : for negative exponent

---

Example 4.1   Represent: −37925.79 in terms of a 7-byte code.

Solution

− 37925.79 in scientific notation can be represented as -0.3792579×10^5

Considering the four parts of the representation:

(a) Sign of number is negative. Hence the first byte is 01.

(b) Next 5-bytes represent number in packed form, they are [37 92 57 90 00].

(c) Last byte is exponent with MSB as sign, i.e.,

[0 000 0101] = > 05

Thus, the 7-byte representation of the number is { 01 37 92 57 90 00 05 }.

Example 4.2   Represent: 0.00017859 in terms of a 7-byte code.

Solution

0.00017859 in scientific notation can be represented as 0.17859×10^{-4}

Considering the four parts of the representation:

(a) Sign of number is positive. Hence the first byte is 00.

(b) Next 5-bytes represent the number in packed form, they are:
[17 85 90 00 00]

(c) Last byte is exponent with MSB as sign, i.e.,

[1 000 0100] == > 84

Thus, the 7-byte representation of the number is { 00 17 85 90 00 00 84 }.

4.1.1 ASCII to Floating Point Conversion

Since real-valued data entered through keyboard are in terms of ASCII digits and are easier to store in the memory in the 7-byte form so that the arithmetic operations can be done on them, there is a need to develop an algorithm to do so. This section describes the algorithm of conversion based on the procedure used in the examples of the previous section. The steps are

1. The data is accepted in ASCII code through the keyboard and stored in the memory area defined as ASCII_IN till carriage return <cr> is detected.
2. Let the number in the form of code is stored in an area which is called CODE_IN. Initialize these seven locations to zero.
3. Check if there is a minus sign, if there is, store 01 at the first location in area CODE_IN. For positive value, there is no need to do any operation as 00 is already stored.
4. Count the number of digits before the decimal point or the number of leading zeros before the first non-zero digit. Store this number as power.
5. Carry out BCD packing. This is done by taking a character, and then masking the upper nibble. Take the next character mask upper nibble and combine with the previous masked character and store the result at an appropriate location.

The algorithm is implemented in the following assembly language program:

```
ASCII_CODE:
            mov  r0, ASCII_IN          ; initialize pointer to the storing area
Loop:       acall  Read_key            ; get the ASCII input from keyboard
            cjne a, #13h, OK           ; continue if not <cr>
            sjmp  OVER
OK:         mov  @r0, a
            inc  r0
            sjmp  Loop
OVER:       mov  r4, #07h              ; initialize counter to 7
            mov  r1, CODE_IN           ; initialize data pointer
            xrl  a, a
INIT:       mov  @r1, a
            inc  r1
            djnz r4, INIT
            mov  a, @r0
            cjne a, '-' , Plus_sign    ; fix sign bit
            mov  a, #01h
            mov  @r0, a
```

```
                        inc  r0
        Plus_sign:
                        mov  a, @r0
                        cjne a, '0', Pos_exp
                        inc  r0
                        inc  r0                  ; to skip decimal point
                        mov  a, @r0
        Lead_0:
                        cjne a, '0', Neg_exp     ; count no. of leading 0s
                        inc  r4
                        sjmp Lead_0
        Neg_exp:
                        mov  a, r4               ; correction for neg. exponent
                        orl  a, #80h
                        mov  r1, CODE_IN +6
                        mov  @r1, a              ; save exponent
                        dcr  r0
                        sjmp Pack
        Pos_exp:
                        mov  a, @r4
                        mov  r1, CODE_IN +6
                        mov  r0, ASCII_IN
                        mov  a, @r0
                        cjne a, '-', Skip
                        sjmp Pack
        Skip:           dec  r0
        Pack:           mov  r1, CODE_IN + 1
        Rept1:          inc  r0
                        mov  a, @r0
                        cjne a, '.', Skip1
                        sjmp Rept1
        Skip1:
                        cjne a, #13h, No_END
                        sjmp end1
        No_END:
                        anl  a, #0fh,            ; mask upper nibble
                        swap a
                        mov  r5, a
        Rept2:          inc  r0
```

```
                mov  a, @r0
                cjne a, '.', Skip2
                sjmp rept2
end1:           end
```

4.1.2 Packed BCD to ASCII

For an input/output operation, it is convenient to use the ASCII format. To display a floating point number being entered, ASCII characters give easier visualization as well as representation.

The numbers entered in a normal format can be displayed either in the same format or in floating point format. In this section we will develop a program which carries out an operation which is reverse of operation in the previous section.

Let the number stored in buffer is

Fp_in ; 01 35 49 87 27 87

The result or output of the program should be given as

ASCII_out ; –0.13549873E –07

The program uses the information regarding the 7-byte code representation of a floating point number.

```
ASCII_REP:
                mov  r0, Fp-in
                mov  a, @r0
                cjne a, #00h, neg_num        ; check for negative number
                mov  a, #20h                 ; output blank for {+} sign
                sjmp disp_sgn
neg_num:
                mov  a, '-'                  ; output {–} for sign
disp_sgn:
                acall display
                mov  a, '0'                  ; output {0}
                acall display
                mov  a, '.'                  ; output {.} for decimal point
                acall display
                inc  r0
                mov  r1, ASCII_out
                acall unpack
                mov  r1, ASCII_out
                acall unpack
                mov  r5, #0ah                ; length which is {10}
```

```
disp_digit:
        mov a, @r1
        add a, #30h                    ; display 10 decimal digits
        acall display
        inc r1
        djnz r5, disp_digit
        mov a, 'E'                     ; display {E} for exponent
        acall display
        mov a, @r1
        clr c
        rlc a
        jc neg_expo                    ; check if power is negative
        mov a, '+'
        sjmp expo_sgn
neg_expo:
        mov a, '-'
expo_sgn:
        mov a, @r1
        anl a, #7fh
        cjne a, #64h, not_equal        ; check if it is 100
        mov r2, a
        mov a, '1'
        acall display                  ; display '1'
        mov a, '0'
        acall display                  ; display '0'
over:   acall display                  ; display '0'_exponent is 100
        sjmp end1
not_equal:
        jc less_100
        mov r2, a
        mov a, '1'
        acall display
        mov a, r2
        clr c
        subb a, #64h                   ; sub. {100} for other digits
less_100:
        mov r3, #00h                   ; initialize counter for sec_dgt of
                                       ; exponent
rept1:  cjne a,#0ah, sec_dgt
        sjmp over
```

```
sec_dgt:
                jc last_dgt
                inc r3
                clr c
                subb a, #0ah
                sjmp rept1
last_dgt:
                acall display
end1:           end
```

4.1.3 BCD to ASCII_STRING

This is called unpack operation. Unpacking means the 5-packed BCD mantissa bytes are converted into the separate digits and stored in buffer.

Example 4.3   Convert a packed BCD number:

38 14 52 04 95 into an unpacked number.

Solution

Let us assume that r0 : points to the first byte of input data.

r1 : points to 1st location in

buffer (Buffer → 03 08 01 04 05 02 00 04 09 05)

The program listing is given below. It is simple and self-explanatory.

```
UNPACK_BCD:
                mov r4, #05
rept1:          mov a, @R0
                Swap a
                anl a, 0F
                mov @r1, a
                inc r1
                mov a, @r0
                anl a, 0F
                mov @r1, a
                inc r1
                inc r0
                djnz r4, rept1.
                end
```

4.1.4 Multibyte BCD_Addition

The BCD addition can be done by taking 5-packed BCD bytes and adding them as shown.

$$NUM1 = 00\ 12\ 34\ 57\ 95$$
$$NUM2 = 12\ 38\ 29\ 32\ 10$$
$$Sum = 12\ 50\ 63\ 90\ 05$$

The program can be developed as follows:

Let us assume that the first number is stored from 41h onwards with the MSB first, while the second is loaded at 51h onwards.

Address	NUM 1		Address	NUM 2		Address	SUM
41	00		51	12		61	05
42	12		52	38		62	90
43	34		53	29		63	63
44	57		54	32		64	50
45	95		55	10		65	12

```
Bcd_add10:
           mov  r0, num1 + 5      ; address of LSB of the first number NUM1
           mov  r1, num2 + 5      ; address of LSB of the second number NUM2
           mov  r3, sum + 5       ; address of LSB of the sum
           mov  r2, #05h          ; counter in (r2)
           clr  c                 ; reset C = '0'
rept1:     mov  a, @ r0
           addc a, @ r1
           mov  @r0, a
           dec  r0
           dec  r1
           dec  r3
           djnz r2, rept1
           addc a, #00
           end
```

4.1.5 Additional Operations

In order to carry out arithmetic operations on floating point numbers, the intermediate or additional operations are to be carried out. They are

1. Right shift
2. Adjustment of exponents
3. Increment/decrement of exponent for normalization of result
4. 10's complement

Following programs are developed for the said purpose and will be used as subprograms in basic arithmetic operations:

Right shift   A program to rotate 10-digit packed (BCD) number by 1 position to right.

```
right_BCD:
            mov r0, #05h              ; counter to rotate 5 times
next:       movx a, @dptr            ; get 1st pair
            swap a
            and a, 0fh
            mov r4, a                ; save it
            dec r3
            jz over
            dec r0
            mov a, @r0
            swap a
            anl a, F0h
            orl a, r4
            inc r0                   ; save at right place
            mov r0, a
            dec r0
            jmp next
Over:       mov @r0, a               ; last digits
            ret
            end
```

Adjustment of exponents   It can be used to adjust exponents of two floating point numbers so that it is possible to carry out add/subtract operation directly on them.

Example 4.4   For the two numbers: Num_1 = 0.35E + 06 and num_2 = 0.7E + 4

A direct addition cannot be done since the powers are different, we can trace small number 1 rotate it twice to right so that we get [0.007E + 06] which can now be added to the number:

$$0.35 \ E06 + .007 \ E06 = 0.357 \ E06 \rightarrow answer$$

The logic of the following program chooses the smaller number always which may not be always true. However, a general program can be developed.

```
adj_exp:
            mov r0, [num1+6]
            mov a, @r0               ; exponent of first number
            mov r4, a                ; save it
            mov r0, [num2+6]
            mov a, @ r0              ; exponent of second number
            clr c
            subb a, r4
            jc comp1                 ; num 1 > num 2
```

```
                  jz end1                    ; alignment not required
                  mov r3, a                  ; save difference
                  push r3
                  mov r5, a                  ; initialize rotate counter
                  mov r0, #(num1+5)
rept1:            call right_BCD
                  djnz r5, rept1
                  mov r0, #(num1+6)
                  mov a, @r0
                  pop r3
                  add a, r3                  ; inc exponent and
                  mov @r0, a                 ; save value
                  jmp end1
comp1:            cpl, a
                  inc a
                  mov r3, a
                  mov r3
                  mov r5, a
                  mov r0, #(num1+5)
REPT2:
                  call right_BCD
                  dec r0
                  djnz r5, rept2
                  mov a, @R0
                  pop r3
                  mov a, r3
                  mov @r0, a
end1:             ret
```

Increment/Decrement of exponents The purpose of the program is to add 1 to the exponent and is self-explanatory.

```
Inc-exp:
                  add a,#00
                  djnz a, no-exp
                  inc a
                  sjmp end1
Neg-exp:
                  cjne a, #80, neg-exp
                  mov a, #01
                  sjmp end
```

```
no-exp:
            dec   a
end1:       end
```

Similarly, a program to decrement the exponent by 1 can be written.

10's Complement of multibyte BCD number   This is useful to carry out subtraction so that instead of developing a new program for subtraction, program for addition can be used.

```
10_comp:
            mov   r0,   num_1 + 5
            mov   r4,   #05h          ; counter for 5 pairs of packed BCD
            clr   c
9_comp:
            push  r0
            push  r4
rept1:      mov   a,    @r0
            mov   r5,   a            ; save pair of packed BCD
            subb  a,    #99h
            mov   r0,   a
            dec   r0
            djnz  r4,   rept1
            setb  c
            pop   r4
            pop   r0
            mov   a,    @r0
rept2:
            adc   a,    #00h          ; 10's complement
            da    a
            mov   @r0,  a
            dec   r0
            djnz  r4,   rept2
            ret
            end
```

4.2 | FLOATING POINT OPERATIONS

Here, we will study four basic mathematical operations of floating point number.

1. Addition
2. Subtraction
3. Multiplication
4. Division

The aim is to develop an assembly language program for each operation. Table 3.3 defines the function name and the purpose of the module for these four basic operations:

Function name	Operation	Definition
F_plus	Add two packed BCD numbers that have been placed in the buffers [num_1] and [num_2]. The numbers must have the same sign.	
F_minus	Subtract [num_1] from [num_2]. Both numbers must have same sign.	
F_prod	Multiply the number [num_1] by [num_2].	
F_div	Divides the number [num_1] by the number [num_2]. Let us assume that both numbers have positive exponents and [num_1] is larger than [num_2].	

The result of the operation is stored in [num_1].

Addition It is necessary to remember that two numbers cannot be added directly if their exponents are not same. In case of the exponents being different, it is necessary to shift the number to the right or left as per the requirement, so that both have the same exponent. After shifting, they can be added.

Example 4.5 Add 1045 and 769 using floating point representation.

Solution

Num_1   ; $1045 ==> 0.1045 \times 10^4 ==> 00\ 10\ 45\ 00\ 00\ 00\ 04$
Num-2   ; $769 ==> 0.769 \times 10^3 ==> 00\ 76\ 90\ 00\ 00\ 00\ 03$

          1814

The num_2 must be shifted to the left by one digit to have equal exponent and it can be added to [num_1].

Num_1 ; $1045 ==> 0.1045 \times 10^4 ==> 00\ 10\ 45\ 00\ 00\ 00\ 04$
Shifted ; Num-2 ; $0.0769 \times 10^4 ==> 00\ 07\ 69\ 00\ 00\ 00\ 04$

----

        sum $==> 00\ 18\ 14\ 00\ 00\ 00\ 04$

The answer is $[1814 \times 10^4] = 1814$.

The example shows that the carry flag must be tracked, since it indicates that correction should be done by incrementing the exponent. The algorithm can be represented as

1. Carry out the shifting to make the exponent of two numbers equal.

2. Carry out normal procedure of adding 10-packed BCD digits.

The program listing in assembly language will be as follows:

```
F_plus:

        acall   adjust
        acall   bcd_add10
```

```
                cjne  a,  #00h,  over        ; no rotation required
                mov   r0,  num_1 + 5         ; rotate to adjust exponent
                mov   a,  @r0
                acall  right_bcd
                orl   a,  #10h
                mov   @r0,  a
                mov   r0,  num_1 + 6
                acall  inc_exp
                mov   @r0,  a
over:
                end
adjust:
                mov   r0,  num_1 + 6
                mov   a,  @r0
                mov   r4,  a                 ; save first exponent
                mov   r0,  num_2 + 6
                mov   a,  @r0
                subb  a,  r4                 ; num_1 > num_2
                jc    cmpl
                jz    end                    ; adjustment not required
                mov   r3,  a                 ; save difference
                push  r3
                mov   r5,  a                 ; rotate counter
                mov   r0,  num_1 + 5
rept1:          acall  right_bcd
                dec   r0
                djnz  r5,  rept1
                mov   r0,  num_1 +6
                mov   a,  @r0
                pop   r3
                add   a,  r3                 ; inc exponent and save
                mov   @r0,  a
                sjmp  end1
cmpl:           cpl   a
                inc   a
                mov   r3,  a
                push  r3
                mov   a,  r5
                mov   r0,  num_1 +5
```

The BCD subtraction process is more complicated than addition, and there is no easy way to do it in 8051 assembly; however, similar to binary subtraction (where we obtain the 2's complement and then add), we can use 10's complement and add.

Once the 10's complement of one of the numbers has been found, using program 10_comp, the number can be added to. In using the program developed above F_plus. However, the absence of a carry out of the last phase when the addition is completed means that the sum (or result) is negative, and we have to call 10_comp again to get the result. The revised form, following program shows the F_minus listing.

```
rept2:
          acall  right_bcd
          dec r0
          djnz r5, rept2
          mov a, @r0
          pop r3
          add a, r3
          mov @r0, a
          ret
right_bcd:
          mov r5, #05
next:     movx a, @ro                            ; get first pair
          swap a
          anl a, #0fh
          mov r4, a
          dec r3
          jz over
          dec r0
          mov a, @r0
          swap a
          anl a, #0f0h
          orl a, r4
          inc r0                                 ; save at right location
          mov @r0, a
          dec r0
          sjmp next
over:     mov @r0, a                             ; last digit
          ret
          end
```

Subtraction   The BCD subtraction process is more complicated than addition, and there is no easy way to do it in X51 assembler. However, similar to binary subtraction (where we obtain the 2's complement and then add), we can use 10's complement and add.

Once the 10's complement of one of the number has been found, using subprogram **10_comp**, the number can be added together using the program developed above **F_plus**. However, the absence of a carry out of the last phase when the addition is completed means that the number (result) is negative, and we have to call **10_comp** again to get it back into the correct form. Following program shows the **F_minus** listing:

```
F_minus:
            acall  adjust
            mov  r0, num_1 +5                    ; find 10's comp
            acall  10_comp
            acall  F_plus
            jc  end
            mov  r0, num_1 +5
            mov  a, @r0
            orl  a, #01h                         ; change sign
            mov  @r0, a
            end.
```

Multiplication   The steps to be carried out are as follows:

1. Normalize the number in the standard format.
2. Use repetitive addition for multiplying num_1 by the least significant digit of num_2.
3. Convert the result in standard form by adjusting exponent.
4. Increase the size of num_2 by a power of 10.
5. If num_2 digits are not over, goto step_2 else stop the process.

The process can be better explained by a numerical example.

we wish: to Multiply num_1 = 257 by num_2 = 29 .

The following steps are carried out:

1. Normalize the number into standard format.

   Num_1 = 257 ; becomes 0.257E+3 and

   Num_2 = 29 ; becomes 0.28E+2

2. Carry out multiple additions.

   0.557E+3 is added to itself 9 times. This gives us 2.313E+3, which is automatically converted back to 0.2313E+4 by F_plus.

3. Increase the size of num_1 by a power of 10 to 0.257E+4 and add this twice to our temporary result. We have 0.7453E+4, which is the correct answer.

The following program [F_prod] will take the number in num_1 and multiply it by the number in num_2. Multiplication is done by repetitive addition. The program **F_plus** is useful here.

```
F_pord:
            mov  r0, num_1
            mov  r4, #07h                        ; counter for 7-bytes
            orl  a, a
Init:       mov  r0, a
            inc  r0                               ; load zeros
```

```
                    djnz  r4, init
                    mov   r0, num_1
                    mov   r1, num_2
                    mov   r4, #07h
        load_mem:
                    mov   a, @r0
                    mov   @r1, a
                    inc   r0
                    inc   r1
                    djnz  r4, load_mem
                    mov   r0, num_2 + 1           ;unpack BCD and put in digit_buf
                    mov   r1, digit_buf
                    acall unpack
                    mov   r0, digit_buf + 9
                    mov   r4, 10
        rept1:      mov   a, @r0
                    add   a, #00h
                    jnz   ok
                    dec   r0
                    djnz  r4, rept1
        ok:         mov   r5, a                   ;counter for repeat addition
        rept_add:
                    push  r0
                    push  r5
                    acall add_bcd10
                    pop   r5
                    pop   r0
                    djnz  r5, rept_add            ;over with one digit
        rept_next:
                    djnz  r4, skip
                    sjmp  adj_expo
        skip:       dec   r0
                    mov   a, @num_2 + 6           ;multiply expo.
                    acall inc_expo
                    mov   a, @r0                  ;next multiplier
                    cjne  a, #00h, ok
                    sjmp  rept_next
        adj_expo:
                    mov   r0, num_1 +6            ;get exponent
                    clr   c
```

```
                rlc   a
                inc   pos_expo
                xrl   a, #80h                          ; take 1's comp as neg expo
                cpl   a
                inc   a
        pos_expo:
                mov   r5, a                             ; save it
                mov   r0, num_2 +6
                clr   c
                rlc   a
                inc   pos_expo1
                cpl   a                                 ; take 1's comp
                inc   a
        pos_expo1:
                add   a, r5
                rlc   a
                inc   expo_2
                cpl   a
                inc   a
                rrc   a
        expo_2:
                mov   r0, num_1 + 6
                moc   @r0, a
                mov   r0, num_1                         ; get sign
                mov   a, @r0
                mov   r5, a
                mov   r0, num_2
                mov   a, @r0
                add   a, r5
                djnz  a, end
                mov   a, #01                            ; save minus
                mov   r0, num_1
                mov   @r0, a
                ret
```

Division   The program **F_div** divides the number in NUM1 by the number in NUM2 using repetitive subtraction. It is assumed that NUM1 is greater than NUM2. The algorithm is described with the help of an illustration.

Example 4.6   Divide 535.8 by 17 using repetitive subtraction.

Solution

We first convert to floating-point format:
$$525.8 \Rightarrow 0.5358E+3$$
$$17 \Rightarrow 0.17E + 2$$

1. Subtract 0.1700 from 0.5258 until we go negative:

   $0.5358 - 0.1700 = 0.3658 - 0.1700 = 0.1958 - 0.1700 = 0.0258 - 0.1700$
   $= -0.1442$

   We did this only twice before going negative : thus 3 is the first digit of our result.

2. Subtract 0.01700 from 0.0258 repeatedly:

   $0.0258 - 0.0170 = 0.0088 - 0.0170 = -0.0082$

   We were able to do this operation only once. Our result so far is 31.

This process continues for a long time. Since the two example numbers are not evenly divisible, **F_div** will take the result out to 10 digits to yield [31.51764706]. **F_div** then finds the difference of the exponents (1 in this example) and adds 1 to it. It does this because the first number is always greater than the second number. So, we get a final answer of [31. 51764706], which we know is correct.

4.3 | CODE CONVERTERS

The codes such as hex, decimal, BCD, packed BCD, ASCII, gray, 7-segment, etc. are used in digital systems. It is necessary to convert from one code to another. The section deals with the algorithms for code converter and their implementation in assembly language.

4.3.1 BCD to HEX

Let us assume that the number 48 is to be converted to hex. It is necessary to separate two BCD digits, i.e.,

$$[48]_{10} = 4 \times 10^1 + 8$$

Dig 2   Dig 1

Dig 2 is to be multiplied by 10 and adding the lower digit Dig 1 to the product will give the answer. The steps to be followed are:

1. Separate the two digit of the byte.
   (i) This can be done by making upper nibble and storing as Dig 1.(r_4).
   (ii) Swap the data nibble and store the data as Dig 2 after and in (r_5).
2. Multiply content of r_5 (Dig 1) by 10 and add contents of r_4 (Dig 2) to get the hex value.

```
BCD-HEX:
            movx a, @dptr                      ;get the data
            mov r5, a                          ;save byte
            anl a, #0fh
            mov r4, a
            mov a, r5
            swap a                             ;exchange nibbles
            anl a, #0fh
            mov r5, a
            mov b, #10
            mul ab,
            add a, r4
            inc dptr
            mov @dptr, a                       ;store the result
            end
```

4.3.2 Binary to Gray

Because of typical characteristic of gray code that there is only a 1-bit charge, gray code is used in the input–output operation. The code is used to carry out input/output operation with devices such as shaft encoder. The table shows codes for hex digits.

Hex	Binary				Gray			
	B3	B2	B1	B0	G3	G2	G1	G0
0	0	0	0	0	0	0	0	0
1	0	0	0	1	0	0	0	1
2	0	0	1	0	0	0	1	1
3	0	0	1	1	0	0	1	0
4	0	1	0	0	0	1	1	0
5	0	1	0	1	0	1	1	1
6	0	1	1	0	0	1	0	1
7	0	1	1	1	0	1	0	0
8	1	0	0	0	1	1	0	0
9	1	0	0	1	1	1	0	1
A	1	0	1	0	1	1	1	1
B	1	0	1	1	1	1	1	0
C	1	1	0	0	1	1	1	0
D	1	1	0	1	1	1	1	1
E	1	1	1	0	1	1	0	1
F	1	1	1	1	1	1	0	0

The Boolean expressions for the bits can be derived using either algebra or Karnaugh map. The relations obtained are shown below.

Binary → Gray:

$$G_3 = b_3$$
$$G_2 = b_3 \,[\text{exor }]\, b_2$$
$$G_1 = b_2 \,[\text{exor }]\, b_1$$
$$G_0 = b1_1[\text{exor}]\, b_0$$

Gray → Binary:

$$b_3 = G_3$$
$$b_2 = G_3 \,[\text{exor}]\, G_2$$
$$b_1 = G_2 \,[\text{exor}]\, G_1$$
$$b_0 = G_1 \,[\text{exor}]\, G_0$$

Binary to gray code conversion As per the expression, it can be seen that the code can be obtained by doing XOR operation between two neighbouring bits (Fig. 4.1). For example:

Fig. 4.1 4-bit binary to gray code converter process

For 8-bit gray code converter (Fig. 4.2), we can accept the data byte (binary). The data can be rotated right with carry so that MSB is 0, while other bits are shifted to right by 1-bit.

Fig. 4.2 8-bit binary to gray code converter process

```
Bin_Gray :
        mov a, #data8           ; get binary data
        mov r1, a               ; save in r1
        clr c,                  ; reset carry
        rrc a
        xrl a, r
        mov r2, a               ; r2 = gray-code
        end
```

Gray to binary code converter This code conversion is not so simple as in the previous example of binary to gray, since the output bit is to be ex-or with the next bit (Fig. 4.3). In this algorithm, we will take the help of Boolean or bit processing capability of the microcontroller.

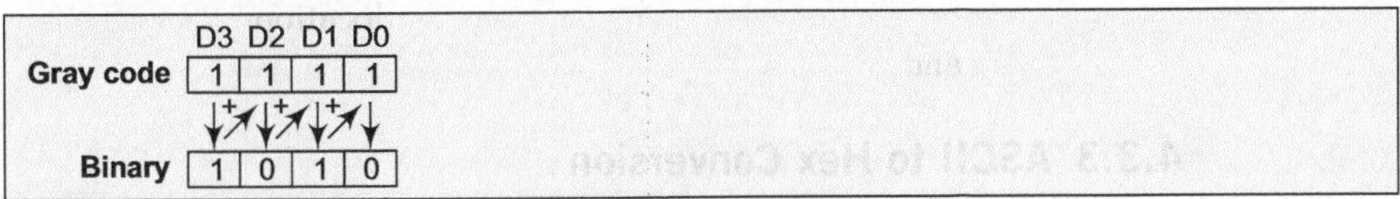

Fig. 4.3   4-bit gray code to binary converter process

The important points to be remembered are

(a) Carry flag is the accumulator for the logical bit operation.

(b) There is no bit ex-or operation supported by the 8051 instruction set. This can be done by using the definition of ex-or logic function:

i.e.,　　a [ex-or] $b = a' . b + a . b'$

The steps followed are

1. Get the gray-code data in the accumulator.

2. Rotate accumulator to the left by 1-bit without carry, so that B_7 becomes the LSB. This is because the first relation says that;

$$B_7 = G_7$$

3. Realise ex-or operation between the MSB and LSB of the accumulator which is having bit-address OE0h and OE7h, respectively. The relation can be as :

$$C = (OE0h) \text{ [ex-or] } (OE7h)$$

i.e.,

$$C = (OE0h) . (OE7h) + (OE0h) . (OE7h)$$

The carry will contain derived binary code bit which can be put in the LSB, by using rotate right with carry.

4. Step 3 repeated 7 times gives the answer in the accumulator, which is the desired byte in binary.

```
Gray_Bin:
            mov   a, #data        ; get gray code
            mov   r5, a           ; save in
            mov   dptr, #2500h    ; (r5) = gray-code
            rl    a,              ; (G7) – (LSB)
            mov   r4, #07h        ; set counter for 7-bit shift
rept1 :     mov   c, OE0h         ; move (LSB) to carry
            anl   c, OE7h         ; c = (OE0) × (OE7)
            mov   temp, c         ; save status of C in location temp
            mov   c, OE0h
            cpl   c
            anl   c, OE7h         ; (c) = (OE0)' (OE7)
            orl   c, temp         ; (c) = (OE0) + (OE7)
            rlc   a
```

```
        djnz r4, rept1
        mov @dptr,a                    ;save in the external memory
                                        location

        end
```

4.3.3 ASCII to Hex Conversion

Hex	8-bit ASCII-code		Hex	8-bit ASCII-code
0	30		8	38
1	31		9	39
2	32		A	41
3	33		B	42
4	34		C	42
5	35		D	43
6	36		E	44
7	37		F	45

The code table shows that desired conversion can be obtained as follows:

1. **Hex to ASCII**

   If digit <= 9

          add (30h)

   Else

          add (37h)

2. **ASCII to Hex**

   If byte <= 39

          sub (30h)

   Else

          sub (37h)

ASCII to hex It is first necessary to check if it is a valid code for a hex digit. This can be done by checking that the byte value is between (30h, 45h) including end values (30h) is subtracted. A checking of result requires additional subtraction of (07h), in case the digit is hex character (A–F).

```
ASCII - Hex:
        mov a, #character              ;enter ASCII code
        cjne a, #30h,high-limit
        jc error                       ;value is less than lower value in
                                       ;valid data

        clr c
        subb a, #30h
        cjne a, #0a                    ;char
```

```
            jc store                        ; value digit 0-9
char:   subb a, #07h
store:  mov r2, a                           ; r2: hex digit
            end
```

Hex to ASCII   The process is just the reverse of the previous case in ASCII-Hex, i.e., instead of subtraction, addition is to be done.

```
Hex_ASCII:
            mov a, # hex digit
            anl a, 0fh                      ; mask the nibble
            add a, #30h
            cjne a, #39h                    ; char
            sjmp store
char:   jc store
            add a, #07
store:  mov r2, a                           ; (r2) will have ASCII code
            end
```

4.3.4 Decimal to 7-segment Code Converter

Since 7-segment LED is the most commonly used output device, there is a need to output a 7-segment code on the output port. To display any decimal digit, the code depends on type of 7-segment module used, i.e., whether the module is using common cathode or common anode configuration.

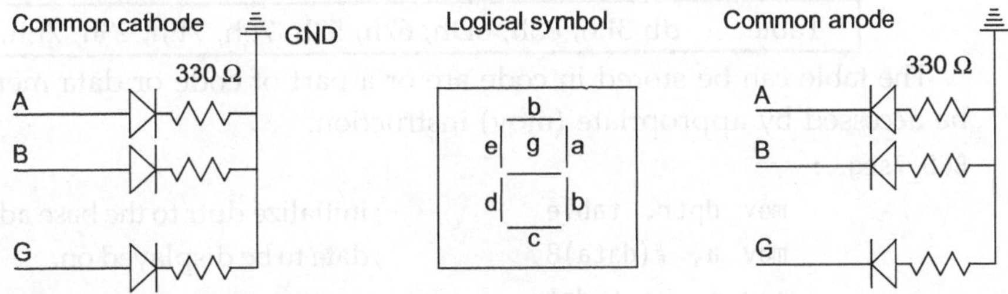

Fig. 4.4   Configurations of 7-segment LED

The Boolean expression for all the output lines can be derived and it can be interrupted using logical instruction.

However, an easier method is to make the use of technique called *look-up table*. The entire code table is stored in the memory. Depending upon the digit to be displayed, the offset is added to the table base. The contents of locations can be read, which gives 7-segment code of the digit to be displayed.

Look-up table for 7-segment code converter

Segment	dp	g	f	e	d	c	b	a	Hex code		Address
Digit	D7	D6	D5	D4	D3	D2	D1	D0	Common anode	Common cathode (complement)	
0	0	0	1	1	1	1	1	1	3F	C0	Table
1	0	0	0	0	0	0	1	1	03	FC	Table + 1
2	0	1	1	0	1	1	0	1	6D	92	Table + 2
3	0	1	1	0	0	1	1	1	67	98	Table + 3
4	0	1	0	1	0	0	1	1	53	BC	Table + 4
5	0	1	1	1	0	1	1	0	76	89	Table + 5
6	0	1	1	1	1	1	1	0	7E	81	Table + 6
7	0	0	1	0	0	0	1	1	23	DC	Table + 7
8	0	1	1	1	1	1	1	1	7F	80	Table + 8
9	0	1	1	1	0	1	1	1	77	88	Table + 9

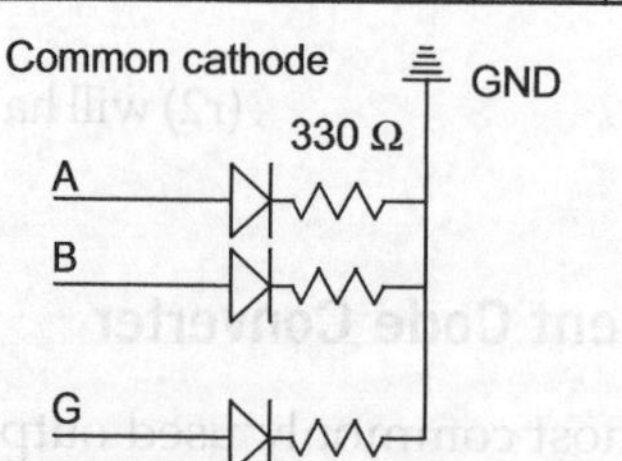

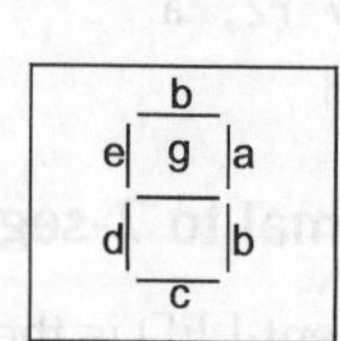

Let us assume common cathode type of module. In this case, the input 'high' segment will glow ON and segment will not glow, OFF, if input is 'low'.

Let us assume table is stored as:

Table	db 3Fh, 03h, 6Dh, 67h, 53h, 76h, 7Eh, 23h, 7Eh, 77h

The table can be stored in code are or a part of code or data memory. It can be accessed by appropriate (mov) instruction.

```
BCD_7seg :
        mov dptr, table          ; initialize dptr to the base address.
        mov a, #(data)8          ; data to be displayed on.
        mov a, @a + dptr
        mov r5, a                ; (r5) will contain seven-segment code
        sjmp  BCD_7seg
```

NOTE ▼ If, the 7-segment LED is common anode instead of common cathode, the entire table has to be prepared with new code value. Instead of preparing a new table, a simple technique can be used. The code, which is read in above program, can be complemented using [cpl] instruction to obtain code for common anode configuration.

The program now becomes

```
BCD-7seg:
            mov dptr, table
            mov a, #0fh
            mov a, @ a + dptr
            mov r5, a                ;common cathode code
            cpl a
            mov r4, a                ;common anode
loop:       sjmp loop
```

4.4 ARRAY HANDLING

Example 4.7 Write a program to find maximum and minimum value in an array of 8-bit numbers.

Solution

    Array db 20, 40, 25, 12, 14

```
Maximum:
            mov dptr, array          ; initialize data pointer, dptr
            mov r3, #05              ; counter for 5 data
            movx a, @dptr           ; 1st element
            mov r4, a               ; assume it to be minimum
            mov r5, a               ; assume it to be maximum
rept1:      inc dptr
            movx a, @dptr
            cjne a, r5, check_min
            sjmp check_max
check_min:
            jc next-check
            cjne r5, a,check_min
next_check:
            cjne a, r4,check_max
            sjmp no_change
check-max:
            jnc no-change
            mov r4, a                ; change max
no_change:
            djnz r3, rept1
            mov dptr, max            ; store max
```

```
                    mov a, r4
                    movx @dptr,a
                    mov dptr, min
                    mov a, r5
                    movx @dptr, a          ; store min
                    end
```

Example 4.8 Write a program to arrange an array in ascending/descending order.

Solution

There are various methods available for the arrangement of an array in order. They are called sorting algorithms. Typical sorting algorithms are

(i)   Exchange sort

(ii) Insertion sort

(iii) Selection sort

(iv) Bubble sort

```
Start:
            mov r3, #data           ; length of an array
rept1:      mov a, r3
            mov r2, a
            mov dptr, # add         ; starting address of array
pass:       movx a, @dptr
            mov r0, a
            inc dptr
            movx a, @dptr
            mov r, a
            cjne a, r0,no_change
            sjmp skip
no_change:
            jc no_xchg
ex_chg:
            mov a, r0
            movx @dptr,a
            clo a
            cjne a, @dptr, skip
            dec dptr
skip:       dec dptr
            mov a, r1
            movx @dptr,a
            inc dptr
```

```
no_xchg:
            djnz r2, pass
            djnz r3, pass
            end.
```

Example 4.9 Write a program to reverse an array. The length of an array is stored in 2500h, while the data is stored from 2501h onwards.

Solution

(This is an example of realization of exchanging data in external memory as there is no direct instruction in the set). The process is to be repeated $L/2$ times where L is the length of an array. Divide by 2, operation can be implemented using a right operation. Let us assume that length of the array is less than 255-bytes.

```
Array_Reverse:
            clr  a              ; clear accumulator
            clr  c              ; clear carry flag
            mov  dptr, #2500    ; get the pointer in dptr
            movx 41h, a         ; 41h will contain the LSB of last address
            rrc  a
            mov  42h, a         ; 42h will contain the loop_count
rept1:
            inc  dptr
            movx a, @dptr
            mov  r3, a          ; r3 temp register for holding data from array
            push dpl
            mov  dpl, 41h
            movx a, @dptr       ; get the data from bottom
            mov  r2, a          ; save it in r2
            mov  a, r3
            movx @dptr, a       ; bring data bottom location
            mov  a, r0
            pop  dpl            ; set the top address
            movx @ dptr, a
            dec  41h
            djnz 42h, rept1
            end
```

4.5 | COUNTERS AND DELAYS

The counters and delays are important techniques commonly used in applications such as traffic signals, digital clocks, process control, and serial data transfer where processor has to wait for a fixed amount of time without doing any useful operation.

Counters are designed to keep a track of events, while delays are important in setting up a reasonable account of timing between two events.

Counters are designed by loading an appropriate number into one of the registers and using inc/dec instructions. A loop is established to keep track and update the count as well as check for the final value. If the final value is reached the process stops. In case of delays, the process is similar to that of the counter except that a count is loaded in a single or more registers depending on value of the desired time delay. The contents of the register are decremented until it reaches zero.

The methods used to generate delays can be grouped into two categories: hardware method and software method.

Hardware Method

In this method, real time timer/counters are used. An extra IC chip is added onto the system to take care of events. Advantages of this method are as follows:

1. Microcontroller is relieved of housekeeping job (updating the register value and checking for final valve). This is done by dedicated chip.

2. They are most accurate and accuracy is of the order of T-state and is determined by system frequency.

3. The ICs can be configured into one of the modes like interrupt driven, programmed IC or status check mode as per convenience.

The disadvantage of this method is that an additional IC requires extra space on the PCB. The hardware changes are required which increases the cost.

Software Method

The logic used in this method is simple. As stated in previous paragraph for delays, one or more registers are loaded with a predefined value. The register is decremented by one until it reaches zero. Advantages of this method are as follows:

1. It is easy to write a code for loading and decreasing the register the predefined number of times using available instructions in the set.

2. Since it requires only a set of instructions to carry out the operation, no hardware changes are required. There is no additional cost except that it will use some memory space for storing the delay program.

The disadvantage of this method is that since the instruction to be executed requires minimum 12 periods, the accuracy is of the order of 12T states.

Depending upon the amount of desired time delay, any one of the three methods described below can be employed.

One-register method This method makes use of one-register. The desired number is loaded in the register and is decremented until it reaches zero. The time taken to carry out the operation gives the desired time delay. The program listing for one register method is

Delay

mov r₄,0ffh ; register r4 can be loaded with desired number, max value is 0ffh

Wait

djnz r4, Wait ; wait for count to become zero

The second instruction requires 24T states when executed once, which is called a *loop time*. Let us denote it by t_L. The total program execution will cause the instruction to be executed [0ffh, 255] times.

Assuming the clock frequency f to be 12 MHz, the clock period (T) can be calculated as:

$$= 1/f = 83.3 \ \mu s$$

Loop time: $\qquad t_L = 24T = 2 \ ms$

Let $(N)_{16}$ be the number loaded in the register.

The maximum value that can be loaded in an 8-bit register is 0ffh, 255d.

$$\boxed{\text{Maximum value of time delay} = [t_d]_{max} = 255 \times 2 = 0.51 \ ms}$$

Hence, the maximum value of time delay using one-register method is of the order of 0.5 ms.

In order to generate the desired value of delay, the value of number to be loaded in the register, termed as delay count must be calculated. The calculations can be carried out as follows:

Let:

$$t_L = \text{Loop time}$$
$$T = \text{Oscillator period} \ [1/f]$$
$$t_d = \text{Desired time delay}$$
$$N = \text{Delay count}$$

Since; $\qquad t_d = N \times t_L$

$$N = t_d/t_L = t_d/24 \times T = [t_d/24] \times f$$

The relation for the count is

$$N = \text{Delay count} \approx \left\{ \frac{\text{Desired delay} \times \text{Frequency}}{24} \right\}$$

Two-register method Two-registers are used to load the 16-bit count. One register is decremented by one until it reaches zero. This loop is included in an outer loop in which another register is decremented until it becomes zero. Program listing for delay using two-register method is as follows:

```
Delay2reg :
            mov r4, # N2        ;N2 loop count for outer loop
rept1:      mov r3, #N1         ;N1 loop count for inner loop
loop:       djnz r3, loop       ;inner loop
            djnz r4, rept1      ;outer loop
            end
```

The total time taken by the program can be calculated as follows:

Inner loop:

```
rept:       mov     r3,    #N1     ;12 clock
loop:       djnz r3,    loop       ;( Max time delay = 0.5 ms)
```

$$(\text{Time delay \_ inner loop})_{max} = 12 \times T + N1 \times 24 \times T$$

$$= N1 \times 24 \times T \ \{\text{If } N1 = 0\text{ffh, MAX} = 0.51 \text{ ms}\}$$

$$\boxed{\text{Maximum delay} = \{(\text{Time delay \_ inner loop})_{max} + 24 \times T\} \times N2}$$

Since maximum value of N2 can also be ffh or 255 :

Maximum value of time delay = $[t_d]_{max}$, using the previously calculated values

$$= 0.13 \text{ s}$$

Hence, the maximum value of time delay using two-register method is of the order of 0.13 s.

Three-register method Two loops of the two-register method are included in the outer loop of a third-register.

```
Delay3reg:
            mov r5, #N3         ;N3 loop count for outermost loop
rept1:      mov r4, #N2         ;N2 loop count for outer loop
loop1:      mov r3, #N1         ;N1 loop count for inner loop
loop:       djnz r3, loop       ;inner loop
            djnz r4, loop1      ;outer loop
            djnz r5 rept        ;outermost loop
            end
```

Maximum delay by three-register method is

$$(\text{Time delay})_{max} = (0.13 + 24 \times T) \times N3 + 12 \times T$$

$$= 32.5 \text{ s } \{\text{if } N3 = 0\text{ffh or } 255\}$$

Thus, three-register methods can be used to generate time delay of the order of 30 s. If the required delay is still more, it is better to use hardware method instead of the software method employing more registers.

Table 4.1 can be used to select number of registers and program module to be used for generating desired time delay.

Table 4.1   Selection of registers for delay generation

No. of registers	Maximum delay generated
One	0.5 ms
Two	0.13 s
Three	32.5 s

4.6 | HANDLING SUBPROGRAMS

Subprograms are small program segment developed to realize a typical operation. There are two types of program modules which are used in software development namely:

1. Subroutine subprogram

   Program codes developed to support modular programming.

2. Interrupt service routine

   ISR is used to support interrupt driven data transfer. The segment exactly tells microcontroller the steps to be carried out if there is an interrupt.

The stack area is a small portion of user RAM which is used by programmer as well as by the microcontroller. The speciality about the handling is that the microcontroller makes use of stack area to save or restore its status. It is known that the information which is at the top of stack can only be accessed. User must take extra care so that only the data saved by the program is accessed and not the information saved by microcontroller, which may lead to conflict.

The address of top of the stack must be loaded in the stack pointer (SP) register. This process is called defining a stack. Since stack grows down word, SP is initialized to the highest address of user RAM.

On power up, SP is initialized to 07, (it is the register R7 in bank). Using stack beyond this may lead to destroying data in the SFR. As already mentioned, the names of SFR cannot be used to initialize the stack pointer.

For example

To initialize stack pointer;

```
        mov SP, #20              ; is invalid since it uses the SFR name
```

The correct way to do is:

```
        mov 81h, # 21h           ; 81h is the byte address of SP
```

In many situations it is advisable to define the stack size also so that it does not interfere with user data in conventional RAM area.

4.7 | HANDLING SUBROUTINE SUBPROGRAMS

Very often, in a program it is necessary to perform a particular task many times. Subprograms are useful in such situations. A subprogram is a single segment of program code, i.e., sequence of instructions which can be employed over and over again, within a longer program. Instead of writing a large program during development, it is always easier to develop and test in the form of small program segments. This makes the development process faster. The advantages of subprograms over single program are as follows:

1. Since the instructions in a subprogram are to be included once, the length of the code is reduced. This reduces the memory space required to stretch the code.

2. Modular or structured programming is a good strategy. It becomes possible to employ programming technique using subprogram modules. The subprogram improves the structure of the program. It becomes easier to debug or correct the modules. The subprogram encourages modular development so the job of main program reduces to controlling the sequence of calling subprograms and coordinating them to implement the algorithm.

3. The flow of logic is well defined in the main modules so the inspection of the functioning of the module can be easily done. The splitting does not cause any loss of capability of examining first delete of the program.

4. A section of program can be developed in the form of subprograms, which can be shared by users from the library. This reduces unnecessary duplication of codes as well as development time.

5. They provides a flexible and convenient way of exchanging information between the application programs. Generally, a list of library modules is provided to the user.

In order to handle the subprogram, the operations required by the microcontroller system are

1. Invoking or calling a subprogram or a procedure

2. Returning from the module back to the calling program

3. Parameter passing

MC5-51 series of microcontrollers have the instructions to call and return from modules, which are described in the previous section.

4.7.1 Calling a Subroutine Module

Depending upon where the program module is located in the address space of the IC, any one of the following instructions can be used:

Module location	Mnemonics	Address size
Within a 2 K page	ACALL	11-bit
Anywhere in address space	LCALL	16-bit

4.7.2 Returning from a Module

Since the controller reads the return address from the stack, no operation should be specified. However, the modules called may be either a normal subprogram or an interrupt service subprogram. The two statements to return from module in the 8051 instruction set are ret and ret1.

ret: It reads the return address from stack and goes back to main.

ret1: In addition to the operation done by ret, it also enables the interrupt structure so as to provide services to the pending interrupt of the same level.

4.7.3 Methods of Parameter Passing

It is necessary to pass parameters to a subprogram when it is called and to return back the information or output from the module. The number of input/output parameters passed to and from the subprograms can vary. Often a small number of parameters need to be exchanged. But in case of array or matrices a large number of parameters are involved in the process. Even the parameters to be passed may be control/status information which may not have a peripheral for data transfer for the module.

Internal registers/RAM can be used to hold the input as well as the output parameter information. This is because the main and subroutine programs use them without any difference.

It is important to take care of the data in the register/RAM before making a call to the module. They must be saved if the required location is modified by the module. This method is adequate as long as the amount of data/information to be exchanged is small.

To exchange large amounts of data, the information is stored in memory and only the address pointer is passed to the module. The output or result from the module can also be obtained in a similar way.

4.8 | HANDLING INTERRUPTS

On configuring the system in an interrupt driven mode, program in general requires to access hardware I/O devices for I/O operations such as

- Read data from keyboard
- Display data on CRT
- Read/write data on DISC

- Send data to printer
- Serial data transfer through EIA232C (485)

Interrupt is a mechanism which causes processor to stop execution of current program and transfer control to ISR/handler and resume current execution on completion of ISR. Interrupt is a forced event, initiated by the device which is transparent to the program execution. It is an *asynchronous event* as it has no relation with system clock.

They can be classified as external, internal, and software interrupts. Interrupts are recognized in the middle of the instruction. In case of repetition, it is allowed in each repetition. If program logic requires disabling, it must be done manually by the programmer.

Internal interrupts are also termed as an *exception*. Exception is the microcontrollers' normal and predictable documented response to a situation detected during the execution of an instruction.

Microcontroller has a two-tire hardware structure to handle interrupt, which can be configured by two SFRs: IP and IE.

The top tier has two priority levels: high and low. Each interrupt source can be assigned to either high level or low level status by setting the appropriate bits in the IP (interrupt priority) register. When two interrupts of different levels are received simultaneously, the high level interrupt is serviced first. If the bit is '0' the corresponding interrupt has a lower priority and if the bit is 1 the corresponding interrupt has the highest priority. This can be configured using interrupt priority (IP) register.

The second tier of priority is used to resolve simultaneous interrupts within the same level. The priority within level ordering from highest to lowest is fixed as: IE0, TF0, IE1, TF1, R1, or T1. This can be done using IE (interrupt enable) register which is bit addressable. Interrupt enable (IE) register may be used.

On RESET, all the interrupts are disabled. EA is set to '1' to enable the interrupt structure. Desired interrupt may be enabled by setting bit corresponding it to'1'.

Example 4.10 1. **Write a program statement to enable all interrupts.**

Since IE is a bit addressable register, sequence of statements using dot convention the same operation can be done as.

```
setb ie.7
setb ie.4
setb ie.3
setb ie.2
setb ie.1
setb ie.0
```

Alternatively, the data (8f) can be loaded to the (IE) register

```
==> mov ie, 9fh
```

2. **Write a program statement to disable all interrupts**

This can be done by loading 00h to the (IE) register ==> mov, ie, 00h

Illustrative examples given in the next section not only clears the above points but also explains use of methods of parameter passing.

4.9 | SAMPLE PROGRAMS

Example 4.11   Write a program to divide 16-bit number by an 8-bit number.

Solution

In performing such divisions, we carry out the division of the upper 8-bit by the 8-bit divisor. The remainder is saved as upper byte of 16-bit data. The 16-bit number is shifted by 1-bit to the left and process is repeated 8 times. In order to simplify the program structure, we will use a subprogram to shift the 16-bit number to the left by 1-bit.

Let us organize the data in the registers for operation as follows:

	MSB	LSB
Dividend (16-bit)	r3	r2
Divisor (8-bit)		r4
Quotient (8-bit)		r5
Remainder (8-bit)		r6
Counter (8-bit)		r0

```
DIV16_8:
            mov 81, #20h              ; initialize SP
            mov r0, #07h             ; initialize counter
            mov r5, #00h             ; initialize quotient
            mov r2, # LSB            ; load dividend
            mov r3, # MSB            ; load dividend
            mov r4, # divisor        ; load divisor
loop1:
            mov a, r3
            mov b, r4
            div ab                   ; a: quotient, b: remainder
            orl a, r5
            rr a
            mov r5, a                ; save quotient
            mov r3, b                ; get reminder in r3
            acall shift_16L          ; shift dividend by 1-bit to left
```

```
                    djnz r0, loop1
                    end                 ; subroutine subprogram to shift 16-bit
                                        ; number to left by 1-bit

shift_16L:
                    mov 40h, r3         ; get the data in the internal RAM
                    mov 41h, r2
                    clr c               ; clear carry flag
                    mov a, 40h
                    rlc a               ; rotate acc. Left with carry, (c) will
                                        ; contain B7
                    mov r2, a           ; save shifted LSB
                    mov a, 41h          ; set MSB in acc
                    rlc a               ; shift the MSB
                    mov 43h, a          ; save shifted MSB at 43h
                    mov a, r2           ; get now shifted LSB in accumulator
                    adc a, 00h          ; add carry
                    mov r3, a           ; r2 will contain LSB
                    ret                 ; r3 and r2 will have 16-bit number shift
                                        ; to left
                    end                 ; by 1-bit
```

Example 4.12   Write a program to find factorial of a given number. Assume that the number is <9.

Solution

Here we will make use of the subroutine for a 16-bit number multiplied by an 8-bit number. This is due to the fact that the answer may be up to 24-bits.

```
Fact:               mov 81h, #20h       ; initialize SP
                    mov r5, number      ; number < 9 whose factorial is required
                    mov r0, ans         ; pointer to store the result
                    push r0
                    mov a, #01h         ; initialize the result
                    mov @r0, a
                    inc r0
                    dec a
                    mov @r0, a
                    inc r0
                    mov @r0, a
                    pop r0
rept1:
                    acall mul16_8
```

```
                    djnz r5, rept1
                    end
         mul16_8:

                    push r0
                    mov a, @r0
                    mov r3, a
                    inc r0
                    mov a, @r0
                    mov r4, a
                    mov a, r3
                    mov b, r5
                    mul ab              ; (a): LSB and (b): MSB
                    mov @ r0, a
                    inc r0
                    mov @r0, b
                    mov a, r4
                    mov b, r5
                    mul ab
                    add a, @r0
                    mov a, b
                    addc a, #00h        ; consider carry
                    inc r0
                    mov @r0, a
                    pop r0
                    ret
                    end
```

Example 4.13   Write a program to find the address of first two RAM locations, which contain consecutive numbers. If so then set carry flag to '1' and '0' otherwise.

Solution

```
cons_no:
                    mov 81, #65h        ; initialize SP
                    mov r0, #20h        ; r0 = lower address
         loop:      a, @r0
                    inc a
                    mov 1fh, a          ; store incremented number
                    inc r0
                    acall check         ; check for last address
                    jnc end1
                    mov a, @r0
```

```
                        cjne a, 1fh, loop
                        setb 0d7h                ; set carry bit
                        mov  a, r0
                        mov  dptr, #found        ; r0 contain the address
                        movx @dptr,a
                        sjmp ok_end
        end1:           clr  c
        ok_end:         end                      ; check subprogram which checks for
                                                 ; the last address
        check:
                        push a
                        clr  c
                        mov  a, #61h
                        xrl  a, r0
                        jnz  skip
                        cpl  c
        skip:           pop  a
                        ret
                        end
```

EXERCISES

4.1 Write a program to edit ASCII characters so as to replace zeros by blanks. The string starts at address [string] specified by [dptr].

4.2 Write a program to multiply multi-byte number.

4.3 Write a program to convert a gray code number into binary.

4.4 Write a program to convert 16-bit number into BCD number.

4.5 Write a program to convert four digits BCD number into hex.

4.6 Write a program to find maximum of an array of ten numbers.

4.7 Write a program to arrange of an array of ten numbers in ascending order.

4.8 Write a program to arrange of an array of ten numbers in descending order.

4.9 Write a program to check if an array numbers are in ascending order or not.

4.10 Write a program to add two multiple word binary numbers whose length in bytes is in accumulator. The starting address of numbers is NUM1 and NUM2. Store the result in memory locations starting from result.

4.11 Write a subprogram to arrange the array of 8-bit numbers in descending order. Write a main program to accept the new data from the port [P1] and insert it at the proper location so that the array remains in the order. Assume that the array is stored in the external memory from the [2501h] onwards, while the length is stored in the location [2500h]. The new data is discarded if it is already in the array.

4.12 Write a subprogram to arrange the array of 8-bit numbers in an ascending order. Write a main program to remove duplicate data items from the array.

4.13 Write a subprogram to arrange the array of 8-bit numbers in an ascending order. Write a main program to convert the array in descending order.

4.14 Write a subprogram to find number of 1's in a byte. Write a main program to find number of 1's in a word.

4.15 Write a subroutine subprogram to convert 8-bit binary data into gray code. Write a main program to convert an 8-bit gray into binary.

4.16 Write a program to add two multiple word binary numbers, whose length in bytes is in accumulator. The starting addresses of the numbers are NUM1 and NUM2. Store the result in memory locations starting from result.

4.17 Write a subprogram to find factors of a given 16-bit number. Write a main program which finds the factors of number, stored in the memory at [2500h] with the lsb first, and stores the result at [2601] h onwards. It also finds the numbers of factors and stores at the address [2600h].

4.18 Write a subprogram to find prime factors of a given 16-bit number. Write a main program, which finds the factors of number stored in memory at [2500h] and stores the result at [2601h] onwards. It also finds the numbers of factors and stores at the address [2600h].

4.19 Write a program to find maximum in a given array of 16-bit numbers. Write a main program to arrange the array in ascending order using the subprogram maximum. The array starts from [2501h] onwards and is terminated by a blank.

4.20 Write a subroutine, which receives the starting address of the fixed length string and displays it on the LCD. Modify the program if the length is unknown.

Chapter 5

External Peripheral Devices

The communication with the microcontroller is done by the input and the output devices connected at the I/O ports. These devices are also called *peripheral devices*. The device used depends on the application. This chapter describes characteristics, operation, and interfacing of the input and the output devices with the microcontroller.

5.1 | SENSORS

In the process control loops, the signals may be nonelectrical. Microcontroller needs input–output devices other than keys or normal display devices to handle such signals. Generally, nonelectrical signals are converted to electrical signals using devices called *transducers*. In this section, we will study commonly used transducers in the microcontroller-based systems.

5.1.1 Transducers

Transducers are also called sensors or detectors. Transducers constitute the first stage of setup for the digital measurement of nonelectrical quantities. A transducer is a device which receives energy in one form and transfers it to an electrical signal. A sensor can be used to sense a mechanical, electrical, optical acoustic, magnetic, thermal, nuclear, chemical, or combinations of more than one form of energy. The electrical signals can be used for further operations on them. A sensor is a transducer, which converts any nonelectrical quantity into an electrical one. Electrical transducers can broadly be divided into two categories:

1. **Active**   They are self-generating devices, their functioning being based on conversion of energy from one form to another. No external source of energy is necessary to excite them. For example, the thermocouple belongs to this category. Based on the principle of operation, they can be classified further as: resistive, inductive, photoconductive, thermoresistive.

2. **Passive**   These transducers do not generate any energy. They must be excited by the application of electrical energy from outside. The extracted energy from the measured value produces a change in electrical state, which is measurable, e.g., photoresistor can be excited by an emf from a cell and the voltage against the photoresistor can be measured. When exposed to a light of certain intensity (measurand) its resistance changes, thus changing the voltage across it.

5.1.2 Signal Conditioning

A sensor detects and generates an electrical signal. This signal cannot be directly used by the microcontroller as it is in an analog form. For operations on the signal like control, comparison, etc. it must be in a digital form. Analog to digital converters must be employed for the purpose.

It is also known that from the accuracy point of view, it is necessary to make use of variation over full range, e.g., for a 10 V ADC, it is necessary that the input received by the ADC from the sensor must be in a range of 0–10 V. The sensor signal may not have this characteristic. Hence it has to be modified. This process of modifying signal is called *signal conditioning*.

Signal conditioning may involve many more processes, which can be broadly divided into either linear, or nonlinear processes. Amplification, attenuation, integration, etc. are examples of linear processes. Modulation, demodulation, sampling, clipping, clamping, etc. are examples of nonlinear processes. Signal conditioning, thus is very important. Once the analog signal is suitably conditioned, it is first converted into a digital signal so that it can be processed by a microcontroller and then analyzed.

The heart of any conditioning circuit is operational amplifier (op-amp). They were originally designed to perform mathematical operations such as addition, sign changing, integration, differentiation, etc. in analog computers and hence the name.

Basically, op-amp is a very high-gain direct-coupled amplifier having high input impedance and low output impedance. The op-amp consists of a number of direct-coupled transistors, diodes, etc. It can be represented by a logical symbol, as shown in Fig. 5.1 with two inputs V_1 (inverting) and V_2 (non-inverting), and an output V_0.

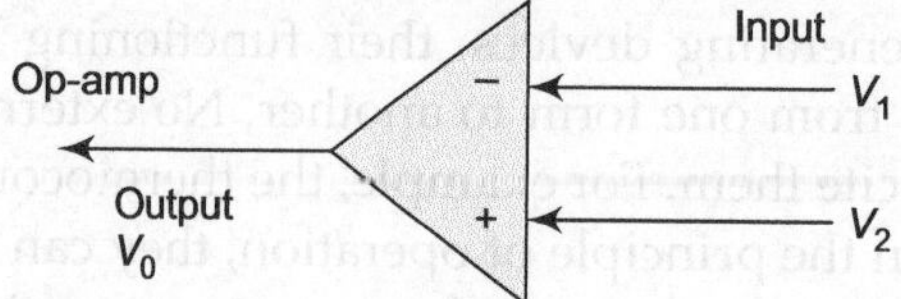

Fig. 5.1   Logical symbol of op-amp

It is understood that all the voltages are with respect to the ground. We will consider different circuits employing op-amps during the project development.

5.1.3 Radiation Transducer

The electromagnetic spectrum is separated into various regions, from radio frequency to X-ray and nuclear. The energy E is given by the following equation:

$$E = \frac{hc}{\lambda}$$

where c is the velocity of light, λ is the wavelength, h is the Planck's constant.

The relation shows that the electromagnetic energy is reciprocally related to the wavelength. The transducers, which are capable of providing electrical signal proportional to such radiation independent of temperature and other effects, are said to be the radiation transducers. We are interested in measuring and detecting light radiation. The relation between light intensity and the strength of source C (in candle power) is

$$L = \frac{[C \times A]}{[D \times D]}$$

where A is the area illuminated and d is the distance from the source to the area.

Photodiodes and Phototransistors

The radiant energy in the form of light to p-n junction of diode can be used to promote an electron from the valence band to the conduction band. Sufficient energy must be supplied by the radiation so that the electron can move from one band to the other. When a photodiode is operated in the absence of light, the reverse current is quite small and the forward current is very large as in the operation of any diode. When light impinges on the photodiode there is virtually no increase in forward current. There is however significant increase in reverse current due to photodiode effect. For successful operation, photodiode must be reverse biased. In such condition the reverse current is directly proportional to the incident radiant energy intensity.

A phototransistor is just logical extension of the photodiode in which the photodiode forms the base–collector junction. The gain of phototransistor is 400 to 500. The typical response curve of phototransistor is shown in Fig. 5.2. Low noise, faster response in terms of µS, and physical isolation between input–output circuits are typical features of the phototransistor.

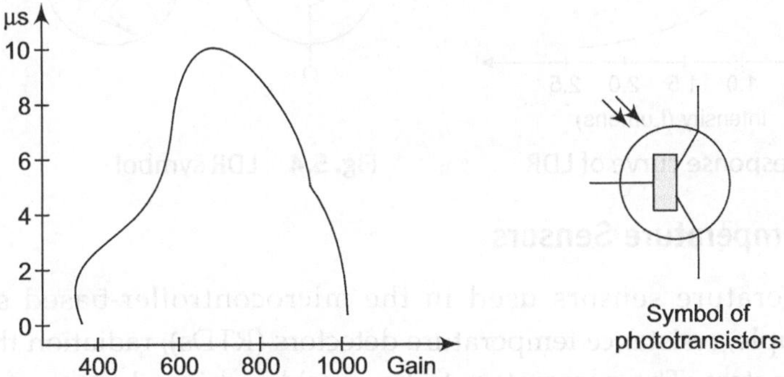

Fig. 5.2   Response curve and logical symbol of phototransistor

Photocells

The extreme interest of such cells is in power generation applications. At MW level, the device is useful in instrumentation and at kW level an array of devices can be used for residential power and space vehicle power.

A photovoltaic cell converts radiant energy into a voltage. The radiant energy produces electron–hole pairs. When the cell is used in the circuit, the current flows as a result of generated electron–hole pairs, due to radiant energy. The voltage developed is directly related to the incident energy.

The properties of the photoconductive cell change when exposed to the light. The conductivity of the material changes. The commonly used materials are the sulphides of cadmium and lead. Photoconductive cells with CdS as the active material have a spectral response curve similar to the human eye. The basic operation of the device is the process of electrons being freed by the radiant energy to move about in the crystal structure decreasing the resistance of the material between the two electrodes. The response curve of the resistance versus intensity is nonlinear as shown in Fig. 5.3. Over a limited range of intensities a linear approximation is valid. The resistance in the absence of light can be in excess of 1 MΩ while the resistance under a given light condition can be as low as 100 Ω. The features of such cells are low cost, small size, and low power consumption. It is used in the light switch type of applications. The symbol and layout of LDR is shown in Fig. 5.4.

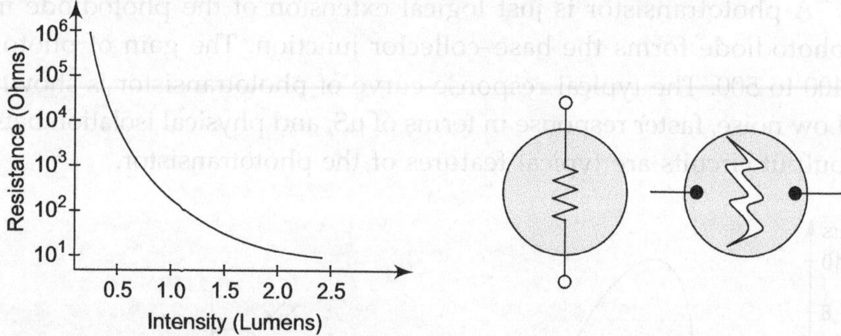

Fig. 5.3 Response curve of LDR   **Fig. 5.4** LDR symbol

5.1.4 Temperature Sensors

The temperature sensors used in the microcontroller-based systems are thermocouple, resistance temperature detectors (RTDs), radiation thermometer, and pyrometers. The parameters to be considered in selection of a particular sensing element are operating range, accuracy and resolution, and medium where it is to be used.

Thermocouple

When a pair of wires made up of different metals is joined together at one end, the temperature difference between the two ends of wires produces a voltage between the two wires. Magnitude of voltage produced depends on the material used for the wires and the amount of temperature difference between the joined ends and the other ends. The combination of metals used in thermocouple are, chrome–constantan (type E), iron–constantan (type J), copper–constantan (type T), chrome–alumel (type K), and platinum–platinum/rhodium (type S).

The junction of the wires of the thermocouple is called the *sensing junction*. Since there is temperature difference between this sensing junction and the other ends, the latter are either kept at a constant reference temperature or at a room temperature.

The advantages of thermocouple over other temperature sensors are as follows:

1. It has rugged construction.
2. Temperature range is from 270°C to 2700°C.
3. Using extension leads and compensation cables, long transmission distance for temperature measurement are possible.
4. Cheaper and offer good reproducibility.
5. Fast response and good accuracy.
6. Calibration checks can be easily performed.

The limitations of thermocouples are as follows:

1. Cold junction and other compensations are essential.
2. They exhibit nonlinear characteristics.
3. Stray voltage pick up may be possible.
4. The output signal is weak and needs to be amplified.

The specifications of a typical K-type thermocouple are as follows:

- Temperature range is –200°C to 1100°C.
- Accuracy is ±3°C or 0.75%.
- EMF outputs – fair emf (30 mV at 750°C).

In order to compensate for nonlinearity of thermocouple characteristics, linearization may be carried out using the analog methods by either linear approximation or series approximation. Alternatively linear approximation can be performed digitally after data conversion. The output signal of thermocouple can be strengthen by appropriate conditioning circuits.

Resistance Temperature Detector (RTD)

Resistance temperature detector commonly employ platinum, nickel or resistance wire element whose resistance varies with temperature. The relationship between the temperature and resistance in the temperature range near 0°C can be calculated using the equation,

$$R_t = R_{ref} (1 + a \times \Delta T)$$

where R_t = Resistance of conductor at a temperature of 1°C.

R_{ref} = Resistance of a conductor at 0°C.

a = Temperature coefficient of resistance.

ΔT = Difference between operating and reference temperature.

The advantages of RTD are as follows:

1. Linear over wide range.
2. Wide operating temperature range.
3. Interchangeable over a wide range than other temperature transducers.
4. Better stability at high temperature.

The limitations of RTD are as follows:

1. Low sensitivity, higher cost.
2. Requires no point sensing.
3. It can be affected by contact resistance, shock, and acceleration.

A recent development in RTD design is the thin film sensor, a small ceramic substrate on which resistance film is deposited. The temperature range for a 100 Ω platinum RTD is from – 200°C to 850°C.

Integrated Sensors

The three terminal IC such as LM335 can be used as the transducers cum sensor. They combine the sensor and the electronic circuit into single monolithic IC package.

It is possible to adjust transducer to maximum accuracy over the range of measurement, using the two-point calibration method. However, the limitations of these transducers are as follows:

1. They can be used up to 200°C.
2. External power supply is required.
3. They are self-heating.
4. Available in limited configurations.

5.2 | INPUT DEVICES

In order to communicate with the microcontroller there are a variety of input devices such as keys, switches, keypad, relays, ADCs, and sensors.

5.2.1 Switches/Keys

Switches or keys are the most common form of input device to be connected to the microcontroller. The keys may be classified as follows:

    SPST : Single pole single throw

    SPDT : Single pole double throw

    DPST : Double pole single throw

    DPDT : Double pole double throw

They can be either TOGGLE type or PUSH button type. Figure 5.5(a) shows the logical symbols for the different keys used in practice.

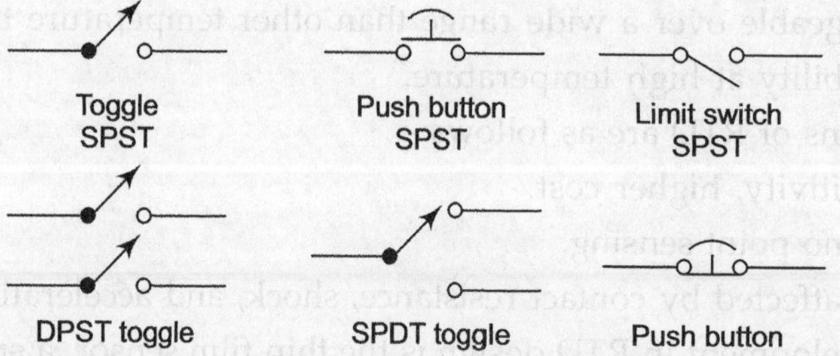

Fig. 5.5 (a) Logical symbols for the different types of keys

5.2.2 Thumbwheel Switches

Thumbwheel switch has an in-built mechanical encoder, which makes contact with either V_{CC} or ground depending on switch setting and generates a BCD outputs corresponding to the decimal number set on the front panel [Fig. 5.5(b)].

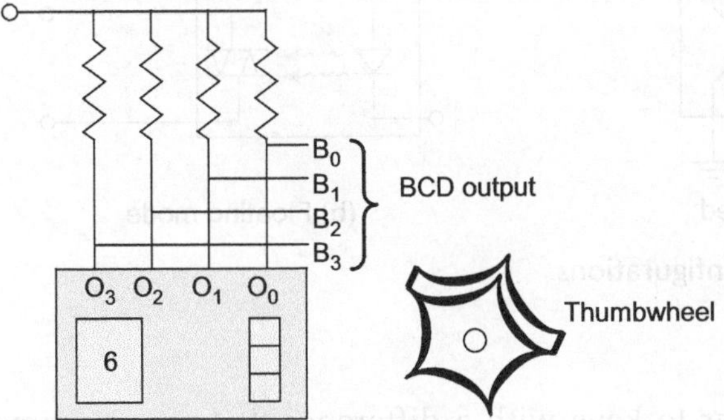

Fig. 5.5   (b) Thumbwheel switch

5.2.3 Transistor Switches

The microcontroller may be required to accept signals from an external transducer, which may be connected as an input device. The voltage levels of the device may not be compatible with the logic levels required at the I/O port. A transistor driver can be used as an input device.

The transistor turns ON and acts like a switch closed to ground. When the voltage is removed from the base, the transistor turns OFF and acts like an open switch. Transistors are much faster than mechanical switches and can last almost indefinitely when used properly.

When transistor is operated in a linear region, it works like an amplifier. A small change in the transducer output may be considered as a change in the input status. This produces bouncing, similar to the one in a mechanical push button switch contact due to spring action. Introducing a time delay, either using an IC device such as Schmitt trigger gate or through some software, can solve the problem.

Isolation is required as a result of this variation of Murphy's law if a direct path exists between the outside world and the microcontroller, a destructively high voltage eventually will be applied to it. The solution is to have a gap in the path through which only signals can be transmitted.

This can be accomplished using optocouplers. An optocoupler uses LED and a phototransistor to transmit signals by light while allowing for separate grounds between voltages in the outside world and those inside the microcontroller-based equipment. These IC chips can withstand a large difference between the input and output ground levels. As shown in Fig. 5.6, they are used in two configurations.

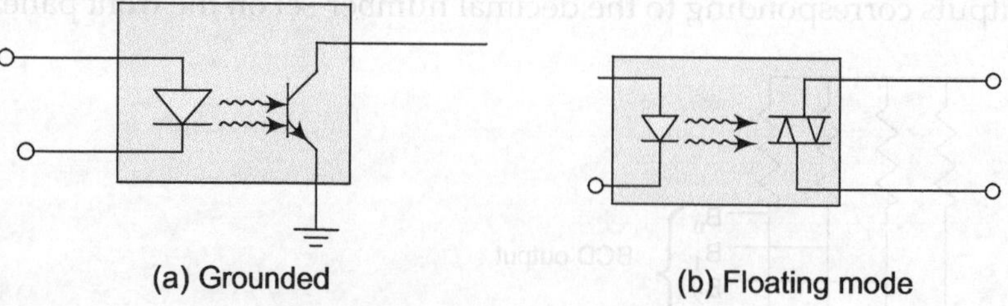

(a) Grounded   (b) Floating mode

Fig. 5.6 Transistor configurations

5.2.4 Relays

The relay are similar to keys with a difference that many a times relays are described in terms of their forms.

1. Form A → Normally open contact (NO)
2. Form B → Normally close contact (NC)
3. Form C → Single pole double arrangement (SPDT)

Figure 5.7 represents the logical symbols used.

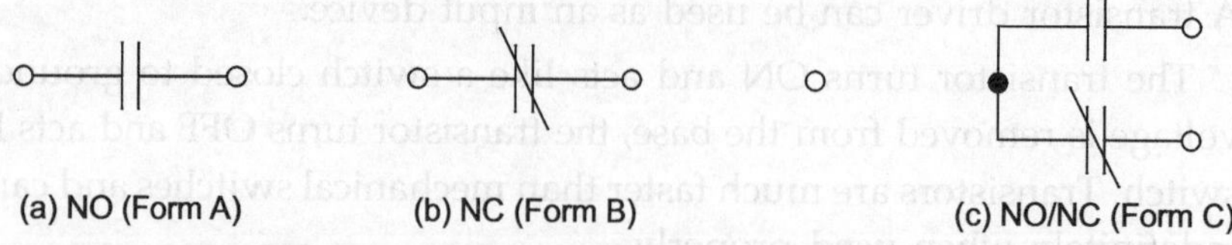

(a) NO (Form A)   (b) NC (Form B)   (c) NO/NC (Form C)

Fig. 5.7 Logical symbols for the relays

5.2.5 Keypad/Matrix Keyboard

If a system uses only a few switches, their status can be read directly through an I/O port. When many switches are used, they are arranged in a matrix form so that the number of input lines can be reduced. The following configurations are commonly used:

  (i) 16-switch hex keypad, as shown in Fig. 5.8.

  (ii) 64-switch typewriter style keyboard.

The problem with them is that operations like key closure detection and key identification along with the code generation are required which are to be handled by the software or hardware.

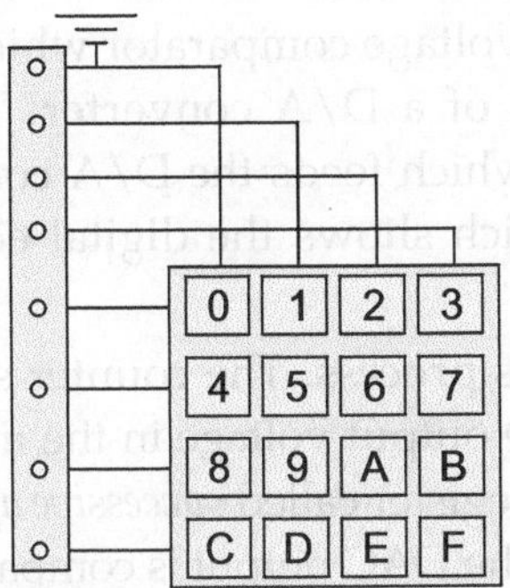

Fig. 5.8 Keypad

5.2.6 Analog to Digital Converters

Analog to digital conversion is a must in most of the systems. This is because of the reason that in practical systems the signals handled are continuous or analog, while for controlling or other purpose microcontroller requires digital signals.

There are many ways of converting analog signals to digital form. Each system has its own advantages and disadvantages and the choice of a particular type depends on the application and many other factors like speed, accuracy, stability, and cost.

The methods used for these conversions can be grouped as follows:

1. Flash converter ADC
2. Ramp type ADC
3. Dual slope ADC
4. Successive approximation ADC

Successive Approximation Type ADC

The successive approximation method is a compromise between speed and accuracy. It is the most popular method and most of the ICs use this technique for conversion.

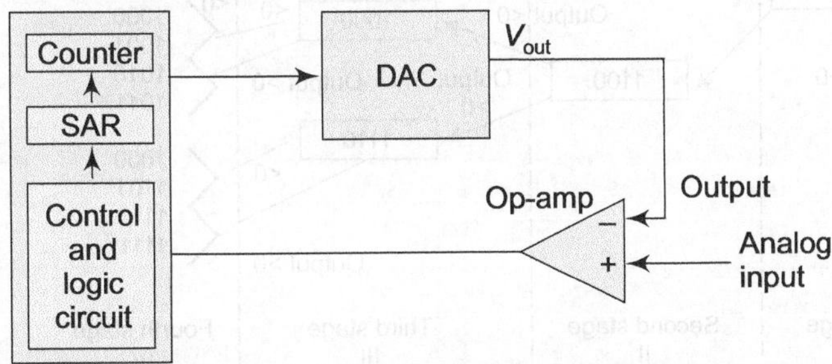

Fig. 5.9 Successive approximation method

As shown in Fig. 5.9, the SAR ADC has a voltage comparator which compares the analog input voltage with the output of a D/A converter. The voltage comparator is followed by a control logic, which feeds the D/A converter. The register is also connected to a display, which shows the digital equivalent of the analog input voltage.

The binary search method is used for this process. The counter starts with a binary input value, which corresponds to the output voltage in the middle of the range, i.e., $V_{mid} = (V_{max} - V_{min})/2$, stored in a register called *successive approximation register (SAR)*. This input is applied to DAC. The DAC output is compared with the applied analog signal. If the output of the op-amp is negative, it means the counter value is less so the actual value is in the upper half range, i.e., between V_{max} and V_{mid}. Value is in the lower half range, if the output of the op-amp is positive. The new mid-value is selected in corresponding range, as shown in Fig. 5.10(a) (illustration for a 4-bit approximation) and binary value of the mid-value of the output voltage as shown in Fig. 5.10(b) is stored in SAR. The process is repeated till output of DAC is zero. The content of counter represents applied analog value in binary.

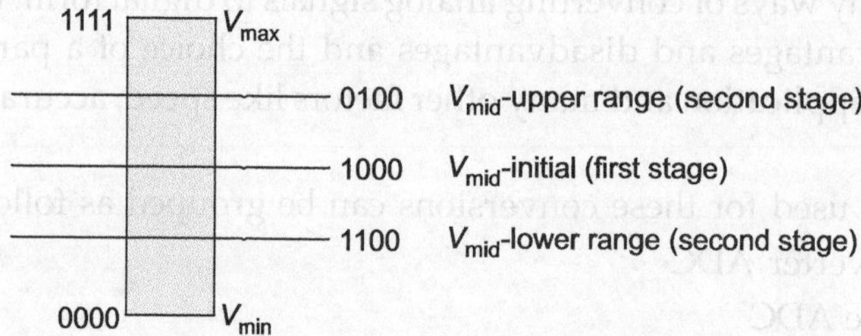

(a) Selection of mid-value in the selected range (binary search)

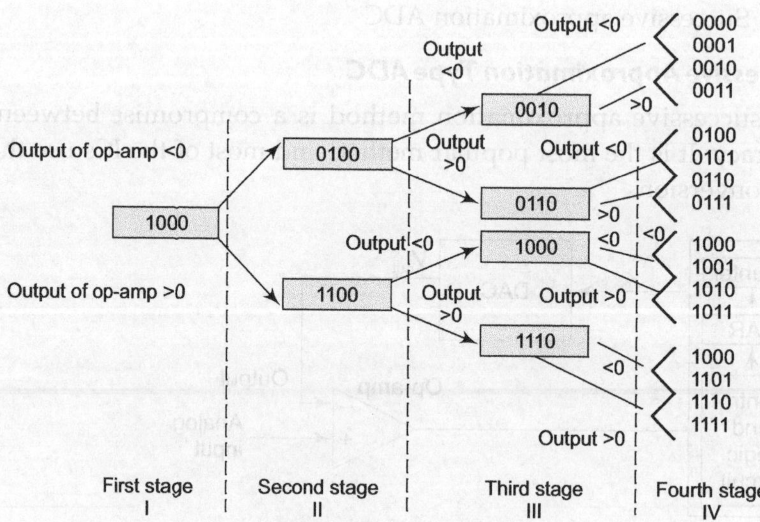

(b) Process of successive approximation for 4-bit approximation

Fig. 5.10

The various sources of errors involved with analog-to-digital converters are

1. Slew rate
2. Offset error
3. Gain error
4. Missing codes

There is a wide range of ADC available in the integrated circuit form with resolution varying from 8-bit to 16-bit. For example,

1. AD7975 and AD7812KN are the ADCs manufactured by the analog devices. AD7975 is superior for digitizing wideband signals at audio frequencies and uses SAR method. AD7812KN is designed with full compatibility with microprocessors and microcontrollers.

2. Motorola offers MC10319P: parallel flash 8-bit ADC, which is fully compatible with TTL logic families. It is available in 24-pin DIP package.

3. GEC offers ZN series of ADC. Most of them are of SAR type like ZN448 and ZN509.

4. ADC0858 from National Semiconductors is 16-bit ADC with power failure detection function also. It is available in 20-pin DIP package.

ADC0809

ADC0809 is an 8-bit microprocessor-compatible successive approximation A/D converter. A clock is built into the ADC, only external components like resistor and capacitor are required.

The detailed block diagram of ADC0809 is shown in Fig. 5.11. National Semiconductors offers ADC080X series fully compatible with microcontrollers. The variations are in speed and accuracy. All are available in 28-pin DIP package.

ADC0809 is an 8-bit SAR-type ADC that includes 8 channels multiplexer with control logic. The device offers high speeds, high accuracy, minimal temperature dependence, and low power consumption.

The device is CMOS 8-bit monolithic and it operates with 5 V supply. The conversion time is 100 μs and consume 0.3 MW of power. There are latched tri-state outputs and addressed inputs.

The device eliminates the possibility of missing codes and need for zero or full scale adjustments. It uses SAR technique. There is a high impedance threshold detector, a sample of switched capacitor array and hold circuit, and successive approximation register. There are tri-state output latches from SAR and latched inputs to the multiplexer address decoder.

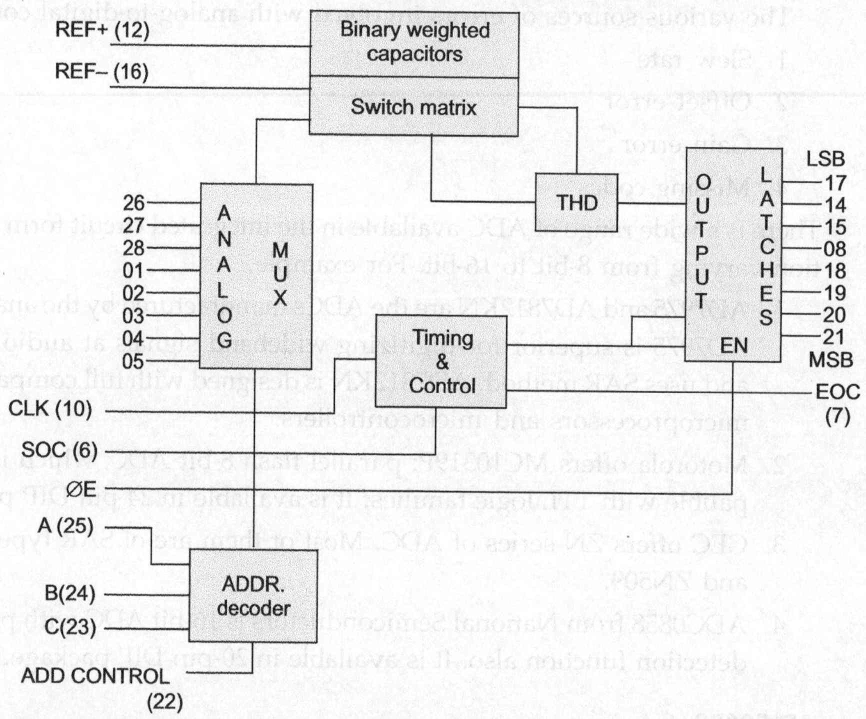

Fig. 5.11 Block diagram of ADC0809

Multiplexer The analog multiplexer selects 1 of 8 single-ended input channels as determined by the address decoder. The address load clocks control the address code into the decoder on a low to high transition. The output latch is reset by the positive going edge of the start pulse and lasts for 32 clock periods. The previous data will be lost, if a new start of conversion SOC command is given before the 64th pulse. Continuous conversion may be accomplished by connecting EOC to the start input. In this mode an external pulse should be applied after power-up to assure start-up.

Converter The CMOS threshold detector in SAR system determines each bit by examining the charge on a series of binary weighted capacitors. The conversion process uses SAR successive approximation register to count and weight the bits from MSB to LSB.

In the first phase of conversion process the analog input is supplied by closing the switch SC and all the ST switches and by simultaneously charging all the capacitors to the input voltage.

In the next phase of conversion process all SC and ST switches are opened and the threshold detector begins identifying bits by identifying the charge on each capacitor relative to the reference voltage. In the switching sequence, all eight capacitors are examined separately until all 8-bits are identified and then the charge convert sequence is repeated.

In the first step of conversion phase, the threshold detector looks at the first capacitor (weight = 128). Node 128 is switched to voltage and the equivalent nodes of all the other capacitors on the ladder are switched to REF. If the voltage at the summing node is less than the trip point of threshold detector, this 128-weight capacitor remains connected to the REF through the remainder of the capacitor sampling (bit-counting) process. The process is repeated for the 64-weight capacitor, the 32-weight capacitors, and so on, until all bits are counted.

The clock required by the ADC may be derived from counter ICs like

1. 7490 / 290 / 390 — Decade (2 × 5)
2. 7492   — Duo-decimal (2 × 6)
3. 7493 / 293 / 93 — Ripple counter (2 × 8)

However, due to synchronization and stability problems, in practice clock signals are derived using the on-chip timers of the microcontrollers or the 8253/8254 timer chip used in Mode 3, i.e., square wave generator mode.

ADC0816/0817

This IC is similar to ADC0809 except that

1. This has a 16-channel multiplexer.
2. There is a provision for connecting Sample and Hold between the output of the multiplexer and the input to the actual (ADC).

A typical block diagram is shown in Fig. 5.12.

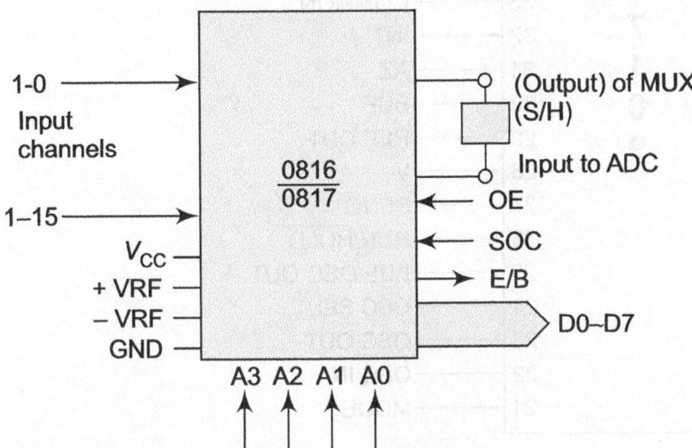

Fig. 5.12   Pin layout of ADC0816/0817

A/D Converter 7109

The A/D converter 7109 is a high performance, CMOS, low power integrating converter designed to interface with the microcontroller. The output of ADC is 12-bit, has polarity and over range features. It may be directly accessed under control of 2-byte active low enable inputs LBEN and HBEN and an active low chip select input CE for a simple parallel bus interface. A UART handshake mode is provided to allow 7109 to work with the industrial standard UART in providing serial data transmission, for remote data logging application. The RUN/HOLD input and STATUS output allows monitoring and control of conversion timing. The 7109 converter provides the user with high accuracy, low noise, low drift, versatility, and economy of the dual slope integrating A/D converter.

The features of 7109 are listed below:

1. True differential inputs.
2. Reference drift of less than 1 µW per degree centigrade.
3. Maximum input bias current of 10 pA.
4. Typical power consumption of 20 mW.
5. Operation up to 30 conversions per second.
6. On-chip oscillator operation with expensive 3.58 MHz TV crystal giving 7.5 conversion per second for 50 Hz rejection.
7. It may be used with RC network oscillator for clock frequencies.

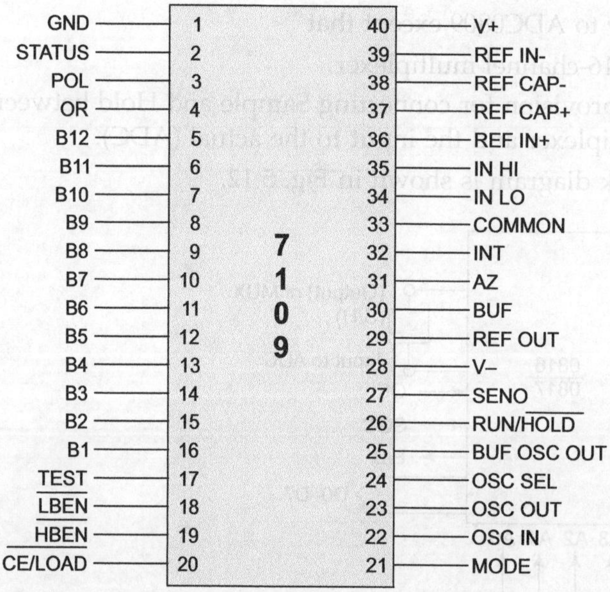

Fig. 5.13 Pin layout of ADC7109

Figure 5.13 depicts the pin layout of the ADC7109. Functions of the important pins are described below:

MODE input The MODE input is used to control the output mode of the converter. When the MODE pin is low or left open, the converter is in its "Direct" output mode, where the output data is directly accessible under the control of chip and byte enable inputs. When the MODE input is pulsed high, the converter enters the UART handshake mode and outputs the data in 2-bytes and then returns to "Direct" mode when the MODE input is left high, the converter gives output in the handshake mode at the end of every conversion cycle.

STATUS output During a conversion cycle, the STATUS output goes high at the beginning of the signal integrate, and goes low one-half period after the new data from the conversion has been store in the output latches. This signal may be used as a "Data valid" flag to drive the interrupt, or monitor the status of the converter.

RUN/HOLD input When the RUN/HOLD input is high or left open, the circuit will continuously perform the conversion cycle, updating the output to the output latches after zero crossing during the denigrating portion of conversion cycle. In this mode of operation, the conversion cycle will be performed in 8192 clock period regardless of the resulting value.

RUN/HOLD goes low at any time during denigrate after zero crossing has occurred, the circuit will immediate terminate denigrate and jump to auto-zero. This feature can be used to eliminate the time spent in denigrate after the zero crossing.

If RUN/HOLD stays or goes low, the converter will ensure minimum auto-zero time and the wait in auto-zero until the RUN/HOLD input goes high.

The converter will begin the integrate portion of next conversion 7 clock periods after the high level is detected at RUN/HOLD.

Direct mode When the MODE pin is left at low level the data output (bit 1 through 8 lower order byte, bits 9 through 12, polarity and over range high order byte) are accessible under control of the byte and chip enable terminals at the inputs. These inputs are all active low. When the chip enable input is low, taking a byte enable input will allow the output of that byte to become active. This allows a variety of the parallel data accessing techniques to be used. It should be noted that these control inputs are asynchronous with respect to converter clock, the data may be accessed at any time. Thus it is possible to access the latches while they are being updated, which could lead to erroneous data. Synchronizing the access of the latches with the conversion cycle by monitoring the STATUS output will prevent this. Data is never updated while status is low.

Oscillator 7109 has a versatile three terminal oscillator. It generates the internal clock. The oscillator may be overdriven or may be operated with an RC network or crystal.

The OSCILLATOR select input changes the internal configuration of the oscillator to optimize for RC or crystal operation. When this is high or left open, the oscillator is configured for RC operation and the internal clock is of the same frequency and phase as the signal at the BUFFERED OSCILLATOR OUTPUT.

When this input is low, a feedback device and the output and the input capacitor are added to the oscillator. The oscillator operates with most crystal in the 1 to 5 MHz with no external components.

It also inserts a fixed divider of 58 circuit between the BUFFERED OSCILLATOR OUTPUT and the internal clock.

Using 3.48 MHz crystal, this division ratio provides an integration time given by

$$T = (2048 \text{ clock period}) \times 58/3.58 \text{ MHz} = 33.18 \text{ ms}$$

The converter operates reliably at conversion rates of up to 30 per second, which corresponds to a clock frequency of 245.8 kHz.

Function cycle of 7109   The function cycle is divided into three phases:

Auto_zero phase   During this phase three operations take place.

1. Input high and low are disconnected from their pins internally and shorted to the analog COMMON.

2. The reference capacitor is charged to reference voltage.

3. A feedback loop is closed around the system to charge the auto-zero capacitor to compensate for the offset voltages in the buffer amplifier, integrator, and the comparator.

Since the comparator is included in the loop, the auto-zero accuracy is limited only by the noise in the system.

Signal integrate phase   The auto-zero loop is opened and the internal short is removed. The internal low and high are connected to external pins. The converter then integrates the differential voltage between IN HI and IN LOW for a fix time of 2048 clock periods. The polarity of the integrated signal is determined at the end of this phase.

Denigrate phase   During this phase the input low is internally connected to analog common and input high is connected across the previously charged reference capacitor. The circuitry within the chip ensures that the capacitor will be connected to the reference capacitor. The circuitry within the chip ensures that the capacitor will be connected with the correct polarity to cause the integrator output to return to zero with a fixed slope. Hence the time for the output to return to zero is proportional to the input signal.

Component value selection   There are two sections in ADC7109, i.e., analog and digital section. It is necessary to select the components for optimum performance from the ADC.

For the analog section, care must be taken in the selection of the values of the components such as: integrator capacitor, integrating resistor, auto-zero capacitor, reference voltage, and the conversion rate. These values must be chosen to suit the particular application.

Integrating resistor   Both the amplifier and the integrator have a class A output stage with 100 μA of drive current with negligible nonlinearity. The integrating resistor must be, large enough to remain in this region over the input voltage range and small enough so that undue leakage requirements are not placed on the board. The relation for selection can be given as:

$$R_{int} = \frac{\text{Full-scale voltage}}{20\ \mu A}$$

Integrating capacitor   The integrating capacitor must be selected to give the maximum integrator output voltage swing with saturating the integrator. The relation is

$$C_{int} = \frac{2048 \times \text{clock period} \times 20\ \mu A}{\text{Integrator output voltage swing}}$$

Auto-zero capacitor   The integrating capacitor must be selected to give the noise of the system. The smaller is the capacitor, lower is the noise.

The auto-zero capacitor, C_{az} cannot be increased without limits since it, in parallel with integrating capacitor forms an RC time constant that determines the speed of recovery from the overloads and more important the overloads, and the error that exists at the end of an auto-zero cycle. A value of C_{az} equal to half of C_{int} is recommended.

Reference capacitor   A 1 μF capacitor generally gives good result in most of the applications.

Reference voltage   The analog input required to generate full-scale output of 4096 count is $V_{in} = 2\ V_{ref}$.

For normalized scale, a reference of 2.048 should be used for a 4.096 V full scale, and 204.8 mV should be used for 0.4096 V full scale. The stability of the reference voltage is a major factor in the overall absolute accuracy of the converter. The resolution of the converter at 12-bits is one part in 4096 or 244 ppm. An extremely high quality reference has to be used where the ambient temperature is not controlled or where high accuracy is needed for the absolute measurement.

For the digital selection   The digital section of the 7109 includes clock oscillator, scaling circuit, 12-bit binary counter with output latches, TTL compatible tri-state output driver, polarity, overage and control logic, and UART handshake logic. Care should be taken to select the parameters.

5.2.7 Sample and Hold (S/H)

The analog signal must be held constant during the conversion period. In case of a time varying signal, an instantaneous value is applied. To keep it constant over a period of conversion, a sample–hold circuit is required. Sample–hold I/Cs are available: e.g., LF198/298/398. These ICs are manufactured by the National Semiconductors and have fast acquisition time. A logical symbol of S/H is given in Fig. 5.14.

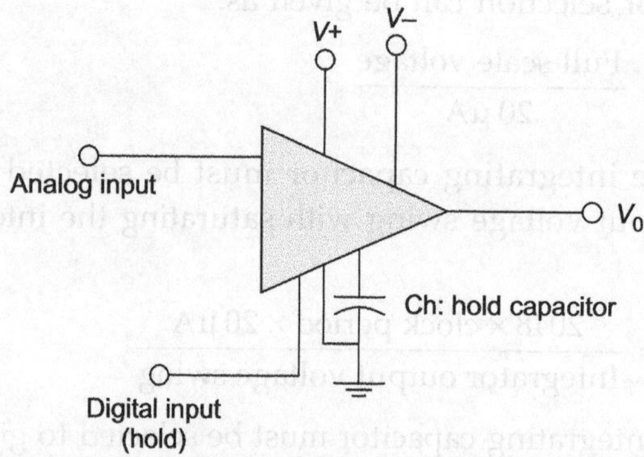

Fig. 5.14 Logical symbol of S/H

The operation of the S/H circuit is simple. When digital input is applied, the capacitor tries to charge up the applied analog input. When the digital input is removed, the capacitor remains charged, as there is no path available for it to discharge. Hence, the voltage is kept practically constant. The capacitors used are made up of dielectric materials like polyester, polypropylene, or Teflon.

Specifications

The specifications of S/H circuits to be considered for an application are described below:

Droop rate It is the rate at which the output of the capacitor discharges or the rate at which output of S/H decreases. The rate also depends on value of the capacitor, e.g., for $C = 1\ \mu F$, it is about 5 mV/min.

Aperture time It is the time delay between the hold command and the instant at which the input is actually applied to hold capacitor. It is in (nsec), e.g., for $C = 0.047\ \mu F$, it is 40 msec.

Acquisition time It is the time required by the S/H circuit capacitor to charge from one voltage level to another.

Small signal bandwidth The frequency at which the held output amplitude is 3 dB below the input amplitude, an input condition of a 100 mV *p-p* sine wave.

Full power bandwidth   The frequency at which the held output amplitude is 3 dB below the input amplitude, under an input condition of a 10 V *p-p* sine wave.

Effective aperture delay   It is the difference between the switch delay and the analog delay of the sample and hold amplifier (SHA) channel. A negative number indicates that the analog portion of the overall delay is greater than the switch portion. This effective delay represents the point in time, relative to the hold command, that the input signal will be sampled.

Aperture jitter   The variations in aperture delay for successive samples. Aperture jitter puts an upper limit on the maximum frequency that can be accurately sampled.

Hold settling time   The time required for the output to settle to within a specified level of accuracy of its final held value after the hold command has been given.

Feed through   The attenuated version of a changing input signal that appears at the output when the SHA is in the hold mode.

Hold mode offset   The difference between the input signal and the hold output. This offset term applies only in the hold mode and includes the error caused by charge injection and all other internal offsets. It is specified for an input of 0 V.

Tracking mode offset   The difference between the input and the output signals when the SHA is in the track mode.

Nonlinearity   The deviation from a straight line on a lot of input versus held output as referenced to a straight line drawn between endpoints, over an input range of –5 V and +5 V.

Gain error   Deviation from a gain of +1 on the transfer function of input versus held output.

Inter channel isolation   The level of correlated between adjacent channels while in the sample (track) mode with a full-scale 100 kHz input signal.

Inter channel aperture offset   The variation in aperture time between the four channels for simultaneous hold command.

Differential offset   The difference in hold mode offset between the four SHA channels.

Power supply rejection ratio   A measure of change in the hold output voltage for a specified change in the positive or negative supply.

Sampled DC uncertainty   The internal rms noise that is sampled onto the hold capacitor.

Hold mode noise The rms noise at the output of the SHA while in the hold mode, specified over a given bandwidth.

Total output The total rms noise that is seen at the output of the SHA while in the hold mode. It is the rms summation of the sampled DC uncertainty and the hold mode noise.

Output drive current The maximum current the SHA can source (or sink) while maintaining a change in hold mode offset of less than 2.5 mV.

Sample and Hold Amplifier (SHA)

The AD684 is a monolithic quad sample and hold amplifier (SHA). It has four sampling channels (Fig. 5.15). Each channel is controlled by an independent hold command and has an independent hold capacitor. The channels are self-contained and require no external components or adjustments. The AD684 is manufactured on a BIMOS process, which provides a merger of high performance bipolar circuitry and low power CMOS logic.

The AD684 is ideal for high performance, multi-channel data acquisition systems. Each SHA channel can acquire a signal in less than 1 μs and retain the held value with a droop rate of less than 0.01. AD684 is an ideal front end for high speed 12-bit and 14-bit ADCs.

The AD684 has self-correcting architecture that minimizes hold mode error and insures accuracy over temperature. Each channel of the AD684 is capable of souring 5 mA and incorporates output short circuit protection. The AD684 is specified for three temperature ranges.

$$J \text{ grade } -40°C \text{ to } 85°C$$
$$A \text{ grade } -40°C \text{ to } 85°C$$
$$S \text{ grade } -55°C \text{ to } +125°C$$

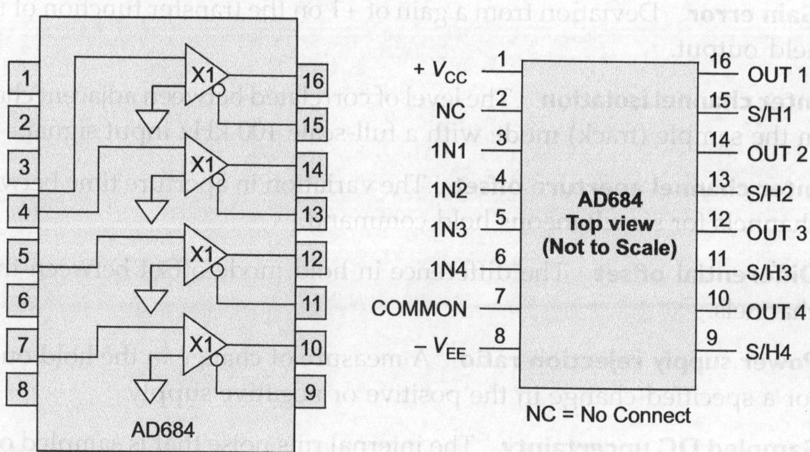

Fig. 5.15 Pin layout and internal organization of AD684

The AD684 is a complete quad SHA capable of high-speed sampling with 12-bit accuracy in less than 1 µs.

The AD684 is completely self-contained, including on-chip hold capacitors, and requires no external components or adjustments to perform the sampling function. Each SHA channel can operate independently, having its own input, output, and sample/hold command. Both the inputs and the outputs are treated as single ended signals, referred to common. The sample and hold circuit of AD684 corrects for internal errors after the hold command has been given by compensating voltage for amplifier gain and offset errors and charge injection errors. In hold mode, the internal circuitry is reconfigured to produce an accurately held version of the input signal.

The features of the AD684 are summarized below:

1. Fast acquisition time of 1 µs and low aperture jitter of 200 ps.
2. Monolithic construction insures excellent inter-channel matching in terms of timing and accuracy as well as high reliability.
3. Independent inputs, outputs, and sample-and-hold controls allow user flexibility in system architecture.
4. Low droop is 0.01 mV/sec and internally compensated hold mode error results in superior system accuracy.
5. The AD684's fast settings time and low output impedance make it ideal for driving high speed analog to digital converters such as the AD674, AD7572, and the AD7672.

5.3 | OUTPUT DEVICES

Output devices are used to get the response or the result from the microcontroller system. Popular devices which are connected to the system are: LEDs, 7-segments, alphanumeric displays, LCD. For controlling the process loop or to generate the control signals rather than only displaying the output requires special devices like stepper motors, printers, data converters, etc. In this section we will study characteristics, operation, and interfacing of the commonly used output devices with the microcontroller.

5.3.1 LED

It is known that LEDs are light emitting diodes. They emit light when it conducts. A diode conducts when it is forward biased. Hence, when the anode of the LED is positive with reference to the cathode, it conducts, emits light and glows. But if it is reverse biased, it does not conduct and does not glow. They are used only as an indicator to keep track of the signal or the logic level in a digital system.

5.3.2 Display Modules

Instead of using a single diode to display the status or the result, the diodes are grouped together to form display modules so that the result can be displayed in a user-friendly manner. Care has to be taken for the device so that the sourcing or sinking capability of device is not exceeded. The current flowing in the module must be enough to drive current in LEDs. The problem is solved by the current drivers/buffers in the form of Darlington pair or special IC such as 75491/75492 are used based on the module configuration.

7-Segment LEDs

These are the most popular output devices. There are 7 LEDs configured to display a decimal digit on the device. It is easy for the user to interpret the result. There are two types of arrangements used in the device.

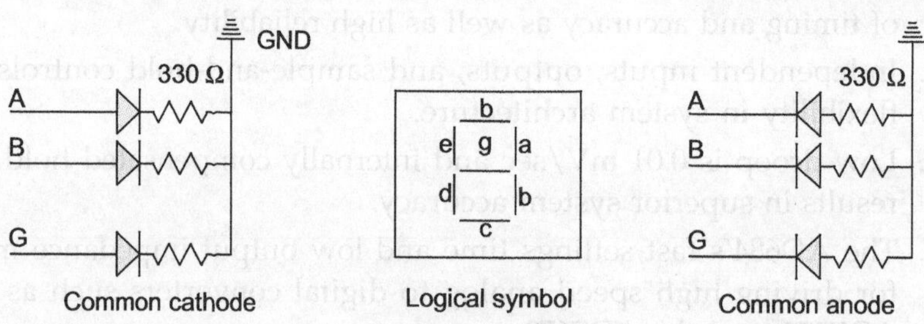

Fig. 5.16 Configurations of 7-segment LED

Common anode As shown in Fig. 5.16, in this arrangement, all the anodes are tied together and connected to V_{CC}. Since for the LED to glow, the cathode of the LED must be connected to ground. These terminals are connected to the port pins of the microcontroller and grounded selectively based on the digit to be displayed. A look-up table for the code is prepared. The code is output on the port pins to display the desired digit.

Common cathode As shown in Fig. 5.16, all the cathodes are tied together and connected to the ground. For the LED to glow, the anode of the LED must be connected to V_{CC}. The anodes are connected to the port pins of the microcontroller and are made logically high selectively based on the digit to be displayed. A look-up table for the code is prepared. The code is output on the port pins to display the desired digit.

16-Segment Display Units

With the 7-segments it is not possible to display alphabets or complete character set. A unit having more segments is required. Such a unit is available in the market. A pin diagram is shown in Fig. 5.17. The arrangement of the display module using 16-segment is shown in Fig. 5.18. Figure 5.19 represents alphabetic character set.

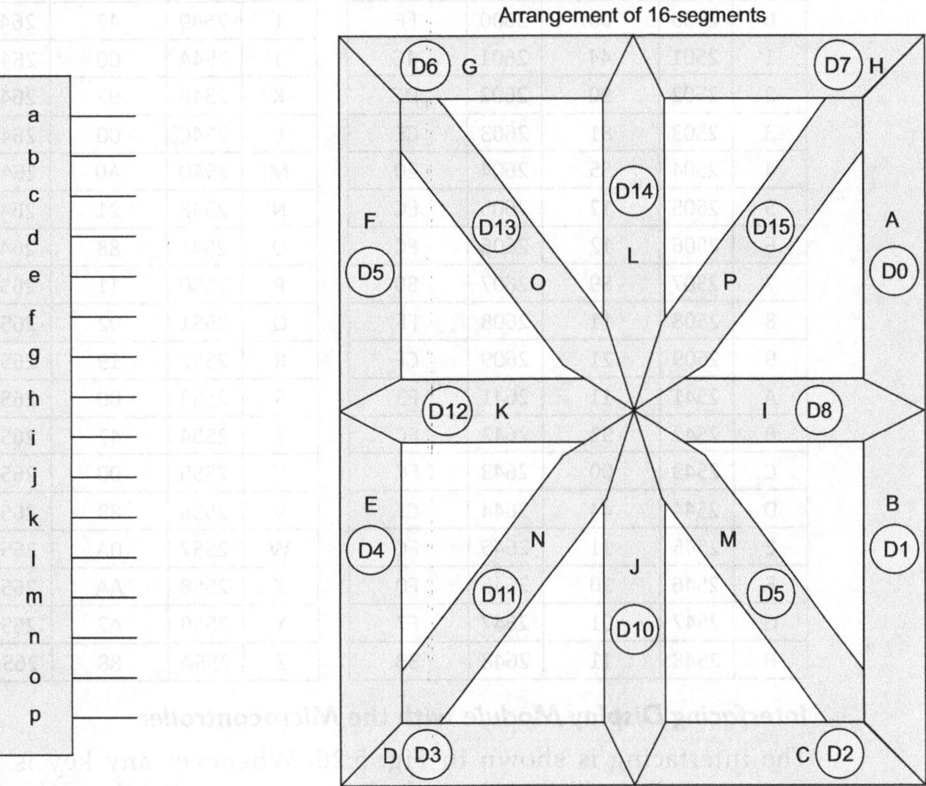

Fig. 5.17 Pin diagram

Fig. 5.18 Arrangement of 16-segment display unit

0	6	C	I	O	U
1	7	D	J	P	V
2	8	E	K	Q	W
3	9	F	L	R	X
4	A	G	M	S	Y
5	B	H	N	T	Z

Fig. 5.19 Set of alphanumeric display characters

The interfacing of these devices is very simple. Since microcontroller has 8-bit ports, two ports must be used to interface the module directly to the microcontroller through drivers. A look-up table should be prepared in memory similar to one for a 7-segment device. Assuming a common cathode configuration of the 16-segments display module, the code is developed and stored in the form of look-up table shown in Table 5.1.

Table 5.1   Look-up table for alphanumeric character display on 16-segment module

Char.	Addr.	LSB Code	Addr.	MSB Code	Char.	Addr.	LSB Code	Addr.	MSB Code
0	2500	00	2600	FF	I	2549	42	2649	CC
1	2501	44	2601	4C	J	254A	00	264A	1F
2	2502	90	2602	DC	K	254B	92	264B	30
3	2503	81	2603	CE	L	254C	00	264C	3C
4	2504	55	2604	20	M	254D	A0	264D	33
5	2505	12	2605	EC	N	254E	21	264E	33
6	2506	12	2606	FC	O	254F	88	264F	FF
7	2507	89	2607	80	P	2550	11	2650	F1
8	2508	11	2608	FF	Q	2551	02	2651	FF
9	2509	21	2609	CF	R	2552	19	2652	F1
A	2541	11	2641	F3	S	2553	00	2653	CE
B	2542	92	2642	FC	T	2554	42	2654	C0
C	2543	00	2643	FC	U	2555	00	2655	3F
D	2544	44	2644	CF	V	2556	88	2656	30
E	2545	11	2645	FC	W	2557	0A	2657	33
F	2546	10	2646	F0	X	2558	AA	2658	00
G	2547	01	2647	F7	Y	2559	A2	2659	00
H	2548	11	2648	33	Z	255A	88	265A	CC

Interfacing Display Module with the Microcontroller

The interfacing is shown in Fig. 5.20. Whenever any key is pressed, the microcontroller will find the code. Let us assume that the subprogram named rd_key is available for finding an ASCII code of the key pressed. Our aim is to write a program to display the hex key pressed on the display unit.

The look-up table is stored in such a way that the ASCII code is the LSB of the address byte. The 16-segment of the character is stored so that the upper byte of the code is stored at address byte [25××], while LSB at the address byte [26××]. The program segment is shown below:

```
disp_16:
        mov 80h, #20h           ; initialize stack
rept1:
        acall read_key          ; call a subroutine for ASCII code
        mov dph, #25h           ; upper byte of address of hex code
        mov dpl, a              ; lower byte of the address of hex code
        movx a, @dptr           ; get upper byte of code
        mov P1, a               ; out upper byte at Port 1
        inc dph
        movx a, @dptr           ; get LSB of the code
```

```
        mov P0, a                       ; out lower byte at Port 0
        sjmp rept1
        end
```

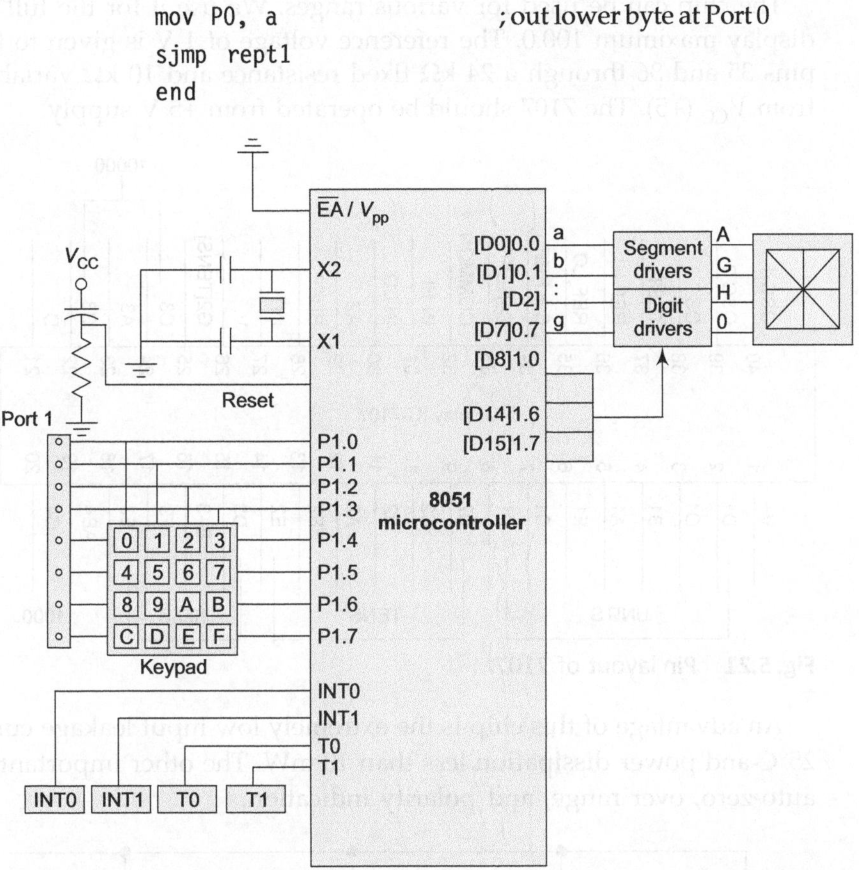

Fig. 5.20 Interfacing keypad and 16-segment display module with the 8051

Digital Panel Meter

It consists of 7-segments LED and IC-7107. The function of IC is analog to digital conversion as well as to display the centigrade on a 7-segments LEDs. For ease of reading, digital read out of any measurement is an essential feature. For case of reading in industrial environment in day time we select LED display rather than LCD display, even though the power consumption is higher.

IC-7107 is used instead of IC-7106 for accuracy and compactness. It is a self-contained ADC, display-decoder, and 7-segment drives for 3½ digit display indications.

Figure 5.21 shows the 7107 from the Inersil. It is a 40-pin plastic DIP single chip that contains all the active circuitry for 3½ digit panel meter. It contains a precession dual slope converter, a BCD to 7-segment decoder display drivers, a clock, and a reference. The general description and features of the IC can be found from the manual.

The chip can be used for various ranges. We use it for the full scale, i.e., to display maximum 100.0. The reference voltage of 1 V is given to the reference pins 35 and 36 through a 24 kΩ fixed resistance and 10 kΩ variable resistance from V_{CC} (+5). The 7107 should be operated from +5 V supply.

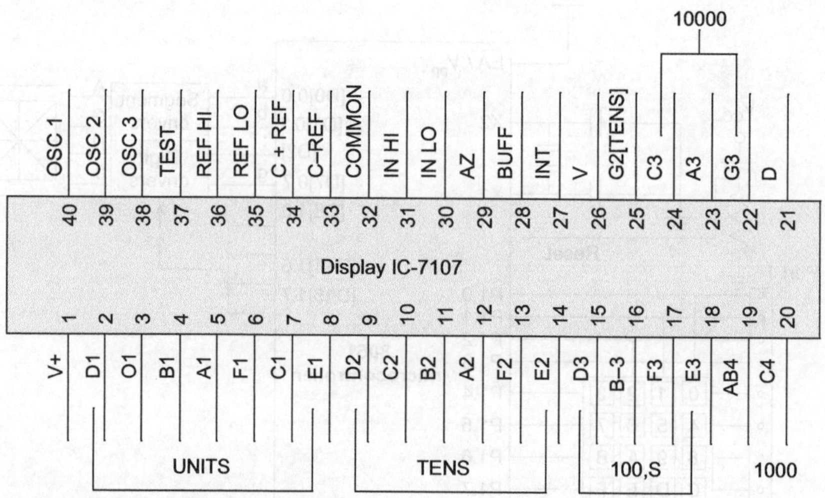

Fig. 5.21   Pin layout of 7107

An advantage of this chip is the extremely low input leakage current 1 pA at 25°C and power dissipation less than 10 mW. The other important features are auto-zero, over range, and polarity indication.

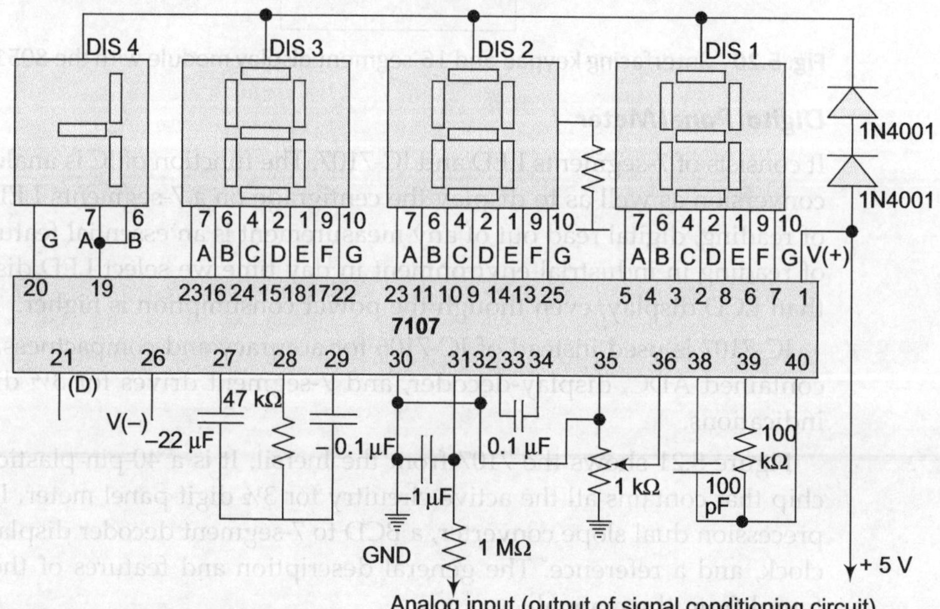

Fig. 5.22   Display panel meter

In the digital panel meters, four FND507 (which is a common anode LED display) are connected to IC-7107 to obtain the reading. Decimal point of second FND507 is permanently grounded. Common anode is given a +5 V supply. Figure 5.22 shows circuit diagram of the panel meter.

Calibration of display   The calibration is carried out as follows:

1. The IC-7107 is tested after the implementation by pulling up the test pin 37 to + V_{CC}, the display must be OFF.

2. Since temperature to be indicated is below 100°C, the last digit is taken as decimal point by grounding input for that point.

3. The total span is to be read up to 100.0 at 100°C and 000.0 at 0°C. Hence, the 1 V is used as reference by adjusting the 10 K potentiometer connected between the pin 36 and pin 35.

4. The readings of mercury thermometer and the display read out are compared by connecting output of signal conditioner block to input of IC-7107.

5.3.3 LCD Displays

These are dot matrix liquid crystal displays used to display alphanumeric, characters, and symbols. The built-in controller and the driver provide convenient connectivity between a dot matrix LCD and the 8-bit microcontroller. All the functions required for dot matrix liquid crystal display drive are internally provided. Internal refresh is provided in the modules. Built-in dot matrix LCD controller has the font 5×7 or 5×10 dots and the display RAM for 80 characters. There is a character generator ROM, which provides 160 characters with font 5×7 dots and 32 characters with font 5×10 dots. These RAM can be read by the microcontroller. The built-in oscillator provides clock. There is a wide range of instructions for controlling display operation such as: clear display, cursor home display ON/OFF, cursor shift, and display shift.

Many times modules do not have conventional CS terminal, the enable signal has to be derived from the CS or I/O select. When connecting the module through a parallel input/output device, the burden of ensuring proper operation falls on the software. Incorrect combination of data bus direction and R/W pin logic can damage both. DB_0 to DB_7 have the capacity of driving one TTL or capacitance of 130 pf. The data bus terminals have three-state construction. When the enable signal is at low level, these data bus terminals will remain in high impedance state. The data bus has pull-up MOS, so when the data bus is open, it produces high output voltage.

At the interface of LCD module, there are three power supply terminals V_{DD}, GND and V_0. The LCD is driven by the voltage determined by $V_{DD} - V_0$. Since optimum voltage of power supply for LCD shifts according to temperature change, voltage at V_0 terminal needs to be adjusted. V_0 also needs adjustment for optimum contrast at the angle at which the module is viewed.

The modules are automatically initialized (reset), when the power is turned ON using the internal reset circuit. The busy flag (BF) holds "*H*" and does not accept instructions until initialization ends. The busy state is 1 ms after V_{DD} rises to 4.5 V. The following instructions are executed in initialization:

1. Display clear
2. Character font: 5×7 dots (F = low)
3. Number of lines: 1-line (N = low)
4. Interface width: 8-bits (DL = high)
5. Address counter: Increment (I/D = high)
6. Display shift: Off (SH = low)
7. Display: Off (D = low)
8. Cursor: Off (C = low)
9. Blink: Off (B = low)

Components of Display Module

The LCD display module is consisting of units such as: oscillator, controller, RAM, and internal reset circuit. The functions, operation, and use of each unit is described below:

Busy flag (BF)   When the busy flag is '1', it indicates that the controller is in the internal operation mode and the next instruction will not be accepted. When R/W pin is '1' and RS is at '0', the status of BF is output from DB_7. The next instruction is written after the busy flag goes low.

Address counter (AC)   The address counter (AC) generates the address for the DD RAM, the CG RAM and for the cursor display. When an instruction code for the RAM address is written to the controller, after deciding that which RAM is addressed (DD RAM or CG RAM), the address information is transferred to AC. After the read/write operation specified by the instruction, AC is automatically incremented (or decremented). The data of the AC is output to $DB_0 - DB_6$ when RS is '0' and R/W is '1'.

Character generator ROM (CG ROM)   The character generator ROM generates 5×7 dot or 5×10 dot character patterns from 8-bit character codes. It can generate 160 types of 5×7 dot character patterns and 32 types of 5×10 dot character patterns. When the 8-bit character code of a CG ROM is written to the DD RAM, the character pattern of the CG ROM corresponding to the code is displayed on the LCD display position corresponding to the DD RAM.

Character generator RAM (CG RAM)   The character generator RAM (CG RAM) is the RAM with which the user can generate character patterns by program. The CG RAM has the capacity to store 8 types of 5×7 dots or four types of 5×10 dots. Programming of these character patterns is explained in CG RAM programming.

Display data RAM (DD RAM)   The DD RAM stores the display data represented in 8-bit hex character codes. It has the capacity for eighty 8-bit characters. The display data RAM (DD RAM) that is not used for display can be used as general data RAM. Depending on the 8-bit character code, that is written into the DD RAM, LCD will select the character pattern from either CG RAM or CG ROM.

Display data with font 5 × 7 dot matrix

Character codes	Operation specified
00h to 07h	Select CG RAM blocks 0 to 7
08h to 0Fh	Select CG RAM blocks 0 to 7
10h to 1Fh	are invalid
20h to 7Fh	Select CG ROM displayable ASCII character
80h to 9Fh	Invalid
A0h to FFh	Select CG ROM Kana (Japanese) and Greek characters

Display data with font 5 × 10 dot matrix

Character codes	Operation specified
00h, 01h, 08h, 09h	Select CG RAM block 0
02h, 03h, 0Ah, 0Bh	Select CG RAM block 1
04h, 05h, 0Ch, 0Dh	Select CG RAM block 2
06h, 07h, 0Eh, 0Fh	Select CG RAM block 3
10h TO 1Fh	Invalid
20h TO 7Fh	Select CG ROM displayable ASCII character
80h TO 9Fh	Invalid
A0h TO FFh	Select CG ROM for Greek characters

Example 5.1   Display the character C.

Solution

To display the character C, the character code 43h has to be written to the display data RAM. Relation between DD RAM address and position on the liquid crystal display depends on the type of display configuration, i.e., one line (80 × 1), two lines (40 × 2), or four lines (20 × 4). The DD RAM address (ADD) is set in the address counter (AC) and is represented in hexadecimal. Displays having more than 80 characters have two enable signals. Each enable signal is capable of controlling 80 characters.

CG RAM Programming

The CG RAM has the capacity to store 8 types of 5 × 7 dots or 4 types of 5 × 10 dots. Programming of these character patterns is explained below:

5 × 7 dot character pattern   CG RAM is organized into 8 blocks having 5 columns by 8 rows pixel format. CG RAM address bits 3-5 correspond to the selection of 1 out of these 8 blocks. Bits 2-0 designate each row in that block. Note that the selection of these 8 blocks of CG RAM requires 7-bit to be '0' and 6-bit to be '1'.

These blocks can be programmed to generate character pattern as required using bits 0-4. Bits 5-7 are not used. Once the address of the block is selected for data pattern generation, data is written using 'Write data to CG RAM' command. The address gets auto incremented after each data is written. '1' for the CG RAM data in the character patterns correspond to dark pixel of the display. The 8th line is the cursor position and display is performed in logical OR by the cursor. Table 5.2 represents the pattern.

Table 5.2   CG RAM data for 5 × 7 dot character pattern

Character codes (DD RAM data)								CG RAM address						Character patterns							
														CG RAM address							
7	6	5	4	3	2	1	0	5	4	3	2	1	0	7	6	5	4	3	2	1	0
Higher					Lower			Higher			Lower			Higher					Lower		
											0	0	0	.	.	.	1	1	1	1	0
											0	0	1				1	0	0	0	1
											0	1	0				1	0	0	0	1
											0	1	1				1	1	1	1	0
0	0	0	0	·	0	0	0	0	0	0	1	0	0				1	0	1	0	0
											1	0	1				1	0	0	1	0
											1	1	0				1	0	0	0	1
											1	1	1								

5 × 10 dot character pattern   CG RAM is organized into 4 blocks having 5 columns by 10 rows pixel format. CG RAM address bits 4–5 correspond to selection of 1 out of these 4 blocks. Bits 0–3 designate each row in that block. Note that the selection of these 4 blocks of CG RAM require 7-bit to be '0' and 6-bit to be '1'. These blocks can be programmed to generate character pattern as required using bits 0-4. Bits 5-7 are not used once the address of the block is selected for data pattern generation, data is written using 'write data to CG RAM' command. The address gets auto incremented after each data is written, '1' for the CG RAM data in the character patterns correspond to dark pixel of the display. The 11th line is the cursor position and display is performed in logical OR with cursor. Since the 12th–16th lines are not used for display, they can be used for the general data RAM. Table 5.3 represents pattern of character code.

Table 5.3   CG RAM data for 5 × 7 dot character pattern

Character codes (DD RAM data)								CG RAM address						Character patterns							
														CG RAM address							
7	6	5	4	3	2	1	0	5	4	3	2	1	0	7	6	5	4	3	2	1	0
Higher					Lower			Higher			Lower			Higher					Lower		
										0	0	0	0	.	.	.	0	0	0	0	0
										0	0	0	1				0	0	0	0	0
										0	0	1	0				1	0	1	1	0
0	0	0	0	·	0	0	0	0	0	0	1	1	1				1	1	0	0	1
										0	1	0	0				1	0	0	0	1
										0	1	0	1				1	0	0	0	1
										0	1	1	0				1	1	1	1	0

(Contd)

(Contd)

	0 1 1 1	1 0 0 0 0
	1 0 0 0	1 0 0 0 0
	1 0 0 1	1 0 0 0 0
	1 0 1 0	0 0 0 0 0
	1 0 1 1	. . . . .
	1 1 0 0	. . . . .
	1 1 0 1	
	1 1 1 0	. . . . .
	1 1 1 1	

Instruction Code

The instruction code is a command set through which the display module is controlled by the controller. Prior to internal execution of the instruction code, the control information is temporarily stored in the internal register. The operation begins on receipt of the instruction code input from the microcontroller. The instructions are grouped together based on the function carried out such as:

1. Designate module functions such as display format, data length, etc.
2. Define internal RAM address.
3. Perform data transfer with internal RAM.
4. Others.

Normally, the third category of instructions are used frequently. Automatic incrementing (or decrementing) of the internal RAM address after each data write. The display shift is performed concurrently with display data write enabling the user to develop systems in minimum time with maximum programming efficiency. Table 5.4 depicts command code table, which represents the instruction codes for a typical display LCD module.

Table 5.4 Command/instruction code table for a typical LCD display module

R/W	D7	D6	D5	D4	D3	D2	D1	D0	Operation
0	0	0	0	0	0	0	0	1	Clear memory, home cursor
0	0	0	0	0	0	0	1	0	Clear and home cursor only
0	0	0	0	0	0	1	R/L	S	Screen action : S = shift cursor/screen [1/0] R/L : right/left shift [1/0]
0	0	0	0	0	1	D	C	B	D : Display ON/OFF [1/0] C : Cursor ON/OFF [1/0] B : Blink/non-blink [1/0]
0	0	0	0	1	S/C	R/L	0	0	S : screen C : cursor
0	0	0	1	DL	N	F	0	0	DL : Data length 8/4 [1/0] N : Number of rows 2/1 [1/0]
0	0	1	Character address						Write to character RAM address
0	1	Display data address							Write to display RAM
1	BF	Current address							BF : Busy/not busy [1/0]

Notations used in Table 5.4 are: S (screen), C (cursor), R (right), L (left), DL (data length; number of characters), N (number of lines; one or two), B (blinking), and the BF (busy flag).

Instruction Set of Typical LCD Module

The operations required for the control of display unit are as follows:

1. Clear display
2. Return home
3. Entry mode set
4. Display ON/OFF control
5. Cursor or display shift
6. Function set
7. Set display RAM address
8. Set CG RAM address
9. Read busy flag and address

Instruction codes can be found from the manual supplied by the manufacturer for the LCD selected. The operating frequency of the controller is around 250 kHz, therefore internal executing time is long. Standard time is 40 µs to 1.6 ms. Since the execution speed of the display module is much lower than its controlling microcontroller; it is necessary to check the busy flag before performing any operation with the module. While internal operation is active, the busy flag is '1'. At this time the only operation that can be performed is a read busy operation.

Clear display

	Rs	R/W	DB7	DB6	DB5	DB4	DB3	DB2	DB1	DB0
Code	0	0	0	0	0	0	0	0	0	1
			Higher order bits		Lower order bits					

When this instruction is executed, the LCD display is cleared and returned to its original status if it was shifted. The cursor goes to the left edge of the display (the left end of the first line if 2-line mode). Character pattern for character code '20', blank pattern is written into all DD RAM addresses. Sets DD RAM address 0 in address counter (AC). Sets I/D = '1' (increment mode) of entry mode.

Return home

	Rs	R/W	DB7	DB6	DB5	DB4	DB3	DB2	DB1	DB0
Code	0	0	0	0	0	0	0	0	1	-
			Higher order bits		Lower order bits					

The cursor or blinks goto the left edge of the display (to the left end of the first line in the 2-line display mode). The display returns to its original status if it was shifted. DD RAM contents do not change. Sets the DD RAM address 0 in address counter.

Entry mode set

	Rs	R/W	DB7	DB6	DB5	DB4	DB3	DB2	DB1	DB0
Code	0	0	0	0	0	0	0	1	1/D	S
			Higher order bits		Lower order bits					

I/D When the I/D is set, the 8-bit character code is written or read to and from the DD RAM, the cursor shifts to the right by 1 character position (I/D = '1'; increment) or to the left by 1 character position (I/D = '0'; decrement). The address counter is incremented I/D = '1' or decremented I/D = '0'. By 1 at this time, even after the character pattern code is written or read to and from the CG RAM, the address counter AC is incremented I/D = '1' or decremented I/D = '0' by 1.

S Shift the entire display either to the right when S is 1; or to the left when I/D = 1 and to the right when I/D = 0. It looks as if the cursor stands still and the display moves. The display does not shift when reading from the DD RAM or when writing into or reading out from the CG RAM when S = 0.

Display ON/OFF control

	Rs	R/W	DB7	DB6	DB5	DB4	DB3	DB2	DB1	DB0
Code	0	0	0	0	0	0	1	D	C	B
			Higher order bits		Lower order bits					

D The display is ON when D = '1' and OFF when D = '0'. When OFF due to D = '0', display data remains in the DD RAM. It can be displayed immediately by setting D = '1'.

C The cursor is displayed when C = '1' and goes OFF when C = '0'. Even if the cursor disappears, the function of I/D does not change during the display data write. The cursor is displayed using 5 dots in the 8th line when the 5×7 dot character font is selected and dots in the 11th line when the 5×10 dot character font is selected.

B The character indicated by the cursor blinks when B = '1'. The blink is displayed by switching between all black dots and display characters at 409.6 ms interval when f_{cp} or f_{osc} = 250 kHz. The cursor and the blink can be set to display simultaneously. The blink interval changes according to the reciprocal of f_{cp} or f_{osc}. It is $409.6 \times 250/270 = 379.2$ ms when f_{cp} = 270 kHz.

Cursor or display shift

	Rs	R/W	DB7	DB6	DB5	DB4	DB3	DB2	DB1	DB0
Code	0	0	0	0	0	1	S/C	R/L	–	–
									No effect	

Shift cursor position or display to the right or left without writing or reading display data. This function is used to correct or search for the display in a 2-line display, the cursor moves to the second line when it passes the 40th digit of the 1st line. Notice that the 1st and 2nd line displays will shift at the same time. When the displayed data is shifted repeatedly, each line only moves horizontally. The second line display does not shift into the first position.

S/C	R/L	
0	0	Shifts the cursor position to the left (AC is decremented by one).
0	1	Shifts the cursor position to the right (AC is incremented by one).
1	0	Shifts the entire display to the left (The cursor follows the display shift).
1	1	Shifts the entire display to the right (The cursor follows the display shift).

Address counter (AC) contents do not change if the only action performed is shift display.

Function set

	Rs	R/W	DB7	DB6	DB5	DB4	DB3	DB2	DB1	DB0
Code	0	0	0	1	DL	N	F	–	–	–
										No effect

DL (set interface data length)   When DL = '1', the data input/output to and from the microcontroller is carried out by means of 8-bits DB7 to DB0. When DL = '0', the data input/output to and from the MPU is carried out in two steps through the 4-bits DB7 to DB4.

N **(set number of display lines)**   The 2-line display mode of the LCD is selected when N = '1', while the 1-line display mode is selected when N = '0'

F **(set character font)**   The 5×7 dots character font is selected when F = '0', while the 5×10 dots character font is selected when F = '1' and N = '0'.

This instruction is executed at the start of the program. From this point the function set instruction cannot be executed unless the interface data length is changed, i.e., software reset is performed.

N	F	No. of display lines	Character font	Duty factor	Remarks
0	0	1	5×7 dots	1/8	-
0	1	1	5×10 dots	1/11	-
1	×	2	5×7 dots	1.16	Cannot display 2-lines with 5×10 dots character fonts

Set CG RAM address

	Rs	R/W	DB7	DB6	DB5	DB4	DB3	DB2	DB1	DB0
Code	0	0	0	1	A5	A4	A3	A2	A1	A0
					Higher order bits		Lower order bits			

Sets CG RAM address into the address counter in binary A5 to A0.

In the 5×10 font mode, A5 and A4 define the CG RAM block number while A3-A0 define the row within the block.

In the 5×7 font mode the CG RAM block is defined by A5-A3 while A2-A0 define the row. Tables 5.2 and 5.3 make this clear.

Set DD RAM address

	Rs	R/W	DB7	DB6	DB5	DB4	DB3	DB2	DB1	DB0
Code	0	0	1	B6	B5	B4	B3	B2	B1	B0
				Higher order bits		Lower order bits				

Sets the DD RAM address into the address counter in binary B6 to B0. Data is written or read from the DD RAM.

When N = '0' (1-line display). B6 to B0 represent: 00h–4Fh

When N = '1' (2-line display). B6 to B0 is 00h–27h for the first line and 40h–67h for the second line.

Read busy flag and address

	Rs	R/W	DB7	DB6	DB5	DB4	DB3	DB2	DB1	DB0
Code	0	1	BF	C6	C5	C4	C3	C2	C1	C0
				Higher order bits		Lower order bits				

Read the busy flag BF that indicates the system is now internally executing a previously received instruction. BF ='1' indicates that internal operation is in progress. The next instruction will not be accepted until BF goes '0'. Check the BF status before the next write operation.

The value of the address counter expressed in binary C6 to C0 is read. The address counter is used by both CG and DD RAM addresses, and its value is determined by the previous instruction.

Write data to CG or DD RAM

	Rs	R/W	DB7	DB6	DB5	DB4	DB3	DB2	DB1	DB0
Code	1	0	D	D	D	D	D	D	D	D
				Higher order bits		Lower order bits				

Writes binary 8-bit data DDDDDDDD to the CG or the DD RAM determined by the previous specification of CG RAM of DD RAM address setting. After write, the address is automatically incremented or decremented by 1 according to entry mode. The entry mode also determines display shift.

Read data from CG or DD RAM

	Rs	R/W	DB7	DB6	DB5	DB4	DB3	DB2	DB1	DB0
Code	1	1	D	D	D	D	D	D	D	D

Reads binary 8-bit data DDDDDDDD from the CG or DD RAM. The previous designation determines whether the CG or DD RAM is to be read. Before entering the read instruction, either the CG RAM or DD RAM address set instruction must be executed otherwise the first read data will be invalid.

The address set instruction need not be executed just before the read instruction when shifting the cursor-by-cursor shift instruction (when reading the DD RAM). The cursor shift instruction operation is the same as that of the DD RAM's address set instruction. After read, the entry mode automatically increases or decreases the address by 1. Display shift is not executed no matter what the entry mode is.

The address counter (AC) is automatically incremented or decremented by 1 after 'write' instructions to either CG RAM or DD RAM. RAM data selected by the AC then cannot be read out even if read instruction is executed.

Interfacing and Initialization of LCD Module

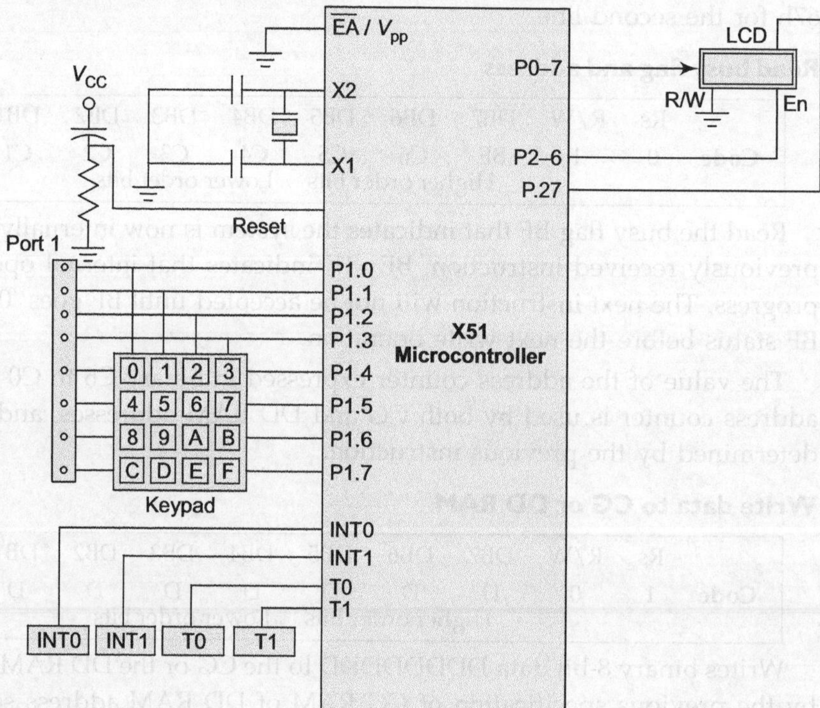

Fig. 5.23 Interfacing LCD module and a keypad with the microcontroller

Figure 5.23 shows the interfacing of a keypad and the LCD module. The port pins P0.0–7 are connected directly to the data pins of LCD module. The control pins of LCD, i.e., RS and En are connected to the port pins P2.6 and P2.7, respectively of the microcontroller. They can be SET/RESET as per requirement of the instruction code of the display operation. The initialization process is shown in the flowchart of Fig. 5.24.

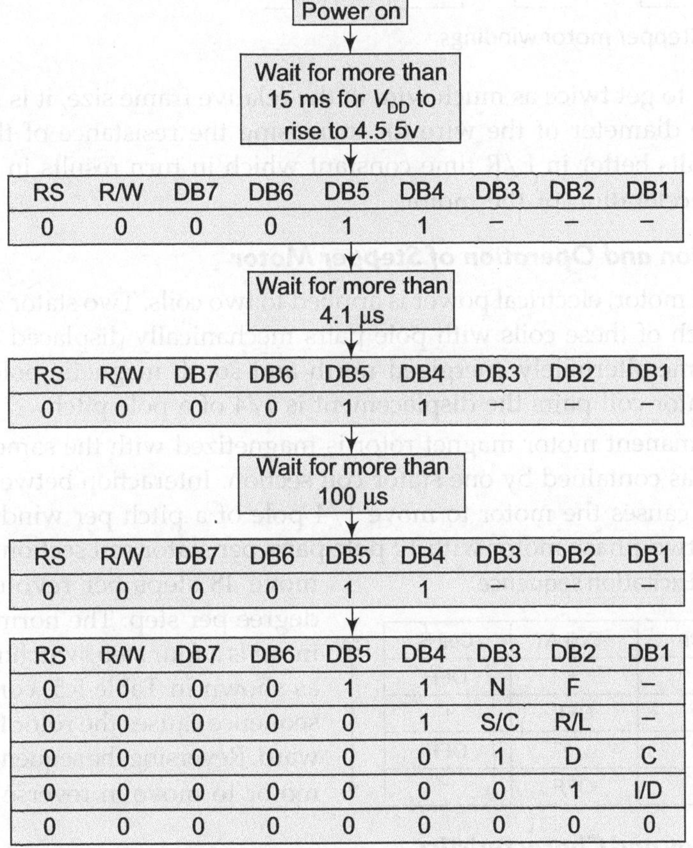

RS	R/W	DB7	DB6	DB5	DB4	DB3	DB2	DB1
0	0	0	0	1	1	–	–	–

RS	R/W	DB7	DB6	DB5	DB4	DB3	DB2	DB1
0	0	0	0	1	1	–	–	–

RS	R/W	DB7	DB6	DB5	DB4	DB3	DB2	DB1
0	0	0	0	1	1	–	–	–

RS	R/W	DB7	DB6	DB5	DB4	DB3	DB2	DB1
0	0	0	0	1	1	N	F	–
0	0	0	0	0	1	S/C	R/L	–
0	0	0	0	0	0	1	D	C
0	0	0	0	0	0	0	1	I/D
0	0	0	0	0	0	0	0	0

Fig. 5.24   Initialization of the LCD display module

5.3.4 Stepper Motors

A stepper motor is stepped from one position to the next by changing the currents through the fields in motor. The stepper motor used in the system has two phases, eight silent poles, toothed iron rotor, and permanent magnet with the windings bound in a bifialer method. The stator cup has two coils A1 and B1 (Fig. 5.25) on the same bobbin as compared to one coil in monofialer scheme.

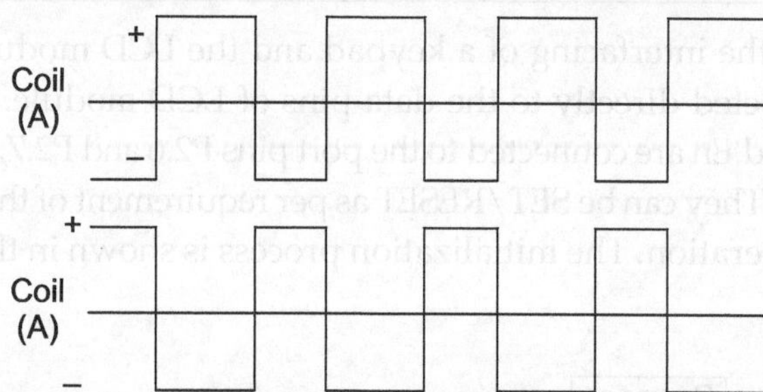

Fig. 5.25   Stepper motor windings

In order to get twice as much wire in the relative frame size, it is necessary to reduce the diameter of the wire for increasing the resistance of the winding, which results better in L/R time constant which in turn results in rapid acceleration/deceleration of the motor.

Construction and Operation of Stepper Motor

In a typical motor, electrical power is applied to two coils. Two stator cups formed around each of these coils with pole pairs mechanically displaced by 1/2 pole pitch become alternately energized north and south magnetic poles. Between the two stator-coil pairs the displacement is 1/4 of a pole pitch.

The permanent motor magnet rotor is magnetized with the same number of pole pairs as contained by one stator coil section. Interaction between the rotor and stator causes the motor to move 1/4 pole of a pitch per winding polarity change. A two-phase motor with 12 pole pairs per stator coil section would thus move 48 steps per revolution or 7.5 degree per step. The normal electrical input is a four step switching sequence as shown in Table 5.5, continuing the sequence causes the rotor to rotate forward. Reversing the sequence will cause motor to move in reverse direction.

Table 5.5   Excitation sequence

Step	Coil A	Coil B
1	+	OFF
2	OFF	+
3	−	OFF
4	OFF	−

Terminology and Characteristics

Residual torque   The non-energized dental torque of a permanent magnet (PM) stepper motor is called *residual torque*. It is the result of the PM flux and bearing friction. It has a value of approximately 1/10th of the holding torque. This characteristic of PM stepper motor is useful in holding a load in the proper position even when the motor is de-energized. The position however will not be held as accurately as when the motor is energized.

Ramping   Acceleration control or ramping is normally accomplished by gating on a voltage-controlled oscillator (VCO) and associated charging capacitor. Varying the RC time constant will give different ramping times. A typical VCO acceleration control frequency plot for an incremental movement with equal acceleration and de-acceleration time is shown in Fig. 5.26. Acceleration may also be accomplished by dividing the frequency. For example, the frequency could start at a 1/4th rate go to a 1/2 rate, finally at 3/4th running rate.

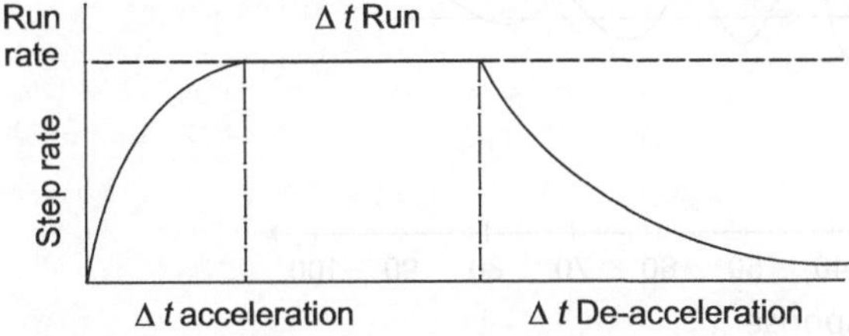

Fig. 5.26   Incremental movement plot

Dynamic torque   A typical torque versus step rate characteristic curve is shown in Fig. 5.27. The start without error curve shows what torque load the motor can start and stop without loss of a step when started and stopped at a constant step or pulse rate.

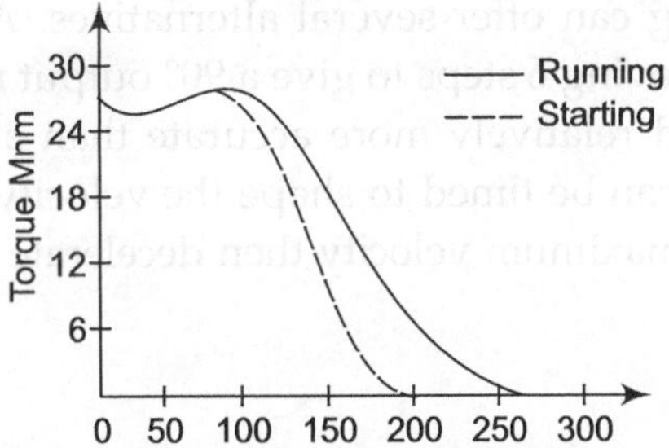

Fig. 5.27   Torque v/s step rate characteristics

The running curve is the torque available when the motor is slowly acce-lerated to the operating rate. It is also called slew curve. It is the actual dynamic torque produced by the motor. The difference between the running and the start without error torque curve is the torque lost due to accelerating the motor inertia. The speed–torque characteristic curves are the key to selecting the right motor and the drive method for specific application.

Resonance   If a stepper motor is operated with no-load over the entire frequency range, one or more natural oscillating resonance points may be detected, either audibly or by a vibration sensor. Some applications may be sensitive to these reso-nating frequencies and therefore, they should be operated with an external dam-ping source, an added inertia, or a software drive. A PM stepper motor however does not exhibit the instability and loss of steps often found in variable reluctance type of motors since the PM has higher rotor inertia and a stronger detent torque.

Single stepping   Figure 5.28 shows typical response of a motor when it moves by one step.

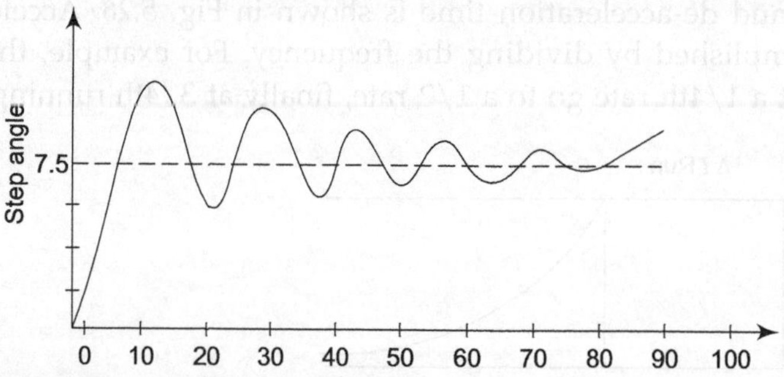

Fig. 5.28 Single step response

The actual response for a given motor is a function of the power input provided by the drive and the load. It can be modified by increasing the frictional load or adding external damping. Mechanical dampers such as slip pads or plates or device such as fluid coupled flywheel can be used but add to system complexity and cost. Electronic damping can also be accomplished.

Multistepping Multiple stepping can offer several alternatives. A 7.5° motor moving 12 steps or a 15° motor moving 6 steps to give a 90° output move would have less overshoot be stiffer and relatively more accurate than a motor with a 90° step angle. Also the pulses can be timed to shape the velocity of the slow motion during start accelerate to maximum velocity then decelerate to stop with minimum ringing.

Drive Method

The normal drive method is 4-step sequence as shown earlier. The following methods are used:

Wave drive In this method only one winding shown in Fig. 5.29 is energized at a time. It produces the same increment same as 4-step excitement.

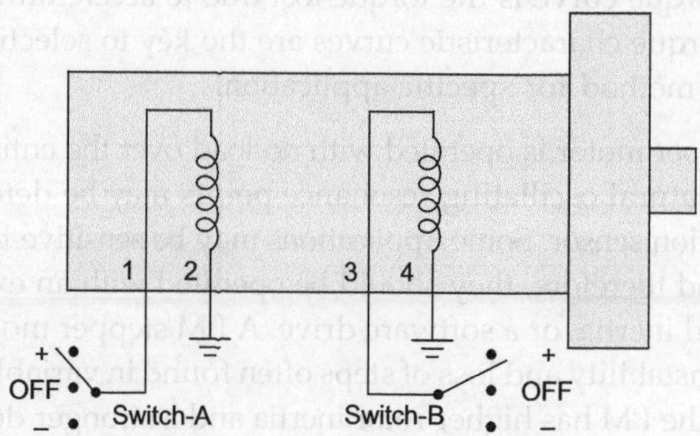

Fig. 5.29 Stepper motor wiring diagram

Table 5.6 Wave drive switching sequence

Step	Coil A	Coil B
1	+	OFF
2	OFF	+
3	−	OFF
4	OFF	−
1	+	OFF

Since only one winding is on the hold and running torque and the rated voltage applied would be reduced to 30% within limits. The voltage can be increased to bring output power back to near rated torque value. The advantage of this type of drive is increased efficiency, while the disadvantage is decreased step accuracy. Table 5.6 describes the wave drive switching sequence.

Half step It is possible to step the motor according to a 8-step sequence to obtain a half step as 0.9 movement for a motor of 1.8 full step rotation with 200 steps/revolution. The holding torque will be energized for a step portion but on the next step two windings are energized. This gives the effect of a strong step and a weak step. Since the windings and flux conditions are not similar for each step when 1/2 stepping accuracy will not be as good as one with the full stepping method. Half step or eight-step switching sequence is shown in Table 5.7.

Table 5.7 Half step switching sequence

Step	Coil A	Coil B
1	+	+
2	+	OFF
3	+	−
4	OFF	−
5	−	−
6	−	−
7	−	+
8	OFF	+
1	+	+

Power Drivers

The signals produced by the logic sequencer are so weak that they are unable to energize the motor winding, power drivers are used to raise the power level sufficiently to energize the motor windings. Power transistors or thyristors are used depending on the required power level.

A motor operated at a fixed rate voltage has a decreasing torque curve as the frequency or step rate increases. This is due to the fact that the rise time of the coil limits the percentage of power actually delivered to the motor. This effect is governed by the inductance to resistance ratio of the circuit L/R. Compensation for this effect can be either increasing the power supply voltage to maintain constant current as the frequency increases or by raising the power supply voltage to maintain constant current. As the L/R is changed more total power is used by the system the series, resistor R are selected for the L/R ratio desired. For $L/4R$ they are selected to be 3 times the motor winding resistance with

$$\text{Watts rating} = (\text{Current per winding}) \times R$$

The power supply voltage is increased to 4 times motor rated voltage so as to maintain rayed current to the motor than a bipolar motor. To minimize power consumption various devices such as a bilevel power supply or chopper may be used.

The equivalent circuit of a motor winding connected to the motor power supply through the power transistor T1 is shown in Fig. 5.30.

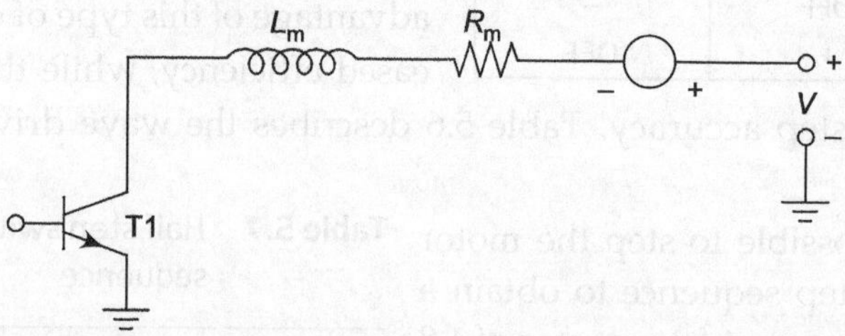

Fig. 5.30   Equivalent circuit of stepper motor windings

This is a typical inductive circuit in which the current rise is governed by the time constant

$$T_e = \frac{L_m}{R_m}$$

Since, the motor torque falls as the stepping rate increases, use of special current forcing circuits is made to cause the current to rise rapidly in the motor windings and thereby to raise the motor torque at high stepping rates.

Current forcing circuits   Four types of current forcing circuits are described in this section.

Series resistance drive   A resistance R_s is connected in series with the motor winding as shown in Fig. 5.31. The winding time constant then reduces to

$$T_e' = \frac{L_m}{R_m + R_s}$$

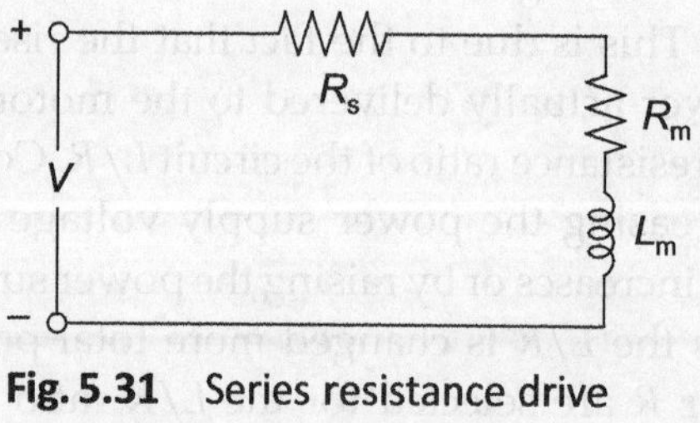

Fig. 5.31   Series resistance drive

The winding time constant is reduced, which causes the current in the winding to rise faster. The disadvantages of this drives are

1. The voltage to be applied to the combination of motor winding and series resistance has to be raised in order to have the same current through the motor winding.

2. There is a substantial power dissipation $= I^2 R_s$ in the series resistance.

3. This drive is, consequently, suitable for small stepper motors.

Bilevel drive   The bilevel drive allows the motor at zero steps/sec to hold at a lower than rated voltage. It is most efficient when operated at a fixed stepping rate. The high voltage may be switched ON through the use of a current sensing resistor, which uses the inductively generated turn OFF current spikes to control the voltage. At zero steps/sec the windings are energized by the low voltage.

In this drive, a high voltage = 4 V to 10 V, where V = rated voltage, is applied to the motor winding in the beginning as shown in Fig. 5.32. The current therefore rises very rapidly towards 4 to 10 times rated current. When the motor current reaches the rated value, the high voltage supply is switched OFF, while the rated voltage supply V is switched ON. Thus, the motor current stabilizes at rated current. The drawback of the scheme is that two regulated power supplies are required.

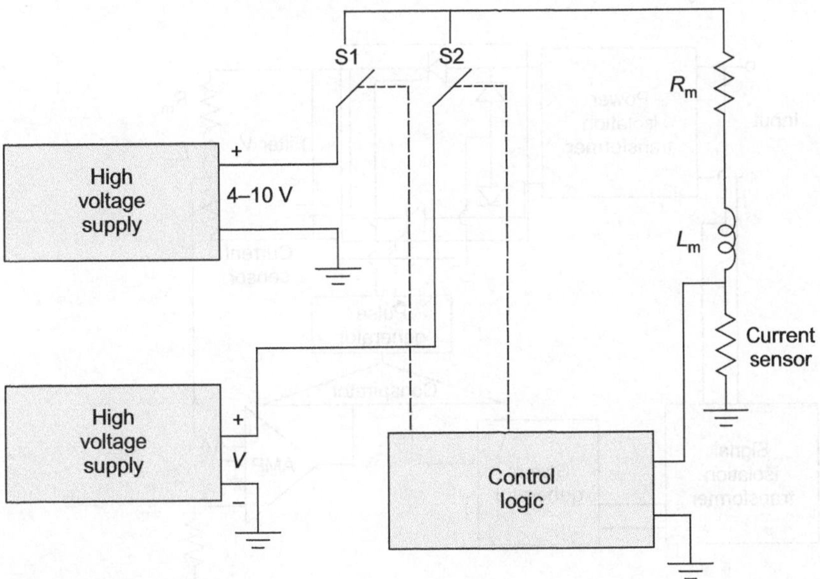

Fig. 5.32   Bi-level drive

Chopper drive   Figure 5.33 shows the chopper drive. It uses a high voltage supply 4 to 10 times the rated voltage, so that the current rises quickly. When the current has reached about $0.99 \times I$, the power supply is again switched ON. The operating frequency of the chopper is of the order of the kHz.

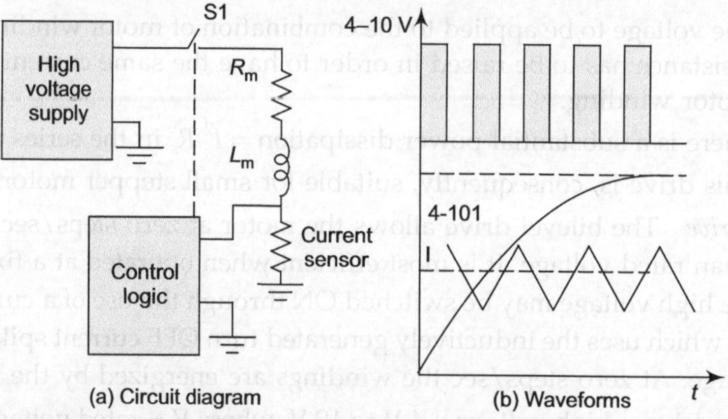

Fig. 5.33 Chopper drive

Constant current drive Block diagram of the constant current drive is shown in Fig. 5.34. In this drive, a controlled converter and a filter is used to produce a DC output, which is applied to the motor winding. The current sensing resistor senses the motor current, compares with the set current to find the difference. The difference is used to adjust the firing angle of the controlled converter in such a way as to maintain the constant current.

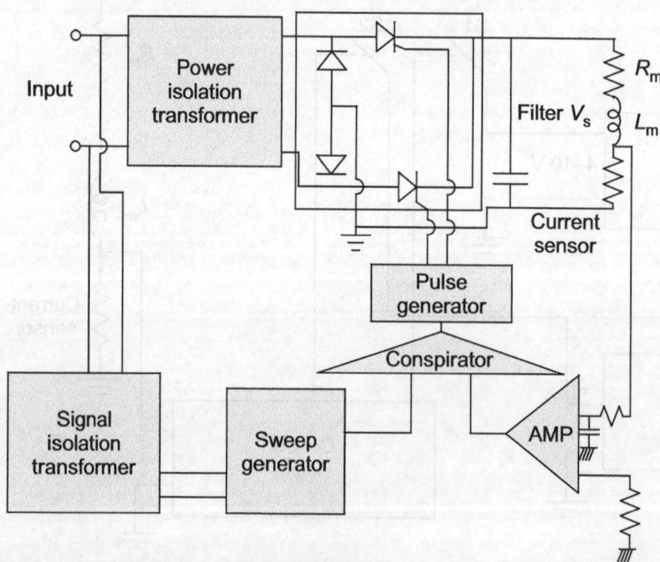

Fig. 5.34 Constant current drive

Current Suppression Circuits

Whenever winding current is turned OFF a high voltage inductive spike will be generated, which could damage the drive circuit. The normal method used to suppress these spikes is to put a diode across each winding. This reduces the torque output of the motor unless the voltage across the switching transistors is allowed to build up to at least twice the supply voltage. The higher this voltage the faster the induced field and current will collapse and thus better performance.

It is important to ensure a rapid of current when the stepper motor winding is switched OFF. The current suppression circuits shown in Fig. 5.35(a) are employed. The freewheeling diode [Fig. 5.35(a)] ensures continuity of current, the rough the stepper motor winding when it is switched OFF.

Connecting a resistance R_f in series with the freewheeling diode as shown in Fig. 5.35(b) causes a more rapid decay of current inasmuch as the decay time constant of the windings.

The best method of ensuring that the current decays to zero in a definite time T_d is to connect a zener diode in series with the freewheeling diode as shown in Fig. 5.35(c).

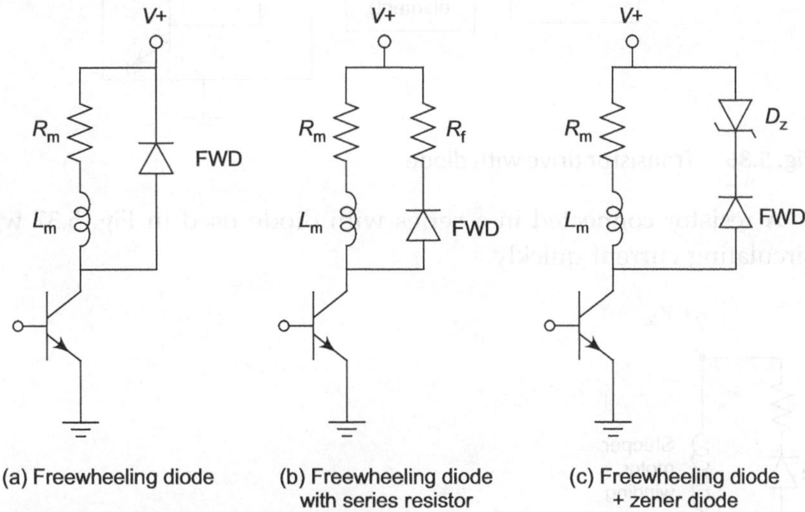

Fig. 5.35   Current suppression circuits

Interfacing Microcontrollers

The motor RPM range required is 30 to 120 RPM, which is a low RPM range. Thus the most suitable drive circuit at low RPM range is an L-R drive in which the output signals from the microcontroller ports are transmitted to the input terminals of the power transistors, which controls the switching ON/OFF of the motor windings.

A problem with this type of switching described above is that when a power transistor is turned OFF, a high-voltage builds up due to $L\,(di/dt)$, which may damage the transistor.

The surge in the voltage is suppressed by connecting a diode in parallel with the winding with polarity shown in Fig. 5.36. There is a flow of circulating current after the transistor is turned OFF and the current will decay with time. The circulating current lasts for a considerable length of time and it produces a breaking torque delivered by the motor.

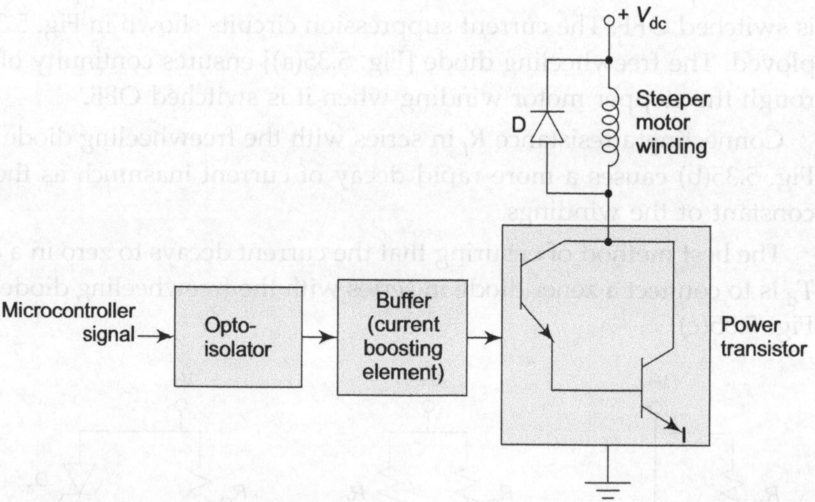

Fig. 5.36 Transistor drive with diode

A resistor connected in a series with diode used in Fig. 5.37 will damp the circulating current quickly.

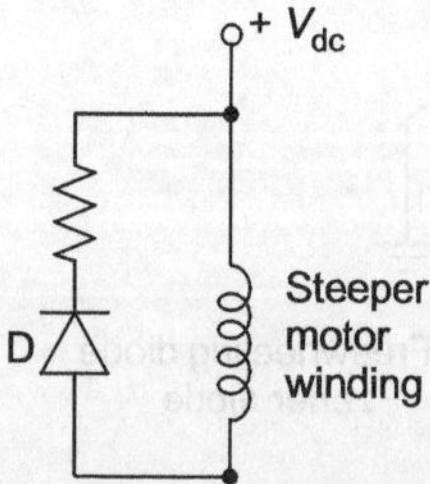

Fig. 5.37 Diode and series resistor

Input/output circuits The input/output circuitry of the system plays a major role in interconnecting the sensor outputs, and the motor power drive and relay inputs with the microcontroller port inputs and output, respectively.

All the input/output lines of the microcontroller are optically isolated using MCT2E which is a NPN silicon planar phototransistor optically coupled to a gallium arsenide infrared light emitting diode. The fact that the system is to be installed on a machine, which will be working in a workshop environment where numerous noise sources like high-voltage lines, electrical noise generated due to switching of other machines, etc. exist, leads to the obvious choice of isolating the I/O lines to ensure safety as well as proper working of the system.

MCT2E is a very popular opto-isolater with high isolation voltage of 2500 V rms for 1 min and surge isolation voltage of 4000 V for 1 s. Figure 5.38 depicts the usage of MCT2E in conjunction with an input digital line.

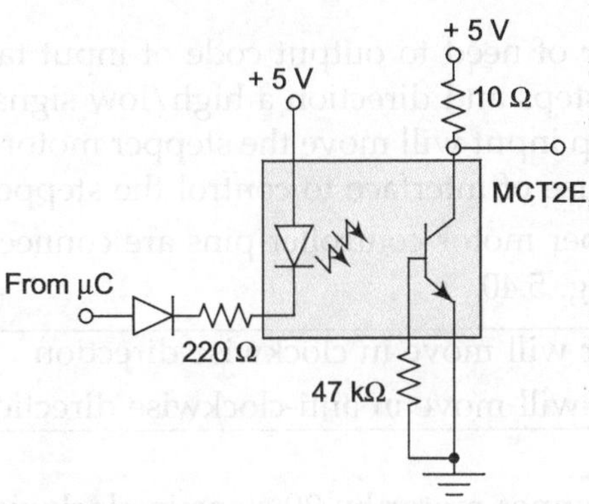

Fig. 5.38 Opto-isolated output line

Figure 5.39 depicts the usage of MCT2E in conjunction with an output digital line.

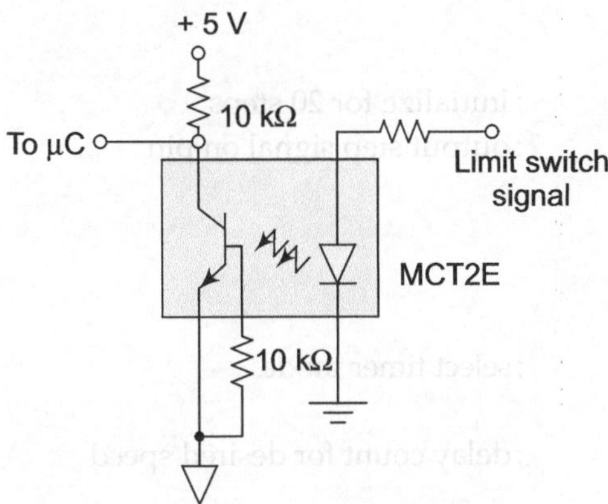

Fig. 5.39 Opto-isolated input line

Stepper Motor Driver Interface

Direct drive method of motor control requires that microcontroller must be dedicated to diving stepper motor. Hardware solutions are available in the form of interfacing an IC 4CN42048. Pin diagram and the interfacing of typical intelligent stepper motorcontroller is shown in Fig. 5.40.

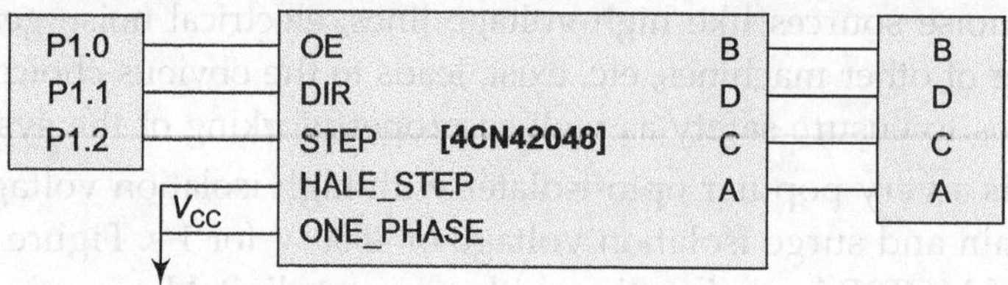

Fig. 5.40 Stepper motor interface

This relieves microcontroller of need to output code of input table. The circuit carries out two functions steps and direction a high/low signal on the pin will control the direction. A step input will move the stepper motor by one step. Sample programs explain the use of interface to control the stepper motor.

Let us assume that the stepper motor controller pins are connected with the microcontroller as shown in Fig. 5.40.

> *DIR pin is high:* The motor will move in clockwise direction
> *DIR pin is low:* The motor will move in anti-clockwise direction

Example 5.2   Write a program to move stepper motor by 20 steps in clockwise direction.

Solution

```
                    dir equ 91
                    step equ 90

                                    ;motor to move in clockwise direction
                    dir_clockwise:
        init:   clr dir
                    mov r0, #14h        ;initialize for 20 steps
        loop:   setb  step             ;output step signal on pin
                    clr step
                    acall  delay
                    djnz r0, loop.
                    end.
        delay:  mov tmod, #21h         ;select timer mode
                    mov tL0, #0Bh
                    mov tH0, #00h        ;delay count for desired speed
                    clr tf0
                    setb tro             ;turn timer on rollover
        wait:   jnb tf0, wait
                    clr tro              ;turn OFF timer
                    ret.
```

Example 5.3   Write a program to move stepper motor by 20 steps in anti-clockwise direction.

Solution

```
        dir equ 91
        step equ 90
                            ; motor to move in anti-clockwise direction
dir_anti_clockwise :
   init: setb dir
         mov r0, #14h        ; initialize for 20 steps
   loop: setb step
         clr step
         acall delay
         djnz r0, loop.
         end.
   delay:
         mov tmod, #21h      ; select timer mode
         mov tL0, #0Bh
         mov tH0, # 00h      ; delay count for desired speed
         clr tf0
         setb tro            ; turn timer on rollover
   wait: jnb tf0, wait
         clr tro             ; turn OFF timer
         ret
```

Example 5.4   Write a program to move stepper motor by 20 steps in clockwise direction using direct interface.

Solution

The motor can use single or half stepping. This may be implemented using a look-up table for the same. The program uses half stepping sequences. The sequence is shown in the following table:

Step	B3	B2	B1	B0
1	1	0	1	0
2	1	0	0	0
3	1	0	0	1
4	0	0	0	1
5	0	1	0	1
6	0	1	0	0
7	0	1	1	0

```
            dir equ 20h
            org 2100h
start:      ljmp init
step_tab:db 0ah, 08h, 09h, 05h, 01h, 05h, 04h,06h,02h
Init:       clr dir                 ; set for clockwise rotation
            mov r1, #14h
            mov dptr, #step tab     ; initialize pointer to the top address of
                                    ; look-up table

rept_step:
            mov r0, #00h            ; counter for table
next_seq :
            mov a, r0               ; get the count
            movx a, @a + dptr
            mov pl, a               ; output the code
            acall delay
            inc r0
            mov a, r0
            cjne a, #08h, next_seq
            djnz r1, rept_step
            end
```

Example 5.5   Write a program to move stepper motor by 20 steps in anti-clockwise direction using direct interface.

Solution

Look-up table of Example 5.4 is traversed in reverse direction.

```
            dir equ 20h
            org 2100h
start:      ljmp init
step_tab:db 0ah, 08h, 09h, 05h, 01h, 05h, 04h, 06h, 02h
init:       setb dir                ; set for anti-clockwise rotation
            mov r1, #14h
rept_setp:
            mov dptr, #step_tab     ; initialize dptr to the base of look-up
                                    ; table
            mov r0, #08h            ; counter for table
next_seq:
            mov a, r0               ; get the count
            movx a, @a + dptr
```

```
mov  pl, a                    ;output the code
acall  delay
djnz  ro, next_seq
djnz  r1, rept_step
end
```

5.3.5 Shaft Encoders

A shaft encoder is a device that converts the position or rotation of a shaft into a digital signal. The shaft encoders can be classified into two groups—incremental and absolute.

Incremental Encoder

A disk is attached to the motor shaft. Disk area is divided into the concentric circular strips. As shown in Fig. 5.41, a typical incremental encoder has three strips.

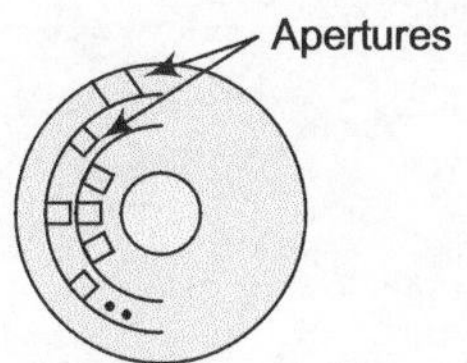

Fig. 5.41   A typical incremental encoder

First strip is called an index track 0. It produces one pulse per revolution. Other two tracks 1 and 2 produce two pulse trains, which are 90° apart. Counting the index pulses per minute gives rpm. Counting pulses on the other tracks with respect to the index pulse gives the position of the shaft. Sensing the time sequence of the pulses on the two tracks (1–2) gives the direction of rotation. The disk is rotated between a setup of an optocouplers provided for all three tracks. Pulse sequence gives desired information. Figure 5.42 shows the timing diagram.

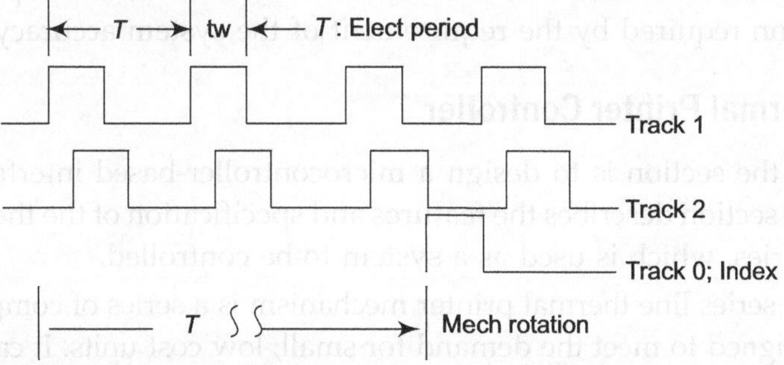

Fig. 5.42   Timing diagram of incremental encoder

Absolute Encoder

Figure 5.43 depicts diagram of an absolute encoder.

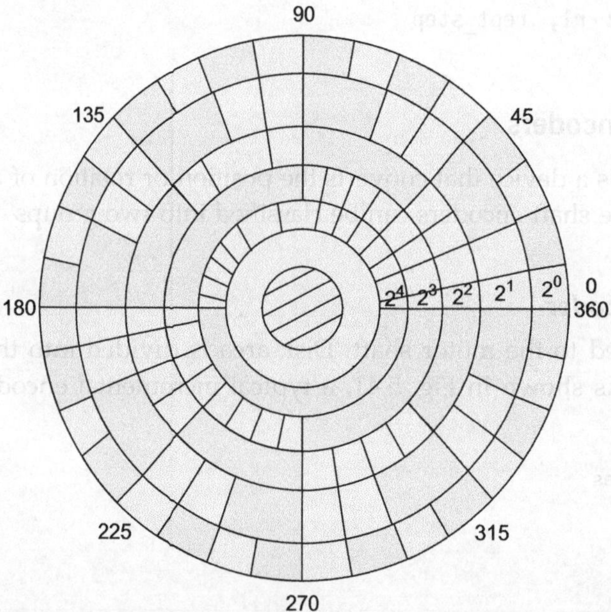

Fig. 5.43   Absolute encoder

Absolute encoders provide a unique binary number corresponding to the position of the shaft using a gray code described in Section 4.3.2. The advantage of gray code is that between any two adjacent positions there is a change of 1-bit, which reduces chances of I/O error.

Figure 5.43 shows a disk divided into five tracks. These five tracks represent a code. The inner track represents LSB of the code, while the outermost tracks represent MSB of the binary value. The tracks are subdivided and marked as black or white indicating Logic 0 or 1, respectively. Thus, any position of the disk is coded as a 5-bit code. This can be extended 6- to 8-bits depending upon the resolution required by the requirement of the system accuracy.

5.3.6 Thermal Printer Controller

The aim of the section is to design a microcontroller-based interface for serial printer. The section describes the features and specification of the thermal printer MTP 201 series, which is used as a system to be controlled.

The MTP series line thermal printer mechanism is a series of compact, thermal printers designed to meet the demand for small, low cost units. It can be used in adding machines, instruments and analyzers, office machines, medical apparatus, and data terminal devices. The MTP series line thermal printer mechanism has the features which include: low cost, maintenance-free, high quality printing, battery drive, silent, non-impact system, and high reliability. They are used in clinics to have practically noiseless printing.

System Specification

Part Number

    Character Printers

    MTP 111-223   1. Base model code

                             2. Characters per line

                             3. Print head type used

    Graphic Printers

    MTP 111-3444   1. Base model code

                             3. Print head type used

                             4. Maximum number of dots per row per line

Motor drive characteristics: The motor is the power source for head movement and paper feed. Through the application of DC voltage to the motor, the head automatically moves back and forth, and paper fed its return: therefore, there is no need reverse the motor.

Tacho-generator output characteristics: The TG is a signal generator connected to the motor that generates false sine waves of cycles per 1 turn of the motor. These waves are converted by a wave from shaping circuit into rectangular wave signals, which are used as a pulse impression-timing signal for the thermal head.

Home switch: The home switch is a push-open type mechanical switch switches "OFF" when the head is at the home position (left end). The switch has two functions. The first is to detect the home position when the head stops there. Upon returning to the home position, the home switch switches from ON to OFF, and brakes are applied to stop the motion to the print head.

The second function is, it is used as reference point for starting the counting of the print timing signals. As the head begins to move to the right from the home position, the home switch switches from OFF to ON, at which point counting of the print timing signal begins.

Thermal print head: Because the thermal print head in the MTP series is a thin film type with high thermal efficiency, it has an excellent print quality. The microcontroller determines which of the vertically placed heating units to energise, and drives them with the thermal head driver (transistor array).

Continuous printing operation: This can be done using home switch signal method or timing signal count method.

System Hardware

Figure 5.44 shows block diagram of a B head mounted printer interface.

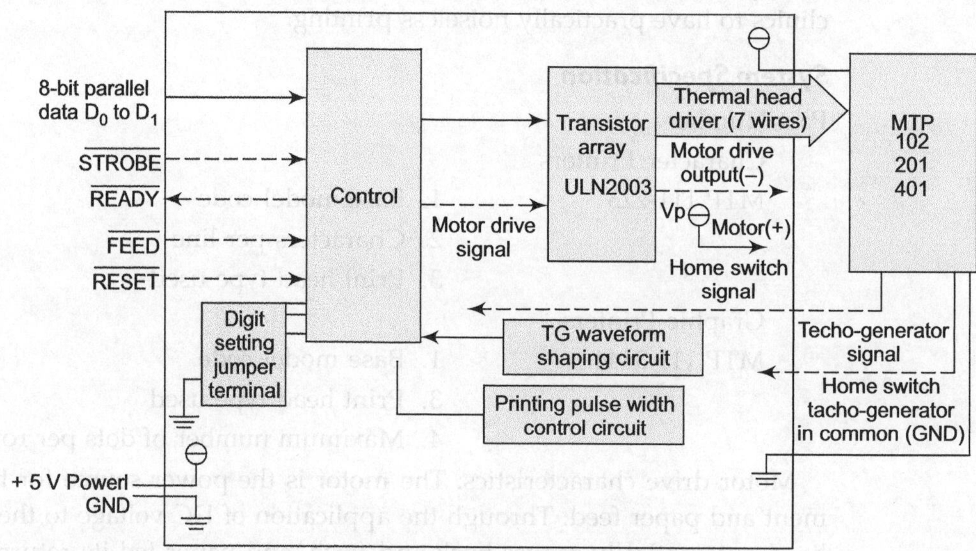

Fig. 5.44   Block diagram of B head mounted thermal printer

It consists of a power supply, controller board, PWC circuit, and the transistor array. We will study each component in details.

Power supply   Figure 5.45 shows a full wave bridge rectifier circuit and regulator, which generate 5 V at the output. Diodes IN4007 (500 mA) are used to build the bridge. Diodes IN5408 are required if the current requirement of the circuit is around 1 A.

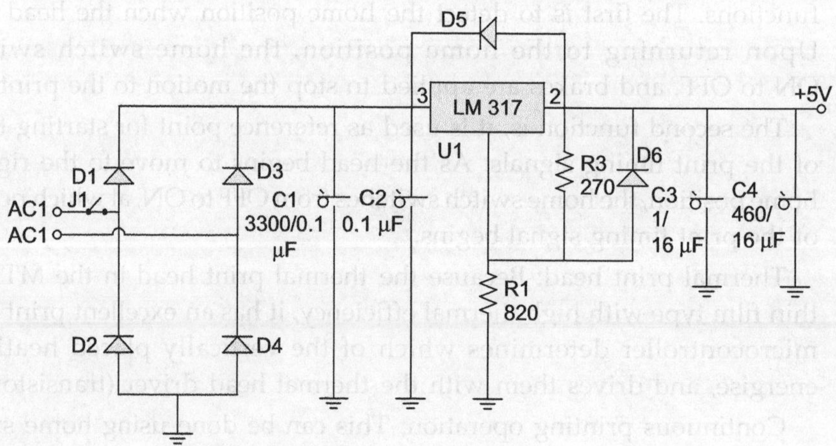

Fig. 5.45   Circuit diagram of the power supply

Since the current requirement during firing of dots and driving the motor is high. Value of capacitor C1 is high, i.e., 3300 μF. The capacitor C2 is for by passing high frequency.

LM317 is an adjustable 3-terminal positive voltage regulator capable of supplying over a 1.2 V to 37 V output range. It is exceptionally easy to use and require only two external resistors to set the output voltage. Normally no capacitors are needed unless the device is situated far from the input filter capacitors. An optional output capacitor C 3 μF is used to improve transient response. LM317 gives variable output voltage depending on value of the resistors R1 and R3.

Diodes N5 and N6 are protection diodes. When external capacitors are used with any IC regulator, it is sometimes necessary to add protection diodes to prevent the capacitors from discharging through low current points into the regulator. When the input is shorted, the output capacitor will discharge into the output of the regulator and can damage the IC. To prevent it, diodes N5 and N6 are used across LM317.

Controller card   The microcontroller 8751 is used. The chip 62C256 has been used as external RAM 32 K. The IC MAX232C provides a serial link. Figure 5.46 shows detail circuit diagram of the controller card.

PWC circuit   Figure 5.47 shows a typical PWC. The circuit generator square wave pulses whose width can be varied using a potentiometer. IC4069 can also be used instead of this circuit.

ULN2003 (7-Transistor array)   Figure 5.48 shows the transistor array used to fires the respective dots: It is a high voltage, high current Darlington transistor array. It has output clamp diode.

74HC240 is an octal inverter. If the microcontroller 8751 remains in reset mode for a longer time, input of ULN2003 will be at a logic level HIGH. In absence of 74HC240, this high on all lines will fire all dots for a longer time and may damage the thermal head.

The pull up resistors R15 to R22 are used so that the I/Ps to ULN2003C will not be floating, i.e., either high or low. A high on the line will cause the firing of the respective dot for a longer time. This may damage the thermal head. However, when we use these pull up resistors in a floating condition, we get lows at I/P of ULN2003 and no dots are fired. This protects the thermal head.

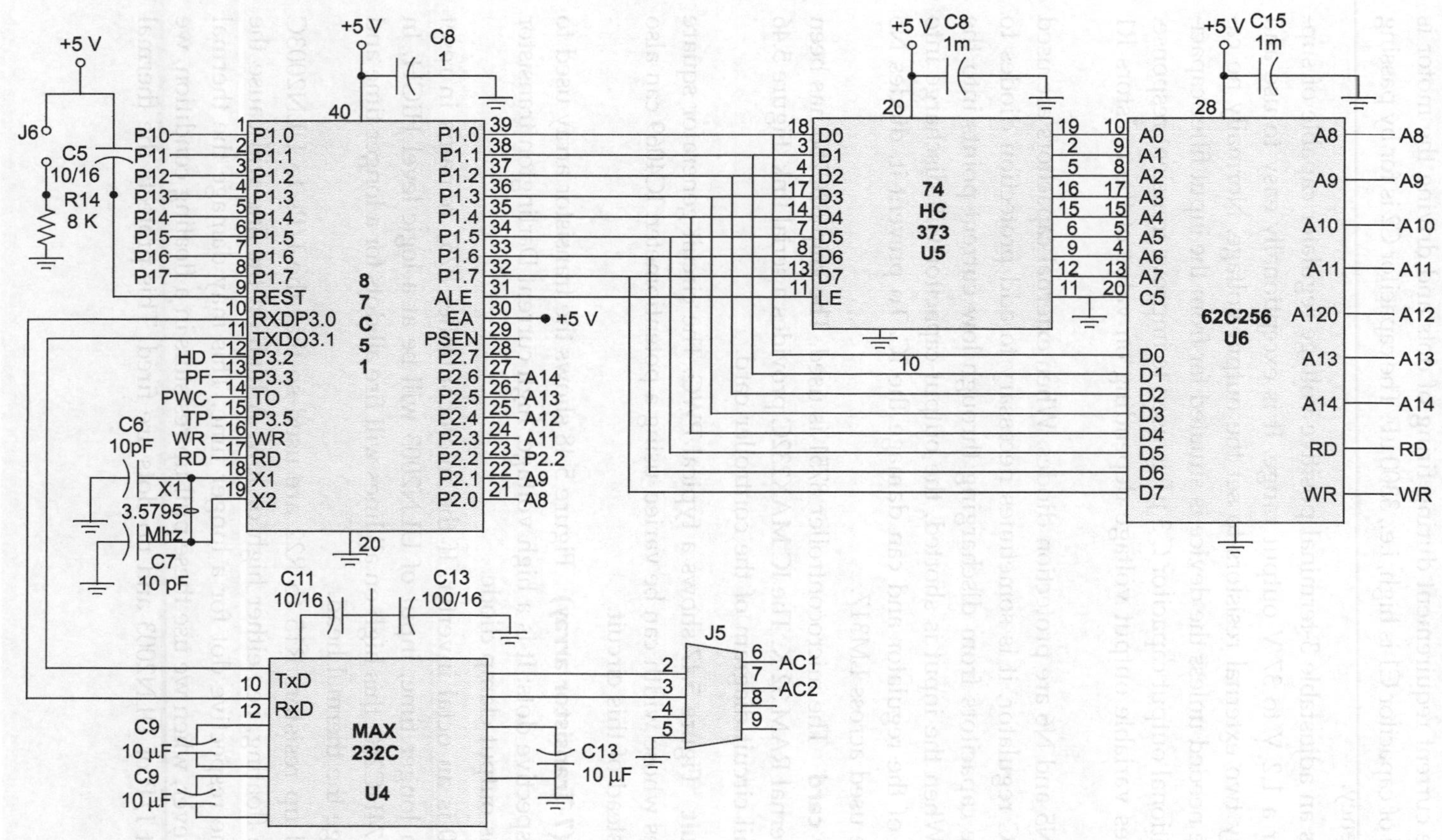

Fig. 5.46 Detail circuit diagram of the controller card with serial link

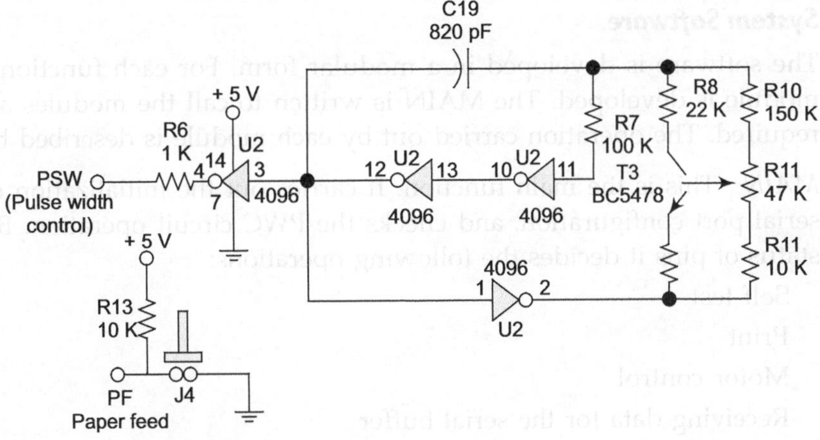

Fig. 5.47 Pulse width control circuit

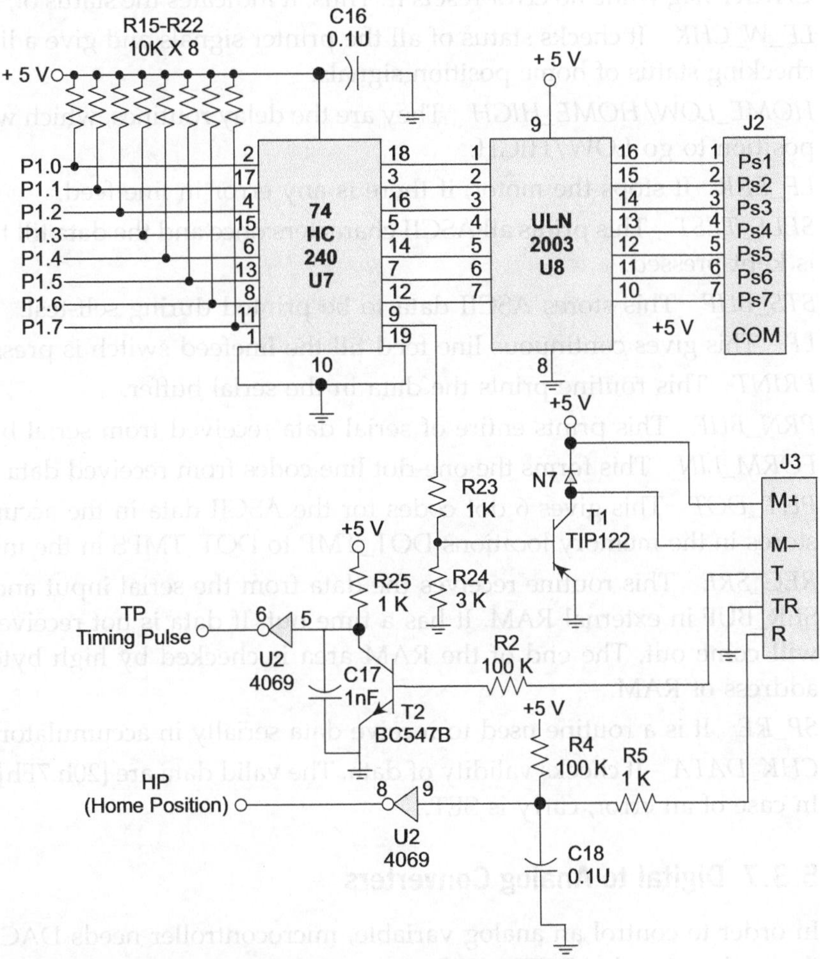

Fig. 5.48 Transistor array interfaced with PWC

System Software

The software is developed in a modular form. For each functional portion a module is developed. The MAIN is written to call the modules as and when required. The operation carried out by each module is described below.

MAIN   This is the main function. It carries out the initialization of the ports, serial port configuration, and checks the PWC circuit operation. Based on the status of pins it decides the following operations:

   Self-test

   Print

   Motor control

   Receiving data for the serial buffer

CHK_PWC   This routine checks if PWC signals are coming or not. Error will set CARRY flag while no error resets it. Thus, it indicates the status of, PWC pulses.

LF_W_CHK   It checks status of all the printer signals and give a line feed after checking status of home position signals.

HOME_LOW/HOME_HIGH   They are the delay routines, which wait for home position to go LOW/HIGH.

LF_ERR   It stops the motor, if there is any error in line feed.

SELF_TEST   This prints all ASCII characters once and the data till the LF switch is kept pressed.

STS_BUF   This stores ASCII data to be printed during self-test.

LF   This gives continuous line feed till the linefeed switch is pressed.

PRINT   This routine prints the data in the serial buffer.

PRN_BUF   This prints entire of serial data received from serial buffer.

FORM_LIN   This forms the one-dot line codes from received data for printing.

PUT_DOT   This gives 6 dot codes for the ASCII data in the accumulator and stores in the memory locations DOT_TMP to DOT_TMPS in the internal RAM.

REC_SRE   This routine receives the data from the serial input and stores it in SER_BUF in external RAM. It has a time out. If data is not received in time, it will come out. The end of the RAM area is checked by high byte of the last address of RAM.

SP_RE   It is a routine used to receive data serially in accumulator.

CHK_DATA   It checks validity of data. The valid data are [20h.7Fh], 0Dh, 0E4h. In case of an error, carry is SET.

5.3.7 Digital to Analog Converters

In order to control an analog variable, microcontroller needs DAC to generate the analog signal. ADC558/557 is a typical 8-bit DAC, which can be connected directly to the 8-bit microcontrollers. Figure 5.49 depicts the internal block diagram of the ADC557/558 DAC chip.

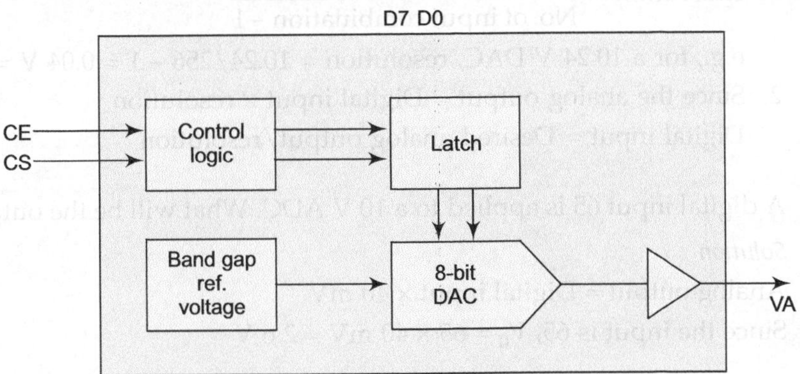

Fig. 5.49 Internal block diagram of DAC 557/558

The built-in latch allows output from the microcontroller to be strobe in. It accepts 8-bit digital pattern and produces the output voltage signal in the range (0–2.56 V). The control logic provides the strobe signal to the latch. The band gap reference is an internal reference voltage; hence there is no need to use high precession power supply. Standard power supply can be used. It can be operated in either uniploar or bipolar mode.

A 1-bit change in the binary input causes the output voltage to change by 2.56/256 = 10 mV.

The important parameters to be considered for selection of DAC chip are:

Accuracy It is the comparison between the actual output and expected output. It is the percentage of full-scale defection, e.g., (10 + 0–2%) % specifies that the maximum error may be 0.002 × 1 = 20 mV.

Linearity It is a measure of how much the output ramp deviates from the line 5 to if the input is varied from 0 to 1. In many applications it is necessary to have better resolution for low output but it does not matter how much output is nonlinear.

Setting time (t_s) When the output is changed, it will change to a new value. This may overshoot the correct value or may ring around desired value. The time required by the output voltage to settle at (+1/2 LSB) of final value is called the *setting time*. This is important as it may not be possible to operate (DAC) above this value.

Output voltage setting on DAC It is possible to calculate the digital input required to produce the desired analog output voltage from DAC, as follows:

1. Compute resolution of the DAC chip.

$$\text{Resolution} = \frac{\text{Voltage range}}{\text{No. of input combination} - 1}$$

e.g., for a 10.24 V DAC, resolution = 10.24/256 − 1 = 0.04 V = 40 mV

2. Since the analog output = Digital input × resolution

   Digital input = Desired analog output/resolution

Example 5.6   A digital input 65 is applied to a 10 V ADC. What will be the output of DAC?

Solution

Analog output = Digital input × 40 mV

Since the input is 65, V_0 = 65 × 40 mV = 2.6 V

Example 5.7   What should be the input applied to the ADC to obtain output of 8 V from the 10 V DAC?

Solution

Analog output = Digital input × resolution

Digital input = Desired analog output/resolution

For a 10 V DAC

Resolution = 10/(256 − 1) H" 0.04 V = 40 mV

Digital input = Desired analog output/resolution

= Desired analog output/0.04 = 8/0.04 = $(200)_{10}$ = C8 h

The digital input pattern to be applied at the input of DAC is 11001000.

5.3.8 Applications of DAC

The DAC can be used to generate waveforms as well as to realize ADC converters.

Waveform Generation

DAC can be used to generate any type of time varying waveform. The hardware setup for the purpose is shown in Fig. 5.50.

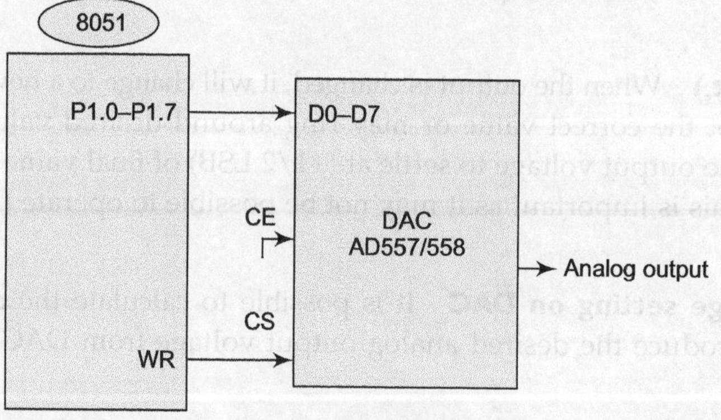

Fig. 5.50   Hardware setup to generate waveforms using DAC chip

The Port 1 pins are connected to the data in DAC. When any input is given to DAC, it will be converted into analog value. Based on the waveform to be generated the required digital value can be calculated. These instantaneous values may be stored in the form of a look-up table. The software and the look-up table changes for different waveforms, hardware is same.

Example 5.8 Write a program to generate square waveform shown in Fig. 5.51

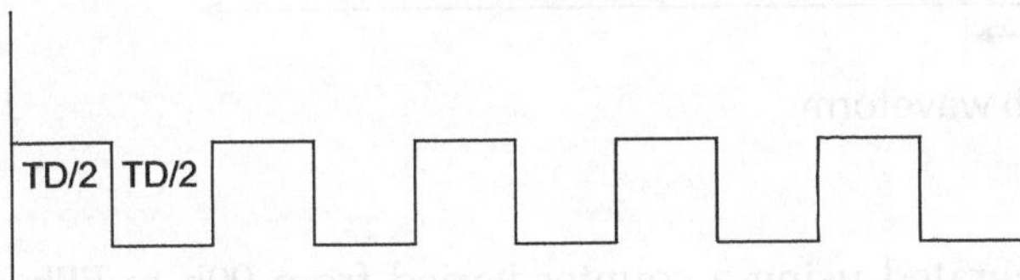

Fig. 5.51 Square waveform

Solution

The square waveform requires that the output is high for half of the period and low for next half. This is to be repeated. Let us assume that a delay subroutine is written, based on the frequency of the waveform required. As shown in Fig. 5.51 the output remains HIGH for $T_d/2$ period and LOW for next $T_d/2$. The delay for a period $T_d/2$ can be written using methods studied in Chap. 4.

The steps are as follows:

1. Load the value FFh in the accumulator corresponding to maximum voltage and output at Port p1.
2. Wait for T_d seconds.
3. Load 00h in the accumulator corresponding to zero (minimum) output, and output at Port p1.
4. Wait for T_d seconds
5. Goto Step 1

This will generate continuous square wave. The program segment is as follows:

```
sqr_gen :
        mov 80, #20             ;initialize stack pointer
rept:
        mov a, FFh
        mov P1, a               ;output Logic 1 for max output
        lcall delay             ;call delay for Td/2 seconds
        mov a, 00h
        mov P1, a               ;output Logic 1 for max output
        lcall delay             ;call delay for Td/2 seconds
        sjmp rept
```

Example 5.9   Write a program to generate saw tooth waveform shown in Fig. 5.52.

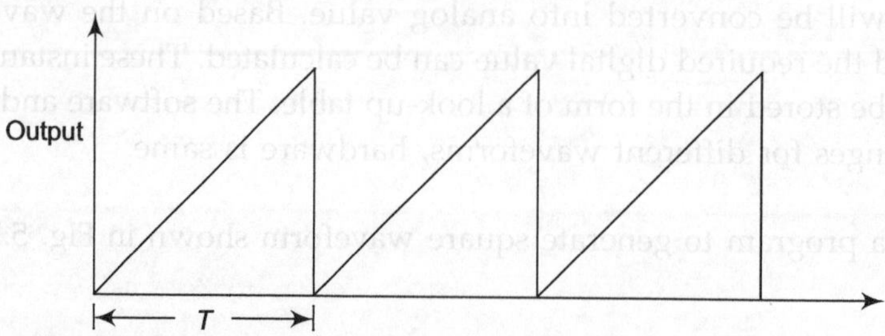

Output

Fig. 5.52   Saw tooth waveform

Solution

This can be generated using a counter varied from 00h to FFh. The counter value is output at Port p1. The counter value can be output after a time delay decided by the frequency of the waveform, e.g., for the waveform shown in Fig. 5.52 for a period of T seconds, counter will count from 00h to ffh. The period is divided into 256 subparts. The required time delay is

$$T_d = [T / 255] \text{ seconds}$$

The procedure steps are as follows:

1. Load 00h corresponding to zero output voltage and output at the Port p1.
2. Wait for T_d seconds.
3. Increment the counter and output the value at the Port p1.
4. Goto Step 2.

This will generate continuous saw tooth wave. The program segment is as follows:

```
saw-tooth_gen:
          mov 80, #20        ; initialize stack pointer
          mov a, 00h
rept1:    mov P1, a          ; output Logic 1 for max output
          lcall delay        ; call delay for Td seconds
          inc a
          sjpm rept1
          end
```

Example 5.10   Write a program to generate triangle waveform.

Solution

This can be generated by using a counter, which varies from 00h to ffh, up to half period and then decrement for the next half. The counter value can be output at Port p1. The counter value can be output after a time delay, which is decided by the frequency of the waveform. For a period of $T/2$ seconds, counter will count from 00h to ffh. The period is divided into 256 subparts. Hence, the delay required is for $T_d = T/512$.

The algorithm is similar to Example 5.4, with a minor modification as given below:

1. Load 00h corresponding to the zero output voltage, and output at the Port p1.
2. Wait for T_d seconds.
3. Increment the counter and output at Port p1.
4. If count is d" ffh

   Goto Step 2
5. Decrement the counter and output at Port p1.
6. Wait for T_d seconds.
7. If count = 00h

   Goto Step 3

   Else,

   Goto Step 5

The program segment is given below

```
trang_  Wave:
        mov 80, #20             ; initialize stack pointer
        mov a, 00h
rept1:
        mov P1, a
        lcall delay             ; call delay for Td seconds
        inc a
        cjne a, #ffh, rept
loop:
        lcall delay
        mov P1, a
        djne a, #00h, loop
        lcall delay
        sjmp rept1
        end
```

Example 5.11   Write a program to generate sin/cosine waveform.

Solution

The important point to be remembered is that the DAC must be configured to generate a bipolar output. The data sheet of the chip to be used should be referred for the setup or configuration.

Increment/decrement operation like previous examples will not give the desired digital input to be applied as an input to the DAC, a look-up table should be prepared. The values can be computed as explained in Example 5.7. A sample look-up table is shown in Table 5.8 for the sinusoidal waveform assuming 36 subdivisions.

Similar table can be prepared for cosine or other regular time varying signals.

Table 5.8 Look-up table for generation of sinusoidal waveform

Address	In_value (Hex)	Address	In_value (Hex)	Address	In_value (Hex)	Address	In_value (Hex)
2500	7F	2513	7F	2526	69	2539	96
2501	FA	2514	FB	2527	02	253A	06
2502	96	2515	69	2528	53	253B	AC
2503	FB	2516	F0	2529	06	253C	0F
2504	AC	2517	53	252A	40	253D	C0
2505	FD	2518	E8	252B	0F	253E	1C
2506	C0	2519	40	252C	2C	253F	D3
2507	E8	251A	DD	252D	1C	2540	2C
2508	D3	251B	2C	252E	1C	2541	DD
2509	DD	251C	D3	252F	2C	2542	40
250A	DD	251D	1C	2530	0F	2543	E8
250B	D3	251E	C0	2531	40	2544	53
250C	EA	251F	0F	2532	06	2545	F0
250D	C0	2520	AC	2533	53	2546	69
250E	F0	2521	06	2534	02	2547	F8
250F	AC	2522	96	2535	69	2548	7F
2510	F8	2523	02	2536	00	2549	FA
2511	96	2524	7F	2537	7F	254A	XX
2512	FA	2525	00	2538	02	254B	XX

The program segment, assuming that the look-up table is stored in the memory from the address 2500h onwards, is given below:

```
sin_wave:
        mov  dptr, #2500h    ;base address of look-up table
        mov  r4, #24h        ;36 intervals
loop:
        movx a, @dptr
        mov  P1, a
        lcall delay          ;delay for [T/36]
```

```
inc        dptr
djnz       r4,loop
sjmp       sin_wave
```

Realization of ADC

In this section we will study how DAC can be used to realize analog to digital conversion process. The popular techniques of converting analog signal into digital one are

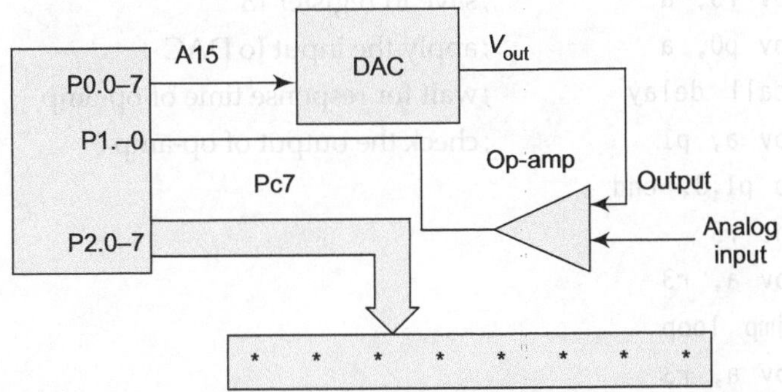

Fig. 5.53   Hardware setup for realization of ADC using DAC

Block diagram of Fig. 5.53 shows hardware setup for realization of ADC using DAC chip. The software changes with the method used for conversion process. We will develop program segments for both the methods.

Counter method   The conversion process starts with application of a reset pulse. On reset, the counter is reset to zero. This counter value is applied to DAC through Port P0. DAC converts this into analog value. The output of DAC is compared with applied input using op-amp. Microcontroller checks output of the DAC. If the output is negative, it means the counter value is less; the counter is incremented by one. The process is repeated till output of DAC is positive. The content of counter represents applied analog value in binary. The counter value is output on display device connected on the Port P2 of the microcontroller. The algorithm steps are described below:

1. On reset is applied, the counter is RESET.
2. The output of the counter is applied to the DAC.
3. DAC generates equivalent analog value which is compared with the applied input.
4. If applied input > output of DAC

   (the output of op-amp is positive)

   Increment the counter and goto Step 2

   Else,

The contents of the counter represent the digital equivalent of applied input signal.

The following program implements the process:

```
adc_cnt :

        mov 80h, #20h       ; initialize SP
rept1:
        mov a, #00h
        mov r3, a           ; save in register r3
loop:   mov p0, a           ; apply the input to DAC
        lcall delay         ; wait for response time of op-amp
        mov a, p1           ; check the output of op-amp
        jb p1.0, end
        inc r3
        mov a, r3
        sjmp loop
end:    mov a, r3
        mov p2, a           ; display the digital value on LED (Port p2)
        sjmp rept1
        end
```

Successive approximation method The method for 4-bit conversion is described in Section 5.6.1. The content of 8-bit counter here represents applied analog value in binary, which is output on display device connected on the Port P2 of the microcontroller.

The algorithmic steps for converting an analog signal into an 8-bit value are as follows:

1. SET the MSB of successive approximation register to '1', load this value to the counter. The counter will start from the mid-point of the range.
2. Apply output of counter to DAC.
3. DAC generates equivalent analog value which is compared with the applied input.
4. If applied input > output of DAC, [the output of op-amp is positive]
   Set next bit to 1 and goto Step 2
   Else,
   If applied input = output of DAC
   The contents of counter represent the digital equivalent of applied input signal. Display the result and STOP
   Else,

RESET current bit, SET next bit, goto Step 2

The process is implemented in the following program:

```
adc_sam:
        mov 80h, #20h           ;initialize SP
rept1:  mov r3, #80h            ;successive digit mid-value
        mov r4, #00h
        mov r5, #08h            ;counter for 8-bit
        mov a, r3
loop:   add a, r4
        mov r4, a
        mov P0, a               ;apply input to DAC
        lcall delay             ;wait for response time
        mov a, P1
        jb P1.0, next_bit
        mov a, r3               ;RESET the current bit
        cpl a
        anl a, r4
        mov r4, a
next_bit:mov a, r3
        rrc a
        mov r3, a
        djnz r5, LOOP
        mov a, r4
        mov P2, a
        sjmp rept1
        end
```

5.4 | MEMORY ELEMENTS

The microcontrollers have memory elements on the chip. For complex applications there may not be sufficient memory and therefore additional elements such as RAM, EPROM, and EEPROM are required. The section describes the characteristics and interfacing of memory elements with the microcontroller. The memory subsystem must have the following characteristics:

1. Any location can be accessed randomly.
2. Each byte has unique address.
3. Data is read/write in one cycle.

5.4.1 Memory Types

Figure 5.54 shows the different types of memories used in a microcontroller-based systems. Even though registers are the fastest storage, due to limited number of registers on microcontroller, memory is used to store both programs and data that determine the operation of the system. The element called main memory includes standard RAM, ROM, EPROM, and EEPROM.

Main memory is directly addressed by the microprocessor. It is possible to access any address randomly. Accessing a memory location means to carry out read or write operation at any desired location. All programs and data in use by the controller at any moment must be stored in the main memory.

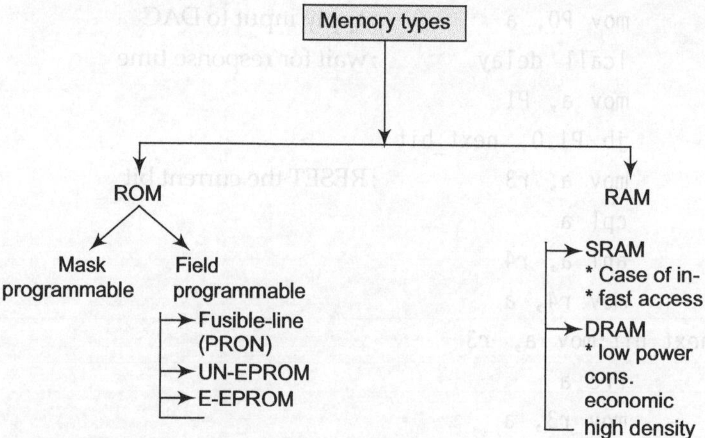

Fig. 5.54   Type of memory elements

5.4.2 Internal Structure of Memory Devices

In the memory device each bit of storage requires one memory cell. This cell may be a flip-flop, a capacitor, or a fuse, depending on the memory type. The memory cells are arranged as an array. The address inputs are decoded to produce the row select and the column select signals to select any memory cell. The memory chip is enabled when the chip select (CS) input is asserted.

For the read operation CS and OE (output enable) are asserted to enable the output three-state drivers. While for the write operation data is provided on the data inputs, CS is asserted and the write enable is driven low. The data in and data out signals may share the same pins or may be on separate pins.

The interfacing a ROM is similar, except there is no write enable (WE) input. Most ROMs and EPROMs have a chip select and an output enable, so the chip select can be connected directly to the address select signal and the output enable to the microprocessor's READ signal.

A key specification of any memory device is its access time. For the READ operation: access time is the time from when a new address is presented to the memory to which the memory provides the correct data at its outputs. While for the WRITE operation, it is the time from when the address is presented to when the WRITE pulse can be terminated and the data removed. Access time from address is the most commonly quoted parameter.

5.4.3 Address Decoding

The task of the address decoder is to generate address blocks for the memory address space of a microcontroller. The desired block size is determined by the capacity of the chip to be interfaced. The range of addresses for which an address select signal is asserted is determined by which address lines are decoded.

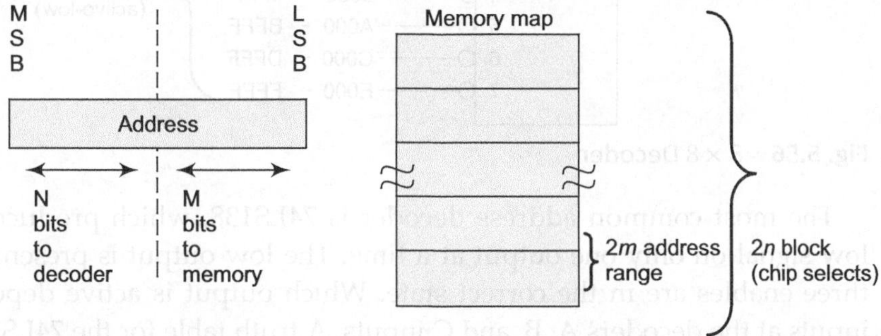

Fig. 5.55   Address decoding

Figure 5.55 shows the top n-bits decoded by the address decoder to produce $2n$ separate address select signals. Each address select signal is asserted for a range of $2m$ addresses, where m is the number of less significant address lines not decoded by the address decoder. For example, if the address decoder decodes A14 through the most significant address bit, the lower 14 address bit A0 to A13 are ignored by the address decoder.

The memory subsystem must include a method to activate the various devices when their assigned addresses appear on the bus. The methods used for the decoding employ logic gates or decoder chips. Decoders are used to divide the memory space into separate memory blocks. Such decoders are termed as *address decoders*.

An address decoder is a circuit, which enables the memory for specified range of addresses and does not overlap the other block. The address space can be partitioned for the devices using following techniques:

1. Single-stage decoding
2. Two-stage decoding

Address decoders reduce the hardware for address decoding and implement the block-wise structure. Figure 5.56 shows an address decoding circuit that generates eight separate address select signals, each of which corresponds to an 8 K block of addresses. One output is asserted at any given time, corresponding to the binary number present at the A, B, and C inputs.

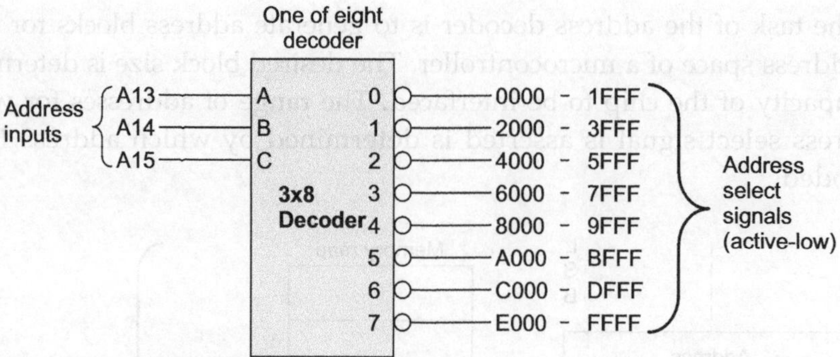

Fig. 5.56 3 × 8 Decoder

The most common address decoder is 74LS138, which produces an active low signal on only one output at a time. The low output is present only if the three enables are in the correct state. Which output is active depends on the inputs at the decoders A, B, and C inputs. A truth table for the 74LS138 decoder is given in Table 5.9.

Table 5.9 Truth table of decoder 74LS138

Input						Output							
G2A	G1A	G1	A2	A1	A0	Y0	Y1	Y2	Y3	Y4	Y5	Y6	Y7
H	X	X	X	X	X	H	H	H	H	H	H	H	H
X	H	X	X	X	X	H	H	H	H	H	H	H	H
X	X	L		X	X	H	H	H	H	H	H	H	H
L	L	H	L	L	L	L	H	H	H	H	H	H	H
L	L	H	L	L	H	H	L	H	H	H	H	H	H
L	L	H	L	H	L	H	H	L	H	H	H	H	H
L	L	H	L	H	H	H	H	H	L	H	H	H	H
L	L	H	H	L	L	H	H	H	H	L	H	H	H
L	L	H	H	L	H	H	H	H	H	H	L	H	H
L	L	H	H	H	L	H	H	H	H	H	H	L	H
L	L	H	H	H	H	H	H	H	H	H	H	H	L

The first three lines of the table clearly show that the outputs will be active only if the enable inputs are correct. The decoder outputs will be active during the entire period when ALE is low. The active output is determined by the input applied at the input pins of the decoder. Using Table 5.9 it is possible to determine the range of addresses over which each decoder output will be active. The range will be the hardware addresses for the device being selected.

I/O devices do not need the entire range of addresses available on any one-decoder output; isolated logic gates can be used to generate single address.

Memory address space decoding may be either full or partial. The full decoding uses all the available address lines. A partially decoded circuit however occurs when one or more address lines is not used in the decoding process. Functionally, the partially decoding system works just as well as the fully decoded system and it usually uses less parts.

Figure 5.57 is a pin diagram of the address decoder 74LS138. It functions exactly as described above, except that it has three enable inputs, two active-low G_{2A}, G_{2B} and one active-high G_1. All the three enable inputs must be asserted.

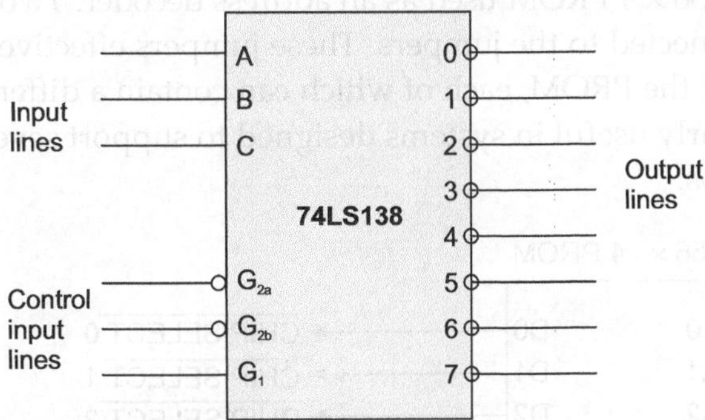

Fig. 5.57   Pin layout of decoder IC 74LS138

Figure 5.58 shows the two-stage decoding scheme to generate more select signals. The low enable inputs can be used to AND necessary control signals. If memory devices without separate output enables are used, one enable on decoders A and B can be connected to READ/WRITE, ensuring that no address select output is asserted unless one of the data transfer control signals is also asserted.

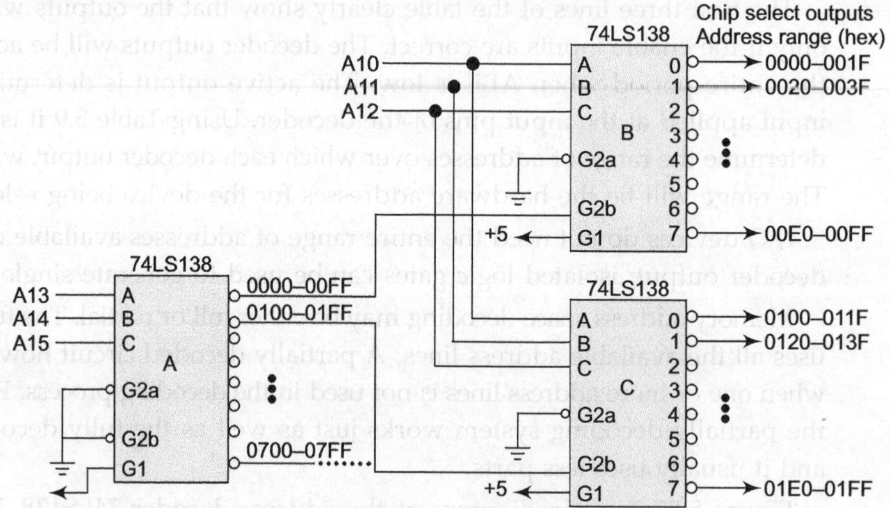

Fig. 5.58   Two-stage address decoding scheme using 74LS138

A specially programmed bipolar PROM is used as the address decoder. Figure 5.59 shows a 256×4 PROM used as an address decoder. Two address bits of the PROM are connected to the jumpers. These jumpers effectively select one of four 64×4 areas of the PROM, each of which can contain a different memory map. This is particularly useful in systems designed to support several different sizes of memory chips.

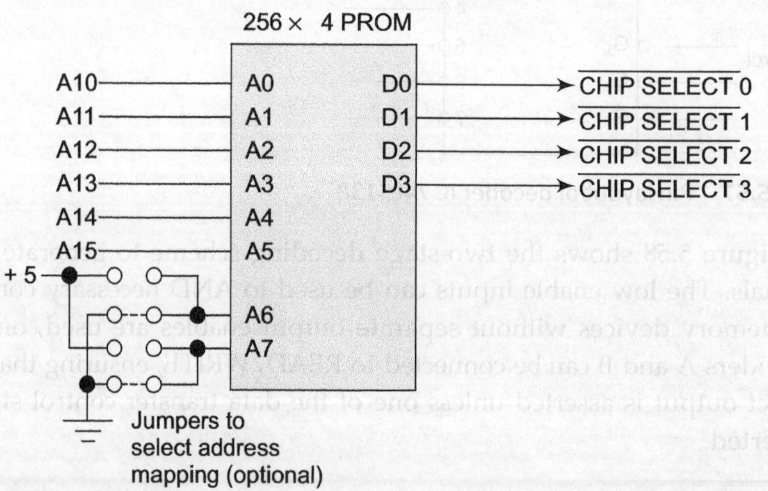

Fig. 5.59   PROM decoder

Address bits A_{10} to A_{15} select one of 64 locations in the PROM. The address bits A_0–A_9 are ignored by the PROM, so that each PROM location is accessed for a range of 1 K addresses. Bipolar PROM is available from many vendors, e.g., MMI 256 × 4 PROM with three-state outputs is the IC63S141. The equivalent device 27S21 from AMD and 74S287 from National Semiconductors.

The address bus is completely decoded; that is, no address bits are ignored. The LSB-bits are decoded by the memory's internal decoder, and the MSB-bits are decoded by the system's address decoder. Decoding logic can often be simplified by ignoring certain address bits.

5.4.4 Basic Bus Interface

Figure 5.60 shows interfacing of the 16 K RAM. The lower 14 address lines are connected directly to the memory chip and select one of the 16 K locations within the chip.

The address decoder decodes the most-significant address lines and provides an address select signal to enable the RAM. The address select signal enables the memory chip to respond to a particular range of addresses. Four such blocks are possible.

The address select output from the address decoder provides the chip select signal for the memory. Since the WE and OE inputs are internally gated with CS, they can be connected directly to the WRITE and READ control signals from the microcontroller.

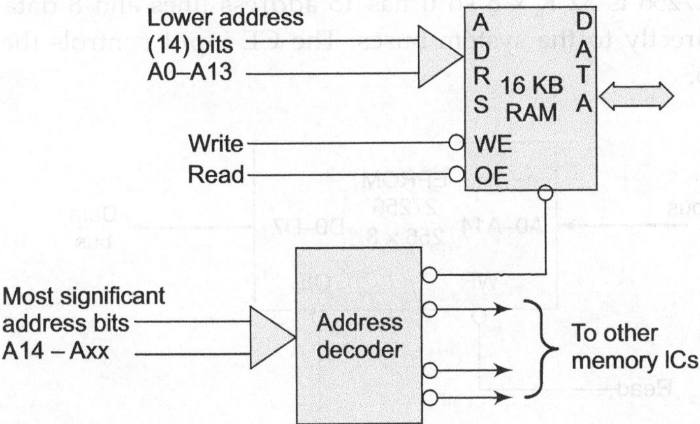

Fig. 5.60 RAM interface circuit

5.4.5 EPROM

Erasable programmable read-only memories (EPROM) are widely used for program storage in digital systems. They are nonvolatile and can be erased and reprogrammed when changes are necessary. All common EPROM are 8-bits wide but the total number of bits determines the size. The 2708 and 2716 were used previously. Now larger and faster devices, such as 2764, 27128, 27256, and 27512 are used. EPROM are designated as 27×××. Last two/three digits of the part number indicate the size in kilo bytes. They are available in NMOS and CMOS. CMOS are more expensive than their NMOS counterparts and are used primarily in systems in which power consumption is critical.

EPROM Programming

EPROM are programmed by a device called an EPROM programmer, and are then inserted into the application system. Programming typically takes from 1 to 5 min, depending on the type of programmer and the size of the memory. The contents can be erased by shining an ultraviolet light into the window in the top of the IC package. The EPROM can then be reprogrammed many times.

To program an EPROM, a programming voltage of typically 12 – 25 V (depending on device type) is applied to the V_{pp} pin. The address pins are driven with the address of the memory location to be programmed, and the data to be programmed is applied to the data lines. Finally, the CE pin is pulsed low, and the data to be programmed is applied to the data lines. Finally, the CE pin is pulsed low, and the data is stored internally.

EPROM Interfacing

Figure 5.61 shows the interfacing of 27256 EPROM to the microcontroller bus. The 27256 is 32 K × 8 so it has 15 address lines and 8 data lines. These connect directly to the system buses. The CE input controls the state of the entire chip.

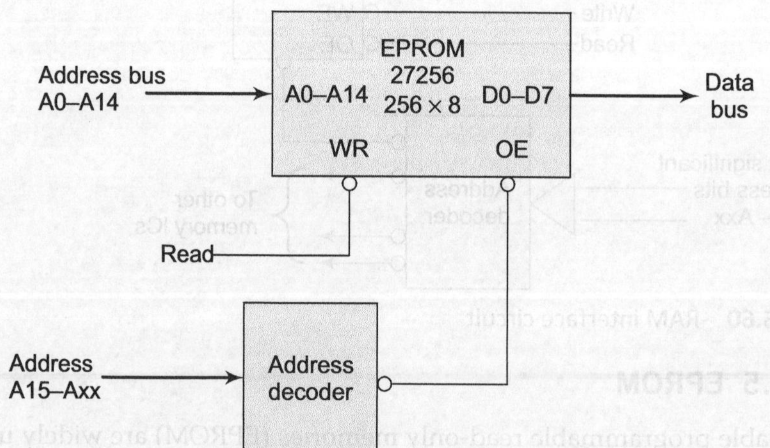

Fig. 5.61　Standard EPROM interface circuit

When the CE is false, the chip is disabled and enters a standby mode. In this mode, power consumption is greatly reduced: the maximum supply current for Intel's 27256 is 100 mA in active mode and 40 mA in standby mode. It is desirable to enable the EPROM only when it is being accessed, to save power. This is achieved by connecting CE to the address select signal from the address decoder. The chip is then selected only when its address is present on the address bus.

The EPROM's outputs are enabled only when the address select signal and READ are both generated.

The 2732A is the largest EPROM available in a 24-pin package. For larger sizes, a 28-pin package is needed for additional address lines. Figure 5.62 shows pin assignment of the 27×××× series of EPROM.

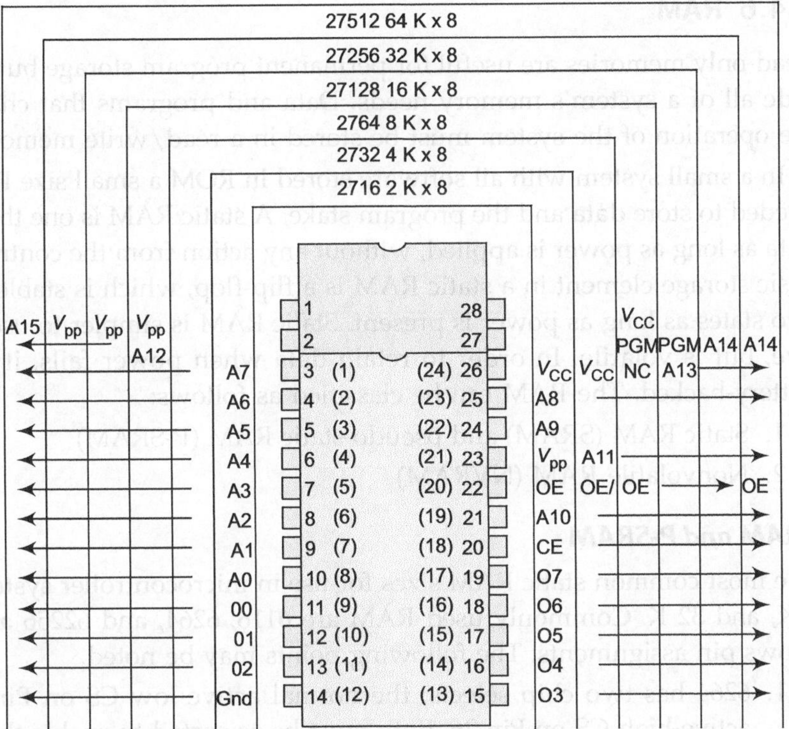

Fig. 5.62   27×××× series EPROM pin assignments

To provide pin compatibility for both package sizes, the pin-out for the 28-pin EPROM is the same as the 24-pin devices in the "lower" 24-pins. A 24-pin EPROM can thus be plugged into a 28-pin socket.

The programming voltage V_{pp}, which shares the output enable pin on the 2732A, is moved to Pin 1 for the larger devices. Pin 28 is assigned to V_{CC} to maintain the standard corner power configuration. On the 2764, Pin 26 is left as a NC so that power can be connected to both Pins 26 and 28 for compatibility with device sizes.

To allow use of all EPROMS including the 27256 and 27512, there is a jumper to allow Pin 27 to be connected to V_{CC} or to the A14. Other than these pin-out changes, the timing and other interface details are similar for all EPROM sizes.

A 27512 EPROM has a capacity of 64 K, which entirely fills the address space of a microcontroller, with a 16-bit address bus. Some address space is needed for RAM and possibly I/O. Mask programmed ROMs, which are programmed during the manufacture of the chip itself, are generally less expensive. Electrically erasable PROM (EEPROM) is also available in the byte-wide standard pin-out.

5.4.6 RAM

Read-only memories are useful for permanent program storage but cannot provide all of a system's memory needs. Data and programs that change during the operation of the system must be stored in a read/write memory.

In a small system with all software stored in ROM a small size RAM may be needed to store data and the program stake. A static RAM is one that retains its data as long as power is applied, without any action from the control logic. The basic storage element in a static RAM is a flip-flop, which is stable in either of two states as long as power is present. Static RAM is simpler to use, cost-effective, but is volatile. In order to retain data when power fails, it can also be battery-backed. The RAM can be classified as follows:

1. Static RAM (SRAM) and pseudo-static RAM (P-SRAM)
2. Nonvolatile RAM (NVRAM)

SRAM and P-SRAM

The most common static RAM sizes for use in microcontroller systems are 2 K, 8 K, and 32 K. Commonly used RAM are 6116, 6264, and 62256 and Fig. 5.63 shows pin assignments. The following points may be noted.

1. 6264 has two chip selects: the normal active-low CS on Pin 20 and an active-high CS on Pin 26. Both must be as sorted to enable the RAM.
2. In a universal byte-wide socket, Pin 26 must have a jumper to connect it to V_{CC} for use with 24-pin devices. This configuration keeps the active-high chip select always asserted, and the active-low CS is connected to the address select signal from the address decoder.
3. 62256 require two additional address lines. Thus, the second chip select input is sacrificed, and Pins 1 and 26 are used for the two most significant address lines.

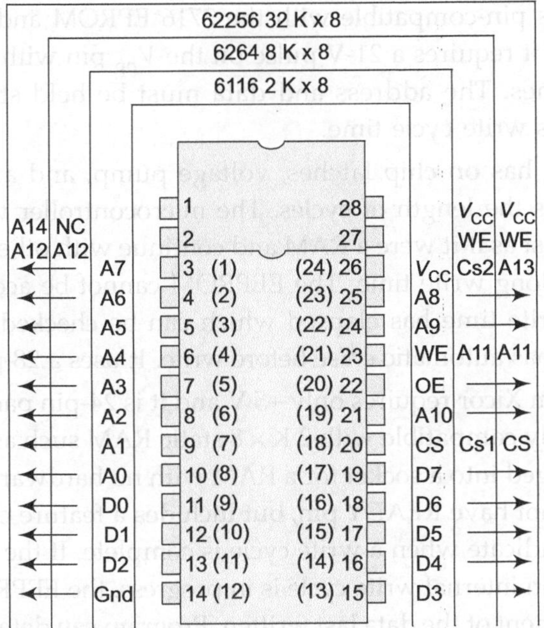

Fig. 5.63   SRAM pin assignments

The pseudo-static RAM is one that uses dynamic storage cells but contains all the refresh logic on-chip so the device appears static to the user. Examples of pseudo SRAM are 42832 from NEC and 65256 from Hitachi.

NV RAM

Current RAM technologies share a common weakness: they are volatile, so all data is lost when power is turned OFF. ROM and EPROM technologies provide nonvolatility but cannot be erased and reprogrammed in the application system. The ideal memory is a nonvolatile RAM, which would perform like a normal static RAM but retain its contents indefinitely without power. The different configurations of the NVRAM are EEPROM, shadow RAM, and battery baked RAM.

1. The most common type of nonvolatile memory is electrically erasable PROM (EEPROM). The contents of EEPROM can be erased and rewritten. The erase and write times are typically 5–10 ms/byte, which is much less than UV-EPROM (erasable on a byte-by-byte basis). They do not require ultraviolet light. EEPROM require only a 5 V supply and include a charge pump on the chip that generates the write voltage from the 5 V supply. Some of the EEPROM used in the microcontroller-based systems are

(a) Intel 2816 is pin-compatible with the 2716 EPROM and equivalent. For writing it requires a 21-V pulse on the V_{pp} pin with specific rise and fall times. The address and data must be held stable for the entire 10-ms write cycle time.

(b) Intel 2817A has on chip latches, voltage pump, and a write timer that controls the length of cycles. The microcontroller writes to the EEPROM just as if it were a RAM and continue with other operations during the long write time. The EEPROM cannot be accessed again until the write time has elapsed which can be checked by READY pin. 2817A has automatic erase before write. It uses a 28-pin package.

(c) X2816A from Xicor requires only +5 V and it is 24-pin package which makes it fully compatible with 2 K × 8 static RAM such as the 6116. It can be plugged into a socket for a RAM with no hardware changes at all. It does not have READY pin, but includes a feature called DATA polling to indicate when a write cycle is complete. If the EEPROM is read while an internal write cycle is in progress, the EEPROM returns the complement of the data last written. Program can determine when the write cycle is complete by reading the location last written and comparing the data read to the data written. EEPROM are also available in 4 K and 32 K size.

2. **A shadow RAM** also called NVRAM, which combines normal RAM cells and EEPROM cells on the same chip. These devices avoid many of the problems of EEPROM at the expense of a much larger chip size and consequently higher cost. One approach to overcoming deficiencies of EEPROM is the shadow RAM, called NVRAM. Nonvolatile RAM are also available with a serial interface.

An NVRAM consists of a memory with one normal RAM cell and one cell, which can be written at full speed. When a power failure is impending, the RAM array is stored in the on-chip EEPROM array. The EEPROM storage cycle occurs only when power fails, not on every write access.

Some of the shadow RAM used in the microcontroller-based systems is

(a) X2210 and X2212 from Xicor have capacities of 64 × 4 and 256 × 4, respectively

(b) 2004 is a 512 × 8 NVRAM from Intel and Xicor. The static RAM array is accessed just like any other static RAM via the address, data, CS, OE, and WR pins.

3. **Battery-backed CMOS RAM** Adding battery backup to a standard CMOS RAM shown in Fig. 5.64 provides an ideal nonvolatile RAM.

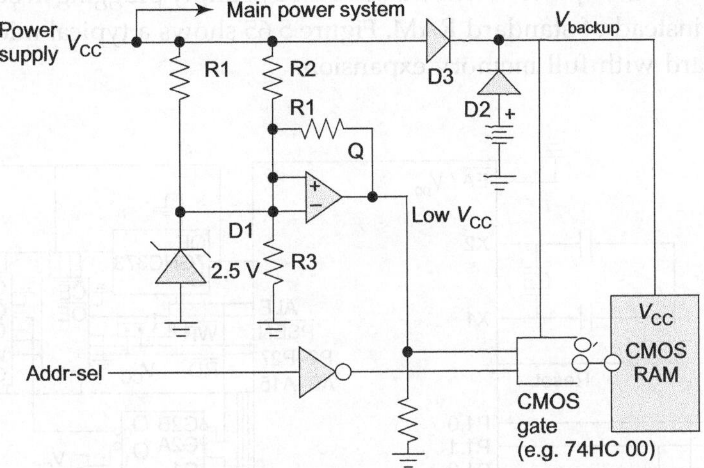

Fig. 5.64   Battery backup RAM

Zener diode D1 provides a reference voltage. Resistors R2 and R3 form a voltage divider. When the voltage produced from this divider is less than the reference voltage, the output of the comparator goes low. R4 provides positive feedback for stabilization during transition. The chip select input to the RAM is gated with LOW V_{CC}, so no accesses can occur when the supply voltage is below the threshold. The RAM enters its low-power standby mode when the battery is providing power. The pull-down resistor R7 ensures that the LOW V_{CC} signal is held asserted if the comparator ceases to drive its output as the voltage drops. Schottky diode D3 isolates the battery supply from the rest of the system. Diode D2 in series with the battery prevents the battery from being charged by the main power supply.

Read and write cycles are fast and unlimited in number, and CMOS static RAM chips are readily available from multiple sources. The RAM chips themselves are less expensive per bit than EEPROM. Batteries have their own disadvantages such as: relatively limited life, less reliable than semiconductor devices and need for extra circuitry to switch the RAM over to battery power when the main supply fails. Since the battery is intended to provide backup power for RAM, it is necessary to isolate RAM power from the rest of the system. It must be inhibited when V_{CC} is below the normal minimum level.

Battery-backed CMOS RAM modules with integral backup control circuitry and lithium battery. On-chip circuitry handles power switching, write protection, and input termination. These modules are typically packaged in a 24- or 28-pin package, which is compatible with a standard static RAM socket. The battery-backed RAM can replace normal static RAM by simply plugging in battery-backed modules instead of standard RAM. Figure 5.65 shows a typical microcontroller-based board with full memory expansion.

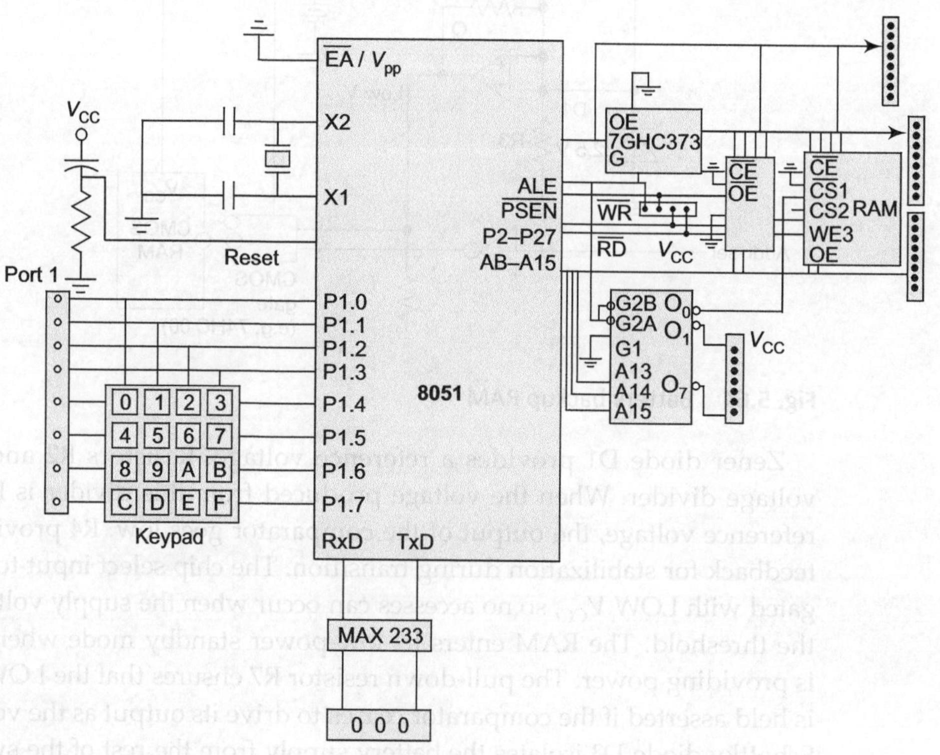

Fig. 5.65 Project board with memory expansion

5.5 | BASIC SYSTEM CONFIGURATION

There are various input/output devices which are connected with the micro-controller. This section describes few general purpose boards with the most common input devices (keys/keypad/thumbwheel switch) and the output devices: LED/7-segment LED/LCD. Anyone of these can be used for developing a small application. User has to write a program module for the application.

5.5.1 Keys and LEDs

Design a scheme to detect the key pressed and output the pin at logic level 1 to make the corresponding LED ON. The hardware setup is shown in Fig. 5.66.

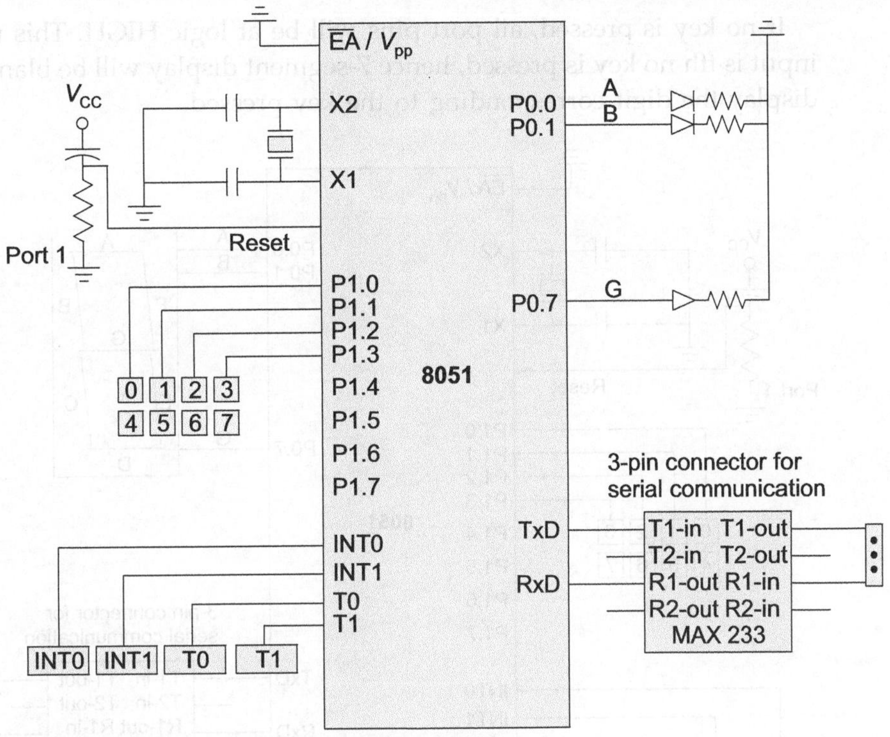

Fig. 5.66   Basic board with keys and LEDs

The operation of the system can be described as follows: If no key is pressed, all port pins will be at logic HIGH, port will read ffh. This means if the input is ffh no key is pressed. Else we will complement the input to make LED corresponding to the key is ON.

```
in_out:
        mov  a, p1          ; read the port and check if any key is pressed
        cjne a, #ff,close
        sjmp in_out
close:
        cpl a
        mov P0, a
        sjmp in_out
        end
```

5.5.2 Keys and 7-segment

Design a scheme to detect the key pressed and output the value on 7-segment display unit connected to Port 0 pins of the microcontroller. The hardware setup is shown in Fig. 5.67. The operation of the system can be described as follows:

If no key is pressed, all port pins will be at logic HIGH. This means if the input is ffh no key is pressed, hence 7-segment display will be blank else it will display the digit corresponding to the key pressed.

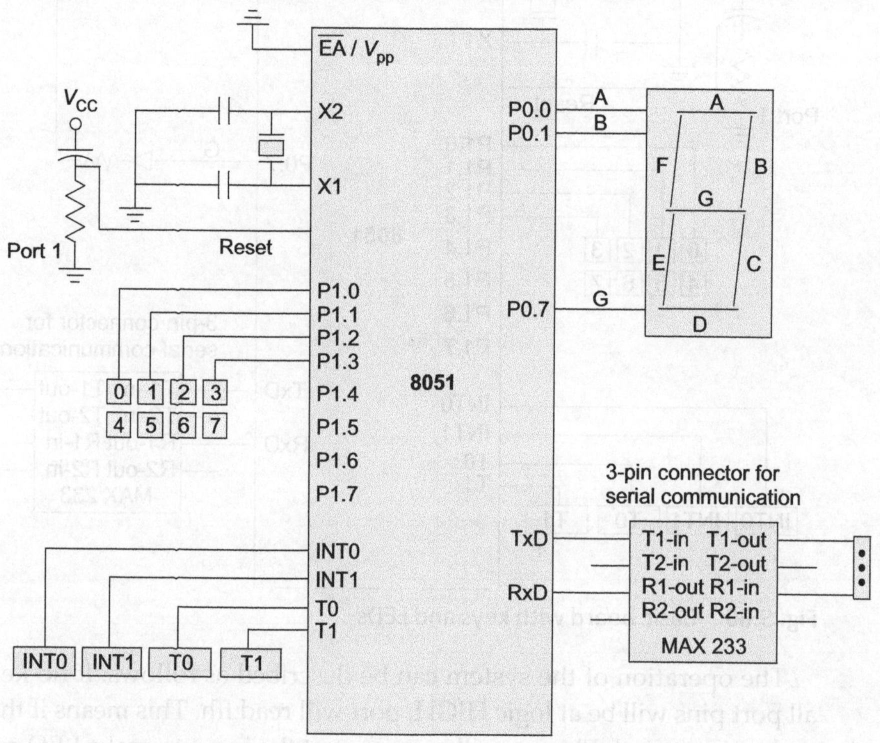

Fig. 5.67 Basic board with key and 7-segment display

Only change in the program with reference to program of Section 5.5.1 is that it is necessary to prepare a look-up table as explained in Chap. 3. The code is to be used to convert 4-bit data to 7-segment code before the data is displayed. First the key pressed has to be identified and then corresponding code is found from look-up table and it is displayed on the 7-segment unit. The program segment is given below:

```
seven_out :

        mov a, p1               ;read the port and check if any key is pressed
        cjne A, #ffh close
        sjmp seven_out
        mov dptr,#lookup        ;pointer to look-up table in the code memory
close:  mov r5, #00h
        rlc a
        jc over
```

```
        inc r5
        sjmp close
        mov a, r5
        mov a, @a +dptr        ; get the code in acc.
        mov P0, a              ; output the code on display unit
        sjmp seven_out
        end
```

There is another method for the above method. Instead of using a look-up table, 4 to 7 decoder IC-7447 can be used. The program segment is given below:

```
decoded-out:
        mov a, p1              ; read the port and check if any key is pressed
        cjne a, #ffh, close
        sjmp decoded-out
close:   mov r5, #00h
close-chk:
        rlc a
        jc over
        inc r5
        sjmp close-chk
        mov a, r5
        anl 0fh               ; mask upper nibble
        mov P0, a             ; output on display unit
        sjmp decoded-out
        end
```

5.5.3 Matrix Keyboard

It is clear that if the keys are connected directly to the port pins such that one key is connected to 1-pin, limited number of keys can be connected. In the form of a 4×4 matrix. One end of the rows is connected to the port pin via key while other end is connected to V_{CC}. Similarly, one end of the columns is connected to the port pin via key while other end is connected to the V_{CC}. There is no connection between the rows and the columns until a key is closed. The column port pins will be logically HIGH, even though the row is grounded, if key is open. When key is closed the column pins will be grounded and are pulled to logic level LOW.

The advantage derived by the configuration is that the number of port pins required for interfacing is reduced. For example, in normal connections, i.e., one pin for each key, 16 keys will require 16-pins. But in the matrix form, only

8 port pins are required. The formula for the required number of port pins is [2.$\sqrt{N}$]. Keypads are readily available which are representative of matrix keyboards. Figure 5.68 shows interfacing keypad with the microcontroller. The limitation of this scheme is that it is necessary to identify the key pressed, which puts overhead on the microcontroller.

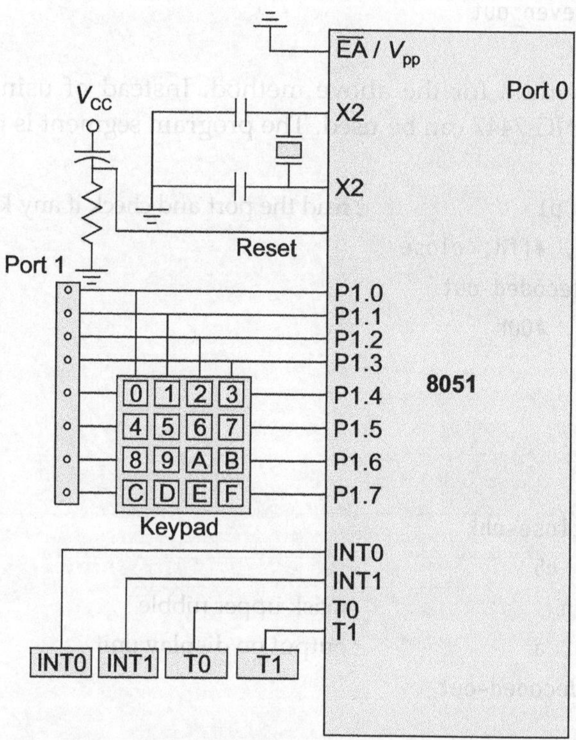

Fig. 5.68 Interfacing keypad with the microcontroller

The identification algorithm for the key pressed is as follows:
1. First ground all the rows by pulling all port pins low.
2. Check the output of the column pins,
3. If output is ×Fh
    goto Step 2
4. Wait for a key debounce period (10 to 20) ms, as some key is pressed
5. Select row r0, by connecting it to ground.
6. If output = ×Fh
    key in the row is not pressed, goto Step 8
7. Select next row and repeat 6 for all four rows.
8. Identify the column in which the key is pressed by rotating column output and find column pin which is grounded.

9. Get the code from the look-up table, once the row and column are identified.

The program segment, which implements the identification algorithm, is given below:

```
rd_keypad:
            mov 80h, #20h          ; initialize SP
            mov E0, #00h           ; ground all rows and
            mov dpl, #00h          ; set column counter to 0
            mov 90h, a             ; output at Port p1
    wait:
            mov a, 90h             ; wait for key closure
            anl a, #0Fh
            cjne a, #0Fh, close
            sjmp wait
    close:
            lcall delay            ; call delay for debounce
            mov a, #7Fh            ; ground one row at a time
            mov r4, #04h           ; row counter
    chk_row:
            rlc a
            mov dph, a             ; save high addr for code byte
            mov 90h, a             ; ground row
            mov a, 90h             ; get column output
            anl a, #0Fh            ; mask upper nibble
            mov r5, #04h           ; column counter
    chk_col:
            rrc a
            jnc over
            inc dpl
            djnz r5, chk-col       ; go to next column check
            mov a, dph
            djnz r4, chk_row
            sjmp start
    over:
            movx a, @dptr
            mov 80h, a
            sjmp rd_keypad
            end
```

5.5.4 System with Keypad and Two-digit Display

Figure 5.69 shows a hardware setup for the system, which uses keypad as input device and 7-segment display units as the output device. Since two-digit display is desired, two display units are used. In order to reduce the software effort of preparing a look-up table for generation of a 7-segment code for the digit display, a hardware decoder IC-7447 is used to interface display unit with the microcontroller. The following programs are developed to carry out specified operation:

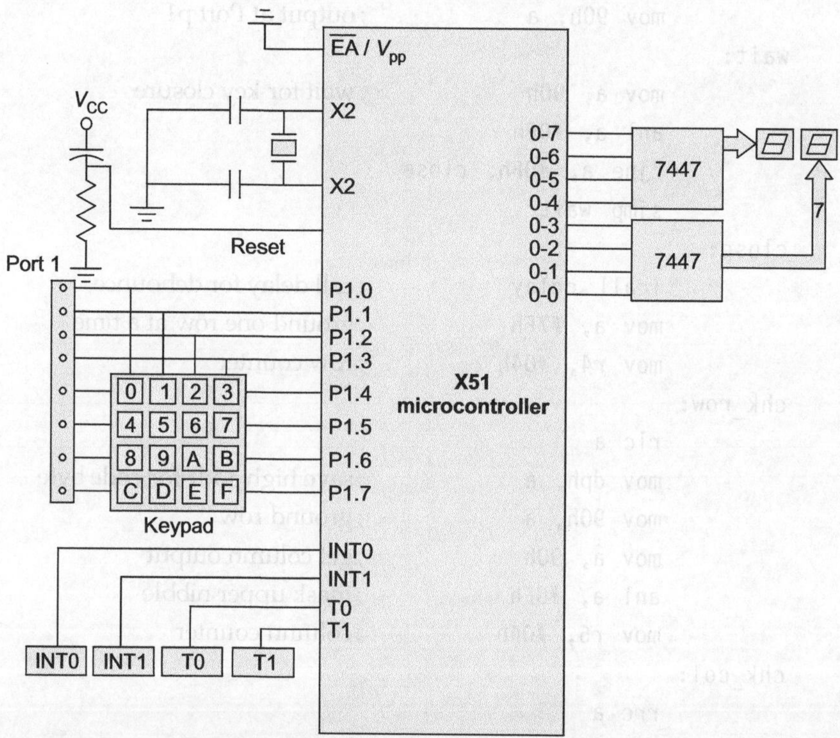

Fig. 5.69   Keypad with two digit display

Example 5.12   Write a program module to accept two-digit number and store it in register r3 in a packed form.

Solution

This program reads two keys from the keypad:

```
rd-keypad:
        mov 80h, #20h          ; initialize SP
read:   acall read_key         ; read first key
        mov r3, a              ; save it in register r3
        acall read-key         ; read second key
        mov b, a              ; save it in register b
```

```
next:      anl a, #0FH
           xch a, r3
           swap a
           anl a, #0Fh
           add a, r3
           mov r3, a                ; save entry in the register R3
           end

                                    ; subroutine for reading a key from the
                                    ; keypad
read_key:
           mov E0, #00H             ; ground all rows and
           mov dpl, #00h            ; set column counter to 0
           anl a, f0h               ; mask lower nibble, as this pins rare connected
                                    ; to columns
           mov 90h, a               ; output at Port p1
wait:      mov a, 90h               ; wait for key closure
           anl a, #0Fh
           cjne a, #0Fh, close
           sjmp wait
close:
           lcall delay              ; call delay for debounce
           mov a, #7Fh              ; ground one row at a time
           mov r4, #04h             ; row counter
chk_row:
           rlc a
           mov dph, a               ; save high address code byte
           mov 90h, a               ; ground row
           mov a, 90h               ; get column output
           anl a, #0fh              ; mask upper nibble
           mov r5, #04h             ; column counter
chk_col:
           rrc a
           jnc over
           inc dpl
           djnz r5, chk_col         ; go to next column check
           mov a, dph
           djnz r4, chk_row
           sjmp read_key
over:      movx a, @dptr
           ret
           end
```

Example 5.13   Write a program module to accept two-digit number, display on 7-segment display unit and also save input in the register r3 in a packed form.

Solution

This program segment uses subprogram for reading a key of Example 5.12

```
rd-display:
        mov 80h, #20h           ; initialize SP
read:
        acall  read-key         ; read first key
        mov r3, a               ; save it in register r3
        acall  read-key         ; read second key
        mov b, a                ; save it in Register b
next:
        anl a, #0FH
        xch a, r3
        swap a
        anl a, #0Fh
        add a, r3
        mov r3, a               ; save entry in the register r3
        mov P0, a               ; display the packed input data on Port 0
        end
```

EXERCISES

5.1   What are the popular devices used as input devices in a microcontroller-based systems?

5.2   Explain operation of a thumbwheel switch.

5.3   State and explain problems associated with transistor switches. How can they be eliminated?

5.4   What is a key bouncing? What are the problems created by it? Describe the methods to avoid bouncing of key.

5.5   Why is it necessary to have isolation between the device and the microcontroller? How is it accomplished in practice?

5.6   How relays differ from keys? Where are they used?

5.7   What are the different methods of converting an analog signal into a digital one?

5.8   Explain successive approximation method of conversion.

5.9   List the sources of errors in ADC conversion.

5.10   With the help of a block diagram, explain the operation of ADC0808.

5.11   Design a scheme to interface ADC0809 with microcontroller in handshaking mode. Write the software for operation of the system.

5.12   Explain the digital section of ADC7109.

5.13   Explain the function cycle of ADC7109.

5.14   Design a scheme to interface [7109] with microcontroller in interrupt driven mode. Write the software for operation of the system.

5.15   What is the need for a sample and hold circuit? When is it useful?

5.16   Define the terms: apperture time, droop rate.

5.17 Describe the operation of the S/H [AD884].

5.18 What are the different types of 7-segment displays used in the microcontroller-based system?

5.19 Use structural organization of 16-segment display to describe display alphanumeric characters.

5.20 Specify the features of LCD units.

5.21 Explain how does LCD operate.

5.22 What do you mean by the instruction code of LCD? How is it useful?

5.23 Prepare a list of operations, which can be controlled using LCD unit.

5.24 What are stepper motors? How they differ from conventional motor? State the advantages of stepper motors.

5.25 Explain various drive methods, used for driving stepper motors.

5.26 Why is it necessary to design an interface for connecting stepper motors with the microcontrollers?

5.27 How can you change the speed of the stepper motor?

5.28 What are the shaft encoders? Why are they used?

5.29 Define the terms: linearity, accuracy, settling time.

5.30 Design a scheme to realize 16-bit ADC using 8-bit DAC chip. The system must display the digital equivalent on four 7-segment displays. Write software for initialization as well as operation.

Chapter 6

Interfacing Parallel Devices

Microcontroller (µC) can be connected with the outside environment through input device, output device, memory, or another µC-based system. In most situations, the µC is not compatible with other systems or devices. The disparity may be in terms of electrical characteristics, logic families, or speed of operation, etc. Additional circuits or networks are necessary for connecting the µC with the peripheral devices. The circuits used to connect µC with the other devices are called *interfacing circuits*.

The disparity or non-compatibility occurs because of the following reasons:
1. Limitations associated with the type of data transfer.
2. Format used for data transfer.
3. Electrical characteristics of the device.
4. Physical characteristics of the device.

The interfacing circuits are subjected to the constraints of the environment and performance. These circuits lack ruggedness if they are to be designed and assembled. However, in practice for most applications, such interfacing circuits are available, which are suitable for most of the µC families. They are in the form of programmable µC chips. The selection of the interface is involved with digital as well as nondigital consideration.

Digital Considerations

The selection of interfacing circuits for the microcontroller as well as peripheral interconnections is based on the consideration of digital parameters such as level shifting, buffering, bus interconnections, serial to parallel as well as parallel to serial conversion and synchronization of digital signals.

Non-digital Considerations

Depending upon the type of signals required by the peripherals, design of interfacing circuits requires consideration of non-digital parameters such as ADC/DAC conversion, type of transducer, need for design of amplifiers, and signal conditioning.

The most important in selection of interfacing circuits is the choice of microcontroller itself. When more than one device is to be connected, several interfaces are required.

The functions expected from a good interface circuit are:

Buffering There is a disparity of speed between microcontroller and peripheral devices. To synchronize the data transfer between them, an intermediate place is required to store the data temporarily during transfer. Interfacing must be capable of providing such a place, termed as buffer. This functionality provided by the interfacing circuits is called *buffering feature*.

Address decoding Since all the devices are connected to the common bus, each and every interfaced device must have a unique address called *device port address*. The selection of I/O device for the data transfer operation requires that the interface must be capable of decoding the address in order to decide if it has been addressed by the controller instruction.

Command decoding In addition to the address decoding, interface must be capable of decoding the instruction so that operation can be identified and executed.

Timing and control Since microcontroller is a synchronous device, operation has to be done at finite instant of time. Due to speed mismatch between peripherals and controllers, there is a need to synchronize the operations with the system clock. This function has to be carried out by an interface.

6.1 | 8255 PROGRAMMABLE PERIPHERAL INTERFACE

It is a parallel programmable peripheral interface, used for parallel data transfer or data acquisition. It is used to interface I/O device with the system data bus. Figure 6.1 shows pin layout of the 8255.

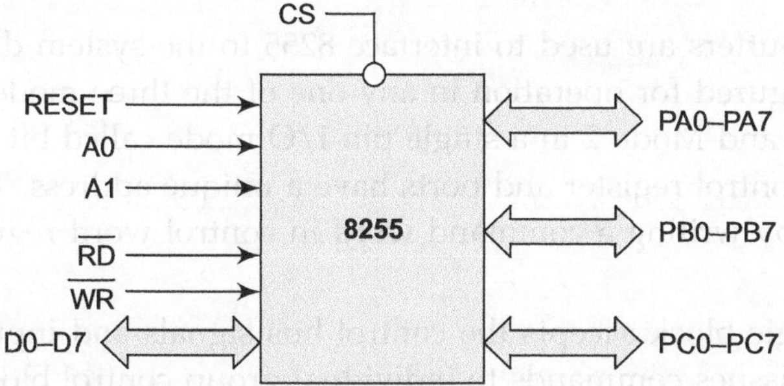

Fig. 6.1 Pin layout of 8255

The 8255 is a general purpose programmable interface, having 24 I/O pins. These pins can be programmed for use as the input or the output in two groups of 12-pins each, as shown in functional grouping in Fig. 6.2.

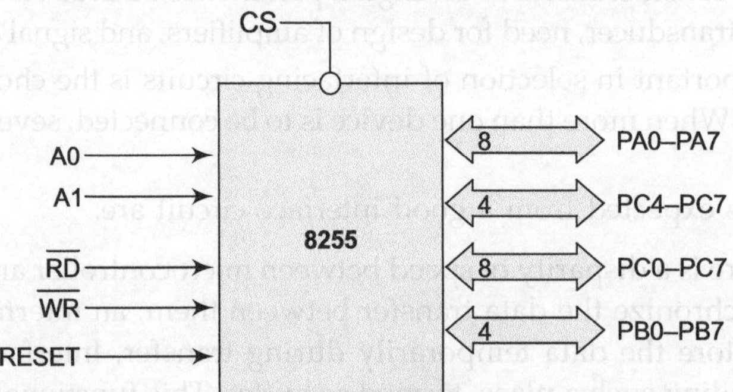

Fig. 6.2 Functional grouping of pins

The simplified internal block diagram of the chip is shown in Fig. 6.3.

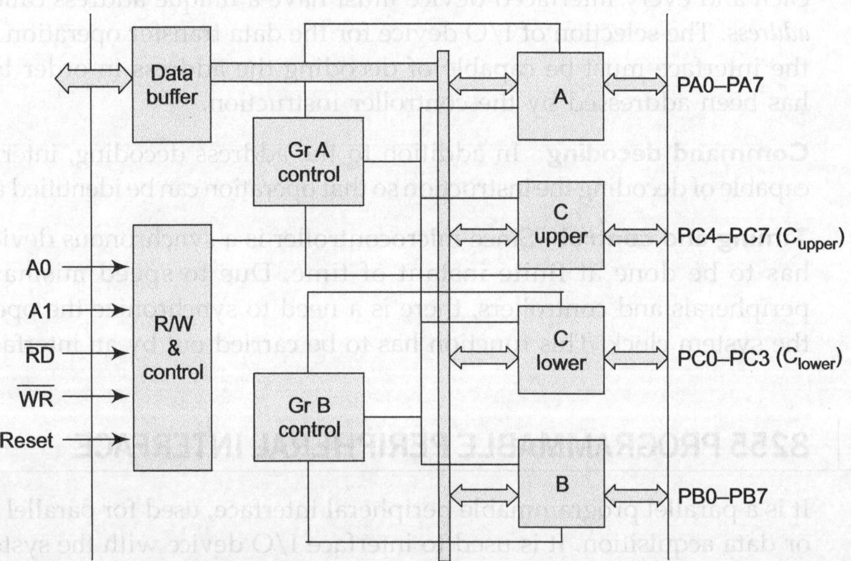

Fig. 6.3 Simplified internal block of 8255

The data bus buffers are used to interface 8255 to the system data bus. The chip can be configured for operation in any one of the three modes termed as Mode 0, Mode 1, and Mode 2 in a single pin I/O mode called bit SET/RESET BSR mode. The control register and ports have a unique address. The chip can be programmed by writing a command word in control word register, termed as CWR.

The control logic block accepts the control bus signals and inputs from the address bus and issues commands to individual group control blocks.

A_1, A_0 pins on 8255 chip corresponds to the internal address of the blocks Port A, Port B, Port C, and CWR. They are used to select the inner blocks of 8255. The four internal blocks A, B, C, and CWR have consecutive addresses in the I/O space. CWR is a write-only register. It shares address with a read-only register called CSR, which indicates status of 8255. Table 6.1 shows selection of inner block, configuration, and direction of flow of signals during the I/O operation.

Table 6.1 Port selection and direction of signal flow

Mode of operation	A_1	A_0	$\overline{RD}$	$\overline{WR}$	CS	Signal direction
Input	0	0	0	1	0	A → Data bus
	0	1	0	1	0	B → Data bus
	1	0	0	1	0	C → Data bus
	1	1	0	1	0	CSR → Data bus
Output	0	0	1	0	0	Data bus → A
	0	1	1	0	0	Data bus → B
	1	0	1	0	0	Data bus → C
	1	1	1	0	0	Data bus → CWR
Disable	X	X	X	X	1	Not selected : High impedance state 'Z'
	1	1	0	1	0	Illegal
	X	X	1	1	0	Not selected : High impedance state 'Z'

6.1.1 Programming 8255

Logic HIGH on RESET pin of 8255 causes all the ports in input mode. All the internal flip-flops are cleared and interrupts are reset. This status is maintained even after RESET signal is withdrawn. It changes when the control word is written in the CWR. The 8255 is programmed by writing a control word in the CWR register.

Port A and C_{upper} pins form group A, while the Port B and C_{lower} pins form group B. Modes of operations of group A and group B can be defined by the control word.

The modes of operations are as follows:

Mode 0   Simple I/O

           Here Port A, B, C are separate ports and can be used as I/O.

Mode 1   Strobbed I/O

           Gr. A and Gr. B can be used in I/O: Port C generates control signals.

Mode 2   Bi-directional I/O

           Allowed only with Port A. This allows bi-directional data transfer on 8-bit data bus.

BSR Mode   This mode of operation is allowed only for Port C. The other modes are inactive. This is used to set/reset Port C pins individually.

The control word format is shown in Tables 6.2(a) and (b). The format to be used is decided by bit B7, called *mode set bit*.

For the bit B7 = 1:

Mode set is enabled and operation can be in Mode 0, Mode 1, or Mode 2.

The pin configurations for these modes are as follows:

Mode 0   All ports PA, PB, PC_u, and PC_l are data ports.

Mode 1   PA and PB are data ports. PC_u and PC_l are control ports used to generate control signals for PA and PB, respectively.

Table 6.2   (a) Control word format for mode set

B7	B6	B5	B4	B3	B2	B1	B0
Mode set		Mode selection bits for Port A	A	PCU	Mode	B Set bit Port B	PCL
'1' Mode active '0' BSR mode		00 Mode 0 01 Mode 1 1X Mode 2		0: Mode 0 1: Mode 1			

B7 = 1   Mode set is active and Mode 0, 1, 2 can be set.

B7 = 0   Mode is inactive and Port C is in bit SET/RESET mode.

In BSR mode, appropriate bits can be SET/RESET as per the requirement by loading a word in CWR. The Port B cannot operate in Mode 2. The control word format for bit SET/RESET BSR mode.

Table 6.2   (b) Control word format for BSR mode

B7	B6	B5	B4	B3	B2	B1	B0
	X	X	X	Port C pin can be selected by 3-bit combination			'0' reset '1' reset

PC0 – 000
PC1 – 001
PC2 – 010
PC3 – 011
PC4 – 100
PC5 – 101
PC6 – 110
PC7 – 111

Any one of the eight combination can be selected for Port C pin

The BSR determines which particular bit from PC0–PC7 is being SET/RESET as per status of the bit B_0. It can be written for each pin of the Port C.

For example:

If PC2 and PC7 are to be RESET and SET: Two words are required.

PC2   00000100 $\rightarrow$ Control word is (04)H

PC7   00011101 $\rightarrow$ Control word is (1D)H

Normally, BSR is used to enable or disable interrupts for Port C, when the 8255 is used in Mode 1 or Mode 2.

6.1.2 Modes of Operations

Figure 6.2 shows block diagram with the grouping of port pins. The port pins can be used for the data or control signals based on selected mode of operation.

Mode 0 Operation

In the control word format MSB, i.e., B7 = '1' indicates that mode set is active.

B7	B6	B5	B4	B3	B2	B1	B0
Mode set	Mode selection bits for Port A		A	PCU	Mode bit Port B	B	PCL
1	0	0	I/O	I/O	0	I/O	I/O

There are various combinations available in which the ports can be used as input or output. This can be set by writing control word in the control word register.

Data bits →	B7	B6	B5	B4	B3	B2	B1	B0
Key code	1	0	0	PA	PCU	0	PB	PCL
80	1	0	0	0	0	0	0	0
81	1	0	0	0	0	0	0	1
82	1	0	0	0	0	0	1	0
83	1	0	0	0	0	0	1	1
88	1	0	0	0	1	0	0	0
89	1	0	0	0	1	0	0	1
8A	1	0	0	0	1	0	0	0
8B	1	0	0	0	1	0	1	1
90	1	0	0	1	0	0	1	0
91	1	0	0	1	0	0	0	1
92	1	0	0	1	0	0	0	0
93	1	0	0	1	0	0	1	1
98	1	0	0	1	1	0	0	0
99	1	0	0	1	1	0	0	1
9A	1	0	0	1	1	0	1	0
9B	1	0	0	1	1	0	1	1

Example 6.1   Design a system to accept a data from an analog device and display the digital equivalent on an LED port using 8255 in Mode 0.

Solution

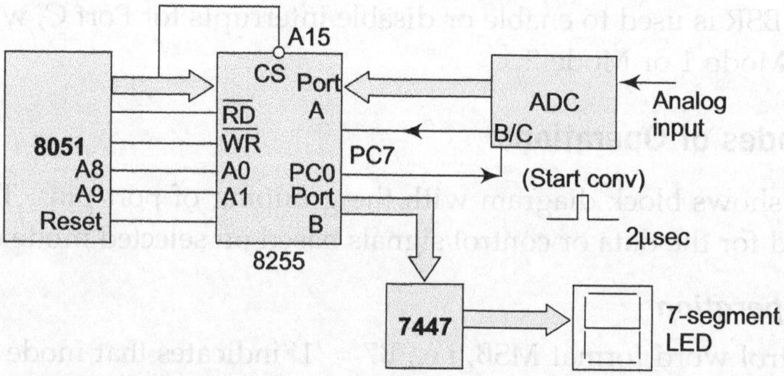

Fig. 6.4   System hardware layout

Figure 6.4 depicts the block schematic of the hardware setup for the desired system. In figure, only those pins which are used in the connection are shown.

The 7-segment LEDs are connected at the Port B through 4–7 code converter IC-7447 so that the look-up table is not required to convert 4-bit output of the port to 7-segment code. ADC is connected to Port A of 8255. Port A is to used in Mode 0. The ADC will start converting the applied analog input signal into the digital, when a start pulse is given. The port pin PC0 is used to generate the start convert pulse using BSR mode. During the process of conversion, B/C pin is logically HIGH indicating that conversion is going on. When conversion is over, the pin will be pulled LOW. This pin connected to PC7. Microcontroller polls the status of B/C pin through PC0 pin. On completion, the data from Port A will be READ and data is displayed on the 7-segment device connected to Port B.

System hardware   The port addresses of 8255 are decided by the connections shown in Fig. 6.4. Using Table 6.1 for the address pins port addresses and desired configuration of ports are

Port Address		
0000	Port A	Input port
0100	Port B	Output port
0200	Port C	C_L = input and C_U = output
0300	Port C_U	CWR

The control word for the configuration is 98h. We are using PC0 for giving start convert, hence

(a) Control word to SET the port pin PC0 using BSR is 01h.

(b) Control word to RESET the port pin PC0 using BSR is 00h.

System software The program segment for operation of the system is given below:

```
start:
        mov dptr, # 0300h
        mov a, 98h
        movx @dptr,a        ; initialize 8255
rept1:  mov a, 01
        movx @dptr,a        ; issue start convert BSR _set PC0
        nop
        nop
        mov a, 00           ; BSR _reset PC0
        movx @dptr,a        ; remove pulse
        mov dptr, #0200h    ; pointer to port-addr Port C
wait:   movx a, @dptr       ; READ Port C
        rlc a               ; wait for conversion to be completed
        jc wait
        mov dptr, #0000h
        movx a, @dptr       ; READ Port A
        inc dptr
        movx @dptr,a        ; WRITE Port B
        sjmp rept1
        end
```

Example 6.2 Design a setup to convert analog signal to digital using the DAC chip, which is interfaced to the microcontroller through the 8255. Assume that 8255 is operated in Mode 0 and counter method is used for the conversion.

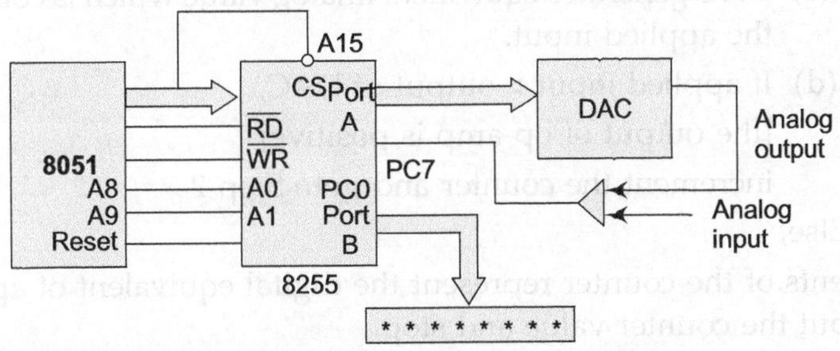

Fig. 6.5 Hardware setup

Figure 6.5 shows hardware setup for realization of ADC using DAC chip interfaced to the microcontroller through 8255. The output of Port A of 8255 is applied as the input to the DAC, which generates equivalent voltage at the output. The output of DAC is compared with the applied analog input. The output of the comparator is connected to port pin PC7. The digital value will be displayed on the LED connected at Port B of 8255.

The conversion process starts with application of a reset pulse. On reset, the counter is reset to zero. This counter value is applied to DAC through Port A. DAC converts this into analog value. The output of DAC is compared with applied input using op-amp, microcontroller checks output of the DAC by polling the status of PC7. If the PC7 is low, it means that the counter value is less than the required value and the counter must be incremented by one. The process is repeated till PC7 is high. The content of counter represents applied analog value in binary. The counter value is output on display device connected on the Port B of the 8255.

■ Hardware

The port addresses of 8255 are decided by the connections shown in Fig. 6.5. Using Table 6.1 for the address pins, port addresses and desired configuration of ports are

Port Address		Configuration
0000	Port A	output port
0100	Port B	output port
0200	Port C	C_U input port
0300	CWR	

■ Software

1. 8255 is initialized in Mode 0 with PA: Output; PB: Output and PC_U: input

2. The steps of conversion algorithm are described below:

    (a) On RESET, the counter is reset.

    (b) The output of the counter is applied to the DAC.

    (c) DAC generates equivalent analog value which is compared with the applied input.

    (d) If applied input > output of DAC,

    [the output of op-amp is positive]

    increment the counter and goto Step 2

    Else,

The contents of the counter represent the digital equivalent of applied input signal. Output the counter value and stop.

The following program implements the process:

```
adc-counter:
                mov 80h, # 20h          ; initialize SP
                mov dptr, # 0300h       ; pointer to the 8255_CWR
                mov A, # 81h            ; 8255 initialization
                movx @dptr,a           ; load 8255 control word
    rept1:      mov a, #00h
                mov dptr, #0000h        ; port addr. Port A-8255
                mov r3, a               ; save in the r3
    loop:       movx @dptr,a           ; input to DAC
                lcall dela             ; wait for response time of op-amp
                mov dptr, #0200h        ; pointer to port-addr C
                movx a, @dptr          ; check output of op-amp
                rrc a
                jnc end1
                inc r3
                mov a, r3
                sjmp loop
    end1:       mov dptr ,#0200h        ; port addr. Port B-8255
                mov a, r3
                movx @dptr,a           ; display result on LED
                sjmp rept1
                end
```

Example 6.3   Design a setup to convert analog signal to digital using the DAC chip, which is interfaced to the microcontroller through the 8255. Assume that 8255 is operated in Mode 0 and successive approximation method is used for the conversion.

Solution

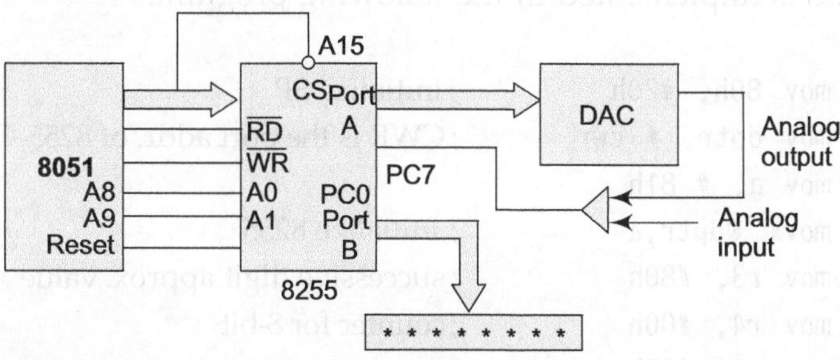

Fig. 6.6   Hardware setup

The hardware setup is same as in Example 6.2.

In successive approximation method, the conversion process is similar to the counter method. Starting value on reset is decided by the binary search method. The counter starts with a binary input value, which corresponds to the output voltage in the middle of the range, i.e., $V_{mid} = (V_{max} - V_{min})/2$, stored in a register called *successive approximation register* (SAR). This input is applied to DAC. The DAC output is compared with the applied analog signal. If the output of the op-amp is negative, it means the counter value is less so the desired value is in the upper half range, i.e., between V_{max} and V_{mid}. Value is in the lower half range, if the output of the op-amp is positive. The new mid-value is selected in corresponding range and stored in SAR. The process is repeated n times, where n is the number of desired output bits. The contents of counter represents the applied analog value in binary. The counter value is output on display device connected on the Port B of the 8255. The process is explained in Figs 5.10(a) and (b) in Section 5.2.6 of Chapter 5.

The algorithm steps for converting an analog signal into an 8-bit value are as follows:

1. Set the MSB of SAR to '1', load this value to the counter. The counter will start from the mid-point of the range.

2. Apply output of counter to DAC.

3. DAC generates equivalent analog value which is compared with the applied input.

4. If applied input > output of DAC,

   [the output of op-amp is positive]

   Set next bit to 1 and goto Step 2

   Else,

   If applied input = output of DAC

The contents of counter represent the digital equivalent of applied input signal, display the result and STOP

Else,

Reset current bit, Set next bit, goto Step 2

The process is implemented in the following program:

```
adc_sar:
            mov 80h, #20h          ; initialize SP
            mov dptr, # cwr        ; CWR is the port addr. of 8255-CWR
            mov a, # 81h
            movx @dptr,a           ; initialize 8255
rept1:      mov r3, #80h           ; successive digit approx. value
            mov r4, #00h           ; counter for 8-bit
            mov r5, #08h
```

```
               mov a, r3
loop:          add a, r4
               mov r4, a
               mov dptr, #PA          ; pointer to Port A
               mov @dptr,a            ; apply input to DAC
               lcall delay            ; wait for response
               mov dptr, #PC          ; pointer to Port C addr.
               movx a, @dptr          ; check output of op-amp
               rrc a
               jc next-bit
               mov a, r3              ; reset the current bit
               cpl a
               anl a, r4
               mov r4, a
Next-bit:
               mov a, r3
               rrc a
               mov r3, a
               djnz r5, loop
               mov a, r4
               mov dptr, #PB
               movx @dptr,a
               sjmp rept1
               end
```

Mode 1 Operation

In this mode of operation, the Port A and the Port B can be used together or individually as 8-bit latched input or the output ports and pins of Port C are used for control/status signals. Since the pin configurations are different in input and output mode, we will study as two different operations, i.e., Mode 1 (input) and Mode 1 (output) for group A as well as group B pins.

Mode 1 (input) Group A can be configured in Mode 1 by writing a control word in the control word register, which selects the Port A for input operation. Port A will accept the data from the parallel input device, to be READ by controller. The format of control word for this operation is

B7	B6	B5	B4	B3	B2	B1	B0
Mode set	Mode selection bits for Port A		A	PCU	Mode bit Port B	B	PCL
1	0	1	1	I/O	1	X	X

Fig. 6.8 Port structure of Port B Mode 1 (input) operation.

X: don't care condition. It indicates that setting of Port A configuration is independent of setting of group B pins which may be configured for Mode 1 input/output operation.

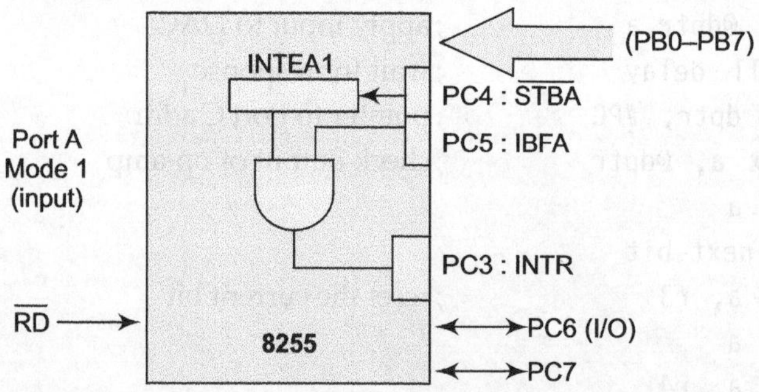

Fig. 6.7   Port structure of Port A Mode 1 (input)

Figure 6.7 shows the internal structure of group A, when configured for Mode 1 (input) operation. Port C_u pins are used for control signals in Mode 1. It is clear from Fig. 6.7, that PC6 and PC7 are free pins and can be configured as input or output ports by setting the bit b3 of control word as: '1' for INPUT and '0' for OUTPUT operation of the Port C free pins.

The peripheral device loads the data in the buffer of Port A with a strobe pulse STB at pin PC4. As a result IBFA signal is generated, which pulls pin PC5 to logic level '1'. IBFA and STBA results in pulling INTR to logic level '1', which is interpreted as an interrupt signal by microcontroller. INTR pin can be used to interrupt the controller. This can be enabled/disabled by INTEA flip-flop. On receipt of INTR signal, controller generates read signal (RD). The falling edge of the RD signal resets IBFA. It goes Low, which indicates that the input buffer is empty and new data can be loaded again.

Figure 6.8 shows the internal structure of group B, when configured for Mode 1 (input) operation. Port C_u pins are used for control signals in Mode 1. It is clear from Fig. 6.8, that all Port C pins are used to carry control signal, there are no free pins in this case.

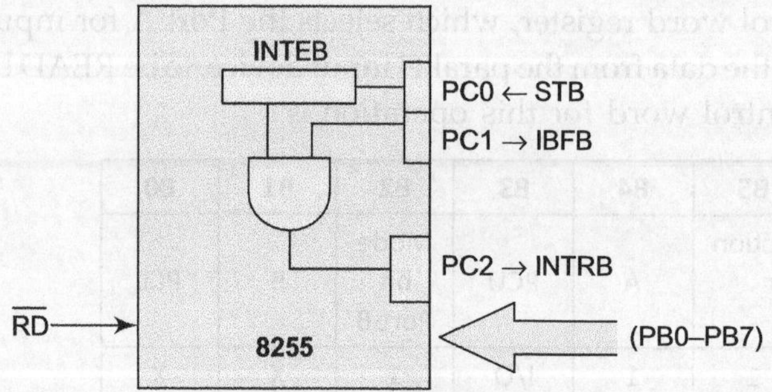

Fig. 6.8   Port structure of Port B Mode 1 (input) operation

The format of status word for Mode 1 (input) operation is

B7	B6	B5	B4	B3	B2	B1	B0
PC7	PC6	PC5	PC4	PC3	PC2	PC1	PC0
I/O	I/O	IBF A	INTE1 A	INTR A	INTR B	IBF B	INTE B

For data transfer, the interfacing of 8255 with microcontroller can be done in two different ways:

1. Using interrupt signal.
2. By polling the status of Port C through the status register using the format shown above.

Mode 1 (output)   Figure 6.9 shows output Port A in Mode 1. Microcontroller sends data to the output device through GR.A pins. The Port A pins are used for data transfer, while the Port C_u pins handle the control signals.

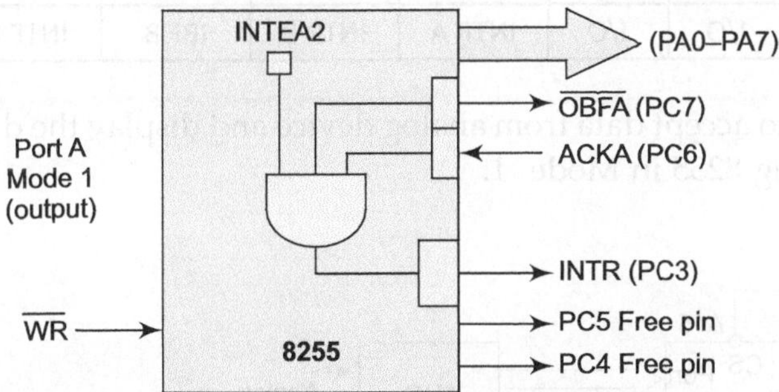

Fig. 6.9   Port structure of Port A Mode 1 (output)

Port C_u pins are used for control signals in Mode 1. It is clear from Fig. 6.9, that PC5 and PC4 are free pins and can be configured as input or output ports by setting the bit b3 of control word as: '1' for input and '0' for output operation of the Port C free pins.

The microcontroller writes data in the output buffer of Port A of 8255 at the falling edge of $\overline{WR}$ signal. The output buffer full $\overline{OBFA}$ signal goes low on rising edge of the $\overline{WR}$ signal, which indicates that there is data in the output buffer of 8255. $\overline{OBFA}$ is used as a strobe to transfer the contents of output buffer of Port A. The peripheral reads the data from buffer. The output device generates acknowledgement signal by pulling ACKA pin low to indicate at the end of operation. This signal is used to reset $\overline{OBFA}$, new data can be loaded by the microcontroller. When ACKA is high, INTRA is made high to interrupt the controller. Port B can also be used in Mode 1 (output). The structure is shown in Fig. 6.10.

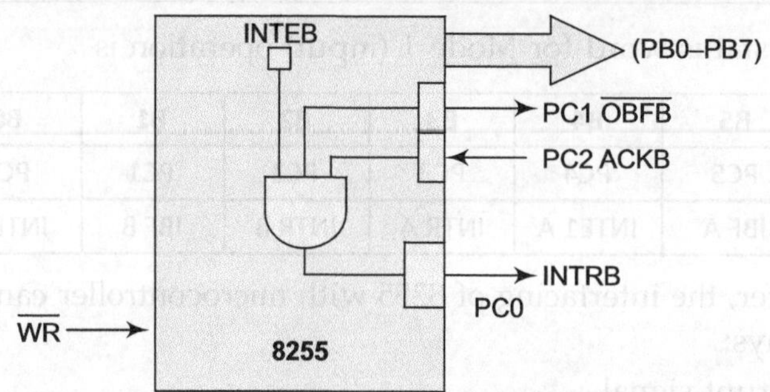

Fig. 6.10 Port structure of Port B Mode 1 (output)

The status word can be read by issuing READ command to Port C. The format of status word for the Mode 1 (output) operation is

B7	B6	B5	B4	B3	B2	B1	B0
PC7	PC6	PC5	PC4	PC3	PC2	PC1	PC0
OBFA	INTE2 A	I/O	I/O	INTR A	INTR B	IBF B	INTE B

Example 6.4 Design a system to accept data from analog device and display the digital output on LED port using 8255 in Mode 1.

Solution

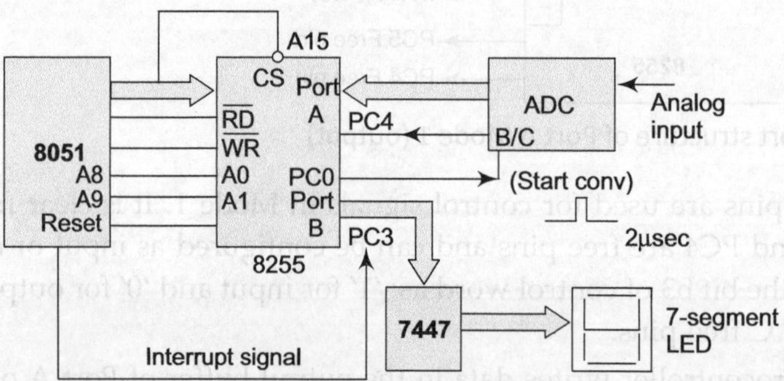

Fig. 6.11 System hardware layout

The system setup shown in Fig. 6.11 is similar to that of Example 6.1, except the INTRA pin is connected to the INT0 pin of the microcontroller. As the ADC is input device, we will use 8255 in Mode 1 (input).

The 7-segment LEDs are connected at the Port B through 4–7 code converter IC-7447 so that the look-up table is not required to convert 4-bit output of the port to 7-segment code. ADC is connected to Port A of 8255. Port A is to used in Mode 1.

The ADC will start converting the applied analog input signal into the digital, when a start pulse is given. The port pin PC7 is used to generate the start convert pulse using BSR mode. B/C pin is connected to pin PC4, which is a strobe pin in Mode 1 input. During the process of conversion, B/C pin is logically high indicating that conversion is going on. When conversion is over, the pin B/C be pulled low, the data will be pushed in the input buffer of Port A. Since buffer is full, an interrupt signal will be generated by 8255 on pin PC3, which is connected to the int0 pin of microcontroller. The data from Port A will be read by controller and will be displayed on the 7-segment device connected to Port B.

Thus when conversion is over, the ADC will input the data into buffer of Port A with a strobe. The buffer being full will generate interrupt on pin int0 of the microcontroller. Microcontroller will read data from the buffer.

- **Hardware** The port addresses of 8255 are decided by the connections shown in Fig. 6.11. Using Table 6.1 for the address pins, port addresses and desired configuration of ports are

  Port

  Address

0000	Port A	Input port
0100	Port B	Output port
0200	Port C_L	C_L: input C_U: output
0300	Port C_U	CWR

- **Software** It is necessary to initialize interrupt structure of 8051, and program the 8255 chip, as per the system requirement. The interrupt source is connected to INT0 pin, we will enable it using IE register.

(i) Enable int0 by writing control word in IE register [address A8]

EA	-	ET2	ES	ET1	EX1	ET0	EX0
1	0	0	0	0	0	0	1

The control word to enable int0 is 81h.

(ii) Interrupt priority level can be defined by IP register.

The format of IP register having address: B8

-	-	PT2	PS	PT1	PX1	PT0	PX0
0	0	0	0	0	0	0	1

The control word to give priority to 01h

(iii) The hardware port addresses of 8255 internal blocks are

    port add: 0000 - A

                         0100 - B

$$0200 - C$$
$$0300 - CSR$$

Let us configure ports as

Port A   -   Input, Mode 1
Port B   -   O/P
C_L -   X
C_U -   O/P

Control word for configuring the 8255 I/O ports in Mode 1 (input)

B7	B6	B5	B4	B3	B2	B1	B0
Mode set	Mode selection bits for Port A		A	PCU	Mode bit Port B	B	PCL
1	0	1	1	0	1	X	X

The control word for the configuration is (b4)h.

(iv) The PC7 is used for giving start convert pulse to ADC. This will be generated by using the BSR mode.

The format of BSR is

0	X	X	X	Port C pin can be selected by 3-bit combination			0 ; Set 1 ; Reset
0	0	0	0	1	1	1	1

To set PC7 control word is 0Eh

0	0	0	0	1	1	1	0

To reset PC7 control word is 0Fh

The program segment for system operation is as follows:

```
main:   mov  dptr, #0300h    ; initialize CWR 8255 pointer
        mov  a, #b4h         ; load control word
        movx @dptr,a         ; write CWR 82553
        mov  a8, # 81h       ; initialize IE
        mov  b8, #01h        ; initialize IP
        mov  a, 0fh
        movx @dptr,a         ; set PC7, BSR mode
        nop
        nop
        mov  a, 0eh          ; reset PC7, BSR mode
        movx @dptr,a
loop:   sjmp loop            ; wait for an interrupt
```

```
0003:     sjmp  int0
 int0:    mov dptr, # 8000h
          movx  a, @dptr       ; read Port A, output of ADC
          inc dptr             ; pointer to Port B
          movx  @dptr,a        ; display the result.
          acall  st_conv       ; issue start convert signal
          ret
  st_conv:
          mov dptr, # 8000h
          mov a, 0fh
          movx  @dptr,a        ; set PC7, BSR mode
          nop
          nop
          mov a, oeh           ; reset PC7, BSR mode
          movx  @dptr,a
          ret
          end
```

Example 6.5 Design a microcontroller-based printer controller to print a line of ASCII characters with the help of a printer interface using 8255 in Mode 1.

Solution

Since printer is an output device, we will use 8255 in Mode 1 (output). We will make use of some of the signals on the Port A. Typical hardware layout is shown in Fig. 6.12.

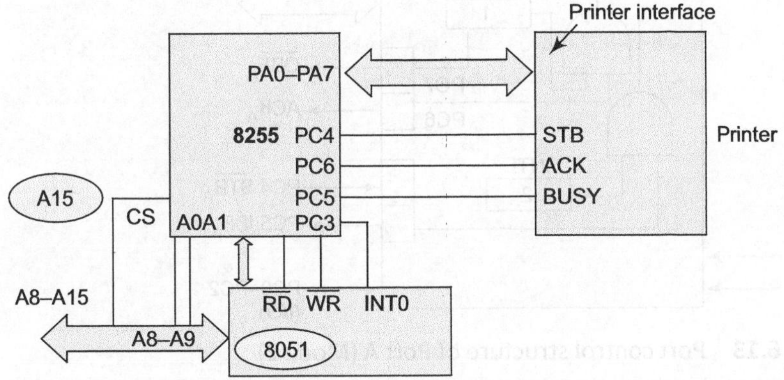

Fig. 6.12 System hardware setup for Example 6.3

The data transfer begins with the system placing ASCII data on the printer input data line. After sometime the STB signal is generated and BUSY signal is generated after it. On completion of data transfer, the ACK is generated so the BUSY signal is withdrawn. As in Example 6.2, the control words can be developed for Mode 1 (output) operation of Port A along with the control signals from Port C. The software, which is similar to Example 6.2 is left to the reader as an exercise.

Mode 2 Operation

In the Mode 2 operation, Port A is used as an 8-bit bi-directional data bus and using the port pins PC3–PC7 for control signals. Port B can be programmed as the input or output for the Mode 0 or the Mode 1 operation. Control word for configuring the Port A of the 8255 in Mode 2 is

B7	B6	B5	B4	B3	B2	B1	B0
Mode set	Mode selection bits for Port A		A	PCU	Mode bit Port B	B	PCL
1	1	x	x	x	x	X	X

Figure 6.13 depicts the port control circuit in Mode 2 operation of the Port A. It can be seen that it is a combination of the Mode 1 (input) and Mode 1 (output). All the pins of Port C in the group A are used by the Port A for control signals. Hence no pin is free for use either as input or output, as in the Mode 1 operation.

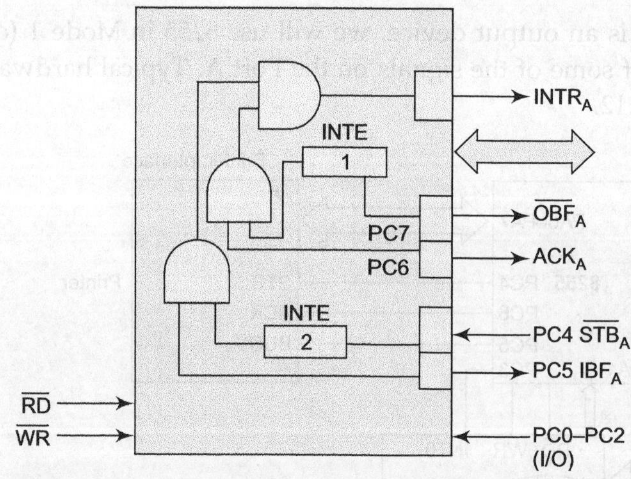

Fig. 6.13 Port control structure of Port A (Mode 2)

The output status signals are

INTE 1, is a flip-flop associated with the output buffer ($\overline{\text{OBF}}_A$)

INTE 2, is a flip-flop associated with the input buffer (IBF$_A$)

The status word format is

B7	B6	B5	B4	B3	B2	B1	B0
PC7	PC6	PC5	PC4	PC3	PC2	PC1	PC0
OBFA	INTE2 A	IBFA	INTE1 A	INTR A	X	X	X

Mode 2 is used to connect two or more microcontroller systems. The advantage of this mode is that the additional I/O ports required to carry out parallel data transfer is reduced and simplifies system hardware.

Distributed Processing

Distributed processing is a typical application of Mode 2. It is economical to have dedicated microcontroller to perform or monitor a specific task, when speed of operation is not critical. Operation of such dedicated microcontrollers performing different tasks can be controlled by a high-speed computer. This concept is known as *distributed processing*. One of the microcontroller works as a master while others operate as slaves. It becomes essential to transfer data between them using handshaking control signals with the property that port has capability of bi-directional data transfer. This can be done by using Port A of 8255 in Mode 2.

Since, speed is important for the master, it is put in interrupt mode, while slave is in use in the status check mode. The steps required in the data transfer operation are as following:

I. **To transfer the data from master to slave**

1. The master reads status of the output buffer by checking the status of the OBF line to verify whether the previous data byte has been read by the slave or not. This is an input function for master.

2. Master writes data in the port buffer and informs slave by causing low signal on the OBF line. This is an output function for the master.

3. Slave checks the status of OBF line for the data from master. This is the output function for slave.

4. Slave reads the data from output buffer of Port A and issues acknowledgement signal by pulling ACK line low. This indicates that data has been read from the output buffer.

II. **To transfer the data from slave to master**

1. The slave checks the IBF signal to find whether the input buffer of Port A is available for data transfer or not. This is an input function for the slave.

2. Slave places a byte in the input buffer with a strobe STB pulse. This is the output function for slave.

3. The master checks the status of IBF line to find whether the data is present in the buffer or not. This is an input function for master.

4. The master reads the data byte from the input buffer. This is an input function for the master.

Hardware setup For a complete bi-directional data transfer scheme OBF and ACK signals are used by the master, while the slave uses IBF and STB signals. This requires three input and two output ports. The Port A in Mode 2 will be sufficient for data transfer, while the port pins on the slave can be used. Figure 6.14 shows typical hardware setup.

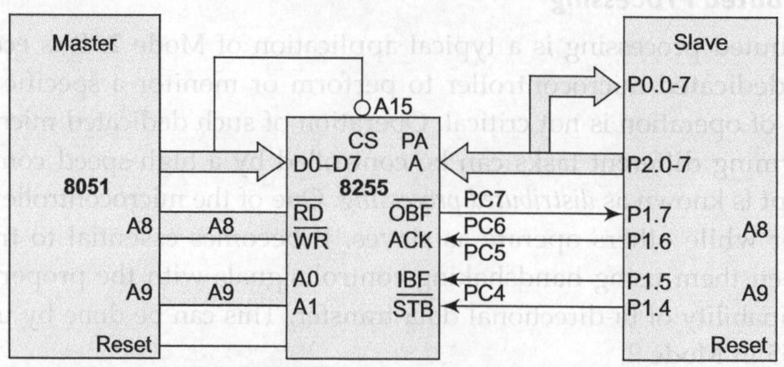

Fig. 6.14 Hardware setup for Port A (Mode 2 operation)

Master 8051 can check the status by reading the status register of the 8255 as it is interfaced with it. The data transfer can be done using Port A of 8255 because it works as a bi-directional data bus.

Slave has to perform read or write operation, the Port 1 bits are used for the handshaking control signals. Port 0 and Port 2 are used for input and output operation, respectively. Slave will read data from the output buffer of 8255, and writes data in the input buffer of 8255.

Software The programs are written in the memory address space of the corresponding controllers. This means that program for the master is written in master's memory address space and same for the slave is written in the memory address space of the slave.

The hardware port addresses of 8255 can be decided based on the connections shown in Fig. 6.14. They are

0000	Port A	Input
0100	Port B	Output
0200	Port Cl	Input
0200	Port Cu	Output
0300	CWR	

Port A of the 8255 is used in Mode 2 and Port B is not used. Control word for this configuration is

B7	B6	B5	B4	B3	B2	B1	B0
Mode set	Mode selection bits for Port A		A	PCU	Mode bit Port B	B	PCL
1	1	x	x	x	x	X	X

Using don't cares as 0, the control word is C0H

The status word format is

PC7	PC6	PC5	PC4	PC3	PC2	PC1	PC0
$\overline{OBF_A}$	INTE 1	IBF_A	INTE 2	$INTR_A$	X	X	X

1. Status of IBF can be checked by AND operation of status register with 20h.

2. Status of OBF can be checked by AND operation of status register with 80h or checking carry after rotate left operation on the contents of the status register after rotate instruction.

Example 6.6 Write a program to transfer 10 data bytes from the master to the slave.

Solution

Slave can read data from Port A by sending the ACK signal. The master has to check for the status of OBF signal, which when high indicates that the data byte has been read and buffer is empty. Slave also checks the status by reading port pin P1.7 of master. In order to issue read instruction, the buffer must be full.

The program segments for both the master and the slave are given below:

1. **Program for the master**

```
master:
        mov  80, #20h
        mov  r0, #addres        ;pointer to the data byte to be transferred
        mov  r4, # 0Ah          ;count, loop counter
        mov  a, #C0h            ;control word for 8255 Mode 2
        mov  dptr, #0300h
        movx @dptr, a           ;write control word
loop:   mov  a, @dptr           ;read status register
        rlc  a
        jnc  loop
        mov  dptr, #0000h
```

```
        mov  a, @r0
        movx @dptr,a
        inc  r0
        djnz r4, loop
        end
```

2. **Program for the slave**

```
slave:
        mov  80, #20h
        mov  r0, #slave_add    ;address where data is to be stored in slave
        mov  r4, #0Ah          ;count for loop counter
rept1:  mov  a,90h             ;read P1
        rlc  a
        jc   rept
        mov  a,80h             ;read P0
        mov  @r0, a
        inc  r0
        djnz r4, rept1
        end
```

6.2 | PROGRAMMABLE INTERVAL TIMERS (8253/8254)

The microcontroller has on-chip timers. But in certain applications they are not enough and there is a need to put extra timers/counters on the system. Due to the stability problem, external timing circuits, like 555, are not acceptable. Since the controller clock is a stable one, it can be used to generate the timing pulses through dividing clock signal by a number to derive desired frequency. The advantage of these timers is that in addition to normal task, they can be used to generate different signals for applications such as: waveforms generator, strobe pulses, interrupt pulses, real-time clock (RTC), event counter, square wave generator, and pulse generator.

The 8253/8254 is a programmable interval timer, available in a 24-pin package. It operates on 5 V power supply. Both the ICs are pin to pin compatible with the small differences in operation, such as:

1. 8253 can operate within a frequency range of 2 to 6 MHz, while 8254/8254-2 can operate up to a frequency of 8 MHz/10 MHz, respectively.

2. 8254 has a read back feature which allows latching of one, two or all the timers simultaneously.

6.2.1 Pin Layout and Block Diagram

Figure 6.15 shows pin configuration as well as the simplified internal block diagram of the 8253/8254 chip. The IC has three timer units and is connected to the microcontroller through an address decoder.

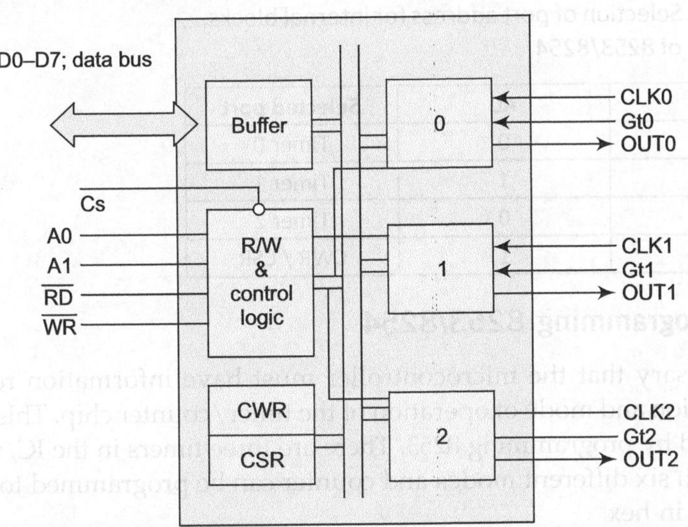

Fig. 6.15 Pin configuration and simplified internal block diagram of 8253/8254

Figure 6.16 represents a typical timer/counter unit. Each unit has a 16-bit counter/16-bit register and can be used in six different modes.

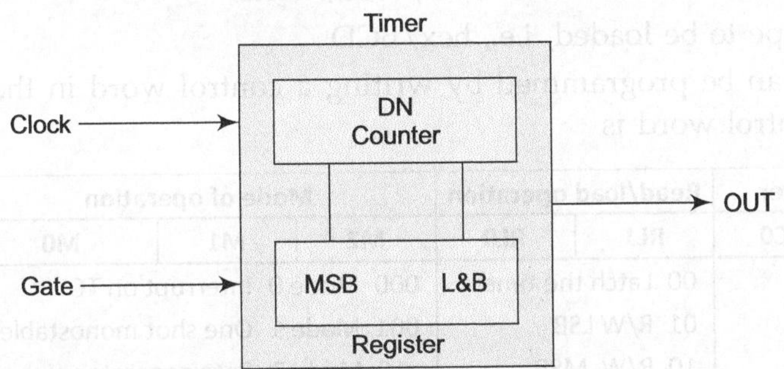

Fig. 6.16 A timer/counter unit of 8253/8254

For the counter operation: 16-bit data count is loaded in the associated register. Data can be loaded in BCD/HEX in the timer. For each clock pulse, the count is decremented by one, till the count is zero. The instant at which the contents become zero is called *terminal count (TC)*. At the end, i.e., when the count is zero, a pulse is generated which is used as an interrupt signal. The output produced at out terminal, depends on the selected mode of operation for the timer. A program control word is written in a control word register (CWR) for mode selection.

CS pin is used to select the chip from various other chips on microcontroller system and decides the hardware address to the IC-8253. The addresses of internal blocks, are decided by address line pins A0, A1. Thus, two pins A0 and A1 decide the port address of timers and CWR/CSR, as shown in Table 6.3.

Table 6.3 Selection of port address for internal blocks of 8253/8254

A1	A0	Selected port
0	0	Timer 0
0	1	Timer 1
1	0	Timer 2
1	1	CWR / CSR

6.2.2 Programming 8253/8254

It is necessary that the microcontroller must have information regarding the configuration and mode of operation of the timer/counter chip. This information is provided by programming 8253. There are three timers in the IC, which can be operated in six different modes and counter can be programmed to count either in BCD or in hex.

Thus, programming 8253 means specifying following information:

1. Timer to be used
2. Mode in which timer is to be operated
3. Operation to be carried out, i.e., read/write
4. Data type to be loaded, i.e., hex/BCD

The chip can be programmed by writing a control word in the CWR. The format of control word is

Select counter		Read/load operation		Mode of operation			Data type
SC1	SC0	RL1	RL0	M2	M1	M0	D
00 Timer 0		00 Latch the timer		000 Mode 0 Interrupt on TC			0 BCD
01 Timer 1		01 R/W LSB		001 Mode 1 One shot monostable			1 Hex
10 Timer 2		10 R/W MSB		x10 Mode 2 Rate generator			
11 Invalid for 8253		11 R/W 16-bits		x11 Mode 3 Square wave generator			
Read back for 8254		(with LSB first)		100 Mode 4 Software triggered strobe			
				101 Mode 5 Hardware triggered strobe			

6.2.3 Interfacing 8253

Figure 6.17 shows a typical schematic diagram of interfacing the PIT chip 8253 with the 8051 microcontroller.

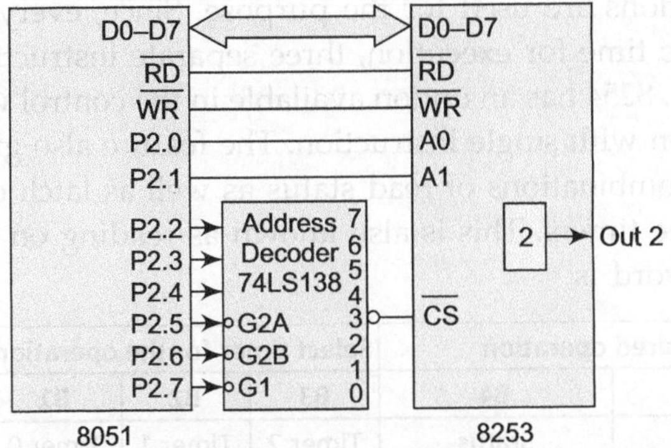

Fig. 6.17 Interfacing of 8253 with microcontroller

The port addresses of the 8253 components can be decided from Fig. 6.17.

P2.7	P2.6	P2.5	P2.4	P2.3	P2.2	P2.1	P2.0
A15	A14	A13	A12	A11	A10	A9	A8
1	0	0	0	1	1	A1	A0

The addresses of the internal components are

8C00	$\rightarrow$	Timer 0
8D00	$\rightarrow$	Timer 1
8E00	$\rightarrow$	Timer 2
8F00	$\rightarrow$	CWR/CSR

We will use this system setup for the examples in this section.

6.2.4 Reading Status of 8253

In order to check the status of the 8253 chip, the status register can be read. CSR is a read-only register, and it reflects status of 8253 during operation.

Out pin status	NC	Read/load operation		Mode of operation			Data type
		RL1	RL0	M2	M1	M0	D
1 High 0 Low	1 Null count 0 Available for counting	Similar to the control word format		Mode of operation set by control word			0 BCD 1 Hex

Null count means that TC is reached. The status of output pin can be polled for event driven applications.

6.2.5 Read Back Feature of 8254

Some applications require the status information of all the three timers at the same instant. 8253 requires three operations to read and latch all the three timers.

Three separate instructions are used for the purpose. Since, every instruction written requires a finite time for execution, three separate instructions are not useful for the operation. 8254 has an option available in the control word, which allows desired operation with single instruction. The feature also gives user an option of all possible combinations of read status as well as latch operation of one, two or all the three timers. This is also known as reading on the fly. The format of the control word is

Select counter		Desired operation		Select timer for the operation			
B7	B6	B5	B4	B3	B2	B1	B0
SC1	SC0	Count	Status	Timer 2	Timer 1	Timer 0	X
1	1	0 : Latch timer 1 : Do not latch	0 : Status Inf. 1 : No status Inf.	1: Select timer for operation 0 : Do not select timer			For future expansion

The bit combinations of B5 and B4 allows four possible operations for the latching and obtaining status information of selected timers. Bits B3, B2, and B1 can be used to select one, two or all three timers for the operations desired as per the combinations of B5, B4.

6.2.6 Modes of Operation

It is mentioned in the previous section that the 8253 timers can be operated in six different modes depending upon the application. The different modes of operations generate different waveforms at the out pin of the timer unit.

Mode 0: Interrupt on Terminal Count

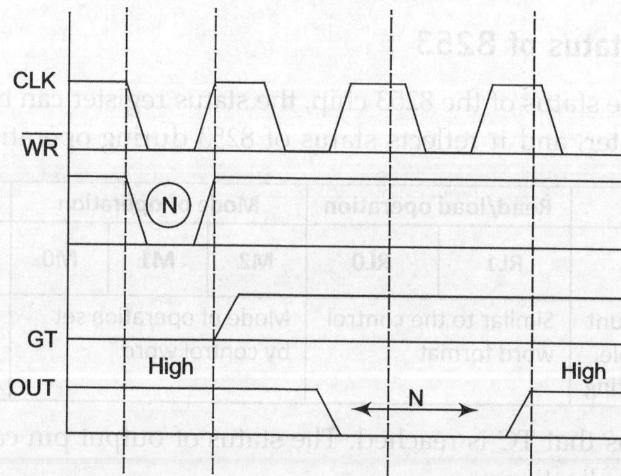

Fig. 6.18   Waveforms on out pin of 8253 timers for Mode 0 operation

This mode of operation provides an output signal at the out terminal of the selected timer, which may be used as an interrupt signal for the microcontroller. The output signal (Fig. 6.18) is generated depending on the number loaded in the register associated with the timer unit.

The Mode 0 operation is selected by writing a control word. A number termed as count is loaded in the timer register. When a clock pulse is received, counter counts down and out pin is pulled at logic level LOW. It is necessary that the gate input must be high. If the gate is at logic level LOW, the counter will stop counting. When TC is reached, the out pin will be pulled high. Thus, out pin can be made high after desired time interval decided by the count. This mode of operation is suitable for the applications which interrupt signal. Example 6.7 uses Timer 2 in Mode 0.

Example 6.7 Design a setup to generate 1 s signal for real time clock, using 8253 in Mode 0.

Solution

Let the frequency be 12 MHz

Since, the microcontroller has 12 pluses in one cycle, the clock period T is

$$T = 12 \,/ 12 \text{ MHz} = 1 \text{ } \mu s$$

It is not possible to directly generate the delay of 1 s by loading 8253 timer register with the count, as the count value is larger than 16-bits.

Program using subprogram We will develop a subroutine subprogram for 50 ms, which when called 20 times generates desired 1 s signal.

The delay count for generating a delay of 20 ms, can be calculated as follows:

$$N = 50 \times 10^{-3} \,/\, 10^{-6} = [50{,}000]_{10}$$

N is a large number and cannot be loaded in the 16-bit timer register, N_{10} should be converted to hex value before loading. Conversion into hex results a value N_{16},

$$N_{16} = [C350]_{16}$$

For generating second signal for the real time clock, following procedure is carried out:

1. Let us initialize 8253 assuming that we select

    Timer 2

    Mode 0 operation   ; interrupt on TC mode of operation

    data is loaded in hex

    Load operation: 16-bit simultaneously with LSB first

The control word bits setting for above initialization is

SC1	SC0	RL1	RL0	M2	M1	M0	Data
1	0	1	1	0	0	0	0

The control word formulated is → (B0)

2. We have to wait for the contents of the Timer 2 counter to be zero. Since we are using status check, it is necessary to latch the timer, read the value to find if TC is reached or not.

The control word bits setting for latching the timer is

SC1	SC0	RL1	RL0	M2	M1	M0	Data
1	0	0	0	0	0	0	0

The control word for latching the Timer 2 → (80)

(a) The subroutine subprogram for 50 ms, will carry out the following steps, when called by the main program.

Load control word, load register of the selected timer, with the count for 50 ms, which is $C350_{16}$

Check if the TC is reached or not. This is carried out by reading content of latched timer register.

On reaching terminal count return to Main program.

(b) Main program will initialize the stack pointer and load count value for the number of times the subprogram has to be called which is $[14]_{16}$ in one of the microcontroller register. When the counter register is zero it indicates that a delay of 1 s is generated and can be used to signal RTC.

Assuming hardware setup of Fig. 6.17, a segment of program listing is given below:

```
                                        ; this is a subroutine to generate delay of 50 ms
Delay_50:
          mov dptr, #8F00h
          mov a, #B0                    ; select the Timer 2, Mode 0
          movx @dptr,a
          mov a, #50                    ; load timer register LSB count
          mov dptr, #8E00h
          movx @dptr,a
          mov a, #C3h                   ; load timer register MSB count
          movx @dptr,a
loop:     mov a, #80h                   ; latch the timer
          mov dptr, # 8F00h
          movx @dptr,a
          mov dptr, #8E00h
```

```
                movx a, @dptr          ; read LSB of latched value
                mov r3, a              ; save in r3
                movx a, @dptr          ; read MSB of latched value
                orl a, r3
                jnz loop
                ret
```

```
                                       ; main program segment to generate 1 s pulse
main:           mov 80, #20h           ; initialize stack pointer
                mov r4, #14h           ; set loop counter to 20
loop:           acall Delay_50
                djnz r4, loop          ; 1 s delay when loop is over
                end
```

Simple program: other approach   Assuming hardware set in Fig. 5.17, it is possible to write a simple program without using the subprogram feature. A program segment is given below:

```
main:           mov 80, #20h           ; initialize stack pointer
sec:            mov r4, #14h           ; set loop counter to 20
loop:           mov dptr, #8F00h
                mov a, #B0h            ; select the Timer 2, Mode 0
                movx @dptr,a
                mov a, #50h            ; load timer register LBS count
                mov dptr, #8E00h
                movx @dptr,a
                mov a, #C3h            ; load timer register MSB count
                movx @dptr,a
wait:           mov a, #80h            ; latch the timer
                mov dptr, # 8F00h
                movx @dptr,a
                mov dptr, #8E00h
                movx a, @dptr          ; read LSB of latched value
                mov r3, a              ; save in r3
                movx a, @dptr          ; read MSB of latched value
                orl a, r3
                jnz wait
                djnz r4, loop
```

Example 6.8   Design a scheme to control PCB making microcontroller in the factory. We wish to keep track of the finished board and keep a count so that check on any board lost in machine can be made.

Solution

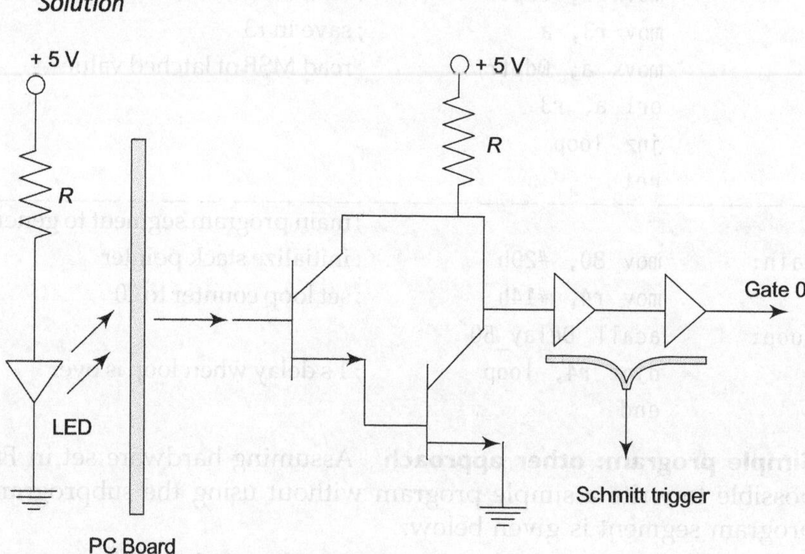

Fig. 6.19   System setup to generate gate pulse for Timer 0 of 8253 operated in Mode 0

Figure 6.19 shows a setup for generation of gate pulse for the 8253 Timer 0. The LED supplies the base drive to the phototransistor *T*. *T* is ON, so the collector of *T* will be at logic level LOW. When a board passes between the diode and *T*, there is no base drive to the transistor *T*. Hence *T* is OFF and collector is at logic level HIGH. This logic signal at collector of *T* is used as a gate signal for Timer 0 of Fig. 6.20.

It is known that in Mode 0, operation gate must be high for countdown, otherwise counter stops counting down. The counter register of Timer 0 can be preloaded with the number, say 10,000. When all boards are over, the counter must reach terminal count. If it does not, it means that the some boards are lost in the machine.

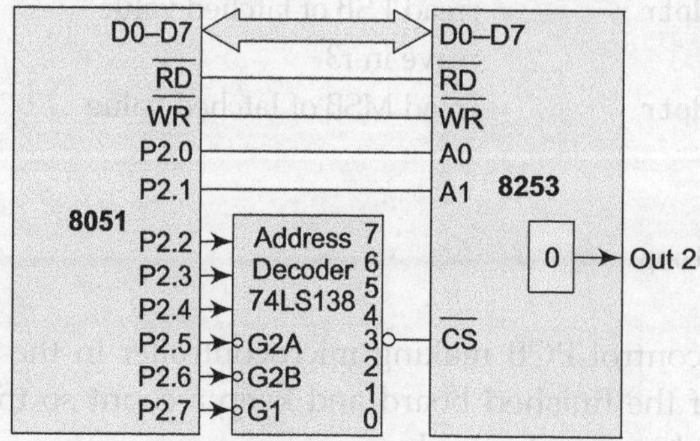

Fig. 6.20   Interfacing of 8253 with microcontroller

The programming part requires initialization of 8253 timer and check for TC. It is left as an exercise for the readers.

Mode 1: Retriggerable Monostable Multivibrator

In this mode, 8253 timer behaves like a re-triggerable monostable multivibrator. The gate input serves the purpose of a trigger. The operation of the timer is to make output pin logically low when counting is going on. On reaching TC, the output pin is made at logic level HIGH. Figure 6.21 shows waveform at the output pin of the timer.

8253/8254 Mode Operation

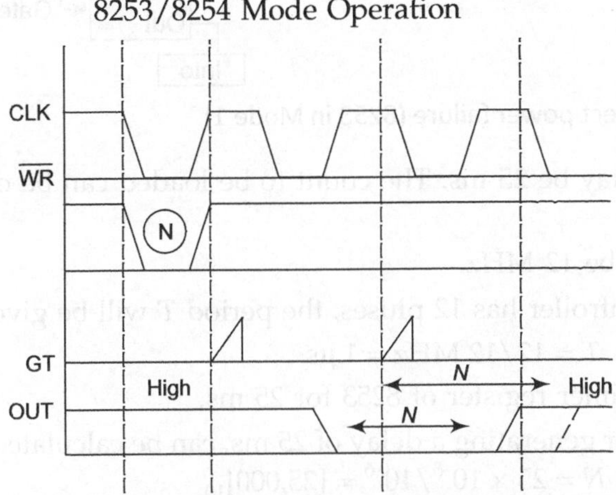

Fig. 6.21 Waveforms on output pin of 8253 timers for Mode 1 operation

In the application, timer is loaded with count which depend on the period for which the output is to be kept low. The trigger is applied in the form of an edge. The counter starts counting down, puts output low, till it reaches TC. If a trigger is applied at the gate of timer, the counter register will be reloaded with the original count and timer starts all over again.

Example 6.9 Design a scheme to detect the power failure.

Solution

The monostable IC 74LS121, 122 whose pulse width depends on associated, R/C network may be used. From view point of stability 8253 in Mode 1 is a better choice. Figure 6.22 depicts a setup for detection of the power failure.

The periodic time of input frequency, i.e., 50 Hz is 20 ms. When the power is present, a trigger will be applied at a regular interval of 20 ms to the gate of timer, 8253 register is loaded with original count and the output remains low as long as power is ON. In case of power failure, due to absence of the trigger pulse at the gate, counter will reach TC and produces high signal at output pin of the timer, which can be used as a signal for a power failure. For safety purpose, the timer is loaded with the value greater than 20 ms.

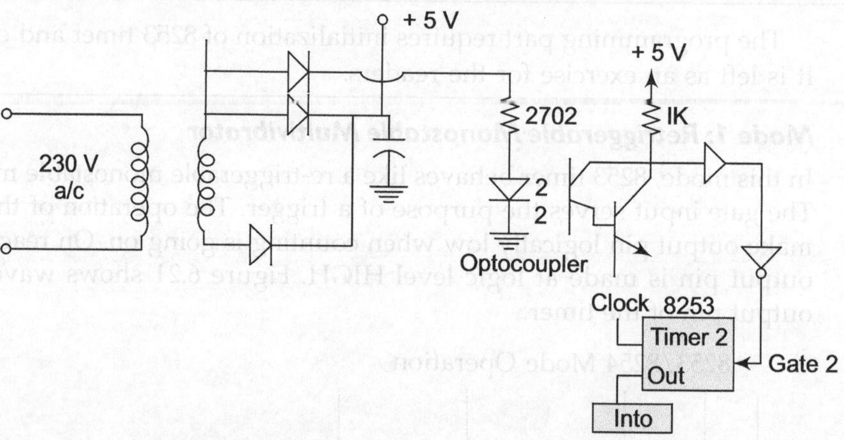

Fig. 6.22 Setup to detect power failure (8253 in Mode 1)

Let T_d; desired delay be 25 ms. The count to be loaded can be calculated as follows:

Let the frequency be 12 MHz.

Since the microcontroller has 12 pluses, the period T will be given by

$$T = 12/12\ \text{MHz} = 1\ \mu s$$

We will load the timer register of 8253 for 25 ms.

The delay count for generating a delay of 25 ms, can be calculated as follows:

$$N = 25 \times 10^{-3}/10^{-6} = [25,000]_{10}$$

Since, N is a large number which cannot be accommodated within a 16-bit timer register, it has to be converted into hex. On conversion, we get

$$N_{16} = (61A8)_{16}$$

Let us initialize 8253 assuming that we select

Timer 2

Mode 1; Re-triggerable monostable multivibrator

Load data in hex

Load 16-bit simultaneously with LSB first

SC1	SC0	RL1	RL0	M2	M1	M0	Data
1	0	1	1	0	0	1	0

The control word is B2

A segment of program listing is given below for operation of setup shown in Fig. 6.20.

```
Mode_1:
        mov dptr, #8F00h
        mov a, #B2          ;select the Timer 2, Mode 1
        movx @dptr,a
```

```
              mov a, #A8h           ;load timer register count LSB
              mov dptr, # 8E00h
              movx @dptr,a
              mov a, #61h           ;load timer register count MSB
              movx @dptr,a
loop:         sjmp loop
              end
```

When power fails, output of 8253 will be high and the system will jump to interrupt 0, where a subprogram may be written for the operation to be performed during power failure.

```
              0003        SJMP        ISR0
ISR0:         ****
              ****
```

Mode 3: Squarewave Generator

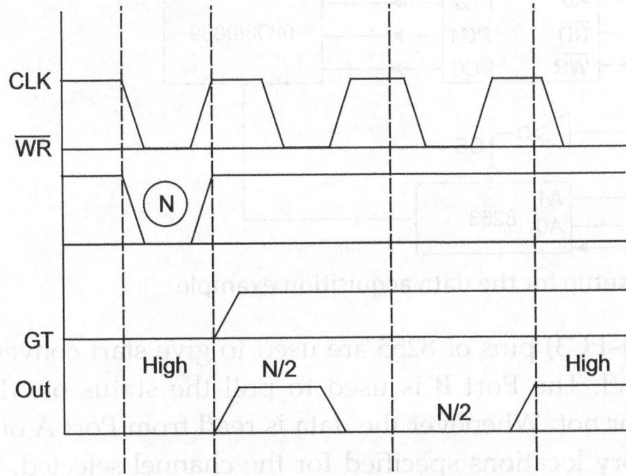

Fig. 6.23 Waveforms on output pin of 8253 timers for Mode 3 operation

In this mode of operation the output remains high for a period of $N/2$ and low for remaining period $N/2$. Once the cycle is over, the count N will be reloaded and operation is repeated. A continuous square wave will be produced at the output pin. The waveforms are as shown in Fig. 6.23.

It is to be noted that if N is even, a square wave with 50% duty cycle is generated. If N is odd, the output is high for $(N + 1/2)$ and low for $(N - 1/2)$, thus square wave will not be symmetrical.

i.e., duty cycle i = 50%.

Example 6.10 Design a data acquisition system, which accepts data from four variables. In a process control loop each variable is received through a channel, which is sampled 15 times, using 0809. The data is to be stored in external memory in the following locations:

Data samples of ch.0 : 30 , 31, 32, - - - - , 3F

Data samples of ch.1 : 40 , 41, 42, - - - - , 4F

Data samples of ch.2 : 50 , 51, 52, - - - - , 5F

Data samples of ch.3 : 60 , 61, 62, - - - - , 6F

Solution

The hardware setup is shown in Fig. 6.24.

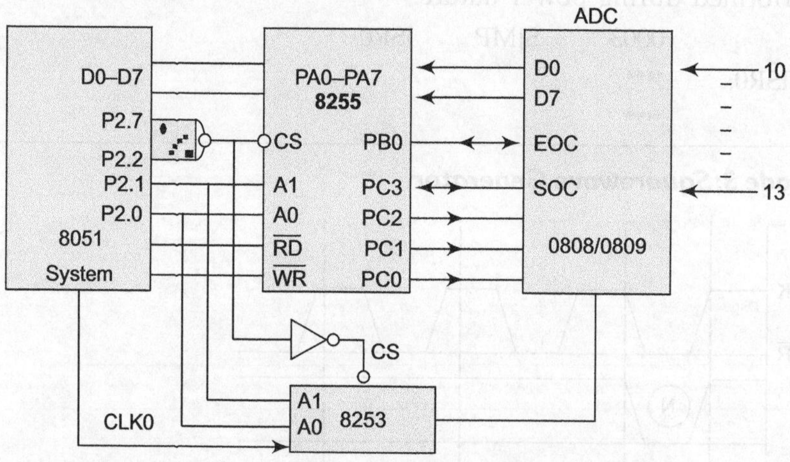

Fig. 6.24 Hardware setup for the data acquisition example

The Port C_L (PC0–PC3) pins of 8255 are used to give start convert as well as to select the channel. The Port B is used to poll the status of ADC, i.e., the conversion is over or not. Whenever the data is read from Port A of 8255 and is stored in the memory locations specified for the channel selected. The process is repeated for each channel 16 times. The start convert pulse must be given for about 100 ns. Average conversion time is of the order of 100 ns. Drawback is that there is no provision for connection of S/H circuit and hence 8 (S/H) are required. The hardware port addresses of the chips are

8255 :		Port Addr.	8253:		Port Addr.
Port A	→	FC00	Timer 0	→	0000
Port B	→	FD00	Timer 1	→	0100
Port C	→	FE00	Timer 2	→	0200
CWR	→	FF00	CWR	→	0300

The control words are to be formulated for initialization of 8255 and 8253 chips.

The software for the operation is listed below:

```
main:       mov 80, #20h        ; initialize SP
            mov dptr, #FF00     ; pointer to 8255 port add
            mov a, #CWR_55      : control word 8255
            movx @dptr,A        ; initialize 8255
            mov a, #CWR_53      : initialize 8053, Timer 0
            mov dptr, #0300h    ; in Mode 2
            movx @dptr,a

start:
            mov r0, #30h        : add for ch-0: sample 1
            mov r5, #10h        : no. of samples

push1:
            push r0
            mov r3, #00         : channel 0 is selected

loop1:
            mov a, #08          : send start convert to ch-0
            mov dptr, #FE00h
            movx @dptr,a
            nop
            mov a, #00
            orl a, r3
            movx @dptr,a        : remove start convert
            mov dptr, #FD00h

wait:       movx a, @dptr
            rlc a
            jnc wait
            mov dptr, #FC00h
            movx a, @dptr       ; read data from Port A
            mov @r0, a
            nop
            nop
            nop
            add r0, #10
            inc r3
            cjne r3, #04, loop1
            pop r0
            inc r0
            djnz r5, push1
            sjmp start
            end
```

Mode 4: Software Triggerred Strobe

In this mode, operation is similar to Mode 0, except that the output will be high when counting is going on. When TC is reached output will be pulled low for one T-state, and again it will be high. Thus, it produces a strobe pulse at the instant decided by the value loaded in the register.

Example 6.11 Design a scheme to provide a 1 s software triggered strobe at the output of the counter 2 of 8253.

Solution

Assuming the crystal frequency of 12 MHz $\rightarrow$ $T = 1$ ms

Hence the count to be loaded in the register for a second pulse is

$$n = 1/10^{-6} = 10,00,000$$

This number cannot be accommodated into the 16-bit timer register. Hence, we use two counters to load the values N_1 and N_2 such that $N_1 \times N_2 = 10,00,000$. We select:

$$1 \quad - \quad N_1 \quad = \quad 1000$$
$$2 \quad - \quad N_2 \quad = \quad 1000$$

The hardware setup is shown in Fig. 6.25.

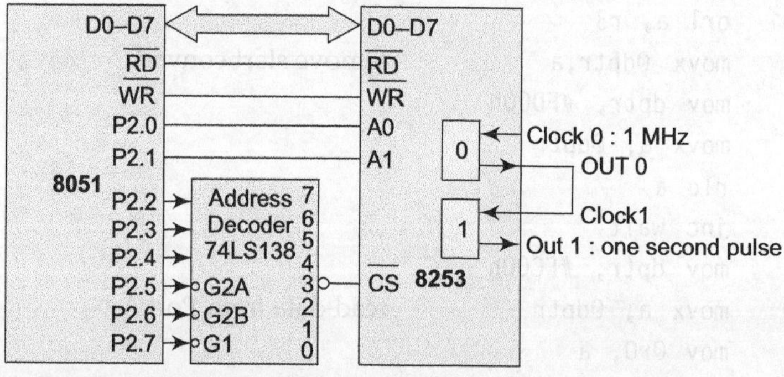

Fig. 6.25   System setup for generating 1 s strobe pulse

For proper operation of the system

1. Timer 0 is used in Mode 2, which is a divide by N counter mode. The input frequency is divided by $N_1 = 1000$ so that a clock frequency of 1 kHz will be produced at the Out 0 terminal of the Timer 0.

2. This output is used as a clock by Timer 1 in Mode 0 operation, generates a software triggered strobe at 1 s, as $N_2 = 1000$ is loaded in the register.

The programming efforts require initialization of Timer 0 and Timer 1 in Mode 2 and Mode 4, respectively, and loading of the counts N_1, N_2 in registers of the timers. The listing is given below:

```
main:       mov dptr, # 8F00h        : initialize memory pointer
            mov a, #35h              ; select Timer 0, Mode 2, loading data in BCD
            movx @dptr,a             ; load CWR
            mov a, #00h              ; load Timer 0 register
            mov dptr, #8C00h
            movx @dptr,a
            mov a, #10h
            movx @dptr,a
            mov dptr, #8F00h
            mov a, #79h              ; select Timer 1, Mode 4, data in BCD
            movx @dptr,a
            mov dptr, #8C00h
            mov a, #00h              ; load Timer 1 register
            movx @dptr,a
            mov a, #10h
            movx @dptr,a
loop:       sjmp loop
```

The program when executed will generate a software triggered strobe at every 1 s at the output pin of Timer 1, as long as program is running.

6.3 | REAL TIME CLOCK WITH RAM: DS1287

The DS1287 real time clock plus RAM is designed to be a direct replacement for the MC146818. A lithium energy source, quartz crystal, and write-protection circuitry is contained within a 24-pin dual in-line package.

The functions include a nonvolatile time-of-day clock, an alarm, a one-hundred-year calendar, programmable interrupt, square wave generator, and 50-bytes of nonvolatile static RAM. The real time clock plus RAM is distinctive in that time-of-day and memory are maintained even in the absence of power.

6.3.1 Features

Summary of the features of the RTC with RAM is given below:

- Pin compatible with the MC146818 A.
- Totally nonvolatile with over 10 years of operation in the absence of power.
- Self-contained subsystem includes lithium, quartz, and support circuitry.

- Counts seconds, minutes, hours, days, day of the week, date, month, and year with leap year compensation.
- Binary of BCD representation of time, calendar, and alarm.
- 12- or 24-hour clock with AM and PM in 12-h mode.
- Interfaced with software as 64 RAM locations.
- 14-bytes of clock and control registers and 60-bytes of general purpose RAM.
- Programmable square wave output signal.
- Bus compatible interrupt signals (IRQ), three interrupts, which are separately software maskable and testable.
- Time of day alarm once/second to once/day. Periodic rates from 122 μs to 500 ms.

The real time clock function will continue to operate and all of the time, calendar, and alarm memory locations in RAM remain nonvolatile regardless of the V_{CC} input. When V_{CC} is applied to the DS1287 and reaches a level of greater than 4.25 V, the device becomes accessible after 100 ms, provided that the oscillator is running and the oscillator countdown chain is not in reset. This time period allows the system to stabilize after power is applied. When V_{CC} falls below 4.25 V, the chip select input is internally forced to an inactive level regardless of the value of chip select (CS) at the input pin. Hence, DS1287 is write-protected.

When the DS1287 is in a write state, all inputs are ignored and all outputs are in a high impedance state. When V_{CC} falls below a level of approximately 3 V, the external V_{CC} supply is switched OFF and an internal lithium energy source supplies power to the real time clock and the RAM memory.

6.3.2 Pin Description

Figure 6.26 shows the internal block diagram of the DS1287 chip, while the pin layout is depicted in Fig. 6.25. GND, V_{CC}-DC power is provided to the device on these pins. V_{CC} is the + 5 V input. When 5 V is applied within normal limits, the device is fully accessible and data can be written and read. When V_{CC} is below 4.25 V the time-keeping function continues unaffected by the lower input voltage.

As V_{CC} falls below 3 V typical, the RAM and timekeeper are switched over to an internal lithium energy source. The time-keeping function maintains switched over to an internal lithium energy source. The time-keeping function maintains an accuracy of ±1 min per month at 25°C regardless of the voltage input on the V_{CC} pin.

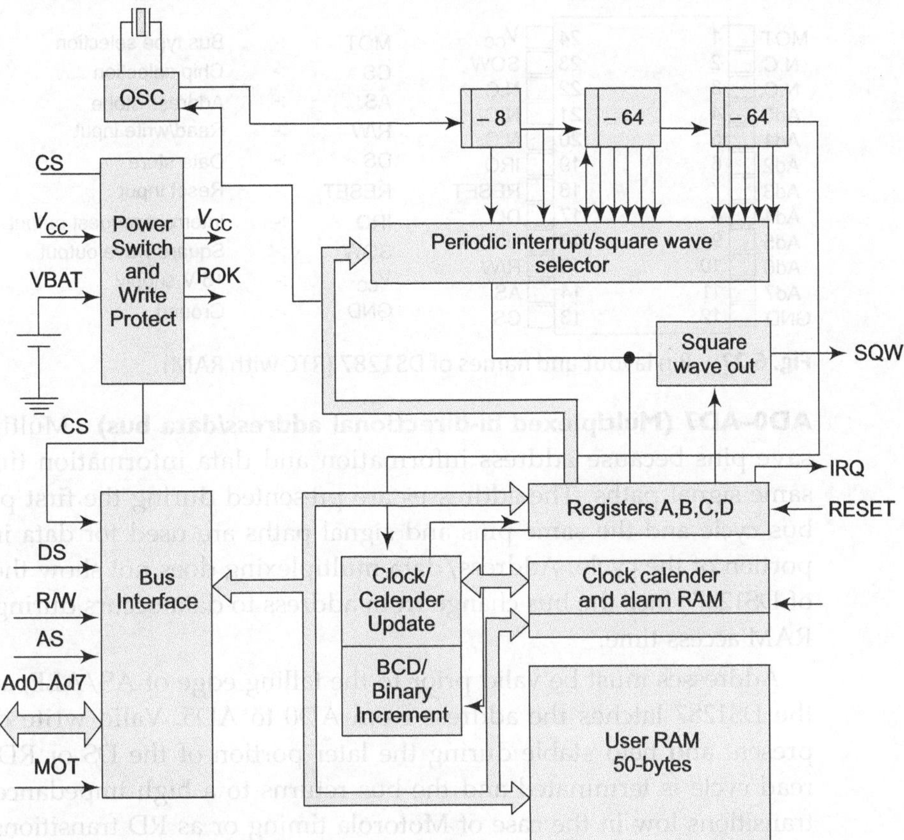

Fig. 6.26   Internal block diagram of DS1287

The details of pin names of DS1287 and their functions are briefly described in Fig. 6.27.

MOT (Mode select)   The MOT pin offers the flexibility to choose between two bus types. When connected to V_{CC}, Motorola bus timing is selected. When connected to GND or left disconnected, Intel bus timing is selected. The pin has an internal pull-down resistance of approximately 20 K.

SQW (Square-wave output)   The SQW pin can output a signal from one of 13 taps provided by the 15 internal divided stages of the real time clock. The frequency of the SQW pin may be changed by programming Register A. The SQW signal may be turned ON and OFF using the SQW bit in Register B. The SQW signal is not available when V_{CC} is less than 4.25 V typical.

MOT	1	24	V_{CC}	MOT :-	Bus type selection
N.C.	2	23	SOW	CS :-	Chip selection
N.C.	3	22	N.C.	AS :-	Address stobe
Ad0	4	21	N.C.	R/W :-	Read/write input
Ad1	5	20	N.C.	DS :-	Data store
Ad2	6	19	IRQ	RESET :-	Reset input
Ad3	7	18	RESET	IRQ :-	Interrupt request output
Ad4	8	17	DC	SQW :-	Square wave output
Ad5	9	16	N.C	V_{CC} :-	+5 V supply
Ad6	10	15	R/W	GND :-	Ground
Ad7	11	14	AS		
GND	12	13	CS		

Fig. 6.27   Pin layout and names of DS1287 (RTC with RAM)

AD0–AD7 (Multiplexed bi-directional address/data bus)   Multiplexed buses save pins because address information and data information time share the same signal paths. The addresses are presented during the first portion of the bus cycle and the same pins and signal paths are used for data in the second portion of the cycle. Address/data multiplexing does not show the access time of DS1287 since the bus change from address to data occurs during the internal RAM access time.

Addresses must be valid prior to the falling edge of AS/ALE, at which time the DS1287 latches the address from AD0 to AD5. Valid write data must be present and held stable during the later portion of the DS or RD pulses. The read cycle is terminated and the bus returns to a high impedance state as DS transitions low in the case of Motorola timing or as RD transitions high in the case of Intel timing.

AS (Address strobe input)   A positive going address strobe pulse used to demultiplex the bus. The falling edge of AS/ALE causes the address to be latched within the DS1287.

DS (Data strobe or read input)   The DS/RD pin has two modes of operation depending on the level of the MOT pin. When the MOT pin is connected to V_{CC}, Motorola bus timing is selected. In this mode DS is a positive pulse during the latter portion of the bus cycle and is called *data strobe*. During read cycles, DS signifies the time that the DS1287 takes to drive the directional bus. In write cycles the trailing edge of DS causes the DS1287 to latch the written data. When the MOT pin is connected to GND, Intel bus timing is selected. In this mode the DS pin is called read (RD). RD identifies the time period when the DS1287 drives the bus read data. The RD signal is the same definition as the output enable (OE) signal on a typical memory.

R/W (Read/write input)   The R/W pin also has two modes of operation. When the MOT pin is connected to V_{CC} for Motorola timing, R/W is a level which indicates whether the current cycle is a read or write. A read cycle is indicated with a high level on R/W while DS is high. A cycle is indicated when R/W is low during DS.

When the MOT pin is connected to GND for Intel timing, the R/W signal is an active low signal called WR. In this mode the R/W pin has the same meaning as the write enable signal (WE) on generic RAM.

CS (Chip select input)   The chip select signal (CS) must be asserted low for a bus cycle in which the DS1287 is to be accessed. CS must be kept in the active state during DS and AS for Motorola timing and during $\overline{RD}$ and $\overline{WR}$ for Intel timing. Bus cycles, which take place without asserting CS will latch addresses but no access will occur. When V_{CC} is below 4.25, the DS1287 internally inhibits access cycles by internally disabling the CS input. This protects both the real time clock and RAM data during power outages.

IRQ (Interrupt request output)   The IRQ pin is an active low output of the DS1287 that is used as an interrupt input to a processor. The IRQ output remains low as long as the bit causing the interrupt is present and the corresponding interrupt-enable bit is set. The IRQ pin normally reads the C register. The RESET pin clears pending interrupts.

When no interrupt conditions are present, the IRQ level is in the high impedance state. Multi devices may be connected to an IRQ bus. The IRQ bus is an open drain out requires an external pull up resistor.

RESET (Reset input)   The RESET pin has no effect on the clock, calendar, or RAM. On power-up, the RESET pin may be held low for a time in order to allow the power supply to stabilize. The amount of time that RESET is held low is dependent on the application. However, if RESET is used on power up, the time must exceed 200 ms to make sure that the internal timer which controls the DS1287 on power-up has timed out.

When reset is low and V_{CC} is above 4.25 V the following occurs:

1. Periodic interrupt enable (PE) bit is cleared to zero.
2. Alarm interrupt enable (AIE) bit is cleared to zero.
3. Update ended interrupt flag (UF) bit is cleared to zero.
4. Interrupt request status flag (IRQF) bit is cleared to zero.
5. Periodic interrupt flag (PF) bit is cleared to zero.
6. The device is not accessible until RESET is returned to zero.
7. Alarm interrupt flag (AF) bit is cleared to zero.
8. IRQ pin is in the high impedance state.
9. Square wave output enable (SQWE) bit is cleared to zero.
10. Update ended interrupt enable (UIE) is cleared to zero.

In a typical application, RESET may be connected to V_{CC}. This connection will allow the DS1287 to go in and out of power failed without affecting any of the control registers.

6.3.3 Address Map

The address map of the DS1287 is shown in Fig. 6.28. The address map consists of 50-bytes of user RAM, 10-bytes of RAM, which contain the RTC time, calendar, and alarm data, and 4-bytes are used for control and status. All 64-bytes can be directly written or read except for the following:

1. Registers C and D are read-only.
2. Bit 7 of Register A is read-only.
3. The high order bit of the seconds byte is read-only.

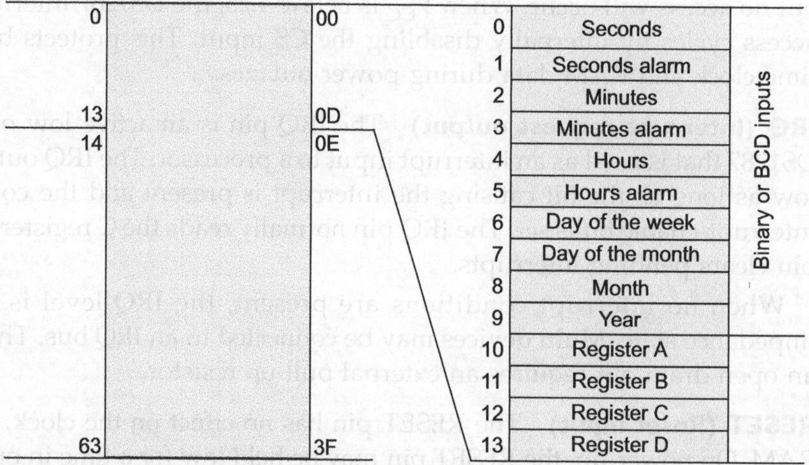

Fig. 6.28 Address map of DS1287

Time, calendar, and alarm locations The time and calendar information is obtained by reading the appropriate memory bytes. The time, calendar, and alarm are set or initialized by writing appropriate RAM bytes. The contents of the ten time, calendar, and alarm bytes may be either binary or binary-coded decimal (BCD) format.

Before writing the internal time, calendar, and alarm registers, the SET bit in register B should be written the internal time, calendar, and alarm registers, the SET bit in Register B should be written to a logical one to prevent updates procuring while access is being attempted.

In addition to writing the registers in a selected format (binary or BCD), the data mode bit (DM) of Register B must be set to the appropriate logic level. All ten time, calendar, and alarm bytes must use the same data mode. The SET bit in Register B should be cleared after the data mode bit has been written to allow the real time clock to update the time and calendar bytes.

Once initialized, the real rime clock makes all updates in the selected mode. The data mode cannot be changed without reinitializing the 10 data byes. Table 6.4 shows the binary and BCD formats of the ten; time, calendar, and alarm locations.

Table 6.4 Time calendar and alarm data modes

Address location	Function	Decimal range	Range	
			Binary MD	BCD mode
0	Seconds	0 – 59	00 – 3B	00 – 59
1	Seconds alarm	0 – 59	00 – 3B	00 – 59
2	Minutes	0 – 59	00 – 3B	00 – 59
3	Minutes alarm	0 – 59	00 – 3B	00 – 59
4	12 Hr. Mode	1 – 12	01 – 0C AM 81 – 8C PM	01 – 12 AM 81 – 92 PM
	24 Hr. Mode	0 – 23	00 – 17	00 – 23
5	Hours alarm - 12	1 – 12	01 – 0C AM 81 – 8C PM	01 – 12 AM 81 – 92 PM
	Hours alarm - 12	0 – 23	00 – 17	00 – 23
6	Day of week Sunday = 1	1 – 7	00 – 07	00 – 07
7	Date of month	1 – 31	01 – 1F	01 – 31
8	Month	1 – 12	01 – 0C	01 – 12
9	Year	0 – 99	00 – 63	00 – 99

The 24/12-bits cannot be changed without reinitializing the hour locations. When the 12-hour format is selected, the high order of the hour byte represents M when it is a Logic 1. The time, calendar, and alarm bytes are always accessible because they are double buffered.

Once per second the 10-bytes are advanced by 1 s and checked for an alarm condition. If a read of the time and calendar data occurs during an update, a problem exists that seconds, minutes, hours, etc. may not correlate. The probability of reading incorrect time and calendar data is low.

When the 12-hour format is selected, the high order of the hour byte represents M when it is at Logic '1'. The time, calendar, and alarm bytes are always accessible because they are double buffered.

Several methods of avoiding any possible incorrect time and calendar reads are covered later in this text.

The three alarm bytes may be used in two ways:

1. First, when the alarm time is written in the appropriate hours, minutes, and seconds alarm locations, the alarm interrupt is initiated at the specified time each day if the alarm enable bit is high.

2. The second use condition is to insert don't care state in one or more of the three alarm bytes. The "don't care" code is any hexadecimal value from

C0 to FF. The two most significant bits of each byte set the "don't care" condition when at Logic 1. An alarm will be generated each hour when don't care bits are set in the hours byte. Similarly, an alarm is generated every minute with don't care codes in the hours and minute alarm bytes. The "don't care" codes in all three alarm bytes create an interrupt every second.

The 50 general purpose nonvolatile RAM bytes are not dedicated to any special function within the DS1287. They can be used by the processor program as nonvolatile memory and are fully available during the update cycle.

6.3.4 Interrupts

The RTC plus RAM includes three separate, fully automatic sources of interrupt for a processor. The alarm interrupt may be programmed to occur at rates from once per second to once per day. The periodic interrupt may be selected for rates from 500 ms to 122 µs. The update-ended interrupt may be used to indicate to the program that an update cycle is complete.

The processor program can select which interrupts, if any, are going to be used. 3-bits in Register B enable the interrupts. Writing a Logic 1 to an interrupt enable bit permits that interrupt to be initiated when the event occurs. A "0" in an interrupt flag is already set when an interrupt is enabled.

When an interrupt event occurs, the relating flag bit is set to Logic 1 in Register C. Setting of these flag bits is independent of the state of the corresponding enable it in Register B. The flag bit can be used in a polling mode without enabling the corresponding enable bits. The interrupt flag bit is a status bit which software can interrogate as necessary.

When a flag bit was last read; however, care should be taken when using the flag bits as they cleared each time Register C is read.

Double latching is included with Register C so that bits, which are set remain stable throughout the read cycle. All bits which are set (high) are cleared when read and new interrupts which are pending during the read cycle are held until after the cycle is completed. 1, 2, or 3-bits may be set when reading Register C. Each utilized flag bit should be examined when read to insure that no interrupts are lost.

The second flag bit usage method is with fully enabled interrupts. When an interrupt flag bit is set and the corresponding interrupt enable bit is also set, the IRQ pin is asserted low. IRQ is asserted as long as at least one of the three interrupts sources has its flag and enable bit both set.

The IRQF bit in Register C is a one whenever the IRQ pin is being driven low. The RTC initiated an interrupt is accomplished by reading Register C. Logic '1' in bit 7 (IRQF bit) indicates that one or more interrupts have been initiated by the DS1287. The act of reading Register C clears all active flag bits and the IRQF bit.

Before the time/calendar data, 244 μs will be changed. Therefore, the user should avoid interrupt service routines that would cause the time needed to read valid time/calendar data to exceed 244 μs.

The third method uses a periodic interrupt to determine if an update cycle is in progress. The UIP bit in Register A is set high between the setting of the PF bit in Register C. Periodic interrupts that occur at a rate of greater than t_{BUC} allow valid time and date information to be reached at each occurrence of the periodic interrupt. The reads should be complete within (TPI/2+t_{BUC}) to insure that data is not read during the update cycle.

6.3.5 Control Registers

The DS1287 has four control registers, which are accessible at all times, even during the update cycle.

Register A

MSB							LSB
Bit 7	Bit 6	Bit 5	Bit 4	Bit 3	Bit 2	Bit 1	Bit 0
UIP	DV2	DV1	DV0	RS3	RS2	RS1	RS0

Periodic interrupt selection (UIP)   The update in progress (UIP) bit is a status flag that can be monitored. When the UIP bit is a 1, the update transfer is quick. When UIP is a zero, the update transfer will not occur for a least 244 μs. The time, calendar, and alarm information in RAM is fully available for access when the UIP bit is read-only and is not affected by RESET. Writing the SET bit in Register B to a "1" inhibits any update transfer and clear the UIP status bit.

Oscillator control bits (DV0, DV1, DV2)   The internal oscillator is turned OFF to prevent the lithium energy cell from being used until it is installed in system. These three bits are used to turn the oscillator ON or OFF and to reset the countdown chain. A pattern of 010 is the only combination of bits which will turn the oscillator ON and allow the RTC to keep time. A pattern of 11X will enable the oscillator but holds the countdown chain in reset. The next update will occur at 500 ms after a pattern of 010 is written to DV0, DV1, and DV2.

Square wave output selection (RS3, RS2, RS1, RS0)   These four rate-selection bits select one of the 13 taps on the 15-stage divider or disable the divider output. The tap selected may be used to generate an output square wave (SQW pin) and/or a periodic interrupt. The RS0–RS3 bits in Register A establish the square wave output frequency. The SQW frequency selection shares its 1-of-15 selector with the periodic interrupt generator. Once the frequency is selected, the output of the SQW pin may be turned ON and OFF under program control with the square wave enable bit (SQWE). The user may do one of the following:

1. Enable the interrupt with the PIE bit;
2. Enable the SQW output pin with the SQWF bit;
3. Enable both at the same time and the same rate; or
4. Enable neither.

The periodic interrupt rate is selected using the Register A bit which select the square wave frequency. Changing the Register A bits affects both the square wave frequency and the periodic interrupt output. Each function has a separate enable bit in Register B. The SQWE bit controls the square wave output. PIE bit in Register B enables the periodic interrupt. The periodic interrupt is used with software counters to measure inputs, create output intervals, or await the next needed software function.

Update cycle   The DS1287 executes an update cycle once per second regardless of the SET bit in Register B. When the SET bit in Register B is set to '1', the user copy of the double buffered time, calendar, and alarm bytes is frozen and will not update as the time increments. However, the time countdown chain continues to update the internal copy of the buffer. This feature allows time to maintain accuracy independent or writing the time, calendar, and alarm buffers and also guarantees that time and calendar information is consistent. The update cycle also compares each alarm byte with the corresponding time byte and gives an alarm if a match or if a "don't care" code is present in all three positions.

The methods used to handle access of the real time clock to avoid possibility of accessing inconsistent time and calendar data are as follows:

1. Use the update-ended interrupt. If enabled, an interrupt occurs after every update cycle, which indicates that over 999 ms are available to read valid time and date information. If this interrupt is used, the IRQF bit in Register C should be cleared before leaving the interrupt routine.

2. Use the update-in-progress bit (UIP) in Register A to determine if the update cycle is in progress. The UIP bit will pulse once per second. After the UIP bit goes high, the update transfer occurs 244 µs later.

Register B

The format of Register B is

Bit 7	Bit 6	Bit 5	Bit 4	Bit 3	Bit 2	Bit 1	Bit 0
SET	PIE	AIE	UIE	SQWE	DM	24/12	DSE

SET   When the SET bit is a zero, the update transfer functions normally by advancing the counts once per second. When the SET bit is written to a one, any update transfer is inhibited and the program may initialize the time and calendar bytes without an update occurring in the midst of initializing. Read cycles can be executed in a similar manner. SET is a read/write bit, which is not modified by RESET or internal functions of the DS1287.

PIE   The periodic interrupt enable (PIE) bit is a read/write bit which allows the periodic interrupt flag (PF) bit in Register C to cause the IRQ pin to be driven low. When the PIE bit is set to 1, periodic interrupts are generated by driving the IRQ pin low at a rate specified by the RS3 through RS0 bits of Register A. A zero in the PIE bit blocks the IRQ output from being given by a periodic interrupt, but the periodic flag (PF) bit is still set at the periodic rate. PIE is not modified by any internal DS1287 functions, but is cleared to zero on RESET.

AIE   The alarm interrupt enable (AIE) bit is read/write bit which when set to a one permits the alarm flag (AF) bit in Register C to assert IRQ. An alarm interrupt occurs for each second that the three time bytes equal the three alarm bytes including a "don't care" alarm code of binary 11XXXXXX. When the AIE bit is set to zero, the AF bit does not initiate the IRQ signal. The RESET pin clears AIE to zero. The internal functions of the DS1287 do not affect the AIE bit.

UIE   The update ended interrupt enable (UIE) bit is a read/write bit, which enables the update end flag (UF) bit in Register C to assert IRQ. The RESET pin going low or the SET bit going high clears the UIE bit.

SQWE   When the square wave enable (SQWE) bit is set to a one, a square wave signal at the frequency set by the rate-selection bits RS3 through RS0 is driven out on the SQW pin. When the SQWE bit is set to zero, the SQW pin is held low; the state of SQWE is cleared by the reset pin. SQWE is a read/write bit.

DM   The data mode (DM) bit indicates whether time and calendar information are in binary or BCD format. The DM bit is set by the program to the appropriate format and can be read as required. This bit is not modified by internal functions or RESET. A one in DM signifies binary data while a zero in DM specifies binary coded decimal (BCD) data.

24/12   The 24/12 control bit establishes the format of the hours byte. A one indicates the 24-hour mode and a zero indicates the 12-hour mode. This bit is a read/write and is not affected by internal functions or RESET.

DSE   The daylight savings enable (DSE) bit is a read/write bit which enables two special updates when DSE is set to 1.

1. On the first Sunday in April the time increments from 1:59:59 AM to 3:00:00 AM.

2. On the last Sunday in October when the time first reaches 1:59:59 AM it changes to 1:00:00 AM.

These special updates do not occur when the DSE bit is zero. This bit is not affected by internal functions or reset.

Register C

The format of Register C is as follows:

MSB							
Bit 7	Bit 6	Bit 5	Bit 4	Bit 3	Bit 2	Bit 1	Bit 0
IRQF	PF	AF	UF	0	0	0	0

IRQF The interrupt request flag (IRQF) bit is set to a 1 when one or more of the following are true:

PF = PIE = 1, AF = AIE = 1, UF = UIE = 1, i.e., IRQF = PF.PIE+AF.AIE+UF.UIE

Any time the IRQD bit is a one the IRQ pin is driver low. All flag bits are cleared after Register C is read by the program or when the RESET pin is low.

PF The periodic interrupt flag (PF) is a read-only bit which is set to a 1 when an edge is detected on the selected tap of the driver chain. The RS3 through RS0 bits establish the periodic rate. PF is set to a one independent of the state of the PIE bit. When both PF and PIE are ones, the IRQ signal is active and will set the IRQF bit. The PF bit is cleared by a RESET or a software read of Register C.

AF A one in the AF (alarm interrupt flag) bit indicates that the current time has matched the alarm time. If the AIE bit is also a one, the IRQ pin will go low and a one will appear in the IRQF bit. A RESET or a read of Register C will clear AF.

UF The update ended interrupt flag (UF) bit is set after each update cycle. When the UIE bit is set to one, the one in UF causes the IRQF bit to be a one which will assert the IRQ pin. UF is cleared by reading Register C or a RESET.

Bit 0 through Bit 3 These are unused bits of the status Register C. These bits always read zero and cannot be written.

Register D

MSB							
Bit 7	Bit 6	Bit 5	Bit 4	Bit 3	Bit 2	Bit 1	Bit 0
VRT	0	0	0	0	0	0	0

VRT   This bit should always be '1' when read operation is done. When it is '0', it indicates an exhausted internal lithium energy source and both the contents of the RTC data and RAM data are questionable. This bit is unaffected by RESET.

Bit 6 through bit 0   The remaining bits of Register D are not usable. They cannot be written and when read, they will always read zero.

6.4   PROGRAMMABLE KBD/DISPLAY INTERFACE (8279)

IC-8279 is designed such that it can be connected directly with the X51 micro-controller bus. It relieves microcontroller of the housekeeping job of refreshing the display and scanning a keyboard 8279 provides an interface for both numeric and alphanumeric displays. It has (16 × 8) display RAM and (8 × 8) FIFO memory.

It performs two major functions:

1. Scans keyboard, detects key closure, identifies pressed key, and transmits the key code along with characteristics of the key pressed to the micro-controller.

2. It puts data from the microcontroller to the display RAM for the use by display devices.

Both the functions are done repeatedly without involving microcontroller, except during the time when the actual data transfer takes place. Thus, microcontroller is relieved of the burden of scanning the keyboard or displaying data.

6.4.1 Modes of Operation

8279 chip can be operated in an input mode and the output mode.

Input Mode of Operation

It can be operated in any one of the following three input modes, depending upon which input device is used. It is also possible to connect any device, whose output is digital as an input device with 8279 directly.

1. Scanned keyboard mode
2. Scanned sensor matrix mode
3. Strobbed I/O mode

Scanned keyboard mode   There can be encoded/decoded scan lines. The key debouncing operation is done automatically. In this mode, when a key is pressed, a 6-digit code is generated which is the characteristics of key position with this position data, SHIFT and CNTL an 8-bit word is formed which is stored in FIFO RAM inside 8279.

D7	D6	D5	D4	D3	D2	D1	D0
CNTL	SHIFT	SCAN			RETURN		

When a byte is received, the IRQ line is raised high and this signal interrupts microcontroller. The interrupt subroutine can be written to read RAM. When FIFO RAM is READ, IRQ line is put Low. IRQ line becomes high again if RAM contains another data. If two keys are pressed within the debounce cycle, it is considered as simultaneous depression. Neither is recognized till one is released. One which remains is considered as a single depression (debounce is 10–20 ms). This multiple key closure problem is solved in 8279 by selecting mode of operation as

Two Key Lockout   When a key is pressed, debounce logic circuit is set. For next two scans, it is checked, whether other key is pressed. At this instant if no other key is pressed then; it is considered as single key depression. The data is taken into FIFO and IRQ is activated. In case FIFO is full, key data is not entered and error flag is set.

Else, if one or more additional keys are depressed, no data is entered into FIFO directly, but the status is resolved as follows:

If the other key is released before the key first pressed, it is a single depression.

If first key is released before the other key, key pressed is ignored.

Thus the end result of this two key lockout mode of operation is that either the first key pressed is entered or ignored, but no other key will be entered.

N-Key Rollover   When key is pressed, the debounce circuit waits for 2–5 s and checks if any key is still pressed. If so, data is entered into the FIFO RAM. There is no limit on how many keys are pressed. In case of simultaneous depression data is entered in order of key pressed.

A special error mode can be set in this option using error mode SET bit. If during a single debounce two keys are pressed, then it sets an error flag, which presents data entry into FIFO. The error flag is READ from the FIFO status word and it is reset by clear command CF = 1.

Scanned sensor matrix mode   In this mode the key is placed in the form of matrix in such a way that:

1. Scan lines forms columns
2. Return lines form rows

The size of the key matrix is 8×8 for encoded keyboard and 4×8 for decoded keyboard. The data in each of the eight return line is entered into eight columns of sensor RAM and each position therefore mapped into a specific sensor RAM. The SHIFT/CNTL are not used. The code generated is as follows:

B7	B6	B5	B4	B3	B2	B1	B0
RL7	RL6	RL5	RL4	RL3	RL2	RL1	RL0

In this mode, instead of a switch other devices can also be connected, e.g., logic circuit. The limitation is that device should be triggerable by scan lines. The output circuit can enter data '1' or '0' on the return line. Here, debounce is not provided. It has an advantage that the microcontroller knows how long the sensor was closed. The IRQ is high, if the sensor value is found to have changed at the end of the scan. When auto increment flag is SET to '0', IRQ is RESET by first data read operation. When it is '1', it is 'RESET' by end INT command.

Strobbed input mode   In this mode, the data from return lines goto FIFO on the raising edge of CNTL/STB line pulses. The data can come from any source, i.e., keyboard or switching matrix. Each line would lead to 8-bit word as follows:

RL7	RL6	RL5	RL4	RL3	RL2	RL1	RL0

Output (Display) Mode

The options available for configuration of the display output are

1. No. of display character 8 or 16
2. Organized in single 8-bit or dual 4-bit group
3. Display format may be:

Left entry mode, also called typewriter mode   In this mode the first character is interred at the extreme left, <CR> advances one step and second character is entered as the second one. At the end of a line, there is a <CR>, but no line feed, then character over writes first character.

In this mode of operation each display position directly corresponds to a single byte/nibble in a display RAM. First entry goes to (add '0'), the first of the 16 addresses to be used for display RAM at the left display position. The 16th entry goes to the last necessary location of RAM, 17th entry fills the left position again (add-0) and the process is repeated. There is a command, which allows writing of the data at any specified address in RAM.

In this mode if carriage does not advance, all characters would appear at the same place (equivalent to this is left entry mode with non auto). The entry is at the same RAM locations and display position.

Right entry mode, also called calculator mode   First character is entered at the right-most position, when second entry is made. The first entry is shifted to the left by one position and the new character is written at the extreme write only. If the new value (entry is in repeated the left most entry is missed after 16 entries). In this mode there is no correspondence with the RAM address and position in display RAM.

6.4.2 Pin Layout and Description of 8279

The pin details of the IC-8279 (Fig. 6.29), their functions, and uses are as follows:

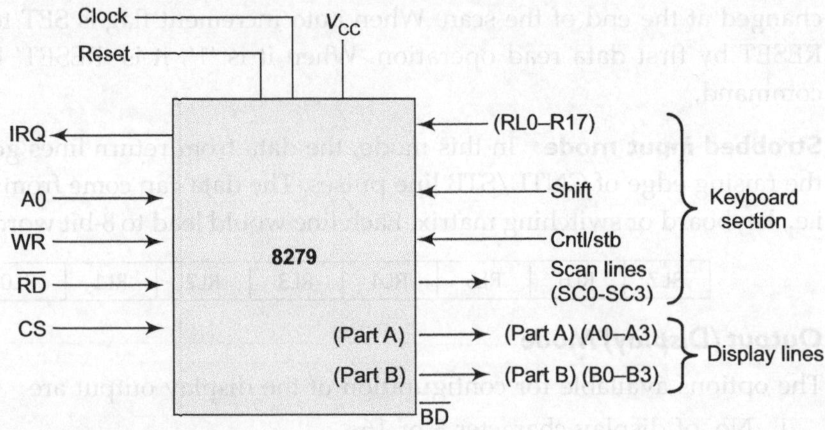

Fig. 6.29   Pin layout of 8279

Scan lines (SC0–SC3)   The scan lines can be used with external decoded to generate 16 lines. Since, it is advised not to use SL3 line for driving decoder, so SL0–SL2 generate 8 lies through the decoder. Combining with return lines RL0–RL7, 8x8 keyboard matrix can be connected to define 64 characters. However with (SHIFT/CNTL) additional lines can be generated. Thus in the encoded mode $64 \times 4 = 256$ keys can be used.

In the decoded mode, the internal decoder of the 8279 is used. Hence, (SC0–SC3) with SHIFT/CNTL gives $4 \times 8 \times 4 = 128$. Maximum keys can be connected to 8279 in decoded mode.

BD   It is useful for blinking the displayed character.

RL0–RL7   They are used with scan lines to connect the keyboard.

PA0–3, PB0–3   They are the 4-bit output ports and can be configured as a single 8-bit output port. The output devices such as LED or 7-segment LEDs can be connected to display the output.

A0   It is the address line. It can be used to select command or data port of 8279.

Clock   This connected to the clock generator so as provide clock signal to 8279. The frequency of clock can be divided if required by the command word. Normal frequency of clock of 8279 is 100 kHz.

IRQ   Interrupt request is generated when there is data in the FIFO RAM of 8279. IRQ pin can be connected to any one of the interrupt pins of the micro-controller.

Reset   Pin is used to reset the chip.

SHIFT, CNTL/STB   In combination these lines are used to connect additional keys.

The details of internal blocks in the 8279 as follows:

Control and timing register   These register store the programmed keyboard. Programming is done by setting address line A0 to '1' and sending write signal. The counter provides scan sequences, i.e., it contains basic timing counter chain.

Scan counters   They can be operated in decoded/encoded mode. It keeps track of the operating sequence.

Return buffer/KBD debounce control   Return lines are latched, buffered by return buffers. When the chip is used in keyboard mode, each line is scanned and key closure is found, a 10 ms debounce is given and if the key is still closed, its status are transferred to FIFO.

Scan mode   The contents of the return lines are directly transferred to FIFO. In strobbed mode contents are transferred to FIFO at rising edge of CNTL/STB pulse.

FIFO and SENSOR RAM   In a strobbed mode, the RAM is FIFO, while in scanned mode it is SENSOR RAM. Each row of sensor RAM is loaded with the corresponding row of sensor in the matrix. IRQ is high, if a change in sensor is detected.

Any key pressed is entered into FIFO through a strobe. The FIFO is an eight-character space in memory. These locations can be considered as a stack of registers with 8-input lines and a control PUSH, 8-output lines and a control line POP. Data is written at top of the memory locations by the push operation. Filling of memory locations in FIFO starts from bottom and up to 8-characters can be stored.

Display access register and display RAM They contain address of the word being READ/WRITTEN by the microcontroller and two 4-bits (nibbles) being displayed. Display RAM can be READ by the microcontroller after command mode and the address is set.

6.4.3 Programming 8279

There are eight commands in the set of 8279 which can be used to program 8279. The commands are used to specify the environment as well as operation to be performed. The format of the command word is

B7	B6	B5	B4	B3	B2	B1	B0
Type of command			Options available with command				

Types of commands It identifies the operation to be performed and is decided by the bit combination as shown below:

Bit combination			Operation specified
B7	B6	B5	
0	0	0	:Keyboard / mode set
0	0	1	:Program clock
0	1	0	:Read FIFO/RAM
0	1	1	:Read display RAM
1	0	0	:Write display RAM
1	0	1	:Display with inhibit / blinking
1	1	0	:Clear
1	1	1	:End interrupt / error mode set

Options Available with Command

With each operation the parametric information is specified by selection of appropriate bit combinations (B4–B0).

Keyboard/Display mode set The format of the display word is

0	0	0	D	D	K	K	K

DD: Combination for configuration of display on output device.

Bit combination	Specified configuration of display
0 0	8-bit character display - left entry mode
0 1	16-bit character display - left entry mode
1 0	8-bit characteristic - Right entry mode
1 1	16-bit characteristic - Right entry mode
K K K :	**Combination for specifying set up of input device.**
0 0 0 :	Encoded keyboard - two key lockout mode
0 0 1 :	Decoded keyboard - two key lockout mode
0 1 0 :	Encoded keyboard - N- key rollover mode
0 1 1 :	Decoded keyboard - N- key rollover mode
1 0 0 :	Encoded sensor matrix
1 0 1 :	Decoded sensor matrix
1 1 0 :	Encoded keyboard strobbed mode
1 1 1 :	Decoded keyboard strobbed mode

Program clock

0	0	1	P	P	P	P	P	P
Command code			Pre-scalar: Any value between 1–31					

This is used to generate internal timing and multiplexing signals. The clock is divided by a pre-scalar value (1–31).

Read FIFO/SENSOR RAM

0	1	0	AI	X	A	A	A
Command code			Auto Incr.		RAM address		

This sets 8279 for a READ operation to read FIFO RAM. AI subsequent READ operations will form the successive locations, if AI = '1', until another command issued. In sensor mode: AAA - Row of sensor RAM if (AI=1).

Read display RAM

0	1	1	AI	A	A	A	A
Command code			Auto Incr.		RAM address		

This sets 8279 for a read operation to read DISPLAY RAM. All subsequent READ operations will form the successive locations, if AI = '1', until another command is issued.

Write display RAM

1	0	0	AI	A	A	A	A
Command code			Auto Incr.		RAM address		

This sets 8279 for a write operation to write the DISPLAY RAM. All subsequent write operations will form the successive locations, if AI = '1', until another command is issued.

Display write with inhibit

1	0	1	X	IW	IW	BL	BL
Command code				Used to mask the nibbles of Port A and Port B 0=Masked 1=Unmasked		Character at A and B are blinked 1=Blink 0=No-blink	

These command sets 8279 for a write operation with inhibit to write the DISPLAY RAM. This mean that the lower of upper nibble of the data byte to be displayed can be inhibited so that it is not displayed on the device. All subsequent write operations will form the successive locations, if AI = '1', until another command is issued.

Clear

1	1	0	CE	CD	CD	CF	CA
Command code			Clear all the raw of displace RAM			Clear (FIFO) Reset IRQ	CF +CD clear all

CE	CD	CD
0-display	0	x - clear all rows with zeros (00h)
1 - enable	1	0 - all rows are filled with blank (20h)
	1	1 - clear all rows with 1's (FFh)

This command is used to clear display RAM as well as clear FIFO flag. When all the locations in FIFO RAM are filled, FIFO flag is set to '1'. No new data is accepted from keyboard or input device unless the flag is cleared. Similarly, the display can be cleared with zeros, ones, or blanks using this command.

End interrupt /Error mode set

1	1	1	E	X	X	X	X

In N-key rollers, if E = 1 chip operates in special error mode. In sensor matrix mode, it clears IRQ line and permits writing of FIFO RAM.

It is possible to find out status of 8279 at any instant of time by reading status register. The format of the status register is

D7	D6	D5	D4	D3	D2	D1	D0
DU	S/E	O	U	F	N	N	N
Display unavailable	Sensor closed/Error mode set	Over run	Under run	FIFO FULL	No. of character in FIFO		

The FIFO status word indicates whether it is under run/over run. It is also possible to have information about the sensor.

D5: Over run It specifies that an additional character is written in the FIFO RAM, when it is full.

D4: Under run Controller is trying to read an empty FIFO RAM.

D7: Display unavailable This bit indicates some data is being displayed on output device.

6.4.4 Interfacing 8279

Figure 6.30 shows interfacing of 8279 with hex keypad and four 7-segment display units. The keys are connected between four scan and four return lines. Since external decoder is not used, the keyboard is a decoded board. Return lines RL0–RL3 are connected to column of the pad and scan line forms row of the keypad.

The display units are connected between the port pins A_{0-3} and B_{0-3} and the return lines. The port pin outputs are connected to the segments (a…g) of the display unit, while the return lines are connected to the base of the transistor drivers. The type of driver used depends on whether the common anode or common cathode display units are used.

When the reset in given clock is set to 31. This provides a scan frequency of about 100 kHz, for a clock of 3 MHz. The 'RESET' also sets chip in mode of 16-character display with 2 key lockout mode of operation.

A15	A14	A13	A12	A11	A10	A9	A8	A7...A0							
0	0	0	1	1	X	X	A0 pin of 8279	X	X	X	X	X	X	X	X
1				8 or 9 depending on A0				0				0			

The data port address is 1800h, while 1900h is the address of the command port.

The 8279 must be initialized by X51 which means specifying input/output configuration of the system. It informs 8279 about the input and output device connected. In order to initialize 8279, control words are to be written at 1900h. Once initialized operation of the 8279 is carried out by specifying the function to be carried out by loading appropriate command word, followed by the data required by the function.

The 8279 read the keyboard column via the return lines RL0–RL2. If a key closure is detected, an internal debounce circuit waits for 10 ms and verifies whether the key is still pressed. If so, a code for this key is moved into FIFO. The 8279 outputs high on IRQ line, indicating availability of a character in the FIFO RAM. IRQ pin is connected to the interrupt pin of the microcontroller. Microcontroller will execute the subprogram for reading the data byte from FIFO RAM.

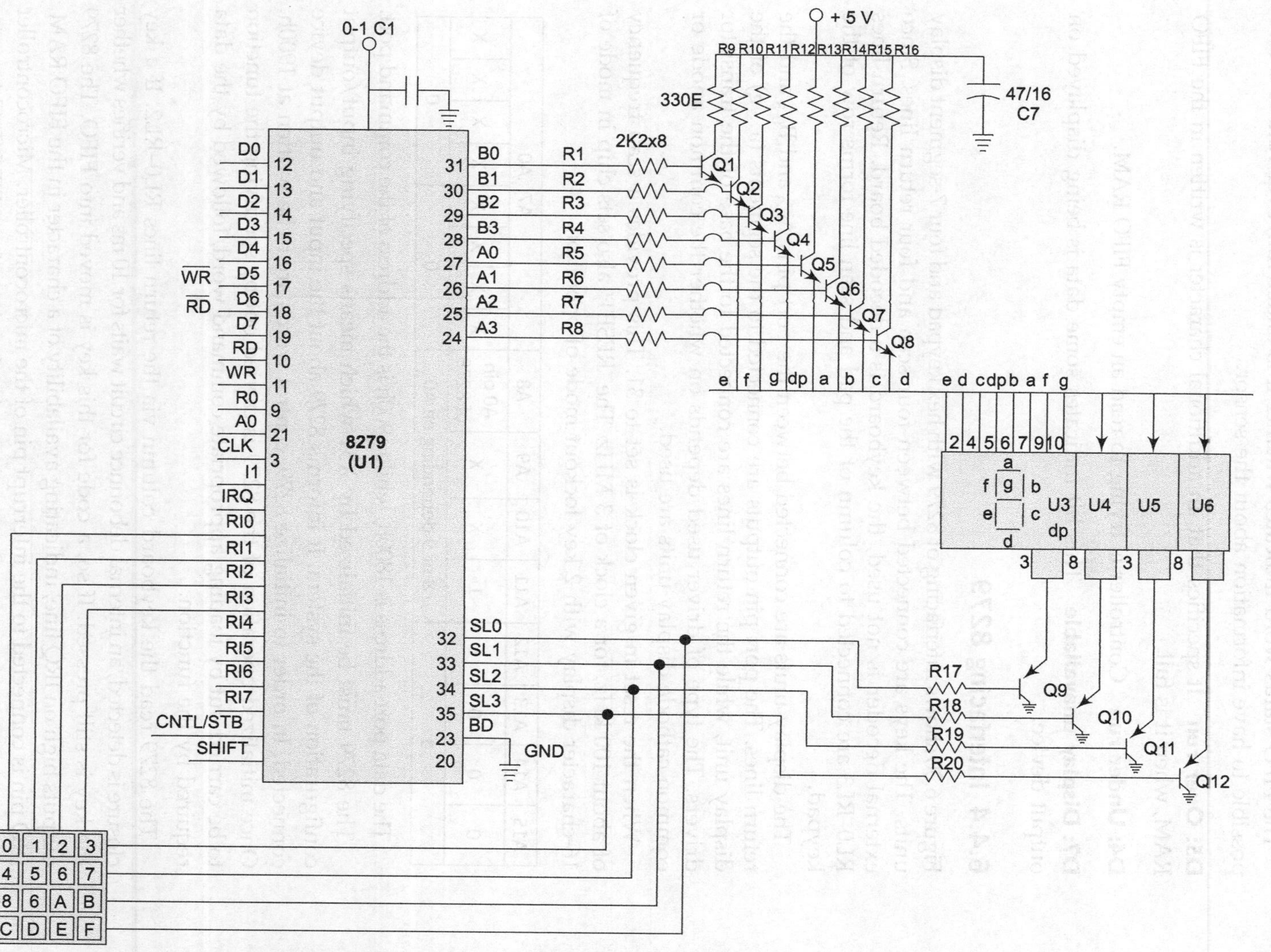

Fig. 6.30   Interfacing keypad and four 7-segment display modules with the 8279

6.4.5 Program Modules

We will study how to realize desired operation using 8279 commands.

Example 6.12   Write a program segment to initialize the 8279 so as to configure the display in 8-character left entry mode. The encoded keyboard is to be configured in two key lockout mode.

Solution

The control word for the initialization is = (OO)H

The program segment for the initialization is

```
Init:      mov 80, #20h
           mov A, #00h
           mov dptr, # 1900h
           movx @dptr,a
```

Example 6.13   Write a program segment to read the data entered via keyboard. Assume that the chip is to be initialized so as to configure the display in 8-character left entry mode. The encoded keyboard is to be configured in two key lockout mode.

Solution

The control word to read a keyboard is

0	1	0	AI	X	A	A	A
Command code			Auto Incr.		RAM address		

$$= 40h$$

When a key is pressed, key code will be entered into the FIFO RAM and interrupt is generated. The interrupt subprogram reads FIFO RAM, by using read command followed by command to read data port. Before the code is stored in memory, it is necessary to strip off the bits indicating the status of CTRL and SHIFT pins.

The program segment is given below:

```
init:      mov 80, #20h        ; stack initialization
           mov a, #00h         ; command for 8279 initialization
           mov dptr, 1900h     ; initialize 8279 to configure keyboard and
                               ; display
           movx @dptr,a        ; load command word
           mov r0, #address    ; initialize the pointer to store data
loop:      sjmp loop           ; wait for interrupt, i.e., key closure
- - - - - - - - - - - - - - - - - - - - - - - - - - - - - - - -
0003:      sjmp rd-key
- - - - - - - - - - - - - - - - - - - - - - - - - - - - - - - -
                               ; interrupt program segment to read FIFO RAM
```

```
            push dpl
            push dph
            mov dptr, #1900h        ; pointer to command port
            mov a, #40h:            ; command to read FIFO RAM
            movx @dptr,a            ; issue read FIFO command
            mov dptr, #1800h        ; pointer to data port
            movx a, @dptr           ; read key code
            anl a, #3Fh             ; mask upper bits
            mov @r0, a              ; load data in internal RAM
            pop dph
            pop dpl
            ret
```

Example 6.14 Write a program segment to display 1 character.

Solution

To display a character, it is written in the display RAM at the location where the character is to be displayed on the device.

The control word to write display RAM is

1	0	0	1	0	0	0	0
Command code			Auto Incr.		RAM address		

$$= 90h$$

It is necessary to initialize the chip. The program segment is given below:

```
init:
            mov 80, #20h            ; stack initialization
            mov a, #00h             ; 8279 initialization
            mov dptr, 1900h         ; initialize 8279 to configure keyboard and
                                    ; display
            movx @dptr,a            ; load command word
            mov r0, #address        ; initialize pointer to store data
disp-ram:
            mov a, #90h             ; command to write display RAM
            movx @dptr,a            ; issue write command
            mov dptr, # 1800h       ; pointer to data port
            mov a, #data8           ; character to be displayed
            movx @DPTR,a            ; write data in the display RAM
            sjmp disp-ram
            end
```

Example 6.15   Write a program segment to clear or blank the display and make display
pointer to point to display unit number 1.

Solution

```
clr_disp :

            push r0
            push a
            mov a, #20h
            mov 24h, #00h              ; pointer to display unit number
            acall disp_ascii
            pop a
            pop r0
            ret
```

                                       ; disp_ascii displays ASCII character in the
                                       ; accumulator at current display position.

```
disp_ascii:

            push dpl
            push dph
            push b
            mov dph, #a
            movx a, @dptr
            mov b, a
            mov a, 24h
            anl a, #07h
            orl a, #90h
            mov dptr, # address
            movx @dptr,a
            mov a, b
            dec dpl
            movx @dptr,a
            inc 24h
            mov a, 24h
            cjne a, #08h, ok
ok:         jc skip
            mov 24h, #00h
skip:       pop b
            pop dph
```

```
                    pop dpl
                    ret
```

Example 6.16 Write a program segment, which waits for a key to be pressed and gives key code in accumulator.

Solution

```
key_in:
            push dpl
            push dph
            mov dptr, #1800h        ; data port pointer
wait:       mov a, @dptr            ; read data from port
            anl a, #0Fh
            jz wait
            mov a, #40h             ; read command
            movx @dptr,a
            dec dpl
            movx a, @dptr           ; read data
            pop dph
            pop dpl
            ret
```

Example 6.17 Write a program segment to accept a 4-digit number from keyboard and display it on the last four display unit. Assume that the packed data is in register pair r4, r3.

Solution

```
disp_4:
            push r5
            push r6
            push r7
            push r1
            mov 09h, r4
            mov 08h, r3
            acall disp_reg
rept1:      mov r3, #00h
            mov r4, #00h
            mov r7, #04h            ; configure for digits
read:       acall key_in
            jnb 0D5h, next
```

```
                mov  r4, 09h
                mov  r3, 08h
                sjmp end_proc
next:           anl  a, #0Fh
                mov  r1, #04h
                xch  a, r3
                swap a
                xch  a, r4
                swap a
                xch  a, r4
                xchd a, @r1
                anl  a, #F0h
                orl  a, r3
                xch  a, r3
                mov  09h, r4
                mov  08h, r3
                acall disp_reg
                djnz r7, read
                sjmp rept1
end-proc:
                pop  r1
                POP  R7
                POP  R6
                POP  R3
                RET

disp_reg :
                acall clr_disp
                mov  a, r4
                swap a
                anl  a, # 0Fh
                add  a, #30h
                acall disp_ascii
                mov  a, r4
                anl  a, #0Fh
                add  a, #30h
                acall disp_ascii
                mov  a, r3
```

```
                    swap a
                    anl a, #0Fh
                    add a, #30h
                    acall disp_ascii
                    ret
```

Example 6.18   Write a program segment to display a message, maximum 8 characters, stored as a string, terminated by 03.

Solution

```
disp_msg:
                    push a
                    mov r5, #08h
                    acall clr_disp
rept:               mov a, @dptr
                    cjne, #03h, disp
                    sjmp end
disp:               acall disp_ascii
                    inc dptr
                    djnz r5, rept
end:                pop a
                    ret
```

EXERCISES

6.1 Design a light intensity controller system having following specifications :

- It should be capable of providing various intensity levels.
- Beep during the set process.
- inc and dec keys to vary set_intensity level.
- Facility to cancel previous settings.
- Four 7-segment displays for output.

6.2 Design a microcontroller-based frequency divider for a square wave applied at the Port B of 8255 and the output is produced at the Port A of 8255. Write an associated software for operation of the system. The system should accept 8-bit number, by which the frequency is to be divided, from thumbwheel switches connected at the Port C of 8255.

6.3 Design an IC tester capable of accepting number of IC under test from Hex keyboard and display the result of testing, i.e., PASS/FAIL/ERROR, on 7-segment display connected via 8279.

6.4 Design an EPROM programmer, capable of accepting number of IC to be programmed, from keyboard and display status of the system Programming (P), End of Process (O), and Error (E), on LCD connected to the Port 1, of the microcontroller.

6.5 Design a scheme to connect Hex [4 × 4] keyboard to a serial pins of a microcontroller.

6.6 Design a scheme to expand the INT0 pin of microcontroller so that eight devices can be connected to it. Write necessary software for operation of the hardware.

6.7 Design a home security system using interrupt driven data transfer scheme making use of [int 1] pin. The sources of interrupts are :

- Door bell
- Rain detector
- Alarm
- Auto ON/OFF lights based on timer
- Mail box
- Fan controller for automatic ON/OFF, based on temperature.

6.8 Design a laboratory function generator to generate square, sinusoidal, and triangular waveforms, with the provisions for

- Adjustment for frequency and magnitude of the waveform.
- Type of the waveform.
- Display for selection.

6.9 Design a scheme to interface two DC motors driving a rolling mill with microcontroller. The motors are coupled using DC tachogenerator which has voltages as: 0.5 V for 250 rpm and 40 V for 1000 rpm. If the speed difference between two motors exceed 100 rpm, an error message must be displayed on LCD.

6.10 Design a digital controller for a chemical process system to keep [pH] of the reagent mix to a constant value. The system should have facility to accept input parameters from keyboard and display status on panel using LCD.

6.11 Design a darkroom timer, to be used by the development of the photographic film, having following facilities:

- Key to set time.
- Two digit display to display SET_ TIME in [seconds].
- Beep during the process.

- inc and dec keys to vary set_intensity level.
- Facility to cancel previous settings.

6.12 Design a scheme to simulate the motion of an elevator and controller for a four storeyed building using 8051 microcontroller. Write the operation assuming that:

- The operation starts with a reset, while lift will wait at the ground floor.
- The elevator waits for 10 ms, at each floor, if there is a request.

6.13 Design a presetable down counter using 8255 in Mode 1. The starting value is input through two thumbwheel switches. Data is entered only when start input is given by pressing key. The hardware address of Port A of 8255 is 00D1h. The counter must decrement by one at every 10 ms, until it is zero. When it reaches zero, an LED must glow to indicate end of operation.

6.14 Design a scheme to simulate an electronic lock. The correct combination will consist in pressing four keys. The LED should glow when combination is correct. Three unsuccessful trials cause an alarm to sound. The operating sequence is as follows:

- Press four keys from the numeric keypad connected to 8279
- If it is correct
  open the lock, by making LED glow.
  Else,
- Combination is not correct, try again.

If three trials are made, sound an alarm. The alarm should sound until "0" key is pressed.

6.15 Design an incubator with following specifications:

- Temperature controlled within [−0.5, 0.5]
- Means to adjust and display set point.
- It must have minimum current drain.

6.16 Design a digital controller for a cyclo_converter to generate 50 Hz AC signal. There should be facility for setup magnitude of the signal via keyboard.

6.17 Design a digital controller for a dc_converter to drive motor load up to 1 kW.

6.18 Design a digital controller for an AC motor using ON/OFF control. There should be facility for setup motor speed from keyboard and display set speed on display unit connected via 8279.

6.19 Design a digital controller for an AC motor using phase control. There should be facility for setup motor speed from keyboard and display set speed on display unit connected via 8279.

6.20 Design a scheme for AC voltage regulator.

Interfacing Serial Devices

In a computer, the data is transferred in parallel because it is the fastest way to do it. For transferring data over long distances, however, parallel data transmission requires too many wires. Therefore, data to be sent over long distances is usually converted from parallel form to serial form so that it can be sent on a single wire or a pair of wires. Serial data received from a distant source is converted to parallel form so that it can be easily transferred to the system bus. Three terms often encountered in literature on serial data system are simplex, half-duplex, and full-duplex. A simplex data line can transmit data only in one direction. Half-duplex transmission means that communication can take place in either direction between two systems, but can only occur in one direction at a time. Full-duplex means that each system can send and receive data at the same time.

7.1 | SOME DEFINITIONS

The various terms used in serial communication are described below:

Data terminal equipment (DTE) The terminals and computers that send or receive data are referred to as data terminal equipment.

Data communication equipment (DCE) Data communication equipment refers to all communication equipments (for the most part modems or their equivalents) that connect to the telephone-line side of DTE. Modems along with multiplexer are the two types of communication equipment that are usually thought of when the term DCE is used.

Bit rate A bit is the abbreviation of binary digit, which can be either 0 or 1. In digital communication, a bit is therefore one basic unit of information. Bit per second or bps (bit rate) are the number of bits that pass a point on a communication channel per second.

Baud rate The term baud rate is used to indicate the rate at which serial data is being transferred. Baud rate is defined as 1/ the time for a bit cell. If the bit time is 3.33 ms, for example, the baud rate is 1/3.33 ms or 300 baud.

The process of writing programs for handling a serial data transfer bit is labourious. This is because the stream of input data bits should be sampled in the middle of the expected bit duration and has to be assembled into a byte.

Since the operation has to be done in real time, it is handled via a universal asynchronous receiver transmitter (USART) chip, which converts the data streams into bytes that can be read by the processor. The processor is relieved from doing all the time-critical jobs of shifting and timing data bits.

The USART is usually a programmable device having necessary hardware circuits for implementing asynchronous serial communication. It consists of four sections:

1. Receiver: Receives serial stream and formulates a word
2. Transmitter: Converts byte into serial bit stream and transmits
3. Modem control: Generates handshaking control signals
4. Logic control: Generates control signals for operation

7.2 | USART: 8251

In a serial communication environment, interface must convert data in serial form for transmission and in parallel form after receiving. It must be capable of inserting and detecting information unique to the techniques used, 8251 fulfills this requirement.

It is a programmable chip, which means that it can be configured to work in various environments. It can be programmed by loading a set of two 8-bit control words known as:

1. Mode set instruction
2. Command word

The control words are used to define the system configuration and parameters such as: baud rate, character length (5, 6, 7, 8), number of stop bits (1, $1\frac{1}{2}$, 2), type of operation synchronous or asynchronous operation, and selection of presence or absence of parity.

8251 can accept a character from the microcontroller for transmission on transmit data (TxD) or accept serial data on its receive data (RxD) lines and stores it for the use by the microcontroller. It is possible to examine the readiness of 8251 to transmit/receive data by examining the status word.

7.2.1 Block Diagram of 8251

Figure 7.1 depicts internal block diagram and pins of the USART 8251 available for interfacing to the microcontroller. The internal structure can be divided into the functional modules as:

  (i) Interface and control logic
 (ii) Modem control section
(iii) Transmitter section
 (iv) Receiver section

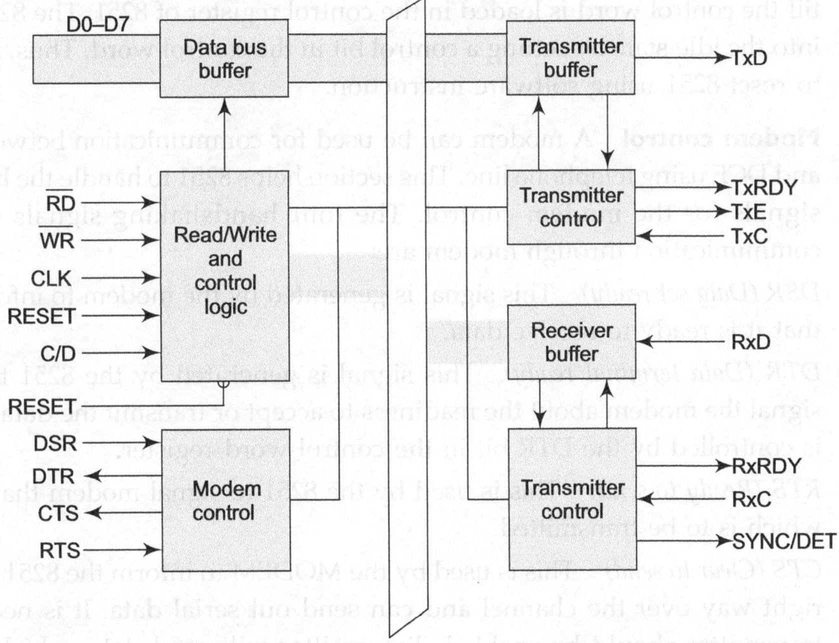

Fig. 7.1 Block diagram of 8251

Read/write and control logic This block accepts signals for the device operation. It contains 8-bit data buffer register, 16-bit control word register (CWR), and 8-bit status register. These registers can be accessed when CS pin is LOW. Depending upon logical status of the C/D, the control or the data port can be accessed.

LOW on the C/D pin indicates the read/write operation at the data port. HIGH on the C/D pin indicates the read/write at the control port. 8251 will read either control word register or status register. Table 7.1 represents the operations performed by 8251 for all combination of the C/D, RD, WR, and CS pin status.

Table 7.1   Summary of the 8251 operations based on status of specified pins

Logical status of the 8251 pins				Operation performed
C/D	RD	WR	CS	
X	X	X	1	Data bus in the High impedance state
0	0	1	0	(data port)=>(data bus)
0	1	0	0	(data Bus)=>(Data port)
1	0	1	0	(Status register)=>(data bus)
1	1	0	0	(Data bus)=>(control register)
X	1	1	0	Data bus in the high impedance state

Clock is used for internal timings and it is about 30 times faster than the baud rate. Reset is used for 8251 to enter into the idle state. It remains in an idle state till the control word is loaded in the control register of 8251. The 8251 can enter into the idle state by setting a control bit in the control word. Thus, it is possible to reset 8251 using software instruction.

Modem control   A modem can be used for communication between the DTE and DCE using telephone line. This section helps 8251 to handle the handshaking signals for the modem control. The four handshaking signals used in the communication through modem are:

DSR (Data set ready)   This signal is generated by the modem to inform the 8251 that it is ready to receive data.

DTR (Data terminal ready)   This signal is generated by the 8251 to inform or signal the modem about the readiness to accept or transmit the data. This signal is controlled by the DTR bit in the control word register.

RTS (Ready to send)   This is used by the 8251 to signal modem that it has data which is to be transmitted.

CTS (Clear to send)   This is used by the MODEM to inform the 8251 that it is the right way over the channel and can send out serial data. It is necessary that transmitter should be enabled. Transmitter will send data, which is there in buffer before disabling command.

Transmitter section   Figure 7.2 depicts the transmitter section with its associated pins. The transmitter buffer will accept the data in parallel form from data bus buffer via internal data bus. This is automatically transferred to the serial output register, when it is empty. The data is shifted out of register on TxD pin, after adding framing bits. Transmitter must be enabled and the C/D pin must be at logic level LOW.

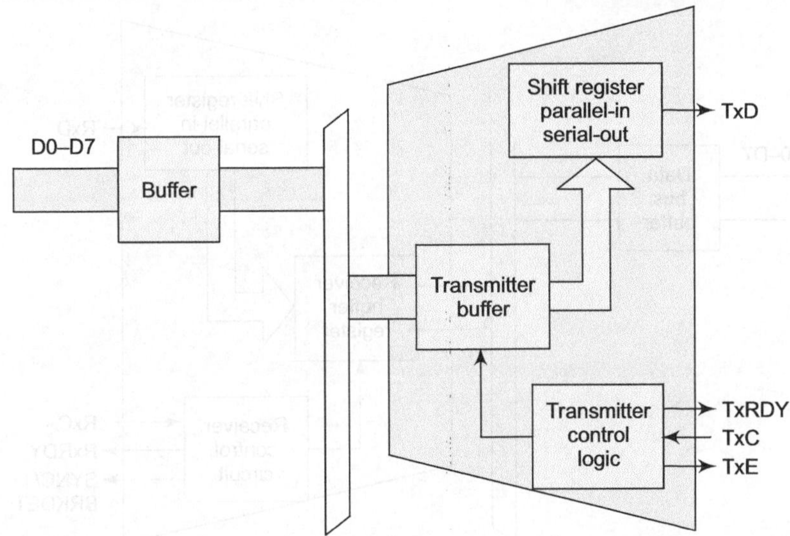

Fig. 7.2   Transmitter section of 8251

TxD   This pin is used for transmitting data output of the serial register. The pin is HIGH when RESET is given or the CTS signal is OFF or transmitter is disabled or empty. Break characters can be sent on TxD by writing a command word.

TxRDY   A logic level HIGH on this pin indicates that transmitter buffer is empty and it can receive a byte from the microcontroller for transmission. It can be used to interrupt the microcontroller directly or its status can be polled by the microcontroller in the status register of 8251. Since, this is not controlled by the control word, the bit will specify that the transmitter is empty, when the transmitter is not enabled. TxRDY is reset, when microcontroller writes a byte into the buffer.

TxC   This pin controls the baud rate at which data is shifted out of the serial output register. In case of a synchronous transmission the baud rate is 1, 1/16, or 1/64 times the transmitter clock. The data is shifted out at the falling edge of the clock signal.

TxE   A HIGH on this pin indicates that the serial output register is empty and no characters are being sent out serially. It is RESET when a character is loaded into the buffer by the microcontroller. This is useful in case of half-duplex system to indicate when they may be turned around.

Receiver section   Figure 7.3 depicts the receiver section with its associated pins. 8251 accepts serial data on its RxD pin. The data is clocked on the rising edge of the receiver clock at RxC. This serial data is converted into a parallel byte and loads into the receiver buffer register after checking the input for the framing bits/character.

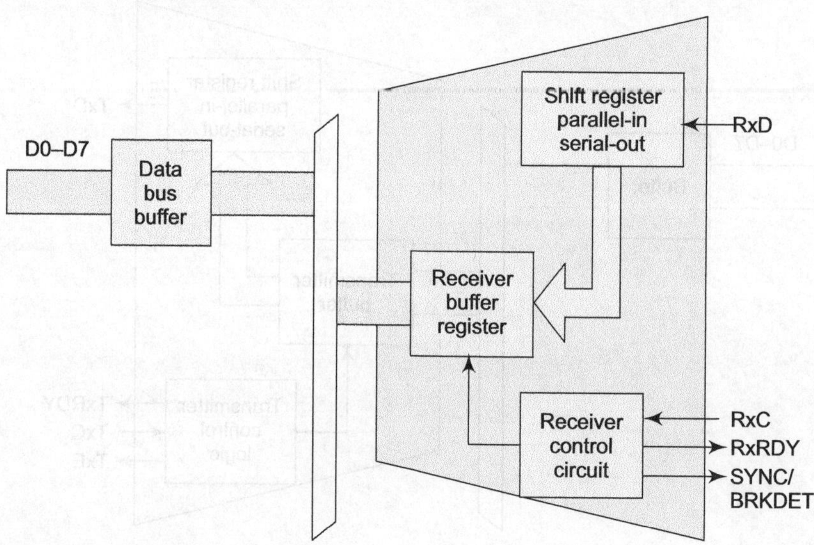

Fig. 7.3 Receiver section of 8251

Control circuit The control section manages all the activities pertaining to the receiver. After initialization, valid HIGH must be detected prior to receiving any data on RxD pin. Once this is detected, the receiver waits for a start bit LOW. This is done by receiver initialization circuit, after master RESET to prevent the receiver from making mistake of unused line RxD as an active RxD line in the break condition. This is done for every master RESET. When RxD line goes LOW, false detection circuit waits for half bit time to ensure presence of LOW bit for a valid start. The control section also sets the framing error status bits in the status word, if there is a parity bit error or absence of a stop bit at the end.

RxC This determines rate at which the incoming serial data is received and it is 1, 1/16, and 1/64 of the clock period. The data is shifted at the raising edge of clock. In synchronous mode the baud rate is equal to the frequency of RxC.

In practice, transmit and receive baud rates are equal, as it is done on a single link. So RxC and TxC pins can be tied together and can be connected to same clock source.

RxRDY This goes high when receiver assembles a character and transfers it to the receiver buffer register. It can be used to interrupt the microcontroller and flag the availability of data or the status of the pin can be polled by polling a bit in the status word register of 8251. In the ASYNC mode, this goes high when receiver is enabled and it has sensed a valid start bit, assembling the character and transferring it to the buffer.

SYNCDET/BRKDET This can be used for detection of the SYNC characters in the case of the synchronous mode of communication and a break character in asynchronous mode of communication.

SYNCDET This can be used as I/O pin in the control word. In output mode, on RESET this pin is low. It goes high when 8251 receives a SYNC character. If 8251 is set for two SYNC then it goes high in the middle of second SYNC character. The status of this pin can be read through the status word. This pin is RESET when the read operation is over.

In input mode, a raising-edge input on this pin prompts 8251 to start assembling the data character on the raising edge of RxC. Once synchronization is achieved, the input signal can be remarked. When an external SYNC is used, the internal SYNC is disabled.

BRKDET In asynchronous mode, this pin is used to detect a break character. It goes high when the RxD pin remains low between two successive stop bit sequences (including parity, data, and start). The pin can be read through the status word. The pin is reset when the RxD pin goes high or on master chip gets a reset pulse or command.

7.2.2 Interfacing 8251

The pins P2.1–P2.6 of the microcontroller are connected to CS using an 8-input NAND gate, Pin P2.0 is connected to the C/D of 8251. This imparts port addresses to 8251. 74LS245 is a bi-directional data buffer.

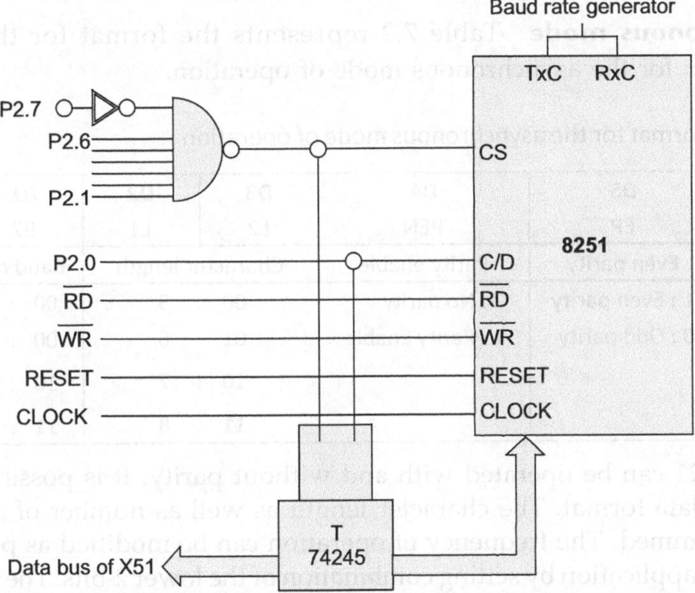

Fig. 7.4 Interfacing scheme to interface 8251 with the microcontroller

The direction line pin T controls direction of the data transfer of the buffer. When the CS is low and the RD pin is low, the pin T is at logic level LOW, direction of the buffer 74LS245 is such that 8251 will put data to the system bus. When the

CS is low and the RD pin is high, the Pin T is at logic level HIGH, direction of the buffer 74LS245 is such that 8251 will get data from the system bus (Fig. 7.4).

7.2.3 Programming 8251

Loading two commands words can program the chip the mode set instruction followed by the command word. The programming of the chip aims at informing 8251 regarding communication parameters. This includes baud rate, no. of data bits, no. of stop bits, and parity information. Once the device has been setup using mode set instruction, actual communication takes place.

For transmission, the microcontroller checks status of the transmitter buffer. A character is loaded into the buffer and sent out via serial line if it is empty. This action is repeated for all the characters to be transmitted.

For reception, the microcontroller checks the status of the receiver register to determine if the receiver has read a character.

Mode Set Instruction

This must follow the RESET operation, which may be external or internal. It defines the general operational characteristics of the 8251. The 8251 support both types of mode of communication: asynchronous and synchronous. The format of control word is different for both the mode of operation.

Asynchronous mode Table 7.2 represents the format for the mode set instruction for the asynchronous mode of operation.

Table 7.2 Mode set format for the asynchronous mode of operation

D7	D6	D5	D4	D3	D2	D1	D0
S2	S1	EP	PEN	L2	L1	B2	B1
No. of stop bits		Even parity	Parity enable	Character length		Baud rate multiplier	
00 : NA		1 : Even parity	0 : No parity	00 : 5		00 : Synch.mode	
01 : 1		0 : Odd parity	1 : Parity Enable	01 : 6		00 : ×1	
10 : $1\frac{1}{2}$				10 : 7		10 : ×16	
11 : 2				11 : 8		11 : ×64	

The 8521 can be operated with and without parity. It is possible to have a variable data format. The character length as well as number of stop bits can be programmed. The frequency of operation can be modified as per the requirement of application by setting combination of the lower 2-bits. The combination 00 of these lower bit indicates that the mode of operation is synchronous.

Synchronous mode Table 7.3 represents the format for the mode set instruction for the synchronous mode of operation.

Table 7.3   Mode set format for the synchronous mode of operation

D7	D6	D5	D4	D3	D2	D1	D0
SCS	ESD	EP	PEN	L2	L1	B2	B1
Single character sync.	External sync. device	Even parity	Parity enable	Character length		Indication of Synchronous communication	
1 : One character 0 : Two characters	1 : External sync. 0 : Internal sync.	1 : Even parity 0 : Odd parity	0 : No parity 1 : Parity enable	00 : 5 01 : 6 10 : 7 11 : 8			

This is similar to asynchronous mode except that instead of stop bits synchronizing characters can be defined as external or internal. The one or two synchronizing characters can also be specified in the command word. The last two LSB bits must be 00 to indicate synchronous mode of operation. If the synchronous mode of operation is specified, 8251 expects the specified synchronizing characters before it can accept the command word for the desired operation from the microcontroller.

Command Word

The command word controls operation of the 8251. It is used to enable 8251 as well as to set the status of modem control. The error flags can be reset or break characters can be sent. The functions of command word are:

1. Enable the transmitter/receiver section
2. Set the modem control
3. RESET error flag
4. Send the break characters
5. Software reset for 8251

Table 7.4 depicts the format of command word.

Table 7.4   Command word instruction format for 8251

D7	D6	D5	D4	D3	D2	D1	D0
EH	IR	RTS	ER	SBRK	RxE	DTR	TxEN

EH (Enter hunt mode)   If this bit is '1', 8251 enters hunt mode. It waits for the synchronizing character. The result will be on the SYNC/BRKDET pin of 8251. This has a meaning only for synchronous mode of operation.

IR (Internal reset)   When power is switched ON, the 8251 enters the idle mode. Initialization has to start from the mode set word. It again enters idle mode on pressing RESET key. Since the pressing RESET key resets all the peripheral ICs interfaced with the microcontroller, this bit helps to give reset instruction only to 8251. A command word with IR bit '1' will force 8251 to idle mode. 8251 will come out from this mode by writing a new mode set word.

RTS (Ready to send)   If this bit is set, it will generate a low signal on the RTS pin of 8251.

ER (Error reset)   This bit is used to reset the error flags which are set due to the communication error of previous byte. RxE and TxEn are used to enable the receiver and transmitter sections, respectively.

DTR (Data terminal ready)   If this bit is set, it will generate a low signal on the DTR pin of 8251.

SBRK   It is used to send break character.

Status Word Format

This word contains information about various processes that may be required by the microcontroller. The status register is read with the C/D pin at the logic level ' 1', i.e., reading the commands port. The status register is not of updated when it is read by the microcontroller. The format of the status register is shown in Table 7.5. It is used to detect the errors in the communication.

Table 7.5   Status word format for 8251

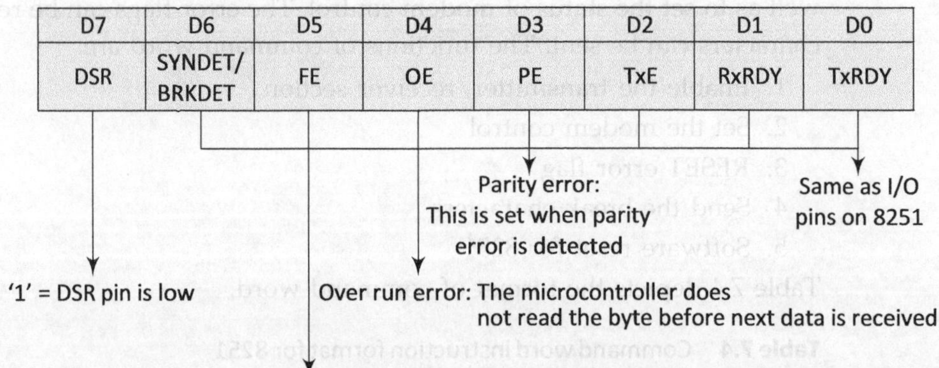

FE (Framing error)   This error occurs when a character is received in a buffer and no stop bits are detected in the stream of data bits. This results in faulty data being received.

OE (Overrun error)   This error occurs when character read on buffer is not read by the microcontroller and the next character arrives. This results in loss of a character.

PE (Parity error)   This error occurs due to the parity mismatch in the data received.

7.2.4 Realization of Additional Serial Port

The 8251 have a serial port, which can be used in different modes of operation. In the complex applications, one serial link is not enough, it is possible to convert the parallel port into serial port using USART. The advantage of this is that it becomes possible to use handshaking signal for the modem, which are available on 8251.

Example 7.1   Design an additional serial port using 8251 to communicate with the terminal. Write a program to transmit the character repeatedly to the terminal.

Solution

The hardware setup as well as software for operation of the system is described below:

Hardware   The setup for additional serial port using 8251 is shown in Fig. 7.5. The RxD and TxD pins of the 8251 provide the desired serial port. The RxC and TxC are connected to common clock generator so that the rate of reception and transmission is same.

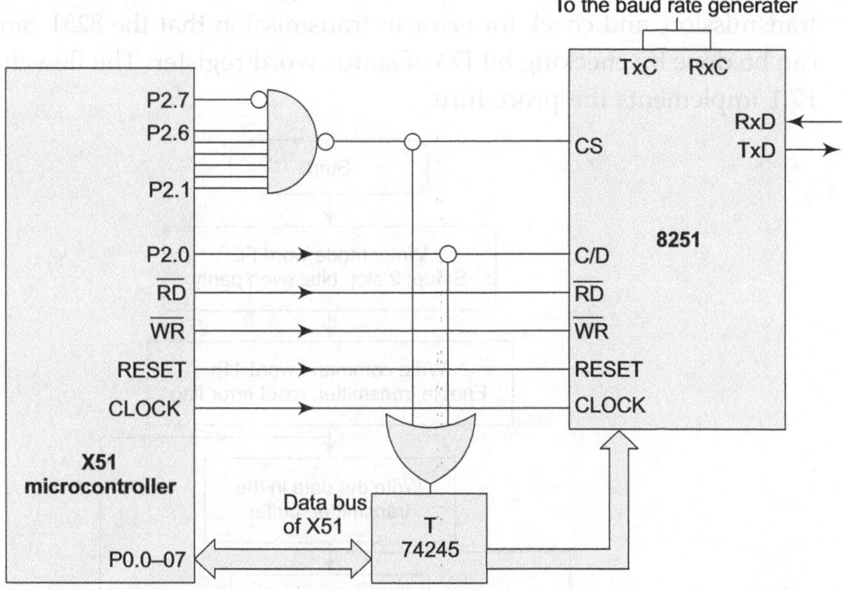

Fig. 7.5   Interfacing scheme to interface 8251 with the microcontroller

Software   It is necessary to load a set of control words in 8251. Microcontroller must write a mode set after the reset based on communication parameters.

Let us assume that the baud rate is 2400 bps, frequency is 16 × baud rate, it has to be asynchronous communication with character length 8-bits, 2 stop-bits, and even parity.

Table 7.6 Mode set word (FEh) for Example 7.1

S2	S1	EP	PEN	L1	L2	B1	B2
1	1	1	1	1	1	1	1

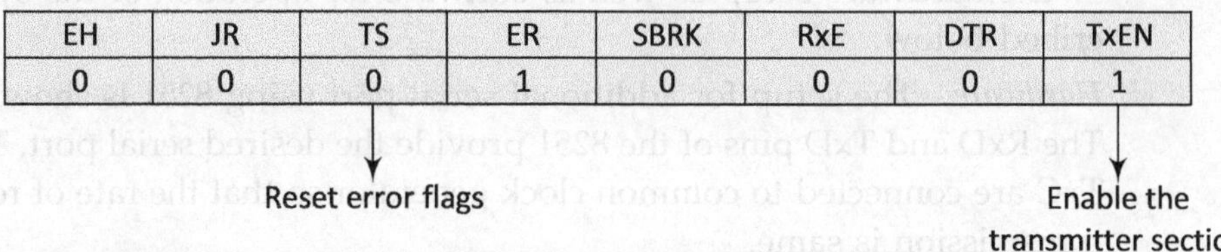

Command word must be written, as asynchronous communication is used, synchronizing characters are not required. The command word should be written to enable the transmitter and reset error flags.

Table 7.7 Command word (= 11h) for Example 7.1

EH	JR	TS	ER	SBRK	RxE	DTR	TxEN
0	0	0	1	0	0	0	1

Now it is necessary to consider status register. It is used to monitor the serial transmission, and check for error in transmission that the 8251 can detect. This can be done by checking bit Do of status word register. The flowchart shown in F7.1 implements the procedure.

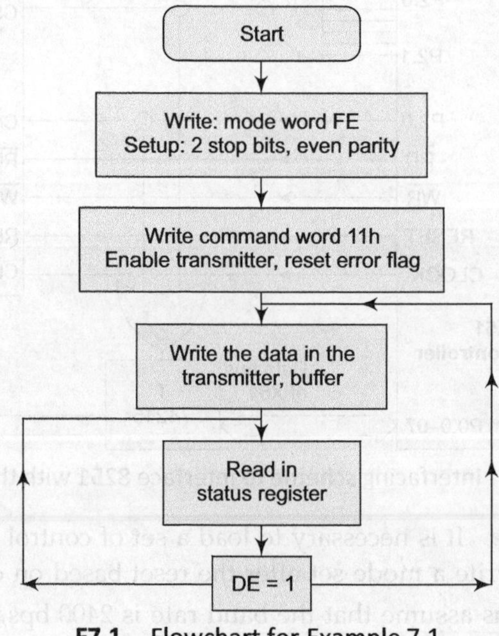

F7.1 Flowchart for Example 7.1

The program segment implementing the flowchart of F7.1 is as follows:

```
Init:
            mov  80, #**            ; initialize stack pointer
            mov  ptr, #007Dh        ; load pointer of command port address
            mov  a, FEh             ; write mode set at command port
            movx @dptr,a
            mov  a, #11h            ; load command word in accumulator
            movx @dptr,a            ; write command word at command port
Restart:
            mov  a, #data₈          ; ASCII data to be transmitted
            mov  dptr,#007Ch        ; pointer to the data port
            movx @dptr,a
            inc  dptr
check:
            movx a, @dptr           ; read the status register
            anl  a, #01h            ; wait for transmitter to be empty
            jz   check
            sjmp restart
            end
```

Example 7.2 Develop a setup to allow echo a character using the 8251 sent by the terminal.

Solution

The hardware is same as Fig. 7.5. There is a change in the software, which requires that the receiver along with the transmitter must also be enabled. The mode set word is same. The formulated command word should enable the transmitter; receiver and reset error flags the mode set word shown in Table 7.8 is 15h.

Table 7.8 Command word (= 15h) for Example 7.2

EH	JR	TS	ER	SBRK	RxE	DTR	TxEN
0	0	0	1	0	1	0	1

Receive operation has high priority so we must write a command to check for receiver to be ready, i.e., full and can receive a character. The command word must check the status register for transmitter to be empty and transmit data. Standard baud rate used in practice are: 110, 300, 600, 1200, 2400, 4800, 9600, 19200 bps.

The program segment for the operation is as follows:

```
init:       mov  dptr, #007D        ; pointer to the command port
            mov  a, feh             ; initialize 8251 mode set word
            movx @dptr,a            ; write mode set word
            mov  a, 15h             ; enable the receiver, the transmitter and clear
                                    ; error flags
```

```
                     movx  @dptr,a          ; write the command word
rec:                 movx  a, @dptr         ; read the status register
                     anl   a, #02
                     jz    rec
mov                  dptr, #007Ch           ; set pointer for the data port
                     mov   a, @dptr         ; receive the data
                     mov   r3, a            ; save received data in register r3
                     inc   dptr             ; pointer to the command port
trans:
                     mov   a, @dptr         ; read the status register
                     anl   a, #01           ; check if transmitter is ready
                     jz    trans            ; wait for transmitter to be ready
                     mov   a, r3            ; load the stored data in the accumulator
                     mov   dptr, #007Ch     ; pointer to command port address
                     movx  @dptr,a          ; write the data in the 8251 buffer
                     Sjmp  rec
                     end
```

7.3 | SERIAL INTERFACE STANDARDS

Serial transmission of data is efficient means for the transmissions of data transfer along long distances. The advantage is that external communication lines/links (letter, telephone, etc. can be used to transfer the data information). There are two types of serial interfaces shown in Table 7.9.

Table 7.9 Serial interface standards

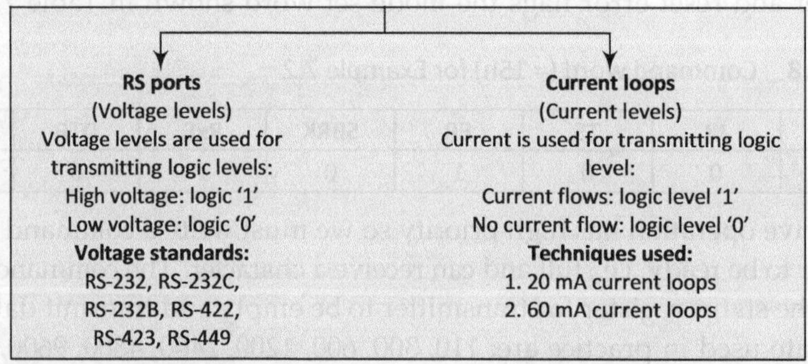

7.3.1 RS-232 Ports

The RS-232 standard was developed in an attempt to ensure that computer hardware shared the same portability characteristics as computer software. In theory, computer equipment that adheres to this standard can communicate with other RS-232 ports with little or no trouble.

The RS-232 interface is a standard interface specified by the Electronics Industries Association (EIA), a trade organization in Washington, DC and is followed by the manufacturers of computers and data communication products. RS stands for the recommended standard. The RS-232C is the popular version of the RS-232 interface. Manufacturers came to a consensus during the 1970s that led to the acceptance of the D-Subminiature 25-pin connector as the standard RS-232 connector.

The RS-232 interface expects a modem to be connected to both the receiving and the transmitting end. The modem is termed as and the computer, terminal, or printer with which the modem is interface is known as DTE. The DCE and DTE are linked via a cable whose length should not exceed 50 ft. Although not reliable it may not affect the communication if the speed of data transfer is reduced when the distance is increased. The DTE has 25-pin D-type male connector and the DCE has 25-pin D-type female connector. However, some manufacturers use 9-pin connectors.

7.3.2 RS-232 Signal Levels

The RS-232 standard follows negative logic. The logical 1 is represented by a negative value, and the logical 0 is represented by a positive value. The 1 (high) varies from −3 to −15 V. In practice, the hardware circuits used for the RS-232 interface maintain the signal level at +12 V (logical 0) and at −12 V (logical 1).

Table 7.10 Pin designation and interface signals of RS-232 port

Pin no.	Signal	Signal name	Source	Destination
1	1	Frame ground	—	—
2	TxD	Transmit data	DTE	DCE
3	RxD	Receive data	DCE	DTE
4	RTS	Request to send	DTE	DCE
5	CTS	Clear to send	DCE	DTE
6	DSR	Data set ready	DCE	DTE
7	SG	Signal ground	—	—
8	RLSD or CD	Received line signal detect or carrier detect	DCE	DTE
20	DTR	Data terminal ready	DTE	DCE
22	RI	Ring indicator	DCE	DTE

Table 7.10 lists the RS-232 interface signals and the pin designations. The TxD is a serial data line carrying the data bits sent by the DTE. The modem receives the TxD signal and uses it for modulating the carrier signal. The TxD is the serial data line from the DCE (modem) to the DTE. The RxD is generated by the modem by demodulating the signal received on the telephone line from the other end of modem.

Before sending the data to the other end, the DTE requests for permission from the modem by issuing the RTS signal. The modem has a method to find out if the telephone line is free and if the other end modem is ready. When the modem finds that the communication path (consisting of telephone line, the other end modem, and DTE) is ready for communication, it issues the CTS signal to the DTE as an acknowledgement (sanction for data transmission) for the RTS. The DTRE issues the DTR signal when it is powered on, error free, and ready for logical connection through the modem. The modem issues a DSR signal to indicate that it is powered on and is error free.

The RI and RLSD signals are used with the dialed modem. When the telephone line is a shared (switched) line, a dialed modem is used and a telephone set is attached to the modem. When a DTE at one end wants to communicate with a DTE at the other end it initiates a dial sequence.

The modem at the sending end sends a dial tone on the telephone line. In response, the called modem issues the RI signal to its DTE and sends an answer tone for 2 s to the calling modem. Then the calling modem sends an 8-ms duration tone on the telephone line. Now the called modem issues CD (carrier detect) signal to its DTE. The CD is an indication to the DTE that it will soon be receiving the data sent by the other end DTE.

7.3.3 Limitations

The RS-232 specification recommends that signals be limited to 20,000 bps. At high speeds, it also recommends that cable length should not exceed 50 ft. These limitations are routinely ignored by RS-232 users. High-speed modems frequently are connected to PCs at 38,400 baud or 57,600 baud with little or no problem. And even high-speed signals such as these will frequently be routed hundreds of feet. If a high frequency digital signal is to be transmitted over very long distances, the channel must be of very large bandwidth and wires have to be laid down which is expensive. However, existing telephone lines can carry analog signals in the range of 40–150 kHz, A device called modem is used to translate digital data in 15 audio tone for transmission on line. At the receiving end this will be converted back into digital data. The data format conversion is done by FSK (frequency shift keying) technique. The logic level representation as the analog frequency signals is as follows:

Logic level		Analog frequency
0	→	2.2 kHzs
1	→	1.2 kHzs

7.3.4 Characteristics of RS-232C

1. At the baud rate of 19,200 baud, the length of the cable should be limited to 50 ft. However, longer cables can be used at low-baud rates.

2. This uses a negative logic

   i.e.,

   $$-12\ V = Logic\ '1'$$
   $$+12\ V = Logic\ '0'$$

3. Noise margin is 2 V, much higher than TTL, which has 0.4 V.

4. To interface with TTL, special line drivers/receivers are required.

   MC1488: TTL $\rightarrow$ RS-232C

   MC1489: RS-232C $\rightarrow$ TTL

   Because of the presence of drivers/receivers on the line, negative logic becomes transparent to the user.

5. It is required that the transition time or settling time must be 4% of transmission rate (1-bit time).

   For example, for a 19,200 baud rate:

   Time $0.04 \times (1/\ 19,200)$ @ 2 ms

   This puts a limit on the length of the cable. For 19,200 baud, the length is 50 ft.

6. The receivers/drivers are single ended and they are reference to the same ground, which puts a limit on the cable length due to the IR voltage drop caused by this common ground current flowing in the line.

RS-232C Connector

It is a 25-pin connector. The pins can be divided into three groups:

1. **Data group** These pins are further divided into two groups: primary and secondary channels. The latter is seldom used, but it allows a path for confirmation.

2. **Timing** They are used for the synchronizing timing signals.

3. **Control** The pins DSR, DTR, RTS, CTS, and DCD are used to establish a protocol between a DCE (modem) and DTE terminal. The remaining control signals pins in this group are for supporting

   (a) a secondary data channel,

   (b) synchronous modem,

   (c) ring detector for autodialing modems, and

   (d) baud is selected for dual-rate modem

   The pin names are defined with reference to the DTE. The 2-pins can be shorter without damaging port. The interfacing is done using the 25-in (socket DB-25S) and (plug B-25P). Since all the pins are not used, care must be taken to check presence of RS port.

Figure 7.6 depicts the equivalent circuit of the RS-232C port.

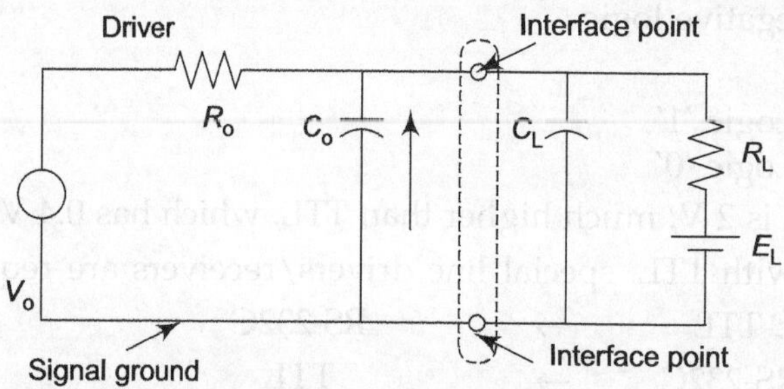

Fig. 7.6   Equivalent circuit of the RS-232C port

7.3.5 Need for Handshaking

Normally, microcontroller is synchronous to character rate of USART by testing (TxRDY) flag. The flaw is that the readiness of the data receiver is not being tested. For example, while interfacing a VDT, the sequence (1B,7A) (ESC, '7') will clear the screen. This operation requires several milliseconds. However, a new character is sent at the rate of 9600 baud, i.e., every milliseconds.

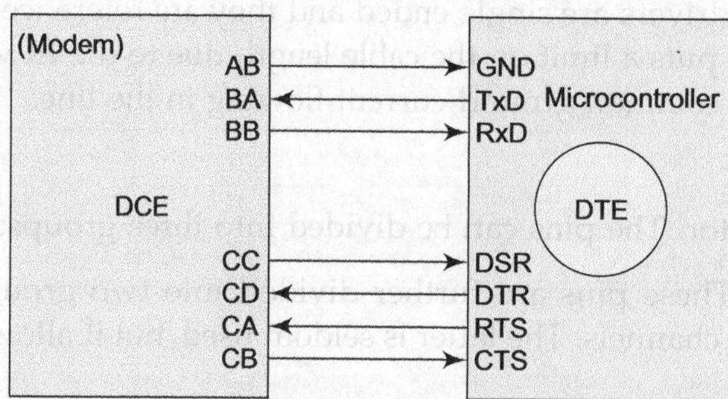

Fig. 7.7   Interfacing DTE and DCE using handshaking signals

As shown in Fig. 7.7, four modem control signals are used to allow a handshaking protocol to be established.

DSR   (data set ready) This is output by DCE (modem) to indicate that modem is ready (i.e., logic '1').

DTR   (data terminal ready) This is output by DTE to indicate that it is ready for communication, the signal is used to switch ON the modem.

RTS   (ready to send) This is output by DTE to indicate that it is ready to transmit data.

CTS   (clear to send) This is an acknowledgement of the RTS signal. It indicates that DCE is ready for transmission.

The handshaking process can be explained as follows:

The signal (DSR/CC) when ($V > 3$), indicates that the modem is connected to the transmission line and is ready to handle data. CD/DTR indicate that the modem should be prepared for transfer of data from DTE, by connecting it to the communication line. CA/RTS is sent to modem when both terminals are ready and DSR is ON. This signal is acknowledged by CB/CTS. At this point DTE is permitted to transmit data onto the line through the modem. The DCD signals DTE that a carrier is present. The ring detector is used in connection with automatic answering machine. The process is described by the flowchart.

The flowchart shown in F7.2 depicts the handshaking process, when RS-232C port data is exchanged with the modem.

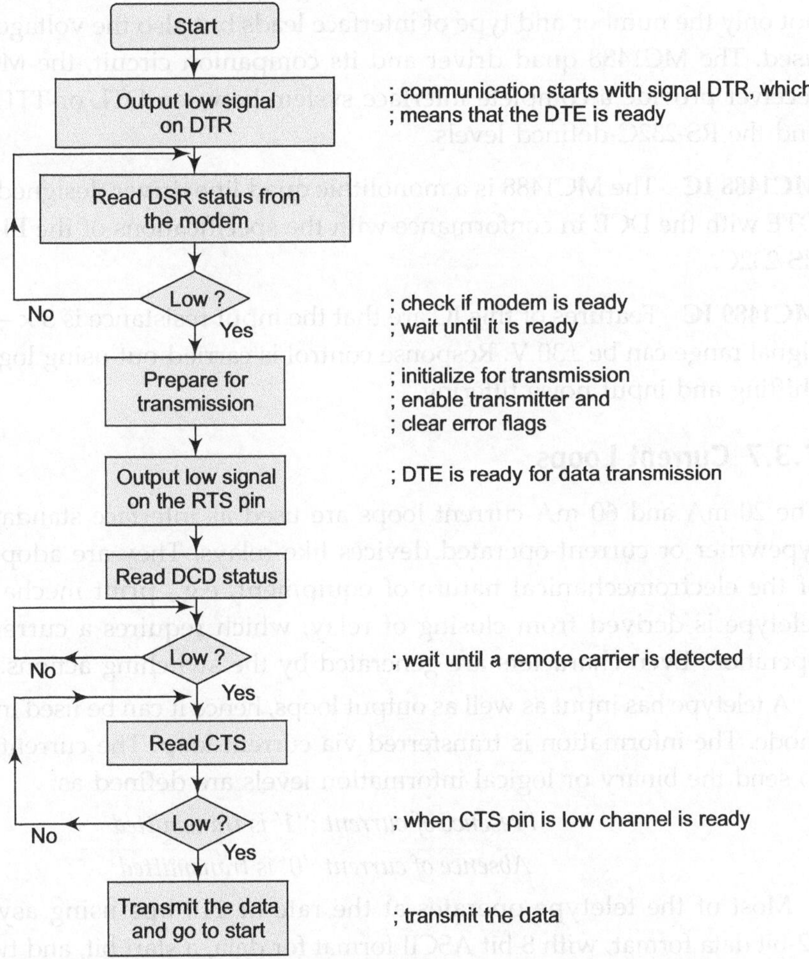

F 7.2 Flowchart for the handshaking process between DTE and DCE

In practice a subset of these signals is used. For example

1. In full-duplex mode of operation, the modem uses only DSR and DCD signals. The microcontroller polls DSR signal to verify that the modem is

ready and not in the test mode. After placing the call, the distant DCD is polled to verify that a modem is present at the other end.

2. In half-duplex mode of operation, DSR and DCD serve the same purpose as in full-duplex mode RTS is used to request the turn around so that controller can transmit. CTS acknowledge this request.

7.3.6 RS-232 Quad Line Driver (MC1488), Receivers (MC1489)

The EIA has released the RS-232C specification detailing the requirements for the interface between data and processing equipment. This standard specifies not only the number and type of interface leads but also the voltage levels to be used. The MC1488 quad driver and its companion circuit, the MC1489 quad receiver provide a complete interface system between DTL or TTL logic levels and the RS-232C defined levels.

MC1488 IC The MC1488 is a monolithic quad line driver designed to interface DTE with the DCE in conformance with the specifications of the EIA standards RS-232C.

MC1489 IC Features of this IC are that the input resistance is $3\,k - 7\,k\Omega$. Input signal range can be ±30 V. Response control is carried out using logic threshold shifting and input noise filtering.

7.3.7 Current Loops

The 20 mA and 60 mA current loops are used as interface standards for tele-typewriter or current-operated devices like relays. They are adopted because of the electromechanical nature of equipment, e.g., print mechanism in the teletype is derived from closing of relay, which requires a current drive for operation. Even characters are generated by the switching actions.

A teletype has input as well as output loops, hence it can be used in full-duplex mode. The information is transferred via current loop. The current when used to send the binary or logical information levels are defined as:

Presence of current : '1' is transmitted

Absence of current :'0' is transmitted

Most of the teletype operates at the rate of 110 bps using asynchronous, 12-bit data format, with 8-bit ASCII format for data, a start bit, and two stop bits. The device to start the mechanical cycle so that it is ready to accept and interpret signals received uses the start bit. Using USART or software can generate the format.

Since the microcontroller is a voltage-operated device, an interface is required to convert data signal from voltage to current form before transmission and current to voltage form on reception of signals. Figure 7.8 shows the TTY with the interface which can be connected to the microcontroller using parallel port or serial port. The send and receive currents loops are separate. The system can operate in a full-duplex mode.

A character is sent to RxD pin from the TTY keyboard by opening or closing a switch. The switch is normally closed so transistor Q1 is ON, this drives RxD pin HIGH. When the switch is open, Q1 is OFF and the RxD pin is at logic level LOW. Thus, the presence or the absence of current used is referred as '1' or '0' in the ASCII character format.

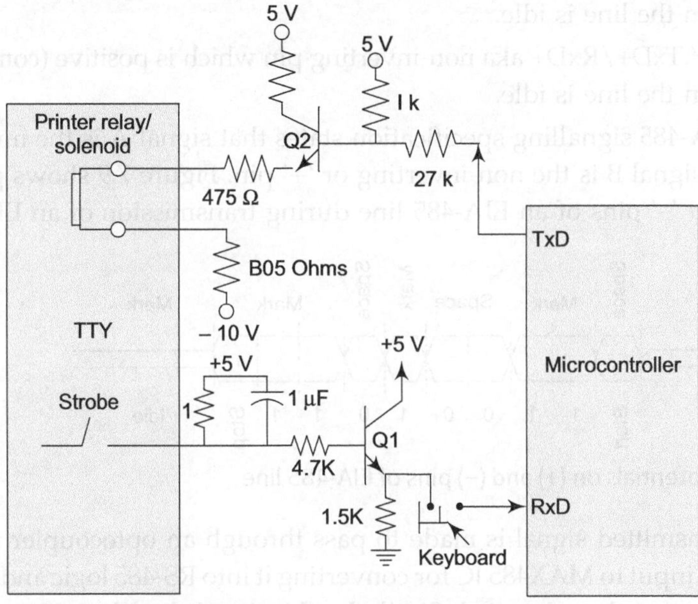

Fig. 7.8   Interfacing TTY with the microcontroller

In order to output the data, microcontroller sends a low on TxD pin, which makes the transistor Q2 ON. This sets a current of 20 mA in a relay coil. A HIGH on TxD pin makes Q2 OFF and the current flows in the loop. Thus, '1' on '0' may be transmitted.

7.3.8 EIA-485

EIA-485 (formerly RS-485 or RS-485) is an OSI Model physical layer electrical specification of a two-wire, half-duplex, multipoint serial connection. The standard specifies a differential form of signalling. The difference between the wires' voltages is what conveys the data. One polarity of voltage indicates Logic '1' level, the reverse polarity indicates Logic '0'. The difference of potential must be at least 0.2 V for valid operation, but any applied voltages between +12 V and −7 V will allow correct operation of the receiver.

EIA-485 only specifies electrical characteristics of the driver and the receiver but does not specify or recommend any data protocol. It enables the configuration of inexpensive local networks and links with high-data transmission speeds (35 Mbps up to 10 m and 100 kbps at 1200 m). It uses a differential balanced line over twisted pair (like EIA-422). It does not specify any connector.

The EIA-485 differential line consists of two pins:

- A: '−' TxD-/RxD− aka inverting pin which is negative (compared to B) when the line is idle.
- B: '+' TxD+/RxD+ aka non-inverting pin which is positive (compared to A) when the line is idle.

The EIA-485 signalling specification states that signal A is the inverting or '−' pin and signal B is the non-inverting or '+' pin. Figure 7.9 shows potentials of the '+' and '−' pins of an EIA-485 line during transmission of an EIA-485-byte.

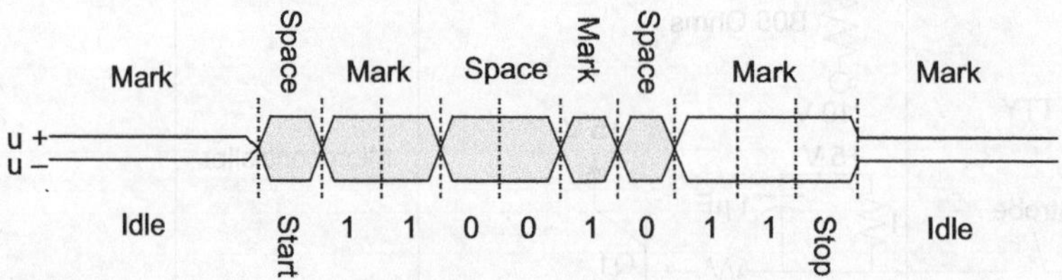

Fig. 7.9   Potentials on (+) and (−) pins of EIA-485 line

The transmitted signal is made to pass through an optocoupler for isolation before it is input to MAX485 IC for converting it into RS-485 logic and transmitted for reception at the other end. Similarly, the signal that the computer sends to the microcontroller is sent in RS-485 logic and is converted to TTL logic before it is received by the microcontroller at the RxD pin via optocoupler.

7.3.9 Serial Peripheral Interface (SPI)

The serial peripheral interface bus or SPI (often pronounced "spy") bus is a synchronous serial data link standard designed by Motorola that operates in full-duplex mode. Devices communicate in master/slave mode where the master device initiates the data frame. Multiple slave devices are allowed with individual slave select (chip select) lines. The features of SPI are: full-duplex communication, higher throughput than I^{2}C, reduced overhead, and multi-master protocol. However, it requires more pins.

There is neither support for hardware flow control nor acknowledgement from slave. The SPI full-duplex capability makes it very simple and efficient for single master/single slave applications. The complexity of the interface grows as the number of slave devices increases due to its lack of built-in addressing.

As shown in Fig 7.10 the SPI bus specifies four logic signals.

- SCLK—Serial clock (output from master)
- MOSI/SIMO—Master output, slave input (output from master)
- MISO/SOMI—Master input, slave output (output from slave)
- $\overline{SS}$—Slave select (active low; output from master)

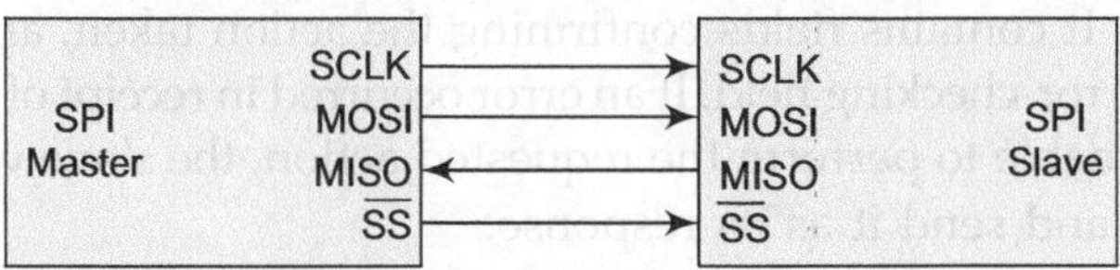

Fig. 7.10 SPI communication

To begin a communication, the master pulls the slave select low for the desired chip and waits for predefined time (if required). The master generates a clock frequency less than or equal to the maximum frequency the slave device supports, configured in the master on a per transmission basis. A full-duplex transmission occurs during this clock:

1. Master sends data on the MOSI line; the slave reads from the MOSI line.
2. Slave sends data on the MISO line; the master reads from the MISO line.

Not all transmissions require all four of these operations to be meaningful but they do happen. Generally, a transmission happens in single byte increments and a master can initiate multiple 8-bit transmissions if it wishes/needs. Other slaves on the bus ignore the clock, MOSI, and MISO lines.

7.3.10 Modbus Protocol

Controllers can be networked directly or via modems. Controllers commu-nicate using a master–slave technique, in which only one device (the master) can initiate transactions (called 'queries'). The other devices (the slaves) respond by supplying the requested data to the master, or by taking the action requested in the query. The master can address individual slaves, or can initiate a broadcast message to all slaves. Slaves return a message (called a 'response') to queries that are addressed to them individually.

Modbus protocol defines a message structure that controllers will recognize and use. It describes the process a controller uses to request access to another device, how it will respond to requests from the other devices, and how errors will be detected and reported. It establishes a common format for the layout and contents of message fields.

The Modbus protocol establishes the format for the master's query by placing into it the device address, a function code defining the requested action, any data to be sent, and an error-checking field.

Modbus message is placed by the transmitting device into a frame that has a known beginning and ending point. This allows receiving devices to begin at the start of the message, read the address portion, and determine which device is addressed (or all devices, if the message is broadcast), and to know when the message is completed. The slave's response message is also constructed using Modbus protocol. It contains fields confirming the action taken, any data to be returned, and an error-checking field. If an error occurred in receipt of the message, or if the slave is unable to perform the requested action, the slave will construct an error message and send it as its response.

Query The function code in the query tells the addressed slave device what kind of action to perform. The data bytes contain any additional information that the slave will need to perform the function.

Query message: master			
Device address	Function code	8-bit data	Error-check

Response If the slave makes a normal response, the function code in the response is an echo of the function code in the query. The data bytes contain the data collected by the slave, such as register values or status.

Response message: slave			
Device address	Function code	8-bit data	Error-check

If an error occurs, the function code is modified to indicate that the response is an error response. The error-check field allows the master to confirm that the message contents are valid. Responses are not returned to broadcast queries from the master. Examples of Modbus functions are read holding/input register; preset single/multiple register.

There are two serial transmission modes used for Modbus frames: ASCII and RTU.

ASCII mode When controllers are setup to communicate on a Modbus network using ASCII mode, each 8-bit byte in a message is sent as two ASCII characters. This mode allows time intervals of up to, 1 s to occur between characters without causing an error.

The format for each byte in ASCII mode is

START	ADDRESS	FUNCTION	DATA	LRC	END
1 chars	2 chars	2 chars	0 to 2x252 chars	2 chars	2 chars : CR,LF
ASCII frame format					

RTU mode When controllers are setup to communicate using RTU mode, each 8-bit byte in a message contains two 4-bit hexadecimal characters for greater character density and has better data throughput than ASCII for the same baud rate. Each message must be transmitted in a continuous stream. The format for each byte in RTU mode is

START	ADDRESS	FUNCTION	DATA	LRC	END
≥ 3.5 chars	8-bits	8-bits	N x 8-bits	16-bits	≥ 3.5 chars
RTU Frame Format					

Modbus uses two kinds of error checking:

Parity checking (even or odd) can be optionally applied to each character.

Frame checking (LRC or CRC) is applied to the entire message.

In RTU mode, messages include an error-checking field that is based on a cyclical redundancy check (CRC) method. The CRC field checks the contents of the entire message. It is applied regardless of any parity check method used for the individual characters of the message.

7.3.11 Universal Serial Bus (USB)

The Universal Serial Bus (USB) specification defines the mechanical, electrical, and protocol layers of the interface. USB defines two types of hardware: hubs and functions. Up to 127 devices may be connected together in a tiered star topology. This tiered star topology allows monitoring of power to each device can be monitored and can be switched OFF if an overcurrent condition occurs without disrupting other USB devices (up to 127 devices). The USB specification was first released in 1994. The current USB standard, Revision 2.0 was released in 2000.

USB supports plug'n'plug with dynamically loadable and unloadable drivers. The loading of the appropriate driver is done using a PID/VID (product ID/ vendor ID) combination. Some organizations provide an extra VID for non-commercial activities such as teaching or research. The user simply plugs the device into the bus. The host will detect this addition, interrogate the newly inserted device, and load the appropriate driver. The end user needs not worry about terminations, terms such as IRQs and port addresses, or rebooting the computer. User can simply plug the cable out; the host will detect its absence and automatically unload the driver.

Wire segments are point-to-point between a host, hub, or function. Transition from hub-to-hub represents a tier and may be connected to a USB bus which is a differential bidirectional serial interface cable bus.

NRZI data transmission may be isochronous or asynchronous at three different rates over a maximum cable length of 4 m (over 4 wires, 2 wires carry data on a balanced twisted pair) from 10 kbps to 400 Mbps in one of three speed modes.

- Slow-speed: 10 kbps to 100 kbps (USB keyboard or USB mouse)
- Full-speed: 500 kbps to 10 Mbps (Devices). It is 2.8 V (high) to 0.3 V (low). Operation at high-speed mode is at 400 mV +/− 10% (high) to 0 V +/− 10 mV (low). Cable impedance for both modes is 90 Ω +/− 15% (differential).
- High-speed [USB 2.0]: ≤ 480 Mbps, with a speed range of 25 Mbps to 400 Mbps. Transmission at the high-speed mode requires the addition of 45 Ω termination resistors between each data line and ground. Operation at control, interrupt, isochronous, and bulk are the packet protocols. Each exchange contains three packets: token packet (address), data packet (data), and a handshake packet (termination).

NRZI produces a change in the signal indicating a Logic 0, no change indicates a Logic 1. Bit stuffing is used with NRZI to stop the signal remaining in the steady state condition; if more than 6 ones are transmitted (no change in the signal) a zero is inserted to produce a transition. NRZI, with bit stuffing is self clocking, allowing the receiver to synchronize with the transmitter.

USB port uses a 4-pin connector, type A or type B (Fig. 7.11), termed as: VBUS(red), D–(white), D+(green) and ground(black). A/B Jacks are used on the chassis side, and the A/B plugs are used on the cable ends. Type A jacks connect to type A plugs, and type B jacks connect to type B plugs. Cables have an A plug on one end and a B plug on the opposite end. The connectors have Pins 1, 4 longer than 3, 4, so power and ground mate first which allow devices to be hot-swappable. Type A connectors point to the hub, while type B connectors point to the function. Normally, a cable will have a type A connector on the computer side [Hub] and a USB type B connector on the far [function] side, to a USB device.

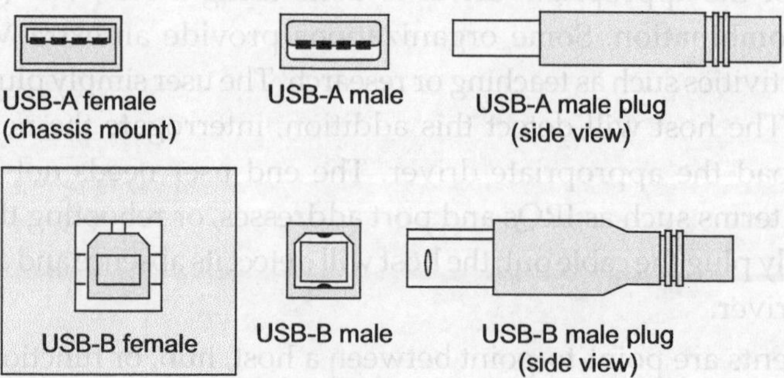

Fig. 7.11 USB port connectors

The USB to RS-232 and USB to RS-422/485 converters all have XP compatible drivers. The RS-232 connections provide I/O to match 9-pin/25-pin serial port. A USB converter must connect to a USB host port. The USB connector (type A male) only fits upstream device connectors (type A female) such as a host port. USB device (slave) devices which don't have a cable attached have a downstream (type B female) connector. USB connectors are designed so that proper connections can be made between the USB host and slaved devices. Two USB devices can communicate with the host, not directly with each other.

The USB is host controlled. There can only be one host per bus. The specification in itself, does not support any form of multimaster7 arrangement. USB 2.0 has introduced a host negotiation protocol introduced by USB 2.0 limited to single point-to-point connections such as a mobile phone and personal organizer. The USB host is responsible for undertaking all transactions and scheduling bandwidth. Data can be sent by various transaction methods using a token-based protocol.

The USB host controllers have their own specifications. With USB 1.1, there were two host controller interface specifications.

1. **UHCI** (universal host controller interface) developed by Intel which puts more of the burden on software (Microsoft) and allowing for cheaper hardware.

2. **OHCI** (open host controller interface) developed by Compaq, Microsoft, and National Semiconductor which places more of the burden on hardware (Intel) and makes for simpler software.

USB 2.0 a new host controller interface specification was needed to describe the register level details specific to USB 2.0. The EHCI (enhanced host controller interface) was born. Significant contributors include Intel, Compaq, NEC, Lucent, and Microsoft so it would hopefully seem they have pooled together to provide us one interface standard and thus only one new driver to implement in any operating systems.

7.3.12 USB to RS-232C Interface

The FT232R is a USB to serial UART interface with optional clock generator output with chip-ID security dongle feature. In addition, asynchronous and synchronous bit bang interface modes are available. USB to serial designs using the FT232R have been further simplified by fully integrating the external EEPROM, clock circuit and USB resistors onto the device. As shown in Fig. 7.12 it is a 28-pin SSOP package.

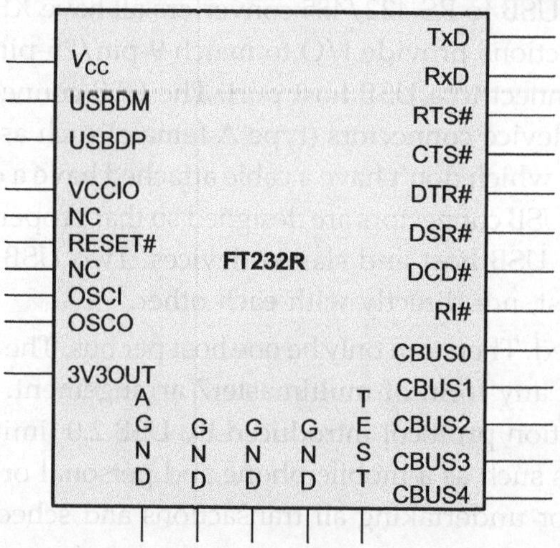

Fig. 7.12 FT232R: Pin configuration

The internally generated clock (6, 12, 24, and 48 MHz) can be used to drive a microcontroller or external logic. A unique number (the FTDI Chip-ID™) forms the basis of a security dongle be used to protect application software.

Fig. 7.13 USB to RS-232C converter

A TTL to RS-232 level converter IC on the serial UART of the FT232R carries out RS-232 level conversion. MAX213CAI (Maxim) and ADM213E (analog devices) may be for communication at up to 115,200 baud, while MAX3245CAI handles communication up to 1 M baud. Figure 7.13 shows a sample converter.

7.4 │ MODEMS

Modem stands for modulator demodulator. The function of a modem is to convert a binary signal into an analog signal of fixed frequency. Since the normal telephone lines have a limited bandwidth, digital signal which requires larger bandwidth cannot be transmitted directly. The process of converting binary 0 or 1 signal into an analog frequency of 2.2 kHz or 1.2 kHz, respectively is called *modulation*. The reverse operation is called *demodulation*.

Modems are classified by the standard protocols that define the baud rate, the method of modulation, and the specific manner in which high levels are represented. The modems are said to be compatible if they both employ same protocols. Many modems support multiple protocols. They automatically configure themselves as per protocols of the originating modem. Bell 103 are the oldest modem standards protocols that employ frequency modulation (FM) to convert signals between analog and digital environment.

A user can initiate the conversation by employing a modem and a terminal to dial up a remote system. The local modem is called originator while the other is operated in auto-answering mode. Bell protocols define two frequencies for originator and two for auto mode. Thus, four audible frequencies support fully duplex communication.

New modems have the capacity to operate as originate or answer mode. In addition to bells protocol, Racal Vadic protocol is also used which employs phase modulation (PM). Most of the modems connected to communication port of computers support both types of protocols. The data transfer between two systems can be done with the help of Kermit, X-Modem, or file transfer protocols.

Most of the modems are packed into a small rectangular box. The indicators provide status information regarding data flow, handshaking signal, and the status of remote carrier. Because the carrier detect is known by multiple names it is marked as CXR. The modem is connected to the serial port of the computer through a standard RS-232C cable.

7.4.1 Smart Modems

The word smart indicates that the modem has in-built intelligence in the form of a microprocessor or microcontroller and firmware. This relieves the user from connecting a telephone line to the system. The numbers can be stored in the memory and based on the command. The number from the list can be selected and dialed. The smart modems have dial-up utility programs. The utilities provide services like sorting, creating, and searching of telephone lists stored on system. Hayse standard instructions are supported by most of the communication software to have autodialing environment.

7.4.2 Microcomputer Modems

The modems to be used with microcomputers and microcontrollers are designed as:

Stand-alone modems They can be used with any system having RS-232 port. The advantage of these modems is that they have indications, which can be used for detection of malfunctioning. They are portable, can be connected without need for an extra slot on the system.

Internal modems They are designed on a single PCB so that they can be inserted in the slot on the system. Thus they do not increase the desk space, commonly termed as the footprint of the system, used by the system. They do not require an RS-232 port and make use of the on board speaker to hear modem dial and available connections. No extra cable for power or connection is required. They are much cheaper than stand-alone ones.

7.4.3 GSM Modem

A GSM modem is a wireless modem that works with a GSM wireless network. A wireless modem behaves like a dial-up modem. Dial-up modem sends and receives data through a fixed telephone line while a wireless modem sends and receives data using radio waves. A GSM modem can be an external device or a PC Card/PCMCIA Card. It may be connected externally using a serial cable or a USB cable. It needs a SIM card from a wireless carrier in order to operate. GSM modems support a common set of standard and extended set of AT commands. The main function of modem is to read, write, delete, send, and receive SMS messages. Messages can also be stored and sent after some fixed time to the owner. GSM modem can also receive messages from the owner and display that on the LCD. Figures 7.14(a) and 14(b) show typical GMS modem while Table 7.11 depicts specifications of a typical GSM modem.

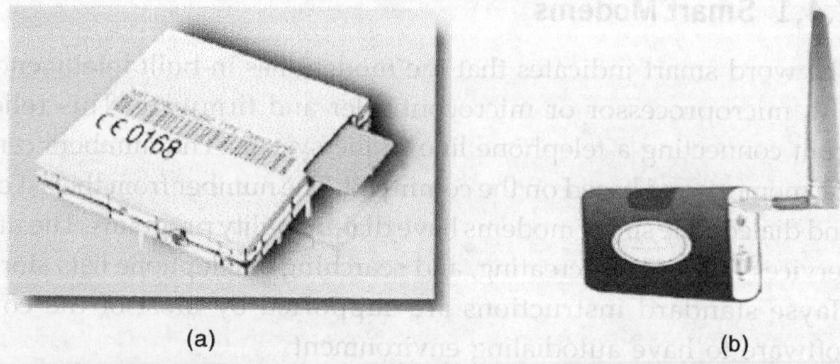

(a)   (b)

Fig. 7.14   GSM modem

Table 7.11   GSM modem: technical specifications

	GSM Modems
Band	Quad Band 850/900/1800/1900
Compliant	ETSI GSM Phase2+
GPRS	Type B, Multi-slot Class 10
Antenna	2.5 dBi, 50&!, SMA (Female) connector Omni directional, Fixed or Roof-top type with flexible cable of 3meters.
Com Port(RS-232C)	DB9 Male
Transmission Power-RF	2W (850/900 MHz band) 1W (1800/1900 MHz band)
Sensitivity	Better than -106 dBm
	External Adaptor (power supply)

Readers may visit website: htpp://www.developershome.com/sms/ GSMModemIntro.asp for more details.

7.4.4 Setting Up Modems

The setup can be simple or complex depending on the modem used. However, two important points need to be considered.

1. Since most modems are disabled until an active level is sensed on the DTR line, care must be taken to inform the modem to either ignore status or always enable all lines, if it cannot assert the DTR signal. Using jumper or DIP switches provided on the modem can do this.

2. The receiver portion of the UART is not enabled until an active low appear on DCD. Many times user and smart modems communicate prior to connection, e.g., commands entered by the user must be echoed on the terminal. This can be done by using DCD assert jumpers or DIP switches provided on the modem.

Most of the modems are capable of performing self-test to check the malfunctioning of either the analog or the digital part of the modem.

7.5 | TELE-LINKS

In order to implement remote control operation, a telephone line can be interfaced with the microcontroller. The telephone instrument can be used as an input device for the operation.

7.5.1 Telephone System

The most familiar instrument to the user of the telephone system is the subscriber set or basic telephone. The telephone is the most extraordinary element of the telecommunication system because of its being synonymous with the telecommunication network. A subscriber set performs the dialling, ringing, transmitting, and receiving functions required at each user location. A typical subscriber set consists of seven main components:

1. Receiver
2. Transmitter
3. Speech network
4. Hook switch
5. Ringer
6. Dialler
7. Bridge rectifier (primarily used with electronic diallers)

In the subscriber set, the hook switch may be in either of the two positions, 'On-hook' or 'Off-hook'. These conditions correspond to idle and busy circuits, respectively. The Off-hook condition is normally activated by lifting the handset. When the handset is lifted, a current sensing device prepares to send the receiver voice communication. This hook switch connects the telephone line to the ringer in an 'On-hook' position and to the speech network in an 'Off-hook' position. In the 'Off-hook' position, subscriber circuit, circuitry receives a DC bias from the power supply from the switching centre. And the 'On-hook' position, a ring signal may be initiated by a caller. An electrical signal of about 80 V_{rms} and 20–30 Hz is typically generated at the switching centre to actuate the ring as at a subscriber set.

Basic Telephone Circuit

A telephone works on the principle that variation in the line current is proportional to the frequency of sound. Using variable resistance does this. The resistance of transducer varies proportionally to pressure exerted on it by sound waves. The speaker generates the sound waves. The transducer, which converts sound waves to electrical signals is called a 'microphone' and the one which does the reverse function is called speaker/earphone. The range of communication depends on the applied voltage and sensitivity of the receiver.

Functions of Telephone

A telephone is designed to perform some definite functions, which are of two types:

(a) The speech functions

1. Electrical signals produced by the microphone
2. Conversion of electrical signals to acoustic signals done by the earphone

(b) The signalling functions

    1. Conversion of electrical signalling information from exchange to acoustic signals for inviting subscribers' attention, which is done by the electrical bell.

    2. Informing the exchange that the telephone is Off-hook, whenever the subscriber lifts the hand set. Closing the DC path between the exchange and subscriber does this.

    3. Informing the called party's address to the exchange for setting up the call, which is done by sending coded signal using dial.

    4. Normally alternating voltage of low value is used for signalling (or ringing)

Operation

Telephone network needs only one pair of wire and yet give an impression of working in a duplex mode, but actually works in a half-duplex mode. The exchange senses the closing of DC loop, whenever the handset is lifted, and disconnects the ringing voltage. Similarly, exchange senses the DC loop, whenever the handset is lifted and sends a dial tone.

There are two methods used to transmit the dialling information to the centre (i.e., telephone exchange):

    1. Pulse dialling

    2. Tone dialling

Pulse dialling (Rotary dialler)   Once the dial tone is received, the user dials the required number. When a number is dialled, DC pulses corresponding to the number dialled are generated by the dial and transmitted on the line. These dial pulses are used for sending the address (telephone number) of the called subscriber to the telephone exchange. The rotary dial sends pulses or the dialling frequency at the rate of 10 pulses/seconds (PPS), with a mark/break ratio of 2:1.

Tone dialling   The modern electronic telephones, in which tone dialling is characterised by absence of rotary dial, are commonly referred to as 'Push Button or Touch Tone' telephone. Such telephones use special purpose ICs for dialling ring detection and speech processing. They can work in pulse dialling or tone dialling mode.

Dial tone   When a subscriber lifts his handset to make a call a continuous tone of 400 Hz is heard indicating that the exchange is ready to accept the dialling. The number should not be dialled until the dial tone is heard.

Ring-back tone   An interrupted tone of 1 second ON and 2 second OFF time is heard in the calling subscriber's receiver when the bell of the called party is ringing.

Pulse dialler used a series of pulses to transmit dial signals to the central office, while the tone dialler generates the tone, which is the combination of various frequencies.

The advantages of tone dialling over pulse dialling can be listed as

1. Faster dialling speed.
2. Ability signal over any noise-grade transmission path.
3. It uses solid-state circuit for the tone generation and detection.
4. After the call has been connected, tone dialling can be used for low speed data transmission.
5. Tone dialling is more compatible and electronically controlled.

DTMF Decoder

A dual tone multi-frequency (DTMF) decoder is necessary in any system using DTMF signalling. The purpose of using this decoder is to decode a valid pair of signalling tones and provide output data corresponding to the DTMF signal received.

DTMF matrix dial (DTMF frequencies)   In tone dialling, which uses distinct voice band frequency signals, each consists of two sinusoidal signals, one from a 'Low Group' and one from a 'High Group' of frequencies. The details are shown in matrix form in Table 7.12.

Table 7.12   DTMF frequencies

Low group frequencies (Hz)	Normal high group frequencies (Hz)			
	1209	1336	1447	1633
697	1	2	3	A
770	4	5	6	B
852	7	8	9	C
941	*	Ø	#	D

If one dials a number '5', then two tones of 770 and 1336 Hz are transmitted. These tones are sensed and decoded by the exchange as the dialled digit, '5' in this case. The column pertaining to tone 1633 Hz is used for special facilities like flash, pause, etc. Figure 7.15 shows DTMF dial arrangement.

The choice of code for touch-tone signalling should be such that imitation of code signals by speech and music should be difficult. Simple single frequency structures are all prone to easy imitation as they occur frequently in speech and music. Hence, some form of multifrequency code is required. Selecting a set of N frequencies and restricting them in a binary fashion to being either present or absent in a code combination easily derives such codes.

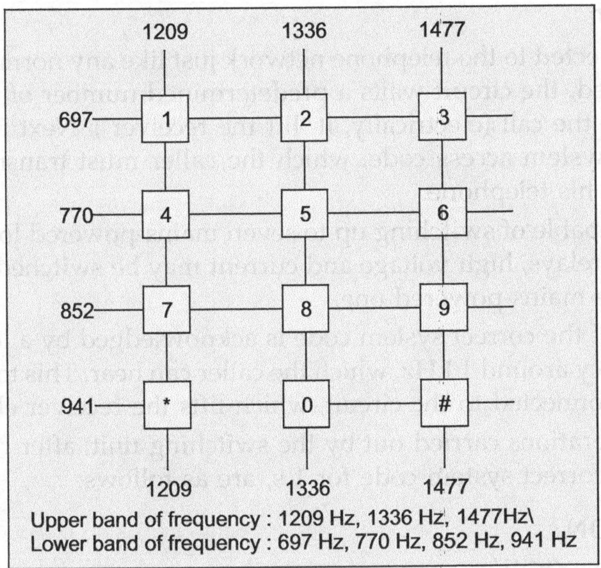

Fig. 7.15   DTMF dial arrangement

Transmitting N frequencies simultaneously involves N-hold sharing of a restricted amplitude range and hence to keep as small as possible the number of frequencies to be transmitted for any valid code word. These factors lead to the consideration of P out of N code.

Before attempting to determine the two specific frequencies at the receiver end, band filtering can be used to separate the frequency groups. This renders determination of specific frequencies simpler. Each frequency component can be amplitude.

7.5.2 Remote Control Using Telephone

The aim of the system described here is to design a switching circuit, which is capable of controlling up to seven loads with the help of a command received via telephone. It is assumed that the commands reach the unit via line in the form of DTMF codes generated on a set. The receipt of the correct code is acknowledged by a short tone which a caller can hear.

The functions expected from the unit after receiving the command are as follows:

1. Unit waits for predetermined number of ring signals and answers the call.
2. Unit switches ON the specified load.
3. It switches OFF the specified load.
4. It switches ON/OFF all loads.
5. Check status of the specified device.

System Operation

The circuit is connected to the telephone network just like any normal telephone set. On being called, the circuit waits a predetermined number of ring signals, and then answers the call (electrically, it 'lift the receiver'). Next it waits for a preprogrammed system access code, which the caller must transmit with the DTMF keypad on his telephone.

The circuit is capable of switching up to seven mains-powered loads ON and OFF. By virtue of relays, high voltage and current may be switched, so that the loads may include mains-powered one.

The reception of the correct system code is acknowledged by a long tone for 1 s of tone frequency around 1 kHz, which the caller can hear. This tone indicates that the caller is connected to the circuit, which lifts the receiver electrically.

The various operations carried out by the switching unit, after receiving the long tone for the correct system code for 1 s, are as follows:

1. **Switching ON**

   Dial $\rightarrow$ 'n' '1'

   First number indicates the device number (from $n = 1 \ldots 7$), while the second number indicates the corresponding device is made 'ON'. System response will be to generate $\rightarrow$ Long Tone

2. **Switching OFF** After ready to accept command,

   Dial $\rightarrow$ 'n' '0'

   First number indicates the device number (from $n = 1 \ldots 7$) and the second number indicates the corresponding device is made 'OFF' System response will be to generate $\rightarrow$ Long Tone

3. **Device status checking** After ready to accept command,

   Dial $\rightarrow$ 'n' '2'

   First number indicates the device number (from $n = 1 \ldots 7$) and the second number indicates check status of device : ON or OFF. System in response will generate:

   (a) Two short tones indicate device is ON

   (b) Three short tones indicate device is OFF

4. **Simultaneous all device ON** After ready to accept commands to make all the devices ON simultaneously.

   Dial $\rightarrow$ '8 '   System will generate $\rightarrow$ Long Tone

5. **Simultaneous all device OFF** After ready to accept commands to make all the devices OFF simultaneously.

   Dial $\rightarrow$ '8' '0'   System will generate $\rightarrow$ Long Tone

6. **Default** If the caller fails, to dial any code within 10 s of hearing any of the above codes, i.e., system access or device access or device acknowledge tones, the system will automatically On-hook the receiver.

Any tone-dialling (DTMF) telephone set or hand-held tone dialer may be used to send commands to the switching unit, and remotely control a wide range of mains appliances in and around the home.

Hardware Setup and Description of System

The block diagram of hardware setup of the switching unit is shown in Fig. 7.16. It includes telephone line. Transformer for inserting tone signal on line, relays, and crystals, etc. The heart of the circuit is 8031 microcontroller. The integrated DTMF decoder (M8870) decodes the tone dialing codes received via telephone line. Interfacing of telephone line consists of two parts:

1. To detect the ring signal and it enable the unit to answer the call at right moment.
2. To receive and transmit tones via the telephone line.

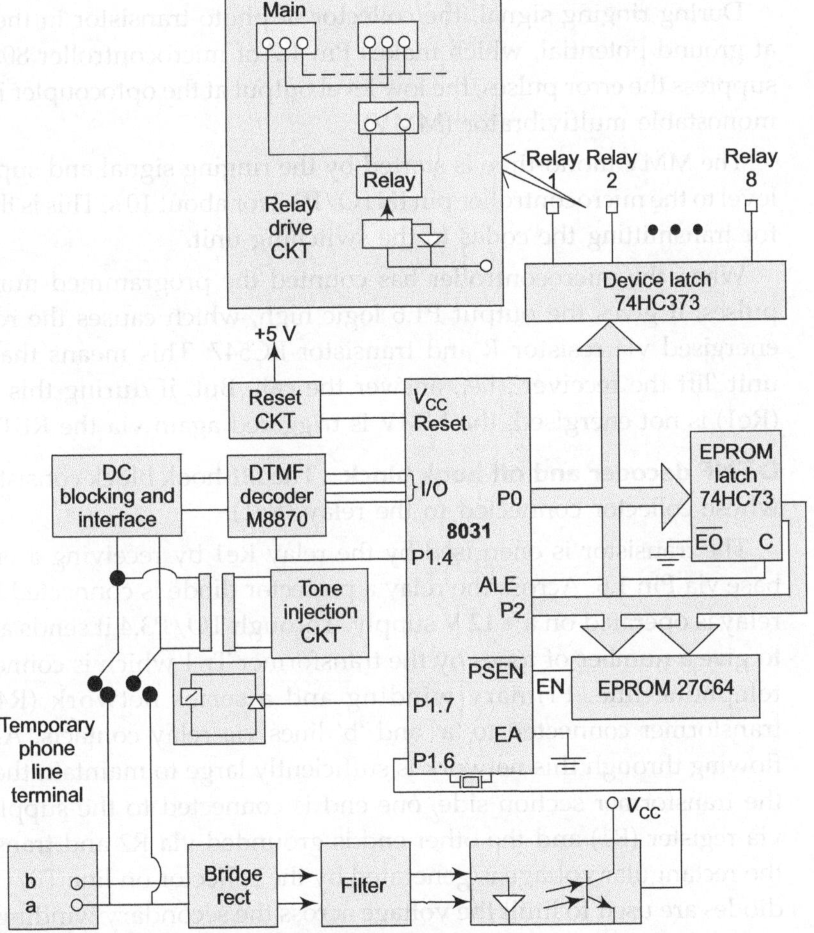

Fig. 7.16 Block diagram of switching unit

Circuit description: The circuit of Fig 7.16 may be divided into five units:

1. Ring detection block
2. DTMF decoder and Off-hook block
3. Reset block
4. Controlling logic block
5. Device switching block

Ring detection block   A bridge rectifier connected to the telephone line which converts the ringing signal of 48 V P-P into pulsating direct current about 15 V DC, with the help of zenner diode, connected across RC filter ($R = 22$ kΩ and $C = 22$ mF). The input of this bridge rectifier, which is coupled with R and C, indicates that the ringing signal detector responds to alternating voltage. Only any direct voltage levels between 'a' and 'b' terminals are ignored. The direct voltage is supplied to LED in optocoupler through register R acting as a current limiter.

During ringing signal, the collector of photo-transistor in the optocoupler is at ground potential, which makes Pin 1.7 of microcontroller 8031 low. Also, to suppress the error pulses, the low level output at the optocoupler is used to trigger monostable multivibrator (MMV).

The MMV mono-time is started by the ringing signal and supplies logic high level to the microcontroller pin INTO/P3.2 for about 10 s. This is the time duration for transmitting the codes to the switching unit.

When the microcontroller has counted the programmed number of ringing pulses, it gives the output P1.6 logic high, which causes the relay (Re1) to be energised via resistor R and transistor BC547. This means that the switching unit 'lift the receiver', i.e., answer the call. But, if during this period, if relay (Re1) is not energised, the MMV is triggered again via the RET connection.

DTMF decoder and off-hook block   The off-hook block consists of a transistor whose collector connected to the relay (Re1).

The transistor is energised by the relay Re1 by receiving a one (high) on its base via Pin 1.6. Across the relay a protector diode is connected in parallel. This relay is operated on a + 12 V supply. Through TO/P3.4 it sends a train of pulses, to give a number of tones by the transformer Tr 1 which is connected across the telephone line. Primary winding and a series network (R4 and R5) of a transformer connected to 'a' and 'b' lines via relay contacts. Also, the current flowing through this network is sufficiently large to maintain the connection on the transformer section side, one end is connected to the supply voltage +5 V via register (R1) and the other end is grounded via R2 and transistor T, so that the rectangular voltage is generated by the collector on line T0/P3.4. Two Zener diodes are used to limit the voltage across the secondary winding for safe levels.

The DTMF decoder (M8870), which is receiving the DTMF code gives the corresponding binary outputs (bit pattern) on output pins Q1–Q4. Pin 15 of M8870 (STD pin) is high, which is detected by 8031 on P1.4, which indicates to read the P1 port. This DTMF decoder IC is configured to its single ended operation. Here, capacitor ($C_{19} = 10$ nF) is kept for over voltage surge protection. The external components to enable the M8870 DTMF decoder are the register, a capacitor, and a quartz crystal (3.579545 MHz).

Reset block   This block consists of retriggrable MMV, whose output is adjusted through timing element R12 and C7 for 10 s. The optocoupler output (Pin 8) is given to the input of NAND gate (IC-7400) whose output is connected to retriggerable input pins of MMV. 4047 is configured in a retriggerable mode. The output from Pin 10 becomes high for 10 s, which generates interrupt 0 (i.e., INT0) for the microcontroller. After 10 s it disables the INT0 pin. The retrigger pin (+T and RET) becomes high either during line detection or during detection of valid DTMF codes (here, STD line is coupled through diode D8 to the input pins +V and RET).

The microcontroller is configured for the edge triggered interrupt; hence when a high to low transition occurs, the system reset on hooking the receiver via relay (Re 1). An LED D10 provides a visual indicator for the call and DTMF signals.

Controlling logic block   The heart of the circuit is the 8031 microcontroller with 8 K memory interfaced through a latch (74HC573) for the multiplexed port P0 for lower order address byte and P2 for higher order address byte. EPROM is enabled through PSEN and latch through ALE from the microcontroller. While writing data on the multiplexed port P0, the latch '1' should be disabled and latch '2' is to enable the clock. A display unit is added with the board.

Device switching block   The latch '2' connected to the device switching transistor gets the data from P0. Latch '2' gets enabled when A15 is high and WR goes low. The output of the latches is connected to the base of transistors whose collectors are connected to the relays and the LEDs, which operate on +12 V supply.

Each relay is made ON, a 1 (high logic) on the base, the collector output low (0), which energises the relay, making the device ON. If the device is made OFF, a 0 is given on the base which de-energises the relay. Also, there is a fly wheeling diode to protect the switching of the transistor against back emf production by the relay coils when they are switch OFF.

System Software

The software is to be developed for the operation and control the hardware. The software can be developed in the form of modules. These types of program modules are required. It is left as exercise for the reader to develop software modules using the following guidelines.

Main   The function of this module is to carry out initialization, line seizing, data collection, and validity checks, decode the command and indicate the response of checking to user. Check the status of load and carry out operation specified by the command.

Subprograms   Each and every operation desired by the user is developed in the form of subroutines. The main program should coordinate these programs to execute the command given by user via telephone line. A subprogram is required to carry out one function. The required functions include delay, checking OFF/ON status, relay operation, long tone, short tone generation, and for setting/checking bits.

Interrupt service routine   Two interrupt service routines are to be developed for operation after the ring is detected and Timer 1 interrupt.

In the chapter we have studied the serial communication using 8251 USART. The generation of an additional serial port for the microcontroller using the 8251 was described. The different types of modems and characteristics of RS-232 ports were covered. At the end, we have studied the characteristics, operation and use of a telephone line for remote operation of the microcontroller.

EXERCISES

7.1 Can 8051 microcontroller support synchronous communication? How?

7.2 What are the two methods for controlling errors in serial communication?

7.3 Draw an internal block diagram of the USART chip. Explain function of each block.

7.4 Define: bit rate and baud rate.

7.5 What is the delay between two successive bits for a transmission rate of 9600 bauds?

7.6 Define: DCE and DTE.

7.7 What are the special features of USART 8251?

7.8 What is the difference between command word and mode set word in USART 8251?

7.9 State and explain different methods to reset USART 8251.

7.10 How USART 8251 can come out of idle mode?

7.11 With the help of a format definition, explain how USART 8251 can be used for asynchronous communication.

7.12 With the help of a format definition, explain how USART 8251 can be used for synchronous communication.

7.13 Explain status word format of USART 8251.

7.14 Explain the initialization procedure for 8251.

7.15 Explain how an extra serial port can be realized on microcontroller for synchronous communication. Design a setup and write a program to transmit using two synchronizing characters.

7.16 State and explain characteristics of RS-232C.

7.17 When not transmitting signal, what are the voltage levels on RS-232C line?

7.18 Define : FRAMMING ERROR, OVERRUN ERROR.

7.19 What are the limitations of RS-232C ? How they can be overcome?

7.20 What is the number of pins on standard RS-232C port?

7.21 What is the number of pins on non-standard RS-232C port?

7.22 Explain the need for handshaking signals.

7.23 A microcontroller system has to be connected to VDU using RS-232C port. Draw a setup to wire microcontroller as DTE and VDU as DCE.

7.24 A serial printer is to be connected to the micro-controller assuming that both are configured as DTE and DTR, RTS lines are used. Describe the cable connections and pin assignments. Write a program to output passed on to the stack to the printer.

7.25 A self-ticketing machine has an interface with a 24-column serial printer. Design a scheme to enable the machine to print the details of ticket and pass the signal to the printer after scrolling two lines.

7.26 What is a modem? Why are they required?

7.27 Explain the points to be considered while setting up modems.

7.28 What are the basic components of a telephone?

7.29 Explain function and operation of telephone.

7.30 What is DTMF? What are the advantages of DTMF?

7.31 How do you select the DTMF frequencies?

7.32 Explain operation of a DTMF decoder.

7.33 A serial link has to be established which accepts input from a keyboard connected at the Port P1 and transmits it to the terminal using serial. Design a setup and write a routine for initialization and operation of the system. Assume that the communication uses, asynchronous mode, 7 data bits, 1 stop bit and no parity and baud rate of 680 bauds.

7.34 A teletype/teleprinter is to be interfaced with the TxD/RxD line of microcontroller. Draw a circuit and explain how a (20 mA) current loop can be employed for the purpose. Explain working of the circuit and write software for operation.

7.35 Design a setup to read ASCII data from an synchronous link using 5-bit code, with odd parity and external synchronization at a baud rate of 680 bps. Assuming more than one message is not retained in the memory, write a software to carry out following steps in addition to initialization:

1. Do nothing until six characters [password] are received on the link.

2. Match the password stored in memory locations [paas]

3. Receive a message < = 32-bytes, terminated by < . > or

   < cr >, if password is correct, make provision for parity check.

4. If message ends with(.)

   {

store it in RAM at data onwards.

respond to the terminal by sending (ok) on line

goto Step 1.

                                        }

                        Else

                                  {

store 6-bytes of message as a new password.

respond by sending (alt). on line

goto Step 1.

                                        }

Chapter 8

System Design

Microcontrollers provide a single chip solution. This means that it is necessary to use them on the input and output device as per the requirement of the application. The I/O devices can be directly interfaced to the microcontroller chip. Software may be developed for the operation and control of the hardware setup. This chapter describes the setup for simple applications and explains the use of internal on-chip components such as parallel as well as serial I/O ports and timers. At the end of the chapter, we will design an IR remote-operated digital controller, which is suitable for the operations in the industrial environment.

8.1 | ON-CHIP PARALLEL I/O PORTS

This section deals with developing applications using port pins. The pins may be used as input or output ports. The two applications deal with the simulation and the testing of digital IC using microcontroller.

8.1.1 Simulation

The term simulation indicates that instead of an actual system, a setup is prepared, which is similar in operation to the actual system. Microcontroller can be programmed to simulate any digital IC such as logic gates, flip-flops, counters and shift registers, etc. The digital ICs can be broadly classified into two groups, combinational and sequential circuits depending upon whether the output is independent of the previous output or not. We will develop setups and programs for simulation of both the groups.

Combinational ICs

It is known that these ICs have output, which depends only on the applied inputs. The operation of gate is described by a truth table. Since a large number of logical operations are defined in the instruction set, instructions can be used to simulate the combinational IC such as logic gates. Consider a setup of Fig. 8.1.

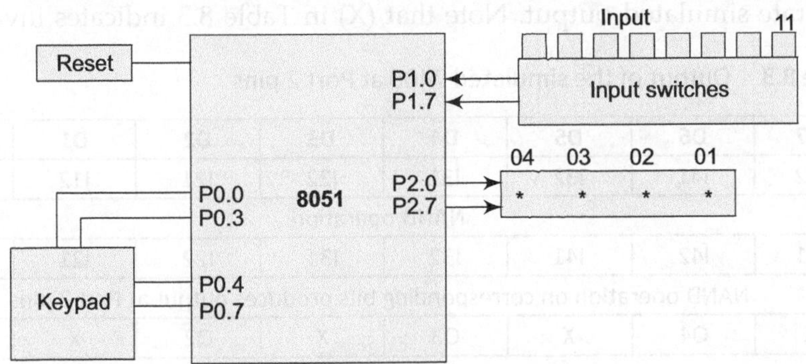

Fig. 8.1 Setup for simulation of combinational IC

A hex keypad is interfaced to the Port 0 pins of the microcontroller. It is used to select the IC number. Switches are connected to the Port 1 pins, which is used to give input to the simulated IC. The program segment will be executed by the microcontroller, based on the number of the combinational IC to simulate the operation and output is generated on the LED connected to the Port 2 pins.

For example, to simulate the IC-7400, which is a quad 2-input NAND gate, a truth table for NAND operation is observed. This can be simulated by an AND operation followed by the NOT operation. It is possible to simulate single gate or all the four gates of 7400. The program logic is instead of doing a logic operation on bit-by-bit basic, to carry out desired operation byte wise. The algorithm is explained below:

The input switches connected to the Port 1 pins. The input from Port 1 is represented by Table 8.1.

Table 8.1 Input to the simulated 7400 gates

D7	D6	D5	D4	D3	D2	D1	D0
P1.7	P1.6	P1.5	P1.4	P1.3	P1.2	P1.1	P1.0
I42	I41	I32	I31	I22	I21	I12	I11
Input to gate 4		Input to gate 3		Input to gate 2		Input to gate 1	

The input data is shifted to the right by 1-bit. The shifted data is shown in Table 8.2.

Table 8.2 Shifted input data 1-bit right shift

I11	I42	I41	I32	I31	I22	I21	I12

Now based on the IC number, logic operation for the gate can be carried between input data and shifted data. For example, 7400 indicates NAND operation is required, which is shown in Table 8.3. The LED connected at the Port 2 pins will generate simulated output. Note that (X) in Table 8.3 indicates invalid output.

Table 8.3 Output of the simulated 7400 at Port 2 pins

D7	D6	D5	D4	D3	D2	D1	D0
I42	I41	I32	I31	I22	I21	I12	I11
NAND operation							
I11	I42	I41	I32	I31	I22	I21	I12
NAND operation on corresponding bits produces output at Port 2 pins							
X	O4	X	O3	X	O2	X	O1

Let us assume that a program module rd-keypad (similar to one developed in Section 5.5.3) is available which reads IC number from keypad at Port 1 pins of the microcontroller. The program segment for the algorithm is as follows:

```
IC-7400:
        mov a, P1              ; accept input
        mov r3, a             ; save it
        rr a                  ; shift right by 1-bit
        anl a, r3            ; AND operation
        cpl a                ; NOT operation
        mov p2, a            ; output at Port 2
        sjmp  IC-7400
        end
```

The program can be modified for the other IC having same base diagram as 7400.

For example:

7408: AND operation, 7432: OR operation, and 7486: EX-OR operation The appropriate output of the gate will be displayed on LED.

```
IC-7408:
        mov a, P1              ; accept input
        mov r3, a             ; save it
        rr a                  ; shift right by 1-bit
        anl a, r3            ; AND operation
        mov p2, a            ; output at Port 2
        sjmp  IC-7408
```

```
IC-7432:    mov  a, P1         ; accept input
            mov  r3, a         ; save it
            rr   a             ; shift right by 1-bit
            orl  a, r3         ; OR operation
            mov  p2, a         ; output at Port 2
            sjmp IC-7432
IC-7486:    mov  a, P1         ; accept input
            mov  r3, a         ; save it
            rr   a             ; shift right by one bit
            xrl  a, r3         ; EX-OR operation
            mov  p2, a         ; output at Port 2
            sjmp IC-7486
            end
```

The procedure is valid for any combinational IC. Care should be taken to connect the input key and the output LEDs at appropriate pins of Port 1 and Port 2.

Sequential Circuits

The basic building blocks of these circuits are flip-flops. The characteristics of these circuits are that the output depends not only on the input combination but also on the previous output. We will study procedure for simulation of flip-flops.

It is known that to understand the operation of a flip-flop, we can use its operation or transition table or derive characteristic equation or draw timing diagram. The second method is suitable for simulation, as it describes behaviour of the sequential circuits by a logical expression, which can be easily implemented using the microcontroller instructions.

For example, to obtain the characteristics of an RS-FF, the operation of RS–FF can be described as shown in Table 8.4 assuming that Y_n denotes current output.

Table 8.4   Operation table of RS-FF

Input		Next output
S	R	Y_{n+1}
0	0	Y_n
0	1	0
1	0	1
1	1	Invalid

To derive the characteristics equation, we consider the current output also as one of the input variable and modify the operation table of Table 8.4. The resulting table is shown in Table 8.5.

Table 8.5 Modified operation table of RS-FF

Input			Next output
S	R	Y_n	Y_{n+1}
0	0	0	0
0	0	1	1
0	1	0	0
0	1	1	0
1	0	0	1
1	0	1	1
1	1	0	Invalid
1	1	1	Invalid

The Boolean expression can be written for $Y_{n+1} = f (S, R, Y_n)$ and any one of the methods, i.e., algebra or Karnaugh map, can be used to simplify the expression. Considering the invalid as a don't care condition, we get characteristic equation for the above function:

$$Y_{n+1} = S + R' Y_n$$

Similarly, the characteristic equations for all types of flip-flop can be derived. They are listed in Table 8.6.

Table 8.6 Characteristic equations for the FF

FF Type	Characteristic equation
D	$Y_{n+1} = D$
T	$Y_{n+1} = T \oplus Y_n$
SR	$Y_{n+1} = S + R' Y_n$
JK	$Y_{n+1} = JK' + K' Y_n$

Example 8.1   Design a setup to simulate JK-FF.

Solution

Block diagram of Fig. 8.2 depicts a JK-FF. It indicates that a JK flip-flop has two input pins J and K, a clock pin, two output pins Yn and Yn', representing current state of the flip-flop. The state is changed in accordance with the operation table of Table 8.5, when a clock pulse is applied at the clock pin.

Consider a system setup shown in Fig. 8.3. Port pins P1.0 and P1.1 represent the J and K pins of the simulated flip-flop. The port pins P1.6 and P1.7 represent the Yn and Yn' of the flip-flop.

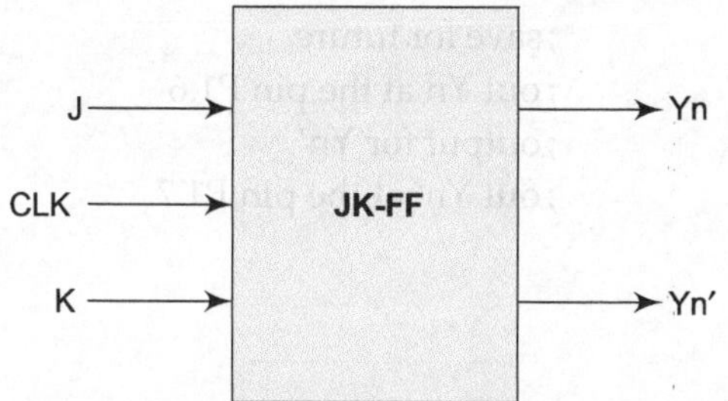

Fig. 8.2   Pin configuration of JK flip-flop

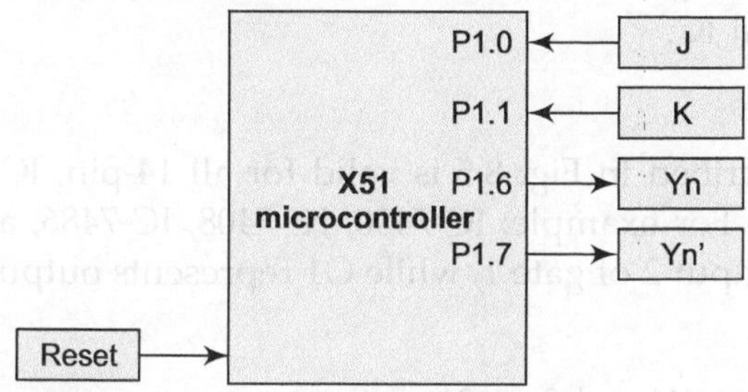

Fig. 8.3   System setup for simulation of JK flip-flop

On reset, the output is cleared, which is to be used for deciding future output. Let us define bit addresses for the application as:

P1.0 (24h): status of 'J' input

P1.1 (25h): status of 'K' input

Previous output (26h): present output (stored at 26h)

(27h): temporary result

The program accepts the status of the J, K from the corresponding port pins and based on the characteristic equation, it decides the next output the program realizes:

$$Y_{n+1} = JK' + KY'_n$$

Important point to remember is that, since all operations are bitwise, ice carry flag will be the accumulator.

```
reset:
            clr  26h                    ; reset output
            mov  c, p1.0
rept1:      mov  24h, c                 ; set status of J
            mov  c, p1.1                ; set status of K
            mov  25h, c
            anl  /26h
            mov  27h, c                 ; save intermediate result
            mov  c, 24h                 ; set J in 'c'
            anl  /25h                   ; obtain JK'
            orl  27h                    ; get next opp
```

```
mov  27h, c             ; save for future
mov  p1.6, c            ; out Yn at the pin P1.6
cpl  c                  ; output for Yn'
mov  p1.7, c            ; out Yn' at the pin P1.7
sjmp rept1
```

8.1.2 Testing Digital IC

The testing procedure for the combinational and the sequential IC are different because the sequential testing must be done in a defined sequence, which is not essential in combinational IC.

Combinational IC Tester

The hardware setup described in Fig. 8.5 is valid for all 14-pin, IC having pin layout shown in Fig. 8.4. For example: IC-7400, IC-7408, IC-7486, and IC-7432. In Fig. 8.5, I12 indicates input 2 of gate 1, while O1 represents output of the first gate.

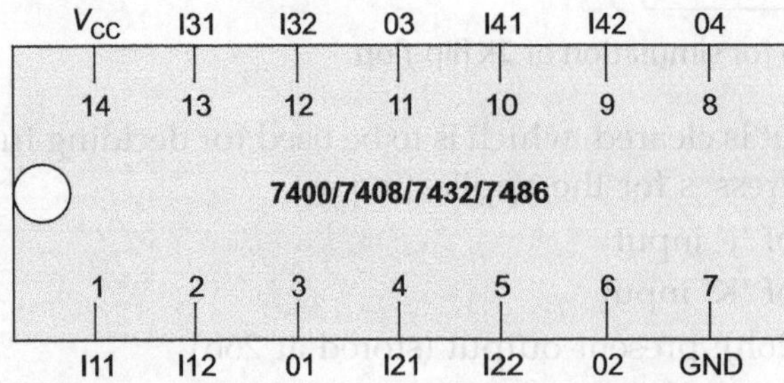

Fig. 8.4   Pin layout for a typical 14-pin digital IC

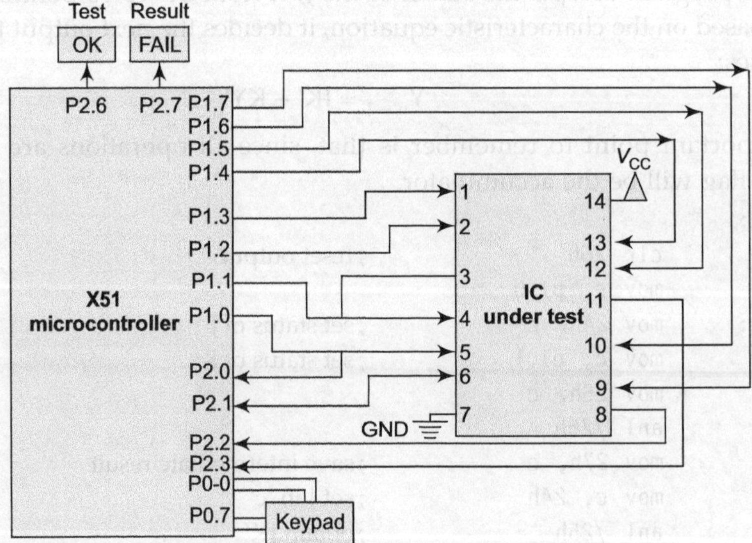

Fig. 8.5   Hardware setup for the IC tester

The number of the IC under test is entered from the keypad. Port P1 is used to give inputs to IC under test, while Port P2 accepts the output from it. The testing sequence is stored in a look-up table. The entries in the look-up table contain the input to be given to the IC under test and the expected output in hex.

On reset, the system waits for an enter key. When the key is pressed the key code is read, which is loaded as the pointer to the look-up table whose base is initialized in the dptr register. Table 8.7 is the look-up table.

Table 8.7   Input–output relation for the IC-7400 (gates) under test

Hex code	Gate 4 I42 I41	Gate 3 I32 I31	Gate 2 I22 I21	Gate 1 I12 I11	Expected output				Hex code
					Q4	Q3	Q2	Q1	
00	00	00	00	00	1	1	1	1	OF
55	01	01	01	01	1	1	1	1	OF
AA	10	10	10	10	1	1	1	1	OF
FF	11	11	11	11	0	0	0	0	00

Similarly, the expected output for other ICs can be determined and the look-up table can be prepared. They can be stored at the base address represented by the IC number. For example, base address for the IC-7400 is 2500, for the IC-7432 it is 2532, for the IC-7408 it is 2508, and for the IC-7486 it is 2586. Table 8.8 depicts the look-up table.

Table 8.8   Look-up table for a 14-pin IC tester

IC-7400	Data	IC-7432	Data	IC-7486	Data	IC-7408	Data
Base address 2500	00	Base address 2532	00	Base address 2586	00	Base address 2508	00
	OF		00		00		00
	55		55		55		55
	OF		OF		FF		00
	AA		AA		AA		AA
	OF		OF		FF		00
	FF		FF		FF		FF
	00		OF		00		OF

The main program waits for an interrupt at ext0 in the form of test key. On return, the status of the memory location OK can be displayed. Only the ISR is given here. Reader should write the main program.

```
isr-ext0:
            mov   dph, #25          ;upper byte of look-up table
            mov   dpl, in_key       ;move input no. to dpl
            mov   r4, #04h          ;counter for input comparison
```

```
rept:       movx a, @dptr
            mov p1, a
            acall delay          ;apply test input
            mov p2, a            ;wait for 10 ms
            anl a, #ofa,         ;mask upper nibble
            mov r3, a
            inc dptr
            cjne a, r3, not-ok
            inc dptr
            djnz r4, rept
            serb ok
            sjmp end1
not-ok:
            clr ok
            end1 reti
            end
```

Sequential IC Tester

The algorithm for testing a sequential IC differs from the testing procedure of a combinational IC. The algorithm for the testing procedure developed in this section can be extended to test any sequential IC.

Let us design the tester for IC-7473 (dual JK-FF). The simplified block diagram of the IC-7473 is depicted in Fig. 8.6.

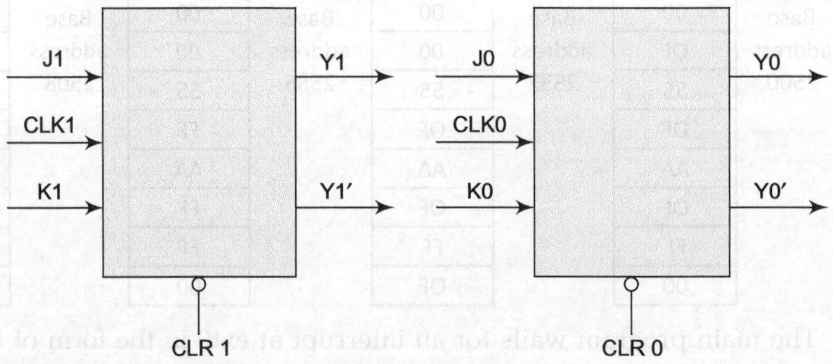

Fig. 8.6 Simplified block diagram of IC-7473

It is known that the 7473 have two JK flip-flops (FF). The suffix 0 indicates FF-0. There are two inputs J, K to each FF. The clear CLR input sets the output pin Y at logic level low at the trailing edge of the clock pulse applied at the clock pin. The Y' pin represents the complimented output. In order to test the IC, three operations must be tested.

1. Operation of CLR pin
2. Operation of CLOCK pin
3. Normal operation of JK-FF depicted in Table 8.9.

Table 8.9 Operation table of JK-FF

Inputs		Next output
J	**K**	Y_{n+1}
0	0	Y_n
0	1	0
1	0	1
1	1	Y_n'

Consider the hardware setup for the sequential IC tester shown in Fig. 8.7.

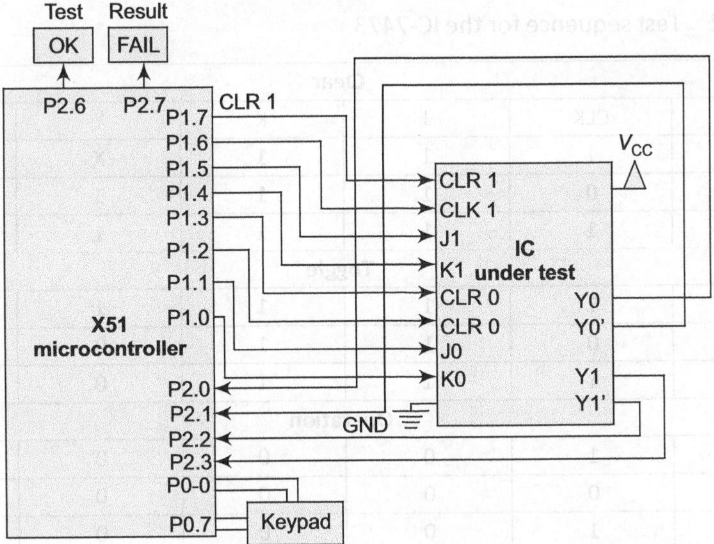

Fig. 8.7 Hardware setup for the IC tester for a 14-pin digital IC of Fig. 8.6.

Port 1 pins P1.0–P1.7 are connected to the input pins and Port 2 pins P2.0–P2 are connected to the output pins of the JK-FF in the IC-7473 under test. The LED connected at the port pins P2.6 and P2.7 of the microcontroller indicate the result of testing. Table 8.10 shows the input data format while Table 8.11 depicts the output data format.

Table 8.10 Input data format for testing IC-7473

P1.7	P1.6	P1.5	P1.4	P1.3	P1.2	P1.1	P1.0
CLR1	CLR1	J1	K1	CLR0	CLK0	J0	K0
Input to the second flip-flop of 7473				Input to the first flip-flop of 7473			

Table 8.11   Output data format for testing IC-7473

P2.7	P2.6	P2.5	P2.4	P2.3	P2.2	P2.1	P2.0
X	X	X	X	Y1'	Y1	Y0'	Y0
Not used				Output of 2nd flip-flop		Output of 2nd flip-flop	

The three operations to be tested for the IC under test are clear, clock, and the operation. The input applied to the flip-flop and expected output is stored in the look-up table. The input is applied to the IC, after allowance for the response time of FF, output from the IC is compared with the expected output. If outputs from the IC for all possible inputs of look-up table agree with the expected outputs then the IC is OK, otherwise it is declared as faulty. Table 8.12 depicts the test sequence.

Table 8.12   Test sequence for the IC-7473

CLR	CLK	J	K	Y	Y'
Clear					
1	1	1	1	X	X
0	0	1	1	1	0
1	1	1	1	1	0
Toggle					
1	1	1	1	1	0
1	0	1	1	0	1
1	1	1	1	0	1
Operation					
1	1	0	0	0	1
1	0	0	0	0	1
1	1	0	0	0	1
1	1	0	1	0	1
1	0	0	1	1	0
1	1	0	1	1	0
1	1	1	0	1	1
1	0	1	0	0	1
1	1	1	0	0	1

The sequence is applied to both the FF' and the look-up table with the test input and expected output is prepared. The look-up table is stored in the memory from 2500h onwards. The alternate entries in Table 8.13 represent the input to be applied to the IC under test and the expected output.

Table 8.13   Look-up table: test sequence of the IC-7473

Address	Input	Address	Expected output
2500	FF	2501	00
2502	33	2503	0A
2504	FF	2505	0A
2506	FF	2507	0A
2508	BB	2509	05
250A	FF	250B	05
250C	C0	250D	05
250E	88	251F	05
2510	C0	251A	05
2512	DD	251B	05
2514	99	2,75	0A
2516	DD	2517	0A
2518	EE	2519	0A
251A	AA	251B	05
251C	EE	251D	05

The program is similar to one developed for the combinational tester. It is left as an exercise to the reader.

8.2 | ON-CHIP TIMERS

Real time counters are extremely useful in applications like event counters, measuring time intervals, elapse time measurements, etc. They are also useful in multitask processing. The counters can be realized using software methods. However, the on-chip timers available on X51 family chips can be used.

When used as a counter: The register is incremented on a 1 to 0 transition (a negative edge) applied to the appropriate input pin T0 or T1. Since, T0 and T1 are alternate functions of the port pins, it takes two complete machine cycles for the 8051 to see the transition. The input must be held high for at least one cycle and low for at least one cycle.

It is known that there are two timer/counters, which can be used as a timer or counter in different modes of operation. They can be programmed and operated using TCON TMOD registers. This section describes applications, which makes use of them in all possible modes of operation.

8.2.1 Mode 1 Operation

This is a 16-bit mode. In Mode 1, the timer is configured as a 16-bit counter that the 8-bit count is in the THX register (TH0 or TH1) and 8-bits of the TLX register (TL0 or TL1) are used for 16-bit operation as shown in Fig. 8.8.

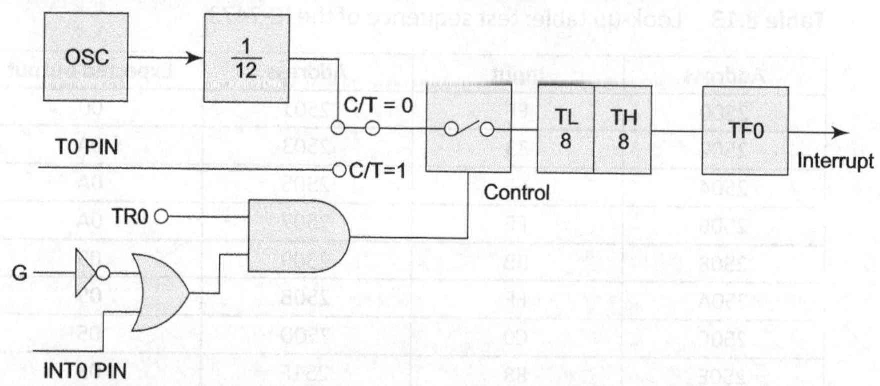

Fig. 8.8 Configuration for Mode 1 operation of timer/counter

When the count in TL and TH goes from all '1' to all '0', the timer interrupt flag (TF0 or TF1) is set. The timer interrupt is the connection between the counter hardware and the program software. Counter/timer (C/T) bit in TMOD is used to select the source of input to the counter. The counting process can be turned ON and OFF (enabled or disabled) independent of the input.

It is clear from the figure that for the counting process to be turned, the TR bit (TR0 or TR1) in the TCON register, has to be at logic level 1. Also, either the appropriate GATE bit in the TMOD register is a 0, or the appropriate intX pin has to be held low. The use of GATE or TR allows software controlled counting while the use of INT allows external hardware controlled counting.

Example 8.2   Design a setup for a 16-bit binary counter, using Timer 0 in Mode 1. The system should count the number of times the count pulse is given by pressing a COUNT key connected to the port pin P3.4 of the microcontroller and display the count on four 7-segment display connected at Port 0 and 1.

Solution

Consider a setup shown in Fig. 8.9. The 7-segments are connected to port pins via BCD 7-segment decoders so that programming becomes simple. Note that only the necessary connections are shown in Fig. 8.9. Whenever the count key is pressed, count will be incremented by 1.

Let us use timer TC0 in Mode 1 which is 16-bit mode. The timer mode can be set by loading a control word in TMOD register. The control word formulation is shown in Table 8.14. Control word is 05h and it is loaded at the byte address 89h.

Table 8.14 Formulation of TMOD control word

D7	D6	D5	D4	D3	D2	D1	D0
GATE	C/T	M1	M0	GATE	C/T	M1	M0
X	X	X	X	0	1	0	1

The steps carried out to program the timer are as follows:

1. Load the TMOD register which selects the mode and operation (timer/counter).

2. Initialize timer registers TL0 and TH0, here it is 00h and 00h.

3. Turn ON the timer/counter operation by setting bits in TCON.

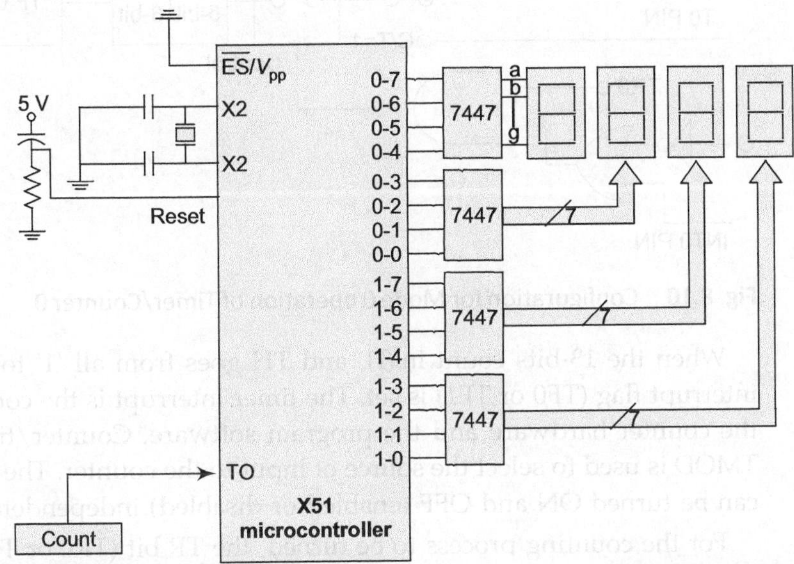

Fig. 8.9 System setup for a 16-bit counter using Timer 0 in Mode 1

The program segment is given below:

```
counter:    mov  8c, #00h        ; load 00h in the timer TH0
            mov  8a, #00h        ; load 00h in the timer TL0
            mov  89, #05         ; load the TMOD register
            setb 8c              ; turn ON Timer 0 (set TR0=1)
rept1:      mov  90, 8a          ; read the counter TL0
            mov  P0, a           ; display at the Port P0
            mov  90, 8c          ; read the counter TH0
            mov  P1, a           ; display value at Port P1
            sjmp rept1
            end
```

8.2.2 Mode 0 Operation

In Mode 0, the Timer 0/1 is configured as shown in Fig. 8.10, a 13-bit counter. It can be considered as an 8-bit counter preceded by a 5-bit divide-by-32 pre-scalar the 8-bit count is in the THX register (TH0 or TH1), depending on the counter to be used. The 5-bits of the TL register are used for count while the upper 3-bits of the TL register are ignored.

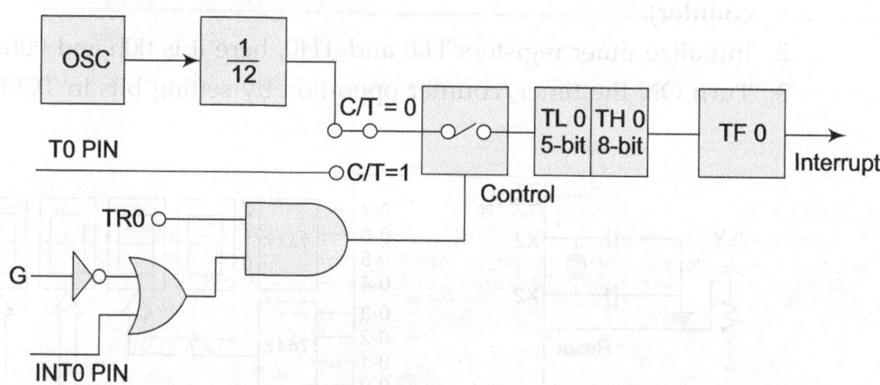

Fig. 8.10   Configuration for Mode 0 operation of Timer/Counter 0

When the 13-bits count in TL and TH goes from all '1' to all '0', the timer interrupt flag (TF0 or TF1) is set. The timer interrupt is the connection between the counter hardware and the program software. Counter/timer (C/T) bit in TMOD is used to select the source of input to the counter. The counting process can be turned ON and OFF (enabled or disabled) independent of the input.

For the counting process to be turned, the TR bit (TR0 or TR1) in the TCON register, has to be at logic level 1. Also, either the appropriate GATE bit in the TMOD register is a 0, or the appropriate INTX pin has to be held low. The use of GATE or TR allows software controlled counting while the use of INT allows external hardware controlled counting.

Example 8.3   Design a setup for a 8-bit binary counter using Timer 0 in Mode 0. The system displays the number of times the COUNT key connected to the Port P3.4 pin of the microcontroller is pressed and displays the count on two segment displays connected at Port 1.

Solution

Consider the setup of Fig. 8.11. This system operation is similar to that of Example 8.2 except that the counter will be incremented only when 32 pulses are given by pressing COUNT key 32 times. Pre-scalar in 8051 can be used to divide the number of elements if used as a counter, by 32. Here 32 events will be recorded as one event.

Let us use timer TC0 in Mode 0 which is 13-bit mode. The timer mode can be set by loading a control word in TMOD register. The control word formulation is shown in Table 8.15. Control word is 04h and it is loaded at the byte address 89h.

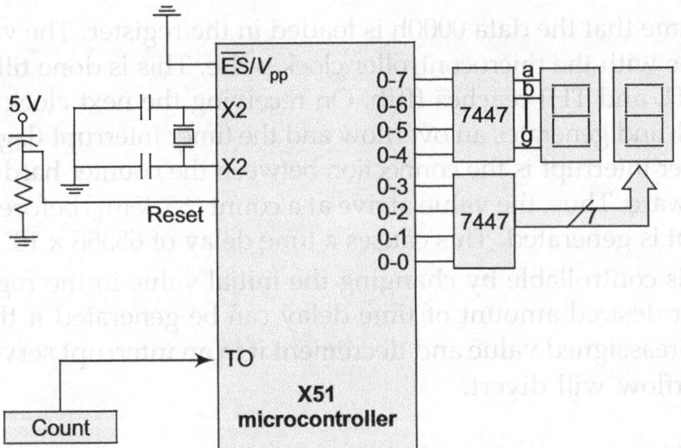

Fig. 8.11   System setup for a 8-bit counter using Timer 0 in Mode 0

Table 8.15   Formulation of TMOD control word

D7	D6	D5	D4	D3	D2	D1	D0
GATE	C/T	M1	M0	GATE	C/T	M1	M0
X	X	X	X	0	1	0	0

The steps carried out to program the timer are as follows:

1. Load the TMOD register with 04h which selects the mode and operation (timer/counter).
2. Initialize timer registers TL0 and TH0. TL0 must be loaded with 5 zeros. As the event proceeds the THO records every 32nd event. The register IT must be read to display every 32nd count here it is 00h and 00h.
3. Turn ON the timer/counter operation by setting bits in TCON.

The program segment is given below:

```
32-count:   mov 8c, #00h        ; load 00h in the timer TH0
            mov 8a, #00h        ; load 00h in the timer TL0
            mov 89, #04         ; load the TMOD register
            setb 8c             ; turn ON Timer 0 (set TR0=1)
rept1:      mov 90, 8c          ; read the counter TH0
            mov P0, a           ; display the value at the Port P0
            sjmp rept1
            end
```

The output will be incremented by one after 32 events.

8.2.3 Time Delay Generation

The internal counter divides the applied frequency by 12. Hence, the effective clock frequency is 1 MHz and the clock period is 1 µs. When used as a timer the register is incremented once per 12-clock period.

Let us assume that the data 0000h is loaded in the register. The value is incremented by one with the microcontroller clock pulse. This is done till the value in the register (TL and TH) reaches ffffh. On receiving the next clock pulse, value becomes 0000h and generates an overflow and the timer interrupt flag (TF0 or TF1) is set. The timer interrupt is the connection between the counter hardware and the program software. Thus, the value arrive at a count $(65,356)_{10}$ before an overflow or an interrupt is generated. This causes a time delay of 65356×10^{-6}, i.e., 66 ms.

The delay is controllable by changing the initial value in the register. A predetermined or desired amount of time delay can be generated if the counter is loaded with preassigned value and decrement it in an interrupt service program, to which overflow will divert.

For example:

Let us load count as 8000h, the counter will start counting from this value, resulting into the delay interval of about 33 ms.

The counter will again start from 0000h onwards after an overflow, hence desired number must be reloaded, if one wishes to obtain desired time interval. This can be avoided by using Mode 2.

Example 8.4   Design a scheme to make LED connected at Port 1 of the microcontroller ON and OFF for 1 s.

Solution

The setup is shown in Fig. 8.12. The LED unit is connected at Port 1. The keys connected at the interrupt pins may be used if necessary.

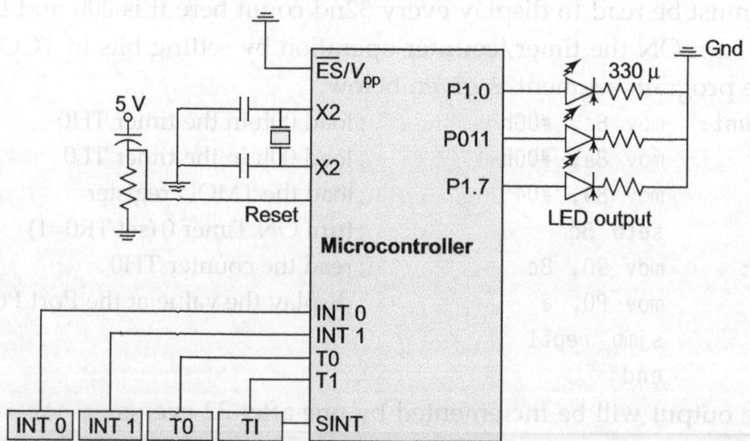

Fig. 8.12   System setup for Example 8.5

Delay generation requires following configuration:

1. C/T bit in TMOD the register is set to 0, indicating timer mode.
2. ET0/ET1 bit at the bit address 0A9h/0Abh is set to '1' to enable interrupt line.
3. Load count for 33 ms delay (8000h) in the TH0–TL0 registers.

Since the maximum delay generated is 66 ms to generate 1 s delay the timer can be used to generate interrupt at a basic delay of either 66 ms or 33 ms the system may wait for 16/32 interrupt signals for a desired 1 s delay.

For example:

If the timer is programmed for a basic delay of 33 ms, waiting for 32 interrupt signals the delay generated is of the order of 33 ms × 32 = 1.056, i.e., about 1 s. Loading r0 with can change the duration other preassigned values. Table 8.16 shows typical values for loading and the resulting net time delay.

Table 8.16   Load value and the net time delay

Delay count r0 = 20h		
Initial value	**Basic delay**	**Total delay**
0000h	66 ms	2 s
8000h	33 ms	1 s
C000h	16.5 ms	0.5 s

The program segment is given below:

```
Delay:
            mov r7, #20         ; count is 20
            mov r0, #20         ; delay (count)
            mov r1, #00         ; initialize r1
            mov 90, #00         ; clear Port 1
            mov 8c, #80         ; (TH0) = (80h)
            mov 8a, #00         ; (TL0) = (00h), (33 ms delay)
            setb 0A9h           ; set bit ET0 in IE register
            mov 89, #01h
            setb 8ch            ; set (TRO), i.e., make Timer 0 ON
wait:       sjmp wait           ; wait for overflow
000b:       sjmp                isr_to
isr-to:
            djnz r0, skip       ; if r0¹0, delay not over
            mov a, r1           ; if r0 = 0
            cpl a
            mov r1, a
            mov 90, a           ; complement output
skip:       mov 00h, 07h        ; reload r0 from r7
            mov 8c, #80h
            reti
            end
```

8.2.4 Mode 2 (Auto-load) Operation

Figure 8.13 shows configuration of Timer/Counter 0 for the Mode 2 (auto-load) operation. Mode 2 operations are similar to Mode 0 except that the TL register is used as an 8-bit counter and the TH register is used to hold a preset number. The contents of TH remain unchanged. The counter can be made to divide the count source by any number from 1 to 255 by means of the automatic reload of TH into TL.

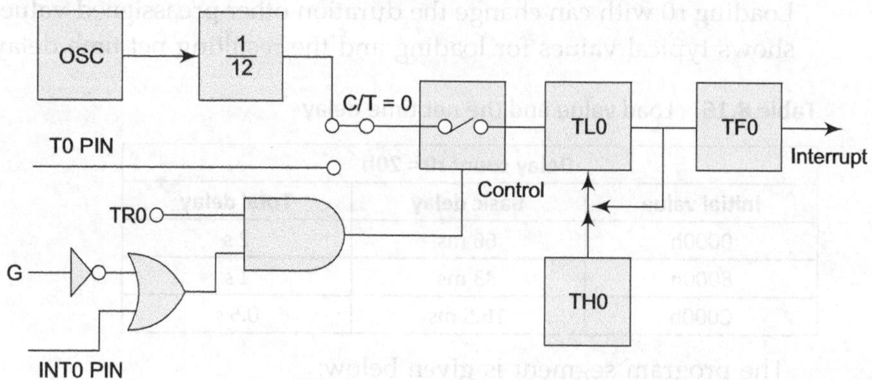

Fig. 8.13   Configuration for Mode 2 operation of Timer/Counter 0

In this mode of operation, an 8-bit value in the TL is loaded automatically the TH whenever the timer/counter passes through FF count. Since this is an 8-bit operation the count in r0 must be increased. The maximum value that can be loaded in the register is ffh. Delay generated is of order of (256 × 256 × 1) ms, i.e., 66 ms.

Example 8.5   Design a scheme to make LED connected at Port 1 of microcontroller ON and OFF for maximum delay time that can be generated by Mode 2 operation.

Solution

The setup is shown in Fig. 8.12. The LED unit is connected at Port 1. The keys connected at the interrupt pins may be used if necessary.

Delay generation requires following configuration:

1. C/T bit in TMOD the register is set to 0, indicating timer mode.
2. ET0/ET1 bit at the bit address 0A9h/0Abh is set to '1' to enable interrupt line.
3. Load count for the maximum delay (0000h) in the TH0–TL0 registers.

The program segment is as follows:

```
delay:
        mov   r7,  #ffh        ; count is 20
        mov   r0,  #ffh        ; delay (count)
        mov   r1,  #00         ; initialize r1
        mov   90,  #00         ; clear (Port 1)
        mov   8c,  #00         ; TH0 = 00h
```

```
                mov  8a, #00              ; TL0 = 00h; ms delay
                setb 0A9h
                mov  89, #02h             ; Mode 2, set ET0, IE register
                setb 8ch                  ; set TR0, i.e., make Timer 0 ON
wait:           sjmp wait                 ; wait for overflow

000b:           sjmp isr_to

isr-t0:         djnz r0, skip             ; if r0¹0, delay not over
                mov  a, r1                ; if r0 = 0
                cpl  a
                mov  r1, a
                mov  90, a                ; complement output
Skip:           mov  00h, 07h             ; reload r0 from r7
                reti
                end
```

8.2.5 Mode 3 Operation

Figure 8.14 shows configuration of timer/counter for the Mode 3 operation. A Mode 3 operation differs from the operations of the other modes. Timer 1 is disabled but holds its count; thus it is essentially frozen.

Timer 0 is split into two separate counters. The first counter is the same as Mode 0. TL0 is used as an 8-bit counter without a pre-scalar. The second counter uses TH0 as an 8-bit counter. Enable control is given by TR1 bit in the TCON register. The first counter sets the TF0 interrupt flag while second counter sets TF1.

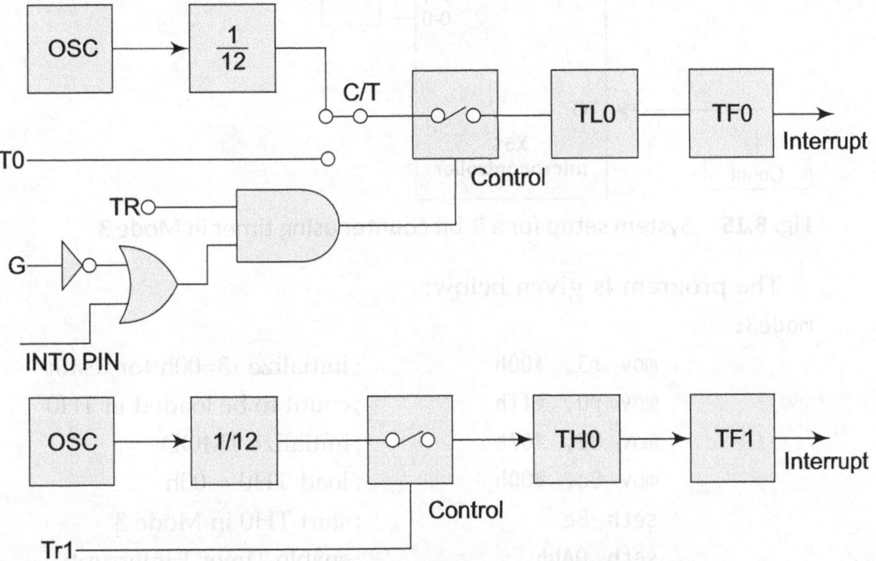

Fig. 8.14   Configuration for Mode 3 operation of timer/counter

Thus in a Mode 3 operation TL0 is used as the timer/counter, while TH0 is used as an 8-bit timer only. The operation of TL0 is controlled by the GATE (TMOD.3) bit having bit address 89h, using instruction [mov 89, #07h] TH0 acts as a timer only and is not affected by the TMOD/TCON bit for TL0. It is enabled by TR1 in TCON register. On overflow, an interrupt is generated at the interrupt branch address 001bh provided the IT1 is enabled by setting 0Abh to '1'.

Example 8.6 Design a scheme to display count of number of times the count key connected to the T0 pin is pressed using Mode 3. Use TL0 in counter mode and TH0 in a timer mode.

Solution

The setup is shown in Fig. 8.15. The LED unit is connected at Port 1. The count key is connected at the interrupt pin T0 (P3.4). System setup requires following configuration:

1. C/T bit in TMOD the register is set to 0, indicating timer mode.
2. ET0/ET1 bit at the bit address 0A9h/0Abh is set to '1' to enable interrupt line.
3. Mode 3 operation TL0 in counter and TH0 in timer mode.

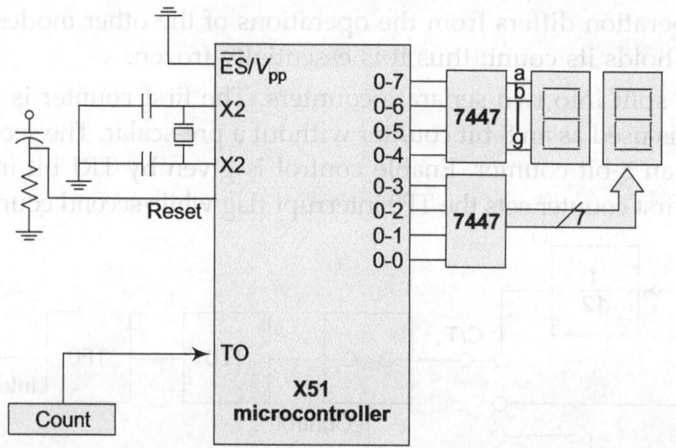

Fig. 8.15 System setup for a 8-bit counter using timer in Mode 3

The program is given below:

```
mode3:
        mov  r3,  #00h        ; initialize r3=00h for result
        mov  r0,  #ffh        ; count to be loaded in TH0
        mov  89,  #07h        ; initialize TMOD
        mov  8c,  #00h        ; load TH0 = 00h
        setb 8e               ; start TH0 in Mode 3
        setb 0Abh             ; enable Timer 1 interrupt
```

```
                 mov  8a, #00h          ; initialize TL0 = 00h
                 sebc 8c                ; turn T0, ON
rept1:           mov  90, 8ah           ; output TL0
                 sjmp rept1
- - - - - - - - - - - - - - - - - - - - - - - - - - - - - - - -
001b:            sjmp isr_t0
- - - - - - - - - - - - - - - - - - - - - - - - - - - - - - - -
isr_t0:          djnz r0, skip          ; if r0¹0, reload Th0 with count
                 inc r3
                 mov r0, #ffh            ; reload r0
skip:            mov 8c, #00h
                 reti
```

8.2.6 Application of Gate Bits

Gate bits in the TMOD register allows the TL0/TL1 in a gating mode. In this mode the counting is done if the interrupt pins are HIGH, and the TR0/TR1 are set. If the gate is LOW, it is independent of the status of INT pins.

Example 8.7　Design a scheme to display the duration for which the interrupt pin is held high.

Solution

Let us select the operation of the Timer 1 in a 16-bit Mode 1. Since the count d8efh is loaded in TH1–TL1 registers, counter will overflow every 0.01 s. The overflowing carry is used to increment seconds. The display can have maximum value 99.99 s. The counter will be frozen when interrupt is made LOW. The setup is shown in Fig. 8.16.

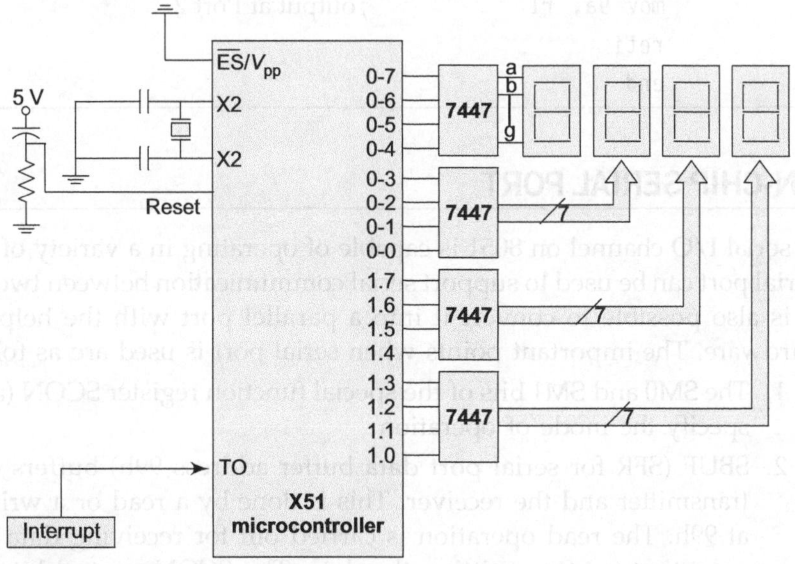

Fig. 8.16　System setup for Example 8.8 (Timer 1 in Mode 1)

The program is given below:

```
start:
            mov r1, #00h          ; register storing seconds
            mov r1, #00h          ; fraction of second
            mov 89, #90h          ; TMOD control word
            mov 8b, #0efh         ; count for 0.01 s
            mov 8d, #08dh
            setb 0abh             ; enable et1
            setb 8eh              ; turn ON t1
wait:       jb wait
001b:       sjmp isr_t1
isr_t1:
            inc r0                ; 0.01 s elapsed
            mov a, r0
            add a, #00h           ; set for DAA
            da a
            mov r0, a
            mov a, r1
            addc a, #00h          ; consider carry spill over
            da a
            mov r1, a
            mov 90, r0            ; output at Port 1
            mov 9a, r1            ; output at Port 2
            reti
            end
```

8.3 | ON-CHIP SERIAL PORT

A serial I/O channel on 8051 is capable of operating in a variety of modes. The serial port can be used to support serial communication between two controllers. It is also possible to convert it into a parallel port with the help of external hardware. The important points when serial port is used are as follows:

1. The SM0 and SM1 bits of the special function register SCON (address 98h) specify the mode of operation.

2. SBUF (SFR for serial port data buffer address 99h) buffers data for the transmitter and the receiver. This is done by a read or a write operation at 99h. The read operation is carried out for receiving data while write operation for transmitting the data. The SCON control bits control the operation sequence, since it is a bit controllable register.

3. The IE register bit ES, must be set or cleared to enable/disable interrupt from RI/TI flags in the SCON register. It is set by hardware when a byte is received, and it is cleared by software after servicing the RI interrupt.

4. The bit IP.4 must be set to give high priority to the serial port.

5. SBUF receiver is double-buffered to eliminate the over-writing of the present received byte by the next byte in the case that the CPU has not yet read the receiver register before the next byte has begun.

6. The REN bit (SCON.4: bit address 9ch received enable control bit) is set or cleared by the software to enable or disable serial data reception. When set by user software, it enables the receiver channel to begin clocking in serial data from the RxD pin on the 8031/51 (P3.0). When the user resets the REN bit, the reception of data is disabled.

7. SM2 (SCON.5) has no effect in Modes 0 or 1. In Modes 2 and 3, if SM2 is called conditional enable because interrupt is generated if RB8 is '1'. This is useful in the multiprocessor environment, where more than one controllers are connected to serial port. It is used to wake up the controllers.

8. The serial I/O of the 8031/51 can handle five types of serial frames. They are shown in Table 8.17.

Table 8.17   Formats of the serial communication frames for serial port on X51

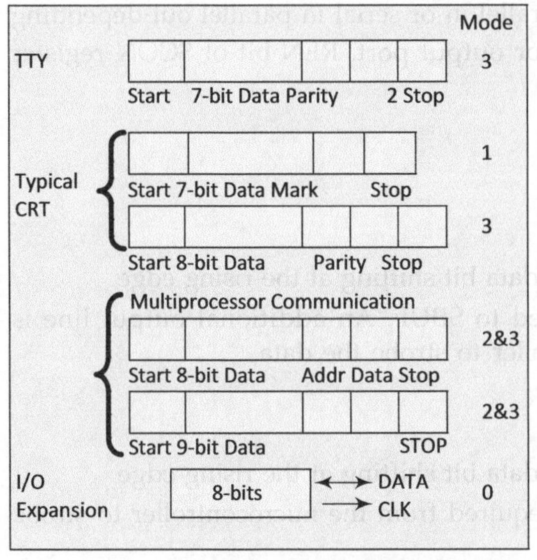

Mode 0   Synchronous I/O expansion using TTL or CMOS shift registers

Mode 1   UART interface with 10-bit frame and variable transmission rate

Mode 2   UART interface with 11-bit frame and fixed transmission rate

Mode 3   UART interface with 11-bit frame and variable transmission rate

Modes 2 and 3 also provide automatic wake up of slave processors through interrupt driven address frame recognition for multiprocessor communications.

We will study the use of the modes of operations of serial port on the microcontroller chip. The serial port has four modes of operation as follows:

1. Mode 0: Half-duplex synchronous operation

2. Mode 1: Full-duplex asynchronous operation

3. Mode 2: Full-duplex asynchronous operation with variable baud rate
4. Mode 3: Full-duplex asynchronous operation with programmable baud rate

8.3.1 Mode 0 Operation

Data is sent and received through the RxD pin in the 8-bit frames with LSB first. The bit rate is fixed at one-twelfth the oscillator frequency. The shift clock is sent out the TxD pin during both transmission and reception and is used to synchronize the receiver with the sender shift clock edge occurs during the valid state of each data bit. Note that the TxD and the RxD are alternate functions of Port 3 pins. This is also called shift register mode of operation.

In complex applications using memory expansion, the parallel ports on chip are used to generate the address bus. To have extra parallel ports, PPI 8255 can be used. To avoid use of an extra IC chip, such as 8255, the serial port can be converted into an additional parallel port with the help of an external hardware used at serial port by using it in Mode 0.

It is already known that Mode 0 operation can be selected by clearing SM0 and SM1 bits in SCON. The serial I/O channel of the 8031/8051, and an external shift register can be used to configure additional input/output ports. The shift registers used is either serial out parallel in or serial in parallel out depending on the port to be realised is input or output port. REN bit of SCON register must be set for expansion.

For example:

1. In Mode 0 input:

   RxD pin is for the input

   TxD pin generates a clock for data bit shifting at the rising edge

   The parallel byte is transferred to SBUF. An additional output line is required from the microcontroller to strobe the data

2. In Mode 0 output:

   RxD pin is for the output

   TxD pin generates a clock for data bit shifting at the rising edge

   An additional output line is required from the microcontroller to strobe the data

Realizing Additional Output Port

The hardware setup used in this experiment for the case of output port expansion is shown in Fig. 8.17.

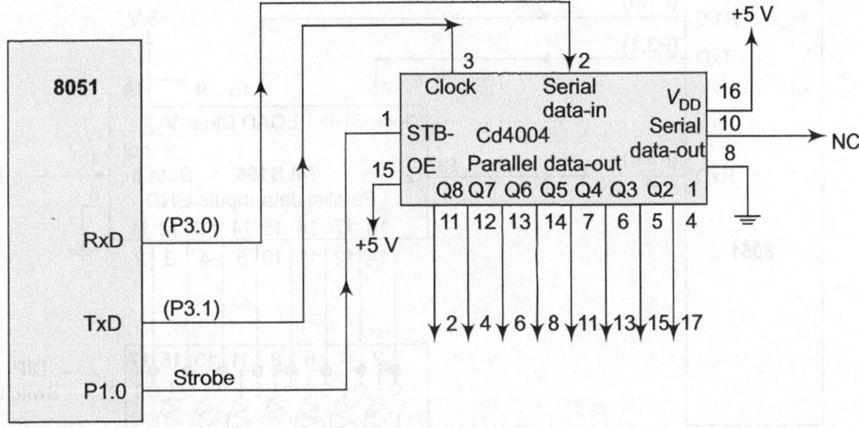

Fig. 8.17 Generation of additional parallel output port

The three pins of the 8031/8051 (RxD, TxD, P1.0) are used to construct an additional output port. P3.1 provides the clock for the shift register and the pin P3.0 outputs serial bits from the 8031/8051 serial channel to the serial-in parallel-out shift register connected at the RxD pin. P1.0 is used to strobe the shifted bits into the latch of shift register.

The software for operation is given below:

```
Start:
          mov  98, #00       ; move 00 into SCON initializes serial
                             ; channel for Mode 0 operation (SM0=SM1=0)
          mov  20, #00,      ; load initial byte into RAM buffer address 20
          mov  r0, #01h      ; R0=>number of expanded ports
          mov  r1, #data     ; move address of data to be output
          clr  99.           ; clear bit T1 in SCON
          mov  99, @R1       ; move contents of RAM to SBUF
wait:     jnb  99, wait      ; wait for data byte transmission
          clr  90            ; lower strobe line
          setb 90            ; strobe shift register's data into their output
                             ; latches

          sjmp start
          end
```

Realizing Additional Input Port

The operation and configuration of Fig. 8.18 is similar to that for an output port except that the shift register to be used is parallel-in, serial-out form. The RxD is used as input line, while the TxD will generate clock pulse. Line P1.0 provides a strobe for load operation.

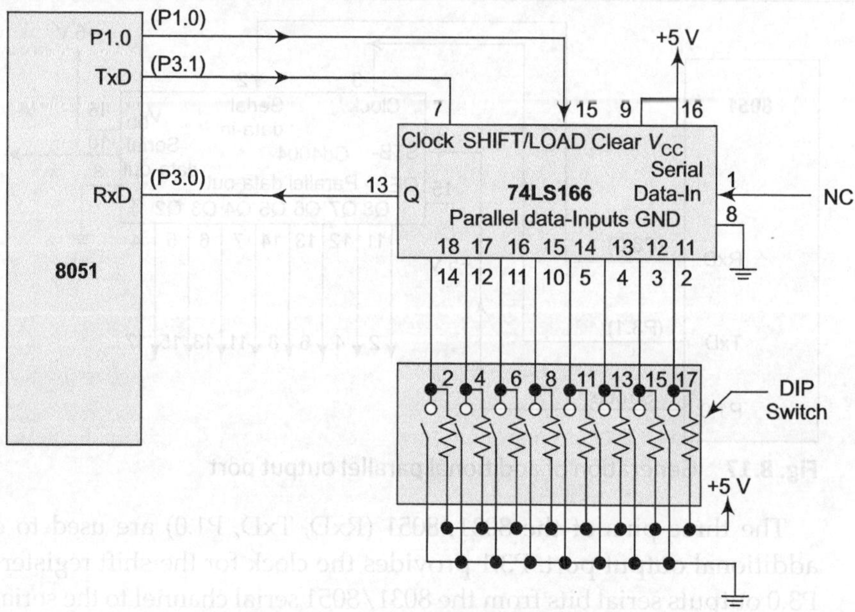

Fig. 8.18   Generation of additional parallel input port

The program for input through input expansion port is shown below:

```
start:
            mov 98, #00         ; loads SCON with Mode 0
            setb 9c             ; sets REN (receiver enable) bit in SCON to 1
                                ; to enable serial data reception
            mov a, #01          ; load ACC with #01
            mov a, 20           ; read RAM buffer 0 data, i.e., PORT' data at
                                ; the switches (Fig. 8.35)
            mov r0, #01h        ; number of input ports input ports
            mov r1, #20h
            setb 90             ; raise shift/load line P1.0 to input Pin15 of
                                ; 74LS166 parallel _serial shift register
            clr 90              ; lower shift/load line
            clr 98              ; reset r1 (Received Interrupt) flag (98) in
                                ; SCON. This causes 8 TxD clocks. The first of
                                ; these latches the data into the 74166's internal
                                ; shift register from its parallel port. The rest
                                ; constitute a dummy cycle "shifting in" 8
                                ; fictitious bits through RxD into SBUF
wait:       jnb 98, wait        ; wait for all 8-bits
            setb 90
            clr 98              ; reset r1 to begin clocking in the real data
                                ; from the 74LS166 into SBUF
```

```
load:       jnb 98, load        ; if r1 is low, wait till all 8 data bits are clocked
                                 ; into SBUF
            mov @R1, 99          ; r1 is high, data byte received by SBUF, write it
                                 ; into RAM buffer

            sjmp start
            end
```

Multiple output ports as well as input port expansion of serial channel can also be obtained. Even it is possible to have input/output port expansion of the serial channel with the help of additional hardware.

8.3.2 Mode 1 Operation: Application as 8-bit UART

Mode 1 is selected by setting SM0 (9Fh) = 0 and SM1 (9Eh) =1 in the SCON register. In this mode the serial I/O port operates as an 8-bit UART with a 10-bit code (an 8-bit data field, a stop bit, and a start bit). A parity bit is not supported by 8051 I/O it must be realized by the software. Normally the bit PSW.0 implements it. A parity checking is part of the program.

Using either an external clock may choose the baud rate clock for the UART in Mode 1 (T1 input through pin P3.5) or the internal oscillator. Baud rate for a frequency of 12 MHz is 31.25 Kbits/s.

Data Transmission

The data byte passed in the accumulator is transmitted via the serial port. The parity flag of the byte is determined from contents of the accumulator, loaded as the 8th bit of the data field of the frame to be transmitted.

We will use an odd parity. The basic requirement is to attach a parity bit for transmission and check it for the received data. Transmitter calculates the parity of the data to be transmitted and appends a corresponding parity bit to the data (8th bit).

Receiver on receiving the data calculates its parity and compares it with the parity used. If the parity does not match, the received data is invalid and discarded. The program segment is given below:

```
trans:
            mov c, 0d0           ; load carry bit with the parity flag (PSW.0)
            cpl c                ; complement the carry to form odd parity
            mov 0e7, c           ; load bit 7 of ACC with carry (parity) bit
wait:       jnb 99, wait         ; test bit 99 in SCON (T1 flag) wait if T1=0,
                                 ; transmission of previous byte is not over
            clr 99               ; clear T1 to prepare for next byte
            mov 99, a            ; load new data in SBUF for transmission
            ret
            end
```

Data Reception

While receiving the data, it is necessary to determine if parity of data is correct or not. If it is correct, it is accepted otherwise it is discarded.

```
receive:
        jnb 98, receive     ; test RI flag (98)
        mov A, 99h          ; read SBUF
        clr 98              ; reset RI flag. Prepare SBUF to receive byte
        mov c, 0d0h         ; load parity of received byte acc.7 transmitted
                            ; byte into carry
        cpl c               ; complement carry to form odd parity
        clr 0e7             ; mask acc.7 (parity bit) for parity test of new
                            ; received byte
        ret
        end
```

I/O Initialization

It is very common to use a timer in auto-reload mode (Mode 2) for generating baud rate. The baud rate is then given by the expression:

$$\text{Baud rate} = \text{Oscillator frequency}/[N\,(256 - (TH1)]$$

where TH1 represents the contents of the register TH1

N depends on the SMOD bit in the PCON register.

For example:

If the SMOD bit is 0, value of N is 384 otherwise value of N is 192.

When selecting a crystal to set the oscillator frequency in a system that will use the serial port, the value stored in TH1 must be an integer number. The oscillator frequency must be an integer multiple of both the baud rate and N in order to obtain an exact baud rate using the above expression.

Let us assume that we are interested in using a baud rate of 300 bauds supplied by Timer 1 used in the Mode 2. This can be set by loading SCON register with the control word 52h, which enables the receiver, and assumes that the SBUF is empty. TMOD is loaded with the value 20h to use Timer 1 in Mode 2 and start baud clock using TCON.

Assembly language segment is as follows:

```
init_srl:
        mov 98, #52h        ; load SCON with #52, select Mode 1, enable
                            ; receiver, T1=1
        mov 89, #20         ; load TMOD with #20. Timer 1, Mode 2
        mov 8d, #98h        ; load TH1 register for 300 baud
        setb 8eh            ; start Timer 1
        ret                 ; return
        end
```

Example 8.8 Write a program to receive 10 data bytes from serial port and store it at memory location specified by register r0.

```
main:     mov r0, #address
rept1:    r1, 0ah              ; loop counter for 10-bytes
          acall init_srl       ; initialize serial I/O channel
wait:     acall receive        ; call routine at to receive data from terminal
          jc wait              ; if parity error; ignore received byte
          mov @r0, a
          inc r0
          djnz r1, wait
          sjmp rept1           ; poll for another byte from terminal
          end
```

Example 8.9 Write a program to transmit 10 data bytes stored in at memory locations specified by register r0 using serial port.

```
main:     mov r0, # address
rept1:    mov r1, 0ah          ; loop counter for 10-bytes
          acall init_srl       ; initialize serial channel
loop:     mov a, @r0
          acall trans          ; call transmit subroutine
          inc r0
          djnz r1, loop
          sjmp rept1
          end
```

8.3.3 Mode 1 Operation: Interrupt Activated Data Transfer

The enable serial (ES) port interrupt enable flag [IE.4] in the interrupt enable (IE) register supports the serial port interrupt. When set/cleared by software, interrupts from T1 or R1 flags are enabled/disabled. (T1 or R1) interrupt, if enabled, vectors to address 0023h. The interrupt service routine there will provide the proper CALLS to support the serial communication.

To incorporate the serial port interrupt, initialization is done as follows:

```
Sr_int-init:
          mov 98, #52h     ; load SCON with #52 ; select Mode 1 and
                           ; enable receiver, T1=1
          mov 89, #20h     ; load TMOD with #20, Timer 1 Mode 2
          mov 8d, #98h     ; load TH1 register for 300 baud
          setb 8eh         ; start Timer 1
          setb 0bch        ; enable high priority for serial port INTR SINT
                           ; IP.4 control bit in IP register
          setb 0ach        ; enable serial port INTR (ES bit in IE)
          ret              ; return
```

Whenever there is a serial interrupt, the interrupt service routine at the address 0023h will be executed. Since RI or TI may generate the interrupt, it is necessary to identify the source and the operation to be done The following program segment carries out the operation:

```
isr-serial:
0023 isr:   push 0d0          ; save PSW
            push 0e0          ; save ACC
            jnb 98, skip      ; test for RI high or not
            acall receive     ; receive byte from terminal
skip:       acall trans
            pop 0e0           ; restore ACC
            pop 0d0           ; restore PSW
            reti              ; return from interrupt to main
            end
```

8.3.4 Mode 2 Multiprocessor System Using Serial Link

The transmission rate of this mode is fixed at a fixed rate of 187.5 K. The 8031/51 hardware supports protocols for communication channel setup and data transfer. The basic operations needed are the well-defined movement of data between multiple microcomputers and the direction of data to and from the appropriate microcomputer.

When the number of processors involved is greater than two, the need to setup the appropriate communication channels arises. The 9th bit plays a vital role. This additional bit provides a convenient means of qualifying the 8-bits of data as either true data or as an address of one of the N microcomputer nodes. It wakes-up desired slave, while leaving the remaining nodes sleeping. The slaves may be addressed to receive data simultaneously or individually. The former case allows a "broadcast" mechanism by which the master sends the same information to all slaves.

Example 8.10 Design a multiprocessor communication system using master-slave configuration. The function of the slave is to monitor and control. The function of master is to load, store, and display sensed data and change control limits. The messages between each microcomputer will consist of 5-bytes of data. The software will handle these 5-bytes as single entity, referred to as MSG.

Solution

It is necessary to initialize serial channel in mode for the master as well as for the slave. A message available flag needs to be defined by the slave processors to signal the receipt of a full message.

```
master-init_mode2:
        mov  0d0, #18          ; select Bank 3
        mov  r7, #XBUF         ; initialize transmit buffer pointer
        mov  r6, #RBUF         ; initialize receiver buffer pointer
        mov  98, #82h          ; set master in Mode 2 with TI=1 , REN = 0 ; to
                               ; prevent reception by the master till it is enabled

slave-init-mode2:
        mov  0d0, #18          ; select Bank 3.
        mov  r7, #XBUF         ; initialize transmit buffer pointer
        mov  r6, #RBUF         ; initialize receiver buffer pointer
        mov  98, #0a2h         ; set slave in Mode 2, TI=1,REN=0. when
                               ; awakened for reception SM2=1 enables
                               ; reception of frames for which RB8=1
                               ; indicating an address and alerts every slave
        setb 0bch             ; set high priority interrupt (IP.4)
        clr  00h              ; initialize "message available" flag
        mov  0a8h, #90         ; enable serial Port INTR(IE), enable T1 and R1
```

Four operations are expected from the system are

1. Connect the controller (EN-LINK): It performs the ADDRESSING of micro-controller before the data transfer.

2. Disconnect the controller (DS-LINK): It performs the disconnection of slave, after the data transfer.

3. Transmit message (TRANS-FIVE): RAM buffer is loaded with the 5-bytes and the TRANS-FIVE routine is used to transmit a 5-byte message.

4. Receive message (REC-FIVE): This routine receives 5-bytes for the RAM buffer.

Master (ISR) The master uses the EN-LINK and DS-LINK routines. The slaves can transmit the messages only when a command is received from the master.

The program listing of different segments to carry out operations by the master are given below:

```
EN_LINK:
wait:   jnb  99, wait         ; wait till transmitter is ready TI=1
        clr  99h              ; transmitter ready. Clear T1
        setb 9bh             ; set TB8 to specify that previous 8-bits in
                               ; data field are an address
        mov  99, a            ; load slave address, (ACC)
        ret                   ; return
        end
```

```
DS_LINK:
no_ds:      jnbB 99h, no_ds        ; wait for transmitter ready TI = 1
            clr 99h                ; clear T1
            setb 9bh               ; set TB8, 8-bit data is address
            mov 99, #0ffh          ; no slave has FF as the address All slaves go
                                   ; back to sleep
            ret                    ; return
            end
TRANS_FIVE:
            push 0d0               ; save PSW
            mov 0d0, #18           ; set RS1, RS0 for Bank 3
            mov r5, #05            ; message counter (5-bytes)
wait:       jnb 99h, wait          ; wait for T1=1
            clr 99h                ; clear T1. Transmission complete
            clr 9B                 ; clear TB8, to indicate that 8-bit DATA
            mov 99, @r7            ; load SBUF with a byte to be transmitted
            inc r7                 ; increment message pointer
            djnz r5, wait          ; decrement byte counter, repeat transmission
                                   ; if not over
            pop 0d0               ; restore PSW
            ret                    ; return
REC_FIVE:   push 0d0              ; save PSW
            mov 0D0, #18           ; select Bank 3
            mov R5, #05            ; message byte counter
wait:       jnb 98h, wait          ; wait for reception to complete
            mov @r6, 99            ; load received byte from SBUF into buffer
                                   ; pointed by r6
            clr 98                 ; reset RI
            inc r6                 ; increment buffer pointer
            djnz r5, wait          ; decrement counter, repeat till over
            pop 0d0               ; restore PSW
            ret                    ; return
```

Slave (ISR) The interrupt service routine (ISR) for the slave processors is
required to respond to the receipt of a frame whose 9th bit (RB8) is set, specifying
transmission of an address. The function of ISR is to compare the 8-bits address
of the frame with the resident slave's address. If they match, SM2 is cleared to
receive all bytes. REC_5 routine is then called to load the receive buffer with
the received message.

```
slave-isr:  0023 push 0d0                     ; save PSW
            push 0e0                          ; save ACC
```

```
            jnb 98h, end             ; rest for RI = 1, return if byte not received
            jnb 9bh, data-rec        ; test for RB8=1 for true data byte received
            mov 0e0h, #SLV-add       ; compare ACC with SBUF byte
            cjne a, 99h,reset-chk    ; goto reset-chk,If not equal, address is
                                     ; not for this slave
            clr 9dh                  ; clear SM2 bit and receive data
data_rec:   mov r7, #addr.           ; load r7 in RB3 with base addr.of message
                                     ; receive buffer
            acall REC-FIVE           ; receive data bytes
            setb 00h                 ; set "message " flag
            sjmp end                 ; carry out end procedure
reset-chk:  mov 0e0, #0ff            ; load accumulator with reset address
            cjne a, 99,end           ; compare it with SBUF byte, do not
                                     ; disconnect if not equal
            set 9dh                  ; disconnect and disable it
            clr 98h                  ; clear R1
end:        pop 0e0                  ; restore ACC
            pop 0d0                  ; restore PSW
            reti                     ; return from interrupt
```

8.3.5 Mode 3 Operation: 9-Bit UART

This mode is also used to handle communication in multiprocessor environment.
It uses an 11-bit frame. A 9th data bit is inserted before the stop bit. Hence, it is
also called 9-bit UART mode. It differs from Mode 2 in selection of baud rate.
In this mode baud rate is programmable and selection is similar to one in the
Mode 1 operation, which uses Timer 1. Reception is enabled when the remote
enable (REN) bit in SCON is set to 1 but the actual reception is initiated by a
high to low transition at RxD pin. Transmission is initiated by writing to SBUF
in any serial mode. Mode 3 frame is characterized by an 11-bit frame {a start bit,
a stop bit Nine data-bits (7 data bits, a parity bit, and the 9th bit [TB8] and
[RB8] of SCON register).

RB8 is set or cleared to indicate the state of the 9th data bit received.

TB8 is set or cleared to determine the state of the 9th bit transmitted in a
9-bit UART mode.

SBUF (99h) will still contain bits 0–7 of the data field. Program has to test/set
the RB8/TB8 bits in SCON (9th data bit) for parity checking and generating. The
program modules for the Mode 3 transmission and reception are listed below:

```
trans_mode3:
            mov c, 0d0h              ; load carry with parity (bit PSW.0)
            cpl c                    ; complement carry to form odd parity
```

```
            mov 9bh, c          ; load TB8 with transmit parity bit
    wait:   jnb 99h, wait       ; loop until transmitter is ready
            clr 99h             ; byte transmitted. Clear TI prior to next byte
                                ; being loaded in SBUF
            mov 99h, a          ; load new byte (bits 0–7) into SBUF
            ret                 ; return
rec_mode3:
    wait:   jnb 98h, wait       ; wait for RI to set
            mov a, 99h          ; read data from SBUF to ACC
            clr 98h             ; reset RI flag to receive next byte
            mov c, 9a           ; load received parity RB8in Cy
            cpl c               ; complement Cy to form odd parity
            ret                 ; return
```

EXERCISES

8.1 Write a program using strobe and acknowledge handshaking. Assume that the strobe is a positive going edge, while acknowledge is a negative going edge.

8.2 Write a code to initialize the Timer 1 to be externally clocked in Mode 0.

8.3 Draw a waveform seen on serial pin when 0x5d is transmitted using 10-bit code. [1 start, 8 data and 1 stop, no parity.]

8.4 How do you select the baud rate for serial communication? What is the number to be loaded if the transmission has to be 2400 bauds?

8.5 What is the pulse width if the count is [0x5b3d] and crystal frequency is 12 MHz?

8.6 What are the problems associated with the push button keys? What will be the effect if they are not taken care of, in the program?

8.7 Assuming that the digit can withstand 18 mA of average current flow, calculate the peak current for a scanned 7-segment display consisting of three digits.

8.8 Design a hardware setup using keyboard, LEDs, and 7-segment displays and write the program to simulate following digital ICs:

    (a) 7432: two input OR gate

    (b) 7410: three input NAND gate

    (c) 7473: dual JK _FF

    (d) 74154: multiplexer

    (e) 74138: decoder

    (f) 74165: shift register

8.9 Design a setup using a keyboard and 7-segment display to test digital ICs [7400/7408/7432/7486]. TEST key is to be connected to [int0]. When the TEST key is pressed, microcontroller should accept the IC number (Last two digits) and display (P) or (F) on the 7-segment display.

8.10 Generate a chip select signal for a memory chip [2K × 8] which occupies the address space (1000 – 1FFF) using

    (a) address decoder 74LS138 and 7400.

    (b) two stage 74LS138 decoder.

8.11 A reverse key is connected to [int0] pin of microcontroller. When the key is pressed it is necessary to reverse an array stored in memory location 2601h onwards, 2600h contains the length of an array. Write a main program and interrupt service routine (ISR) for the [int0] for operation of key.

8.12 A SORT key is connected to int1 pin of microcontroller. When the key is pressed it is necessary to arrange the array stored in memory location [2601] onwards, [2600] contains the length of an array, in descending/ascending order depending on status of port pin P1.0 of microcontroller. If P1.0 is high: descending order else ascending order.

Write a main program and interrupt service routine [ISR] for the [int1] for operation of key.

8.13 A SORT key is connected to [int1] pin of micro-controller. When the key is pressed it is necessary arrange the array stored in memory location [2601h] onwards terminated by [<CR>], in descending/ascending order depending on status of port pin [P1.0]. If P1.0 is high, arrange in descending order else in ascending order.

Write a main program and interrupt service routine (ISR) for the [int1] for operation of key.

8.14 Write a program which accepts the elements of an array, which is terminated by [ffh], from the port [P1]. A DESC key is connected to [int0]. The controller must arrange the array in descending order when the key is pressed.

8.15 Write a program which accepts the elements of an array, which is terminated by [ff], from the port [P0]. An ASC key is connected to [int1]. The controller must arrange the array in ascending order when the key is pressed.

8.16 Write a program which accepts the elements of an array of 8-bits, which is terminated by blank, from the port [P1]. An AVG key is connected to [int1]. The controller must find the average of the array arrange the array, when the key is pressed and store the result in location [2600h].

8.17 Write subprogram to multiply an 8-bit number by 10. Write a main program, which accepts a 4-digit BCD number from port [P1, P2] and converts it into hex.

8.18 An ADD key is connected to the interrupt pin [int0]. Write a main program which accepts two 8-bit data from Ports P1, P2 and waits for pressing of ADD key. When the key is pressed, it adds the data and stores the result at RESULT.

8.19 A DUP key is connected to the interrupt pin [int1]. Write an ISR which deletes a duplicate entry from the array stored in memory. The array starts from [2501] onwards and is terminated by a blank.

8.20 A BCD_IN key is connected to [int1]. When key is pressed, microcontroller allows [int0] to interrupt. It also accepts the hex data from port [P1]. If it is (0 to 9), data is stored as it is at the location INPUT specified by dptr, otherwise it is converted to BCD before storing.

8.21 Write a program to generate a time delay of 1 s using Timer 0.

8.22 Write a program to realise digital clock using Timer 1.

8.23 Write a program to measure the frequency of input signal applied at port pin P1.0.

8.24 Write a program to measure pulse width of a signal applied at port pin P1.1.

8.25 Write a program segment to initialize Timer 0 in external mode and count number of pulses at the interrupt pin T0.

8.26 Write a program to allow serial communication at 2400 bauds using TxD/RxD pin.

8.27 Write a program to communicate with other microcontroller system using serial port in Mode 3 operation.

8.28 Write a program to control power supplied to the microcontroller system by using PCON register control.

8.29 Design a setup and write a program to convert serial port of microcontroller into a parallel port.

8.30 Design a setup and write a program to convert parallel Port 0 of microcontroller into serial port.

Chapter 9

Application I: Smart Energymeters

The chapter introduces the concept of prepaid energymeters and the design of microcontroller-based three phase power meter using the ADE7758. The microcontroller updates its energy registers at a predetermined time by sending the request on the SPI. This module can be expanded to support five ASIC cards and it allows the interface with external CT to increase the current measurement limit. The parameters are software adjustable and can be stored in the internal EEPROM to be passed on to the microcontroller during system initialization. The design and implementation of a prototype module for industrial applications of a prepaid energymeter for single phase electrical system is mainly for the domestic use only. This is one of the three levels of the prepaid metering technology. Installed at the consumer's premises, the meter should continue the power supply when the card is inserted in the meter holder and cut-off the supply when the units indicated on the card are totally used.

Power distribution has been a neglected area for too long. A meter is considered no more than a rusty contraption at the end of the distribution line—just an energy recording device. Besides striving to achieve a higher percentage of metered consumers, better metering equipment is also being experimented within these states. The major deterrents to widespread installation of better meters are their comparatively higher cost. An electronic meter costs five times more than a traditional meter. But the cost evens out in the long run as such meters have a long service life with high accuracy levels. There is a need for firm policies, continuous monitoring, and strict vigilance. Taking a cue from private distribution utilities that have strong vigilance and monitoring units, even the government electricity boards are starting to take electricity theft and pilferage offences quite seriously.

In the current scenario, distribution reforms are likely to be a long-drawn-out process involving systemic overhaul and infrastructure upgradation. Simultaneously, the mindset of the consumers need to change. Prepaid metering technology seems to be one of the most suitable technologies for India. In effect, prepaid metering is just like a debit card. Consumers pay in advance and buy electricity in the form of a card or a token. The electricity equivalent to the worth of token is credited to the consumers account. Subsequently, the account has to be recharged or else the power supply to the consumer would be automatically disconnected. Prepaid metering technologies can be categorized into two types: 'token' and 'token less', i.e., those needing a physical medium to convey credit and those that do not. Token technology mainly utilizes coins, tickets, keys, and magnetic cards.

9.1 | INTRODUCTION

Electromechanical meters have been the standard for metering electricity since billing began. The measurement of electricity supply and subsequent billing to industries has traditionally been achieved through electromechanical meters. Although widely used, this solution has several disadvantages including long term accuracy, cost of calibration, and limited communications. These issues can be overcome by using digital power meters, where it is possible to achieve long term accuracy by removing analog components which are prone to drift over temperature and time and adding value added features such as multiple tariff rates, overload alert, maximum current usage, and improved communications.

In order to accommodate the advanced requirements, manufacturers are adopting electronic solutions. New energy measurement ICs are enabling accurate, dependable, and robust meters with alarm and indications. Digital power meters eliminate problems of distribution companies such as faulty power meters, reactive loading by industries, and power theft and can be configured to provide multi-rate billing, smart billing, and for communication via a network.

9.2 | THREE-PHASE ENERGYMETER

Consider the design of a microcontroller-based Class 0.2 three-phase energymeter to measure active energy as per IS13779/IEC1036 with a PC-based calibration facility using EIA-885 and energy measurement IC having technical specifications of Table 9.1.

Table 9.1 Specifications for 3-phase energymeter

Voltage inputs		Current inputs	
Three phase 3 wire (True RMS measurement)		Nominal current	: 5A AC from external CT
Nominal voltage	: 440 V AC ±10%	Starting current	: 0.4%*basic current Overload
Voltage tolerance	: 50% to 110% of nominal	Capacity	: 4.0 x Nominal continuous
Overload capacity	: 4.0*(Nominal continuous)	Frequency	: 50 Hz ±5%
Accuracy		**Modbus serial communication**	
Class 0.2 as per IS-13779 - IEC 61036		Bus type	: RS-485 2 Wire
Operating temperature	: − 10 °C to 55 °C	Baud rate	: 4800, 9600, or 19200 (user selection)

9.2.1 System Setup

Figure 9.1 depicts a system setup for the three-phase energymeter. The full scale current of 5 A and full-scale voltage of 440 V is stepped down to ±0.5 V peak-to-peak for metering purpose. ASIC ADE7758, energy measurement IC converts the analog inputs into digital form using in-built ADC and updates its various registers accordingly. This is interfaced to controller through SPI in master-slave configuration. ASIC carries out the required calculations and stores the final values in its internal registers. Microcontroller reads ASIC registers using SPI. The data is displayed on LCD or sent to the PC as per request.

The LCD is used in 16 × 4 line character mode. Push button switches with the LCD are used to scroll through the interactive menu and select the parameters to display. The microcontroller communicates with the PC using EIA-485 with Modbus protocol for display as well as for calibration. The SMPS is supplied from the three-phase supply.

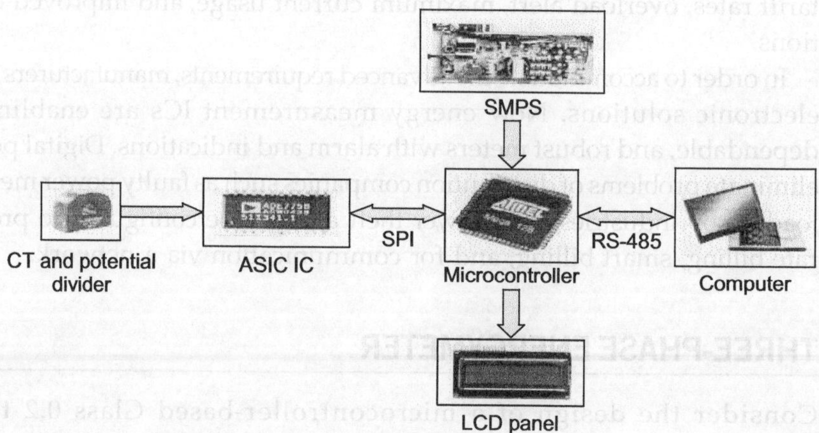

Fig. 9.1 Setup: three-phase energymeter

CT section   Electrohms CT1164 (turn ratio of 1:2500) with class 0.2 accuracy is connected with a burden of 66 Ω via the ground. The main lines are connected to the primary of the CT, which steps down the current in the secondary. Burden resistor across the CT is used to control the voltage. Here it set for a maximum of 353 mV for a full-scale primary current of 5 A.

Potential divider section   The 1 K resistor and the 33 μF capacitor form the anti-aliasing filter. The 1 K resistor along with the others form the potential divider network to divide the full-scale 440 V RMS to 353 mV RMS which is the required input to the ASIC. For the voltage divider, three resistors of 330 K in series with a 1 K resistor in parallel are used to have a ratio of 1:1000 for converting input voltage in millivolts.

9.2.2 ASIC IC ADE7758

ADE7758 has three sections: voltage channel, current channel, and power section.

Voltage channel   The reference voltage is varied by changing the ASIC voltage input (346 mV) using voltage PGA. The reference can be set to 0.5 V, 0.25 V, or 0.125 V. The ADC in ADE7758 is 1-bit sigma-delta ADC. ADC is scaled using the xVRMSGAIN register (default value 1). The offset register for each voltage channel (xVRMSOS), removes the 0 reading offset in the register. Its value is obtained and stored at the time of calibration. RMS output is stored in the xVRMS register (Fig. 9.2).

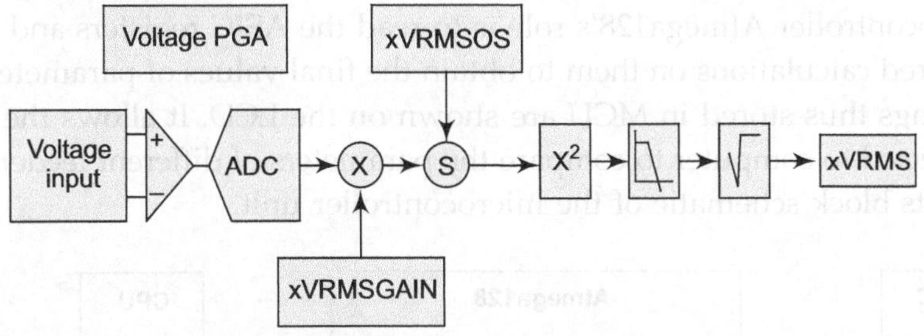

Fig. 9.2   Voltage channel

Current channel   The working of current channel of ASIC is similar to voltage channel with xIGAIN register for scaling xIRMSOS offset register, root mean square module, and a xIRMS register to store the final value. HPF and integrator block after the gain adjustment. HPF is added to remove the ADC offset errors which affect the power calculations. HPF is not implemented on voltage channel because the HPF on the current channel alone is sufficient to eliminate error due to ADC offsets in the power calculation. ADE7758 allows the current channel input from *di/dt* sensors to recover the current signal before it is used, it is necessary to integrate it (Fig. 9.3).

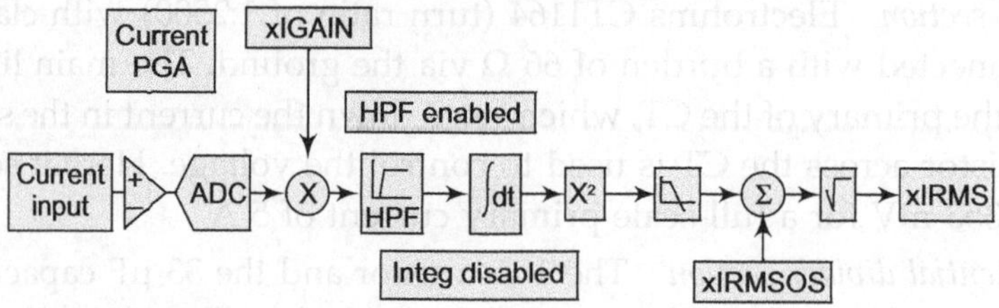

Fig. 9.3   Current channel

Power section   Power section uses the outputs from current and voltage channels. ADE7758 measures three types of power, namely, active, apparent, and reactive. For active and reactive power calculations, ASIC uses original V and I signal values, while for apparent power it multiplies the register values of I_{RMS} and V_{RMS}. Power sections have offset and gain registers. The power register values are in hexadecimal format and the actual value from it has to be calculated. These values are multiplied to obtain the final power readings with constants: Wh/LSB for active power, VARh/LSB for reactive power and VAh/LSB for apparent power (Fig. 9.4).

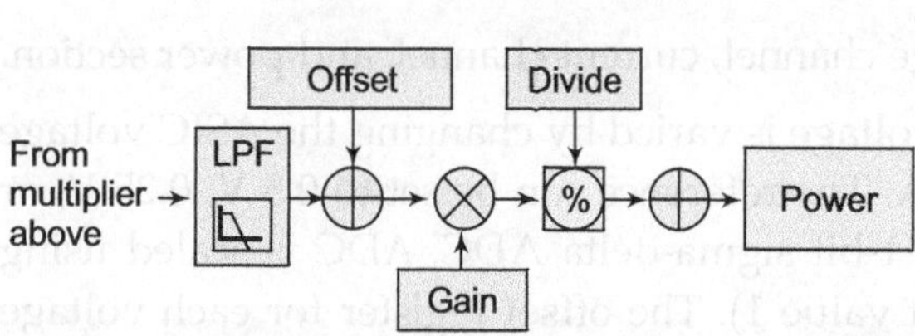

Fig. 9.4   Power section

9.2.3   Microcontroller Atmega128

Microcontroller Atmega128's role is to read the ASIC registers and perform the required calculations on them to obtain the final values of parameters. The final readings thus stored in MCU are shown on the LCD. It allows the meter to be connected to computer to compare the parameters of different feeders. Figure 9.5 depicts block schematic of the microcontroller unit.

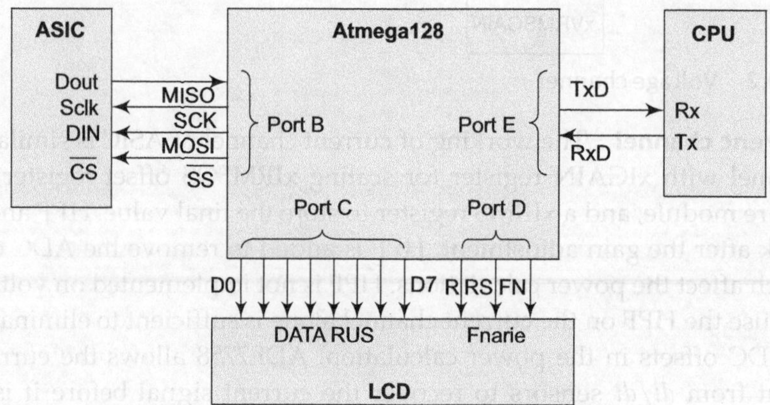

Fig. 9.5   Block schematic: microcontroller interface

The block schematic includes following blocks:

ISP (In System Programming)   ISP is interfaced using a 6-pin connector (Fig. 9.6). The UART pins RxD and TxD together with SCK of SPI are used for interfacing ISP.

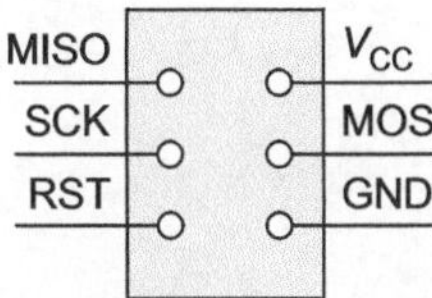

Fig. 9.6   ISP pin diagram

SPI (Serial Peripheral Interface)   SPI is used to communicate between Port B of microcontroller and ASIC.

LCD (Liquid Crystal Display)   6 × 4 LCD is interfaced directly to the Port C for the data and to the Port D for other enable signals. An external potentiometer is used to control the LCD contrast.

UART (Universal Asynchronous Receiver/Transmitter)   Port E is used for communication with a computer via a EIA-485 cable. Modbus protocol is used for this communication, which again is an industry standard. This is explained later.

9.2.4 AVR Studio

AVR studio is an integrated development environment (IDE) for writing and debugging AVR applications in Windows 9x/Me/NT/2000/XP environments. AVR studio provides a project management tool, source file editor, chip simulator, and in-circuit emulator interface for the powerful AVR 8-bit RISC family of microcontrollers. It supports the STK500 development board, which allows programming of all AVR. GUI plug-ins and other modules can be written and hooked to the system. It consists of many views and sub-modules. The major views are described in Fig. 9.7(a).

Project management   All code creation within AVR studio is done as programming projects. The project file that maintains information about the project such as: files included assembler setup, customized views and so forth. It ensures that the project is setup properly and will assemble/compile.

Editor   It is a programming editor with normal window functionality including syntax colour coding. It is used to debug code, break, and trace points.

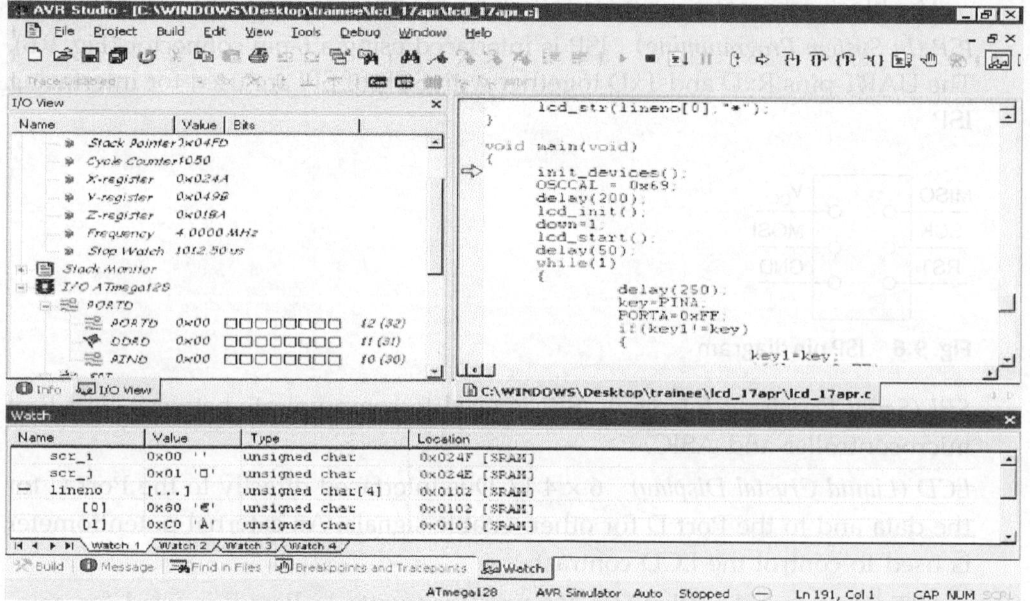

Fig. 9.7 (a) AVR studio 4: screen

Output tabs It is a collection of several views integrated into one tabbed frame. The tabs on the frame select the view from

- *Build:* Output from the compiler/assembler is routed to this window, the result of the compilation/assembly can be read here.
- *Messages:* This window is the common window for all modules of the software to present messages to the user. Messages are colour coded. Most messages are plain messages of no significant priority and with no colour. Warnings signalling potential problems are yellow. Errors are red.
- *Find:* The result of the search is routed to this window.
- *Breakpoints:* Lists all active breakpoints in all modules. Breakpoints can be enabled, disabled, and removed.
- *Workspace tabs:* Workspace view consists of a number of views that are intended to help the developer to debug the code he has written in a systematic way.

Project view The project view lists the files that are included in the code project, if it has been created a code project. It shows the name of the currently loaded object file as well as the source files that the object file refers to.

I/O view

- Register file section: All AVR controllers have a set of 32 general purpose registers that freely can be used by the programmer or the compiler. Every time a simulation or an emulation breaks, the view is updated with the current values of the internal registers in the part. If the value has changed since the last break, the register appears colour coded.

- Processor core register section: The processor registers are updated when the execution breaks.
- I/O register section: Each different AVR device is defined through the peripherals the device supports. An ATmega128 has a completely different set of peripherals compared to an AT90S8515, but the core and the instruction sets of the two chips are quite equal. All peripherals are controlled through 8 or 16-bit registers that can be read and/or written. The I /O view is configured for the selected part and shows all registers and bits logically grouped. Bits and registers can be both read and written when the emulation is in break mode. This view gives total control over the device subjected to debugging.

Info   It has two sections—the interrupt vector section which lists all interrupts and their corresponding interrupt vector addresses, and the available packages section providing list of the available packages of the device with the corresponding pinout of each package. The register view lists all registers from high to low address.

Watch   Debugging high level languages as C/C++ need to watch variables. Click on the variable under watch and drag-and-drop it into the watch window. If the variable is a struct or an array a [+] symbol will occur in front of the variable, indicating that it can be expanded in the view. When the execution breaks, the variable automatically will be updated—if it is in SCOPE.

Memory view   The code executed lies in the program memory, the variables are located in SRAM (mainly), the I/O registers are mapped into I/O-memory area, and the EEPROM is yet another memory area. The memory view is able to display all the different memory types associated with the AVR devices. And as for the watch and the I/O view, the area in display on the screen is automatically updated when an execution break occurs.

9.2.5 ImageCraft C Development Environment (ICCAVR)

The ImageCraft C Development Environment is a program for developing microcontroller applications using the ANSI C language. Its main features are windows integrated development environment (IDE) with integrated editor and project manager. Source files are organized into projects. Editing and building can be done wholly within the environment. Compile time errors are displayed in the status window. The integrated project manager generates a standard makefile. The IDE drives an ANSI C command line compiler that is normally transparent in operation. The compiler is a set of native 32-bit programs and understands long file names. As shown in Fig. 9.7(b), the IDE has three window panes.

Editor and terminal tabs   The top left window pane is the editor and includes the terminal tabs. The editor is capable of syntax highlighting C elements, plus bookmarks, and other features. When opened, the built-in terminal emulator also displays as one of the tabbed windows in the editor pane. It is the main area of interaction between user and the IDE. It contains a list of open files in a tabbed notebook control. If the IDE built-in terminal emulator is invoked, it will be opened in this section as well.

Project manager   The project manager pane contains two tabs: one contains the list of C and assembly files in the project and the other is a "browser" view of the project, listing the defined functions and variables. The project manager allows user to group a list of files into a project and defines compilation options for them. It allows user to break down the program into small modules. On build only source files that have been changed are recompiled. Header file dependencies are automatically generated. The project manager creates a makefile in standard format.

Status window   Status bar displays useful information such as the full file name of the currently active editor, the cursor position, and the full file name of the project file. The status window displays the status information from the IDE, such as from a project build. Compiler error messages start with "! E.." and are tagged with a small red sign on the gutter.

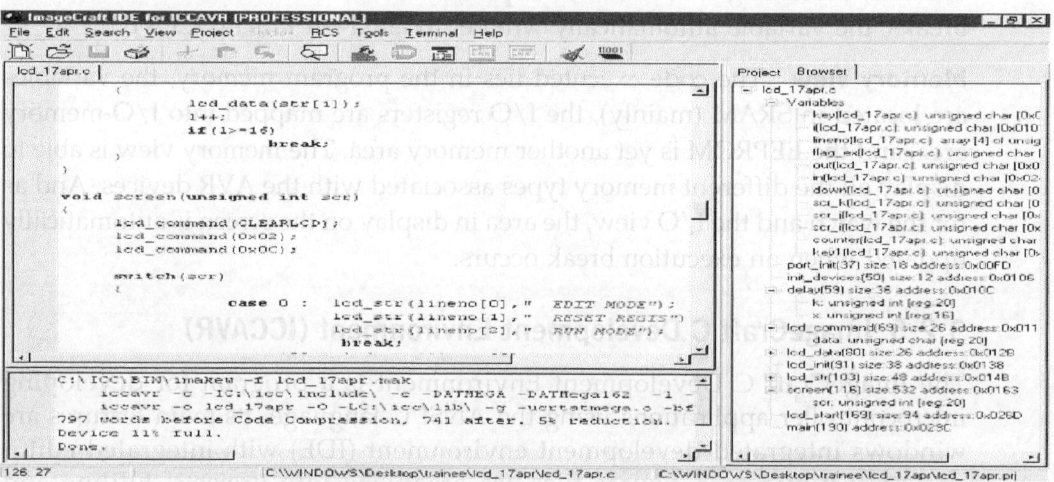

Fig. 9.7   (b) Screenshot: IDE-ICCAVR

9.3 | SOFTWARE DESIGN

The software may be developed in the modular form consisting of modules such as: main, initialization, SPI, LCD display, and UART. The steps to be carried out in each of the module are as follows:

Main Module

1. Define variables, initialize the microcontroller board.
2. Display 'START' message on LCD.
3. Start timer, send first byte on SPDR to start SPI communication with the help of timer interrupt.
4. Wait till all register data through SPI interrupt.
5. If UART receive is complete → Perform UART function.
6. Carry out calculations and display the result on LCD.
7. Goto Step 5.

Initialization Module

1. Disable all interrupts, initialize ports and directions: Port x, DDRx.
2. Initialize SPI in master mode and enable done SPI interrupt.
3. Initialize UART for 9600 buad rate, 9-bit format without parity. Enable transmit and receive interrupts.
4. Initialize timer for the count of 320 ms.
5. Initialize LCD for four line mode.
6. Re-enable all the interrupts.

SPI Module

1. If interrupt from main module, goto Step 3.
2. If interrupt from timer (every 320 ms),
   Start timer again, send first address byte on SPDR.
3. Receive byte in SPDR.
4. Interrupt microcontroller for ISR.
5. Capture received byte.
6. If required number of bytes for register are not present → transmit another byte read, goto Step 7.
7. If all registers are not read → transmit next register address, return.
8. Stop SPI return.

LCD Module

1. Display START screen.
2. Wait for key pressed.
3. Identify key: keyin, keyout, or keydown.
4. Display corresponding Menu Return.

UART Module

1. Receive data in the buffer.
2. Generate receive Rx interrupt.

3. Receive byte.
4. Update byte count according to the function selected.
5. If all byte are not received → goto Step 2.
6. Check the function code.
7. Perform operation as per code and prepare the transmit data.
8. Transmit data on UART.
9. Repeat Step 8 till all bytes are transmitted Return.

9.4 | PREPAID METERING SYSTEM

The concept of prepayment relates to paying in advance for the electricity to be consumed as per customer requirement and concurrence. Prepayment metering systems amalgamate the benefits of electronic meter with the meters of prepayment concept. The consumer purchases electricity and gets a smartcard/key/token receipt with the meter/identification no., electricity credit, and other details of customer on payment. In order to transfer the credit, consumer inserts the card/ key or punches the code on the keypad. The meter reads the data and when the energy has been used up he gets disconnected and has to get credit again from the vending machine.

Prepayment metering has been in use in the world over for over 70 years with variations developing over the years. The earlier electromechanical prepayment meters considered of kWh meter with a switch. Variations have developed over the last 4 or 5 years there have been sea change in the technology and its applications. Magnetic card, smartcard, smartkey, keypad meters are now available. One of the major obstacles to prepayment meters is the cost of a customer site equipment. The installed cost of the meter plus consumer display is around 20 times more than the cost of a conventional watt hour meter.

The basic prepayment system (Fig. 9.8) consists of the following:

- An electricity dispensing unit
- Vending machine
- Smartcard/smartkey/token card
- System master station (SMS)
- Software for vending tariff, generation, and revision

The meters/dispensing unit could have low credit warning, emergency credit and display facilities, customer record, load limit, and control facilities. The sales vending machines are to be networked with the central office of the utility for accounting and control purposes. All components in the system from the card/receipt to the main station are designed and developed to ensure highest possible level of DC current in operation, transmission, storage, and data processing. This level of security is achieved by employing proprietary algorithms, strong data encryption techniques (triple DES), multilevel passwords, software protection, etc.

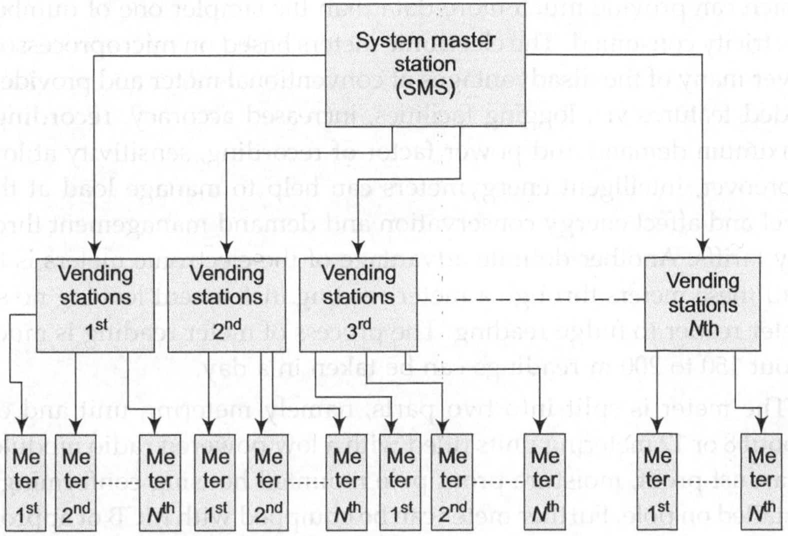

Fig. 9.8 Block diagram: prepaid energy system

A traditional electronic prepayment metering system operates on three different levels.

1. The meter installed at the consumers' premises is at the lowest level.

2. The next level is the vending stations, which are located at the utility's office or at the appointed agent's office. The communication between the vending stations and the meters is in the form of a token, which is used to top up the credit in the meter as well as to transfer or download information to the meter and, in some cases, upload information (depending on the choice of token) back to the vending station.

3. At the top level is the system master station (SMS) or master client, which is necessary to ensure a common database for reporting as well as to provide the total management, administration, financial and engineering control. The SMS communicates with the various vending stations (vending clients) via modem. Information on the consumers, tariff changes, etc. are communicated to the vending station and details of consumer sales is communicated back to the SMS.

It should be remembered that a prepaid metering system is not just an alternative to conventional metering but a complete system that includes the billing and control of the system.

9.5 | ELECTRONIC METERS

Recent advances into the electronics era have ushered in sophisticated technology which can provide much more data than the simpler one of number of units of electricity consumed. The electronic meters based on microprocessor technology cover many of the disadvantages of conventional meter and provide many value-added features viz. logging facilities, increased accuracy, recording of tamper, maximum demand and power factor of recording, sensitivity at low loads, etc. Moreover, intelligent energymeters can help to manage load at the consumer level and affect energy conservation and demand management through time of day tariffs. Another definite advantage of the electronic meters is the ability to read these meters through a meter reading instrument leaving no scope for the meter reader to fudge reading. The process of meter reading is much faster and about 150 to 200 m readings can be taken in a day.

The meter is split into two parts, namely metering unit and display unit. About 8 or 12 metering units filled with a low powered radio module are housed in a dust proof, moisture proof pole mounted housing conforming to IP54 and installed on pole. Further meter can be equipped with MCB of appropriate rating to disconnect the service in the event of any abnormal condition. The user interface fixed with energy recording device/counter is installed in each consumer's premises, which are connected to the pole. A meter reading instrument or computer is provided to meter reader to capture the meter readings. The system is similar to walk by system of remote reading. The advantage of the system is that it effectively controls the theft of energy and at the same time makes the meter reading simple and effective. Though the system is expensive the cost will pay for itself through increases in revenue due to prevention of pilferage of energy.

In the deregulated electricity supply regime, bulk power changes several hands from the generating company to the transmission and distribution companies. Metering assumes great significance at each of these bulk power transfer points. Apart from being highly accurate and reliable, these metering systems must collect, compute, collate, communicatea and present data in a form amenable to analysis and usage. Some key features to be considered while selecting these systems are discussed below.

System accuracy   One of the crucial requirements for metering at bulk power points is the accuracy of measurement. Since the revenue involved is enormous, even small measurement errors can contribute to significant differences in billing. Traditionally, class 0.2 s meters with class 0.2 s potential transformers and class 0.2 s (or at best 0.1 s) current transformers are used for these applications. Meter and instrument transformer inaccuracies, together with the losses on the potential transformer cables, result in an overall system accuracy of about 0.5 percent at full load. In many poorly selected and installed metering systems, the total accuracy may be as low as 1 percent. This may contribute to significant loss of revenue for bulk power transfer points.

Four-quadrant metering and high reactive accuracy It is important to carry out full four-quadrant metering for bulk power exchange points. In many inter-utility bulk power tie lines, conditions of line float are encountered when the active power transfer (import or export) is very low and yet the reactive power transfer (lag or lead) is significantly high. The metering system should be able to measure accurately in all the four quadrants so that a just tariff regime can be agreed upon and applied.

High degree of serviceability The need to have a high uptime for grid metering systems can hardly be overemphasized. The equipment used for such applications is exposed to the harshest environments. It must, therefore, be designed to withstand the high level of surge and switching transients that they are expected to encounter. In case of any eventuality, the meters should be amenable to easy removal and replacement. A modular construction with 'hot pluggability' is the order of the day.

Data communication and data security Comprehensive data storage is a must for managing complex energy flow across networks. All data pertaining to metering, tariff, billing, and demand and load profiling should be aptly communicated for further analysis and use. All control and data transfer functions should be handled through reliable digital data communication methods.

High degree of configurability Different utilities and power producers have different power purchase agreements and tariff structures. The metering system chosen must be amenable to configuration of the definitions of the energy register types, time-of-day registers, rate registers, billing parameters, load surveys, event logging, alarm annunciation, etc.

Self-diagnostics and ease of testing Metering equipment needs continuous monitoring. Self-diagnostic alarm annunciation should therefore be employed. Such alarms may include watchdogs, auxiliary power failure, phase imbalance, etc.

9.5.1 Use of Information Technology in Metering

The SEBs lose as much as 30 to 40 percent of the electricity generated because of power theft. The situation is made even worse by the infrequent meter reading and long lead-time between reading and billing. The chief culprit is the lack of information technology. The SEBs still rely, to a large extent, on manual systems and clerical entry. They can improve their cash flow significantly by computerizing the meter reading, billing, and invoicing process. Meters are looked upon by SEBs as stand-alone devices to measure the energy consumed. Modern metering solutions, in fact, are more than just measurement devices. They can detect theft and tampering which can help to minimize distribution losses. They can also record and store all kind of usage data. Moreover, intelligent energymeters can help manage load at the consumer level.

Meter reading instruments A meter reading instrument (MRI) is a data collector that retrieves usage information from the meter using optical sensors. Typically, the meter reader takes this instrument to the customer site and connects it to the meter. Data is transferred from the meter to the MRI. The meter reader brings the MRI to the billing station. The data is downloaded into the invoicing system. There are several benefits from using MRI instruments. To begin with, there is less scope for SEB staff to fudge the reading. Besides, the actual process of meter reading is much faster because it relies on optical sensors. With the help of an MRI, 150–300 readings can be recorded in a day. There is also almost no time lag between reading the meter and billing. This is so because there is no paper work involved. Information from the meter reader can be downloaded straight into the billing system and invoices can be generated right away.

Remote meter reading If there are good communication facilities, the reading can be done automatically and remotely. In fact, the meter reader does not even need to visit the customer site to collect the billing information. The microcontroller in the meter sends the information to the billing computer which generates the invoice. Tampering and fraud is also curtailed. More importantly, the time between meter reading and billing is minimized. A typical automatic meter reader (AMR) system comprises several main components: meters, meter interface unit (MIU) or gateway device and the host computer. In a typical AMR system, meter data is passed from the meter to a gateway, which is either integrated into or attached to the meter through power lines. From the gateway, data is transmitted over communications medium to either a mobile interrogator unit or a node on a fixed network. From the interrogator, the data is typically stored until the end of the day at which point it is uploaded to a host computer via a telephone modem or serial port connection. AMR meter data can be transmitted over a variety of mediums, including telephone, power lines, satellite, cable and radio frequency.

Spot billing The utilities can go one step further by attaching a printer, portable or otherwise, to the MRI. The bill can be generated on the spot by the meter reader. This can help cut down the billing cycle from a few days or weeks to a few minutes, and thereby expedite revenue collection.

Information technology orientation   The benefits from these devices cannot be fully realised if other areas of SEB operations continue to be IT-deficient. For example, most of the SEBs have well equipped billing and invoicing systems. But the data for the bills is still being fed manually, even though it is being generated by electronic meters. The reason is that many of the MRIs are not compatible with the billing software of the SEBs. Automated billing is thus almost ruled out. In fact, in many instances, the MRI data is uploaded into the computer and a printout is generated. The data is then fed manually into the computer in the data formats acceptable to the billing software. This again underscores the need for a full-fledged systems approach to customer management.

Energymeters are not only the meters used in the distribution system. Other meters perform equally critical functions and also need to be upgraded if SEBs want to reduce T and D losses.

Grid meters   These meters are used by the transmission authorities like the Power Grid Corporation of India whenever there is an interutility transfer or sale. These will also be used by IPPs and SEBs when power is generated by the former and sold to the latter. The accuracy level will become extremely important because it will affect the revenues and the rate of return the IPPs will get.

Audit meters   These are special energymeters programmed to monitor energy losses in a power supply network. For each point of distribution, they provide the energy inflow and outflow. Whenever there is a discrepancy beyond the normal range, the system overseers are informed. It is essential that these meters be installed at every point of distribution for the SEBs to reduce transmission and distribution losses.

Demand meters   Demand meters are intelligent devices which continuously guide the electricity user to avail of its contract demand to the maximum limit even under complex loading conditions without exceeding it. It thus protects the user from the penalties arising from exceeding the maximum demand. This device is equally helpful for the electricity supplier for ensuring that the power system operates at constant loading.

Summation meters   Summation meters are used for error-free summation of four quadrant energy and computation of concurrent demand. The devices are essential whenever there is a bulk power exchange or import/export of energy, like sale of power by IPPs to SEBs or by central units to SEBs.

9.6 | PREPAID ENERGYMETER: SYSTEM OPERATION

Operation of the system is as follows:

1. The load is connected through the meter in the form of bulb. Now the card is inserted in the card holder. And when the supply is switched ON, the bulb receives the supply and it will glow.

2. The input supply is stepped down for metering purpose by the step-down transformer and by using rectifier AC is converted to the DC and for regulated power supply for circuit operation and the relay operation is also provided.

3. Now the supply wire is passed to the load through the current transformer (CT). So the current passing through the CT develops small voltage across its secondary, which is proportional to the current drawn by the load. Hence, if the load is increased the current through the CT will also increase and correspondingly the voltage will increase in the same proportion. But these voltage signals are too small for the metering purpose. So it is necessary to amplify this voltage signals.

4. This amplified voltage signal is given as input to the voltage controlled oscillator. It generates the pulse according to the input voltage signal. Here timer IC is also used for timing and counting.

5. Input of the VCO is also given to the BCD UP/DOWN counter which is used as DOWN counter.

6. The output data in BCD format is given to the BCD to 7-segment decoder, which displays the number of units which are on the 7-segment LED display. The meter starts down counting as the load consumes the energy when the meter indicates the zero then the large current is passed through the card and the card will be fused and hence it will become useless. So the power supply is cut-off. Power supply starts again only when new card is inserted.

7. If the card is taken out before the units become zero then the relay cuts the power supply from the load but the remaining units are still there. They will not become zero. When the card is reinserted, the meter starts counting from where it was before. And at the time of power failure the data is stored with the help of battery backup.

This operation is modified by using the microcontroller, ADC and can be made precise by interfacing the other necessary hardware.

9.7 | SYSTEM HARDWARE

This system mainly consists of four separate cards: power supply card, signal conditioning card, controller card, and display card.

Power supply card This card (Fig. 9.9) consists of step-down transformer of 12–0–12 V, then bridge rectifier, and the regulator ICs to generate the required voltages. The system needs only +5 V, +12 V, –12 V. +5 V is required for the controller card and +12 V and –12 V is needed for the signal conditioning card.

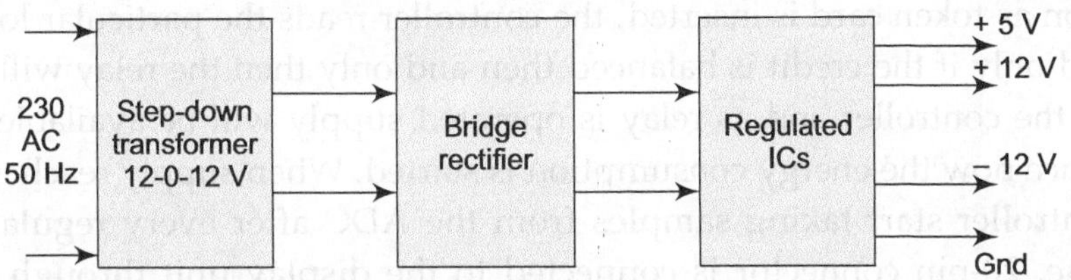

Fig. 9.9 Block diagram: power supply card

Signal conditioning card As shown in Fig. 9.10 it has a zero-cross detectors (ZCD) and precision rectifiers (PR).

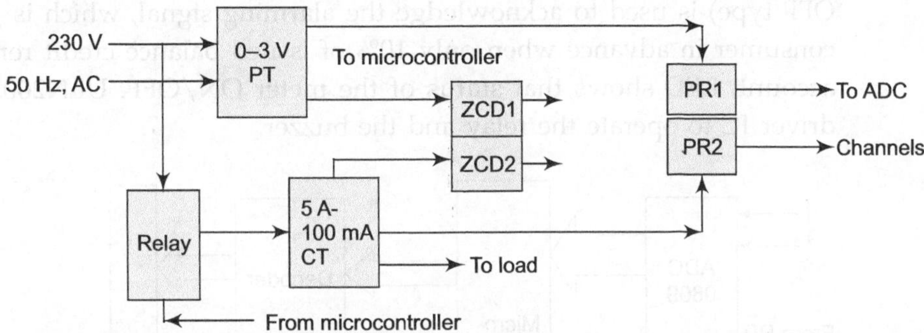

Fig. 9.10 Block diagram: signal conditioning card

One ZCD and one PR is used for voltage signal and the other ZCD and PR is used for current signal. Primary of PT (0–3 V, 1.5 A) is directly connected to the 230 V AC, 50 Hz. Primary of CT is connected to the phase through the +5 V relay and input to these ZCDs and PRs are given from secondaries of PT and CT. Outcome of the ZCDs are given to the microcontroller interrupt mechanism, which can be utilized to find phase difference between two signals. Output of PRs are given to the ADC channels which is mounted on the controller card, which will help to find the consumed energy. Here, phase is connected to the load through relay and CT. Freewheeling diode is connected across the relay for the purpose of the over current protection. Output of ZCDs are rectified only to have the positive pulses and potential divider circuit is also needed to make the signal compatible to deal with the digital ICs (i.e., TTL compatible).

Controller card (Fig. 9.11) This is a main hardware of the project, two separate cards are mounted on a single card. One is a token card and the other is a controller card. It consists of ADC (0809), microcontroller (89C51), decoder (74LS138), inverter (4093), ULN2003, 20-pin connector, 6-pin connector, reset switch (S), two user defined switches (S0, S1), and a buzzer operate over a range from 3 V to 24 V. As token card is also included, it consists of EEPROM also. As soon as token card is inserted, the controller reads the particular location and if and only if the credit is balanced then and only then the relay will be operated by the controller and as relay is operated supply will be available to the load, hence now the energy consumption is started. When supply reaches to the load, controller start taking samples from the ADC after every regular interval of time. 20-pin connector is connected to the display unit through the decoder which provides the necessary signals, refreshing technique is used for the display, 6-pin connector is used to connect the EEPROM which acts as token card. Inverter IC-4093 is used to provide the clock signals to the ADC. Switch S0 (ON/OFF type) is used to display the balance credit when it is pressed. Switch S1 (ON/OFF type) is used to acknowledge the alarming signal, which is given to the consumer in advance when only 10% of issued balance credit remains in the account. LED shows that status of the meter ON/OFF. ULN2003 is used as driver IC to operate the relay and the buzzer.

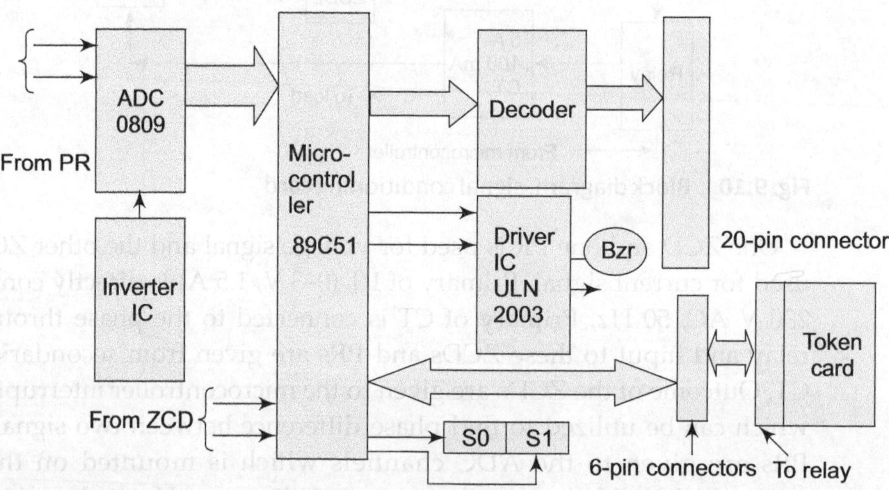

Fig. 9.11   Block diagram: controller card

Display card   This card contains (Fig. 9.12) six 7-segment LED units along with its driver circuit and a 20-pin connector which is connected to the 20-pin connector of the controller card using flat ribbon cable.

The prototype module is designed using microcontroller (89C51), EEPROM (AT 24c16), ADC–0809, Inverter IC-4093, Decoder (74LS 138), ULN 2003, transformers CT/PT, zero-cross detector (ZCD), precision rectifiers (PR), relay, display unit, load, and power supply unit.

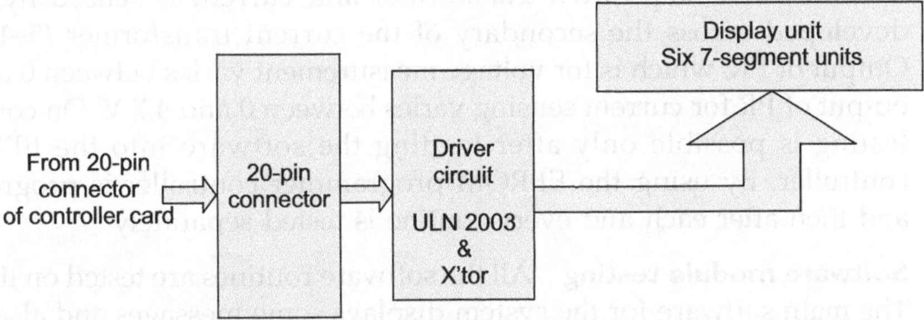

Fig. 9.12   Block diagram: display card

9.8 | TESTING AND TROUBLESHOOTING

In any system design, testing is the main phase. The following steps are suggested for testing the prototype, starting from power supply to final firmware:

Testing of power supply unit

1. First step in any system is to test AC voltage input to the bridge rectifier. Here if the output of 12-0-12 transformer is 13 V then the DC voltage comes out from bridge rectifier is 13 V.

2. In second step one has to check whether the regulator outputs are +5 V, +12 V, and −12 V or not. Here output of 7805 is +5.01 V and that of 7812 is 12.02 V and the output of 7912 is −1.91 V.

3. Checking of each point at which one required to have +5 V, + 12 V, and −12 V on the board by checking each and every points where appropriate voltages are needed. After successful checking all points for +5 V, for +12 V, and for −12 V and if there is any problem then check tracks on the board and jumpers and also check the connections, there should not be any dry solder or any connection should not be left unsoldered. Also check the components whether they are in working condition or not.

4. Transformer should be capable to drive relay without any loading of it. This can be tested directly by testing the relay's ON/OFF operation.

Testing of signal conditioning and controller sections   On signal conditioning, card zero-cross detectors (ZCD) and precision rectifiers (PR) are constructed. The output of ZCD is a square wave having amplitude equal to the operating voltage of the IC LM311, i.e., +10.5 V. Before supplying this output to the controller it must be made TTL compatible by rectifying it using diode which produce +10.5 V DC.

This output is given to the microcontroller which is operated at +5 V only, hence the potential divider circuit is used and then it is connected to the controller card. Input to the precision rectifiers is analog quantities voltage and current. Then output of the PR is given to the ADC channels. Here voltage is sensed from 0 to 3 V step-down transformer and current is sensed by the voltage developed across the secondary of the current transformer (5–1 A, 15 VA). Output of PR, which is for voltage measurement varies between 0 and 3 V. And output of PR for current sensing varies between 0 and 4.7 V. On controller card, testing is possible only after loading the software into the EPROM of the controller. By using the EPROM programmer controller is programmed first and then after each and every routine is tested separately.

Software module testing   All the software routines are tested on the simulator. The main software for the system displays some messages and also checks that the card which is inserted is new, partially used or fused card by checking the balance credit in the EEPROM and accordingly it gives the messages on the display. If balance credit is present then the controller will drive the relay through the relay driver circuit.

Controller takes the samples of current and voltage from the ADC after every 0.1 s and multiply them to find the average power, by using the ADC routine. Another routine is made to find the phase angle (ϕ) between current and voltage using the interrupt mechanism of the controller. Then the program will automatically find out $\cos(\phi)$ which will be useful to find out the consumed energy. This routine performs the following calculations to find out the consumed power:

$$P = V_{rms} \times I_{rms} \times \cos(\phi)$$
$$V_{rms} = (1.11) \times V_{dc}$$
$$I_{rms} = (1.11) \times I_{dc}$$
$$P = (1.11)^2 \, V_{dc} \times I_{dc} \times \cos(\phi)$$

Then consumed power is multiplied with the time to find the energy consumed. One routine is developed to take the security measures, this routine first checks the particular (i.e., defined) location of the EEPROM and if the data stored in this location is matched with the data which is already available in the controller, the token card can be assigned to particular meter. Hence the meter identification is possible. One more software program is developed to give the alarming signal through the buzzer when last 10% of purchased electricity remains as a balance. A program is also developed to indicate that meter is ON at a glance by glowing the LED when the credit is balanced in the token card.

Testing of relay driver section   Relay driver used here is NPN darlington array and inverter, so if we apply logic HIGH (+5 V) to input of driver both transistors of that input becomes ON and output becomes low so particular relay connected to that output becomes ON. This way relay driver can be tested. Even if any problem occur then first check track on the board, if it is also okay than check IC of the driver.

Testing of display and decoder section common anode type   Displays are used in multiplexed configuration. PNP transistors and ULN 2003 are used as digit driver and they are driven by outputs of 74LS138 decoder. Testing of this section is done in two modes—static and dynamic. In static testing, logic LOW (0 V) is given one by one to digit drivers and observing 8 on 7-segment display one by one on respective digits. In dynamic testing, actual display routine is written to display say 1, 2, 3, 4, 5, 6, 7, 8 on six 7-segment display with required all necessary hardware.

Table 9.2 Jumpers and connectors on hardware modules

Hardware module	Symbol	Connector/ Terminal
Main connector blocks	C1	Mains connector
	T1	12-0-12 V transformer for power supply card
	T3	5A/1A-15 VA current transformer
	T2	0–3 V transformer
Controller card	x	J7
	y	J8
	z	J9
	v	J10
Signal conditioning card	R	J4
	S	J5
	T	J6
	u	J7
Display card	J1	20-pin connector to be connected to controller card using flat cable

9.9 | SYSTEM SOFTWARE

This section describes the algorithms for various modules developed for system operation.

9.9.1 Main Module

1. When the prepaid card is inserted, display routine will display a message: 'Prepaid Energymeter'
2. Module checks for the balance credit is present or not in the token card at some secret location defined by the utility.
   (a) If credit is not present, the message
       'card is fused, sorry no credit is present'
       will be diplayed on the display unit.

(b) If any credit is present then this credit balance available in the secret location is stored into some other temp. register. And ease the value from the secret location and store this value in some known location of an EEPROM. Here this is done for decrementing the balance credit after it is used by the consumer.

3. Set counter (ct) to 10 this counter is used to calculate the 1 s. Also initially set the power consumed (P) = 0.

4. The balance credit is present drive the relay so the supply is available to the consumer.

5. Module samples from both the channels of the ADC after every 0.01 s and calculates the average power (P_1) using the relation:

Average power $(P_1) = V \times I = (v_1i_1 + v_2i_2 + v_3i_3 + \cdots + v_{10}i_{10})/10$

6. After receiving the samples of voltage and current after every 0.1 s, they are multiplied and added to the previous value, then decrement the counter (ct) by one.

7. Power consumed is calculated: $P = P + P_1$.

8. If the counter is not zero again read the samples and go for to calculate the average power.

9. If the counter (ct) is zero, after waiting for 1 s, the phase angle (ϕ) between current and voltage is computed using the interrupts and power factor (cos ϕ) is calculated using Taylor's infinite series $(1 - x^2/2! + x^4/4!.)$

10. The consumed power $(P) = V \times I \times \cos ((\phi)$ and consumed energy $(E_c) = P \times$ count

$$\text{One unit } (E_{ut}) = kWH = 1000 \times WH$$
$$= 1000 \times 3600 \text{ W-sec} = 36 \times 10^5 \text{ W-sec is calculated}$$

11. Compare E_{ut} with E_c, if $E_c < E_{ut}$, then again take the samples from the ADC.

(a) if $E_c < E_{ut}$

(b) reduce the credit by one from the available balance credit in the EEPROM (token card).

(c) Goto Step 5 if the balance credit is non-zero.

9.9.2 Status Check

The module checks status of the token card. It identifies the token card whether it is a new card, partially used, or the fused card.

1. It checks the secret location if the credit is present in this location then it will display the message

"New card"

on the display unit and erases contents of this location.

2. If no credit is found in the secret location, it will check other two known locations if any credit is present, the following message will be displayed:

   "partially used card"

3. If no credit is found in any of these three locations then the message displayed will be

   "fused card".

9.9.3 Alarming Signal Module

This routine indicates that last 10% of total purchased electricity is left in the token card by generating the buzzer tone.

1. It checks whether the inserted card is valid or not

   if it is not valid

       no operation will be done

   Else

   it will store the balance credit from temporary register into some other register (R_p).

2. Calculates $X = 10\%$ of the content of the R_p and compares it with the updated value stored in the known location of EEPROM.

   if they are not equal

       Continues comparison

   Else

   It checks whether the acknowledgement switch S1 is pressed (pushed ON) or not, if it is not pressed

   (a) presses it

   (b) supply the logic high to generate the buzzer tone after every half an hour.

   (c) presses S1 (pushed OFF) to off the buzzer tone, which intimates that only last few units have been left in the token card for the user one.

9.9.4 Display Module

It displays the balance UNITs on the display unit having six 7-segment LEDs. It checks whether the switch S0 is closed or not.

If not → presses S0 (pushed ON), will read the updated value from the known location of EEPROM, then display this value.

9.9.5 Security Check Module

This is for the security of meter as well as card. It checks the identification number of the meter given by the utility which is stored at some defined location in the token card. This number in EEPROM and that in the controller ROM must be same for the software to execute main module.

9.9.6 Meter Status

This routine provides an extra feature which indicates that whether the meter is ON or OFF. It checks for the credit balance in the token card or not. In case there is no credit balance, a logic LOW signal will be issued to make LED OFF otherwise LED will glow continuously.

9.10 | OPERATING PROCEDURE

Figure 9.13 shows the system arrangement mounted on the wooden board.

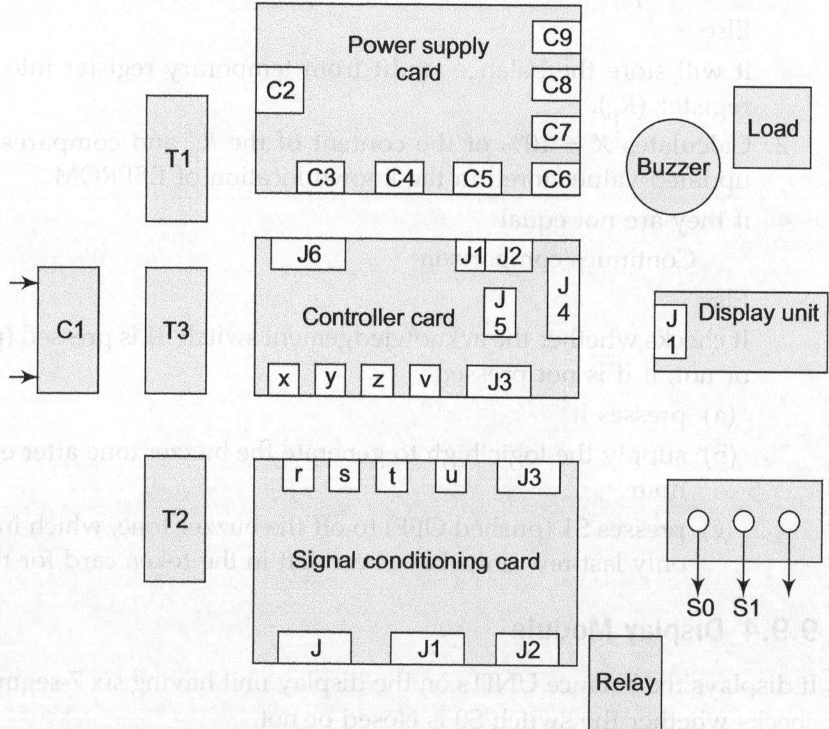

Fig. 9.13 System hardware modules mounted on board

Table 9.3 defines the jumpers and connectors provided on different hardware modules. Following steps describes how to set up and operate the system.

1. Connect the mains cord to the connector C1.
2. Connect the transformer T1 to the 3-pin connector C2 on the power supply card.

3. On the power supply card there are seven connectors, which are C3, C4, C5, C6, C7, C8, and C9 which gives the supply of +12 V, +9 V, +5 V, GND, –5 V, –9 V, and –12 V, respectively.

4. Now connect the C3, C6, and C9 to the connector J2 on the signal conditioning card.

5. Connect the phase and neutral lines to the connector J on the signal conditioning card.

6. The jumpers on the signal conditioning card and that on the controller card should be connected as shown in Table 9.3.

Table 9.3 Jumper connections

Jumpers on signal conditioning card	Jumpers on controller card
J4	J7
J5	J8
J6	J9
J7	J10

7. Connect the connector of 20-pin (J4) on the controller card to the display card.

8. Connect the jumpers J1 and J2 on the controller card to the relay and the buzzer, respectively.

9. Connect the 6-pin connector J3 on the controller card to the token card connector J5.

10. After all these check that whether all the components are mounted or not.

11. Then at last switch ON the power supply from the mains, i.e., 230 V, 50 Hz/ AC.

EXERCISES

9.1 State limitations of electromechanical meters.

9.2 Enlist advantages of electronic meters.

9.3 How do you define accuracy of meters?

9.4 What are IS and IEC standards?

9.5 Explain operations of voltage and current channels of ADE7758.

9.6 Explain the procedure of measuring active, reactive, and apparent power using ADE7758.

9.7 Is it necessary to interface ADE7758 with the microcontroller using SPI port? Why?

9.8 Define: ISP, SPI, UART, ICCAVR.

9.9 State and explain features of AVR studio.

9.10 What is the advantage of using ICCAVR?

9.11 Draw a schematic diagram of interfacing ADE7758 with the microcontroller using SPI.

9.12 What is the significance of IN system programming? Enlist the pins used for interfacing ISP.

9.13 Use algorithm described in Section 9.3, write a program module to initialize the controller card.

9.14 What are the parameters and variables required for initialization of the controller board?

9.15 When does SPI start receiving data?

9.16 Write a program module to identify the key pressed and display corresponding menu.

9.17 Write a program module for communication through UART.

9.18 What is the concept of pre-payment?

9.19 Explain the operation of a prepaid energymeter.

9.20 Discuss levels of pre-payment metering system.

9.21 Discuss the features used in selection of type of metering system.

9.22 Discuss advantages of embedding information technology in metering system.

9.23 Explain role of ZCD in the signal conditioning card of Fig. 9.10.

9.24 Describe test procedure for testing power supply unit.

9.25 Write program module to check present credit balance and display on LCD.

9.26 Write program module to check status of token card of prepaid energymeter.

9.27 Explain need for alarming signal in prepaid systems.

9.28 Write program module to generate appropriate message on display for low balance.

9.29 Identify the signals available on 20-pin connector of Fig. 9.11.

9.30 Modify setup of Fig. 9.12 to interface LCD instead of LEDs.

9.31 The microcontroller 89C51 of Fig. 9.11 is to be replaced by one having in built ADC. Draw modified setup using new microcontroller.

9.32 Write a program module to sample current and voltage channels of ADC at every 0.1 s.

9.33 Write program module to calculate and display consumed power.

Chapter **10**

Application II: Power Line Communication

In the present age of information and technology, the focus is both on creation as well as dispersion of information. The popular technologies currently used to reach the end users for the provision of information include telephone wires, Ethernet cabling, optic fibre, and wireless satellite technologies.

The existing power lines for transmitting data and voice are particularly interesting since these provide customers an alternative to traditional networks. In many countries, power delivery network is ubiquitous and this fact has a tremendous impact on communication market. The advantage of using electric power lines for data transmission is that every building and home is already equipped with the power line and connected to the power grid. The power line communication system (e.g., building a home network) uses the existing AC electrical wiring as the network medium to provide high-speed network access point almost anywhere there is an AC outlet. In most cases, building a home network using the existing AC electrical wiring is easier than trying to run wires, more secure and more reliable than radio wireless systems like 802.11b, and relatively inexpensive as well. For most small office home office applications, this is an excellent solution to the networking problems.

The potential of power lines as a powerful medium to be able to deliver not only electricity or control signals, but even full-duplex high-speed data and multimedia content is being explored. In case of PLC access systems, a low voltage power supply network is used for connection of end users/subscribers to a communication network. The subscribers in a low voltage electrical power supply network share the transmission capacity of PLC network. A high-gross data rate on the medium is necessary to ensure a sufficient quality of service (QoS), making PLC system competitive to the other access technologies. Because of the shared transmission medium, PLC networks have to provide a very good network utilization besides an efficient QoS.

10.1 POWER DISTRIBUTION NETWORK FOR PLC

The terms power line carrier (PLC) communication systems, or residential power line circuit (RPC), or distribution line communication (DLC) systems refer to the low voltage part of the electrical power distribution network. It comprises components/devices attached to the secondary side of the distribution transformer, i.e., the medium voltage (MV) to low voltage (LV) transformer, including the low voltage network within the consumer's/customer's premises and all the loads attached to it. Figure 10.1 shows a typical electric power distribution network for a PLC. Typical characteristics of power lines are:

- Three-phase system, 440 V between phases; loads are typically connected between a phase and neutral. Heavy loads are connected between two phases.

- Operating frequency: 50 Hz.

- Typically 400 houses are connected to a single distribution transformer in a city environment; these houses can be found in a circle with an average radius of 400 m.

10.2 DIGITAL COMMUNICATION SYSTEM

The purpose of any communication system is to transmit information from a source station to a destination through a communication channel. Figure 10.1 shows a very simplified block diagram of communication system which basically comprises information source, a transmission medium, and information destination.

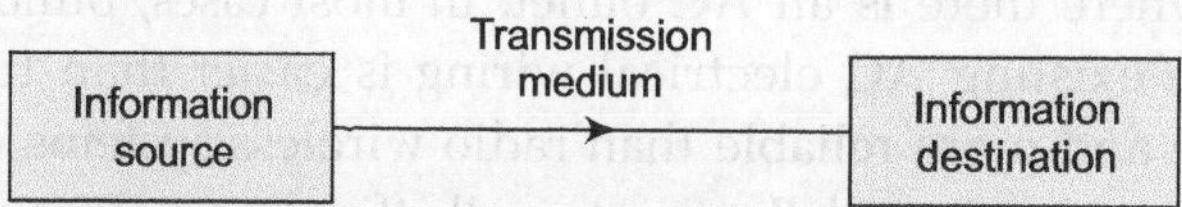

Fig. 10.1 Basic communication block

The source and destination are two separate points and this long distance between them is responsible for the interference/noise signals affecting the desired signal being transmitted (from source to destination). Information may be in the form of symbol that can be analog such as voice, video, music, or digital (discrete in time). Quite often the source information is unsuitable for transmission in its original form and must be converted to more suitable form prior to transmission, i.e., analog information may be required to be converted to digital form for digital communication and digital data may have to be converted to analog form for analog communication prior to transmission.

The term digital communication covers a broad area of communication technique including digital transmission and digital radio. Digital transmission involves transmission of digital pulses from a digital source or transmission of analog signals (after converting these into digital pulses) from analog sources. Digital radio involves the transmission of digitally modulated analog careers between two or more points in a communication system. Digital transmission systems require a physical channel between the transmitter and receiver such as a metallic wire, a coaxial cable, or an optic fibre cable. In digital radio system, the transmission medium could be free space or a physical channel such as a metallic or optical fibre cable.

Figure 10.2 shows the block diagram of the digital communication system. The transmission of information should be as fast as possible, error-free, and accurate. But due to non-ideal channel, noise and interference in the communication system, the received signal will not be same as the transmitted signal, it is always different from source signal.

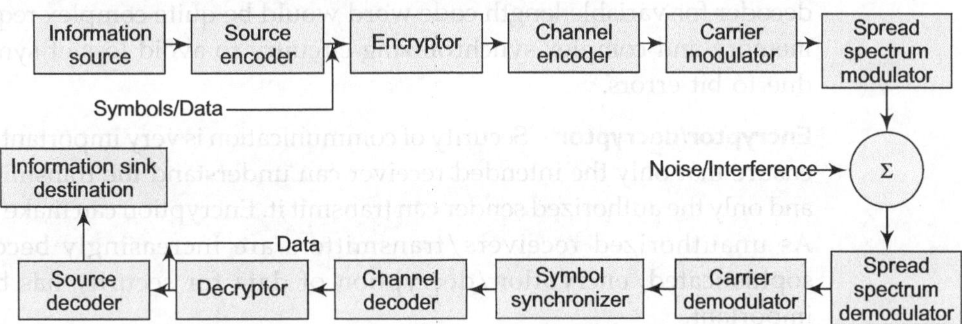

Fig. 10.2   Block diagram of digital communication system

Information source   The source of information can be analog or digital. Analog information source are continuous signals such as human speech, TV, or fax signals. A discrete information source could be a signal generated by personal computer. The analog information is converted into an equivalent digital signal by sampling and quantization. The discrete information sources are characterized by source alphabet and symbol rate. Source alphabet probability and probabilistic dependence of symbol in sequence. The symbol rate is the rate at which the information source produces the symbols/characters. Its unit is symbols/sec.

Source encoder/decoder   The information generated by the source has lots of redundancy which when transmitted as it is results in inefficient utilization of the bandwidths. Source encoding is required to eliminate or reduce the redundancy in the message so as to provide an efficient representation of the information source. The source encoder convert symbol sequence into a binary sequence by assigning code word to the symbols in the sequence. The simplest way in which a source coder can perform this operation is to assign a fixed length binary code word to each symbol of the input sequence. The important parameters used for source coder are block size, code word length, average data rate, and encoding efficiency.

- Block size gives the maximum number of distinct code word that the encoder can represent and depends on the number of bits in the code word.
- The code word length is the number of bits used to represent each code word.
- The average data rate is long-term data rate that can be achieved by the encoder. The coding efficiency is the ratio of actual data rate to the minimum achievable data rate.

The source decoder at the receiver converts the binary output of the channel decoder into a symbol sequence. Memory can be employed to store the code words at the receiver. It is possible to design both fixed length and variable length decoders. Encoders/decoders can be synchronous or asynchronous. The decoder for variable length code word would be quite complex requiring larger memory and complex synchronizing circuitry to avoid loss of synchronization due to bit errors.

Encryptor/decryptor   Security of communication is very important and we must ensure that only the intended receiver can understand the transmitted message and only the authorized sender can transmit it. Encryption can make this possible. As unauthorized receivers/transmitters are increasingly becoming more sophisticated, encryption/decryption of data for security has become very important.

Channel encoder/decoder   The channel encoder provides a different type of communication security than that provided by the encryptor. It increases the efficiency and/or decreases the effects of transmission errors by adding some known redundancy to the message. Error control by channel coding is achieved by adding extra bits to the output of the source encoded/encrypted data. The redundant bits added do not convey any information but enable detection and correction of errors in the message bits at the receiver. Block code and convolution code are the examples of the channel codes.

Carrier modulator/demodulator   The output of the channel encoder is a digital signal composed of symbols. These digital symbols can be transmitted directly on a physical channel (wired channel) after passing it through a line coder (RZ-return to zero, NRZ-non return to zero, Manchester biphase, etc.). In such cases, the transmission is referred to as a base band modulation since the base band signal has been digitized/coded and transmitted at the base band frequency only. However, it is possible that this digital signal is carrier modulated for wireless transmission.

Carrier modulation enables efficient transmission of the signal by minimizing the effect of channel noise, matching the frequency spectrum of the transmitted signal with channel characteristics, and providing the capability to multiplex several signals.

Carrier modulation involves producing an analog waveform corresponding to the discrete symbols at its input. Parameters of modulation are type of modulation (ASK-amplitude shift keying, PSK-phase shift keying, FSK-frequency shift keying, etc.), power level, bandwidth of the modulated signal, and frequency of the carrier signal.

The demodulator performs the reverse operation, i.e., it extracts the message signal from the information bearing carrier signal of the modulator. The receiver is simply a mirror image of the transmitter.

Spread spectrum modulator/demodulator Encryption makes the transmission secure from unauthorized receiver/transmitters. Additional security may be provided by use of spread spectrum techniques where an unauthorized listener is prevented from even distinguishing between the various symbols, often mistaking the received signal for wide band noise. It also provides anti-jamming, multiple access, and interference rejection capability.

The communication channel It is a link between the source (Tx-transmitter) and the destination (Rx-receiver). Due to the non-ideal nature of the channel (physical limitation and finite bandwidth) information signals suffer from amplitude and phase distortion as they travel over the channel. The signal power is further decreased due to attenuation in the channel due to the effect of noise (man-made/natural/thermal add-white Gaussian Noise (AWGN)).

The various channels are
- Wired channel
- Wireless channel (radio channel)
- No wire channel (power line communication)

In wired channel, a physical pair of wire such as co-axial cable, optic fibre is present in between the transmitter and receiver. In wireless channel, a link such as radio channel (satellite, microwave links/free space) is present in between transmitter and receiver. Wireless communication link have multipath distortion effects in the form of fading. One of the ways to minimise the noise is to increase the signal power. However, the signal power cannot be increased beyond the certain limit because the non-linear effect is dominant as the signal amplitude is increased. For this reason, signal-to-noise ratio (SNR) can be maintained at the output of the communication channel.

In no wire channel, no specific wire is employed for communication but the electric supply line (power line) can be used as communication link between transmitter and receiver. Hence, the power line can perform the dual functions; one as electric supply line and another as communication link. In PLC (power line communication), especially design coupling network is required to interface the transreceiver and the power line.

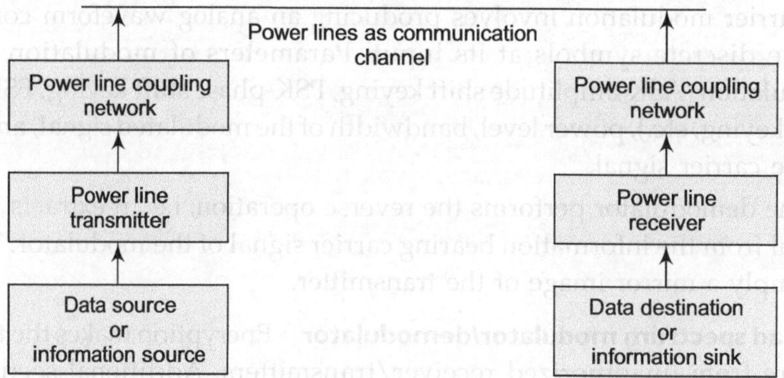

Fig. 10.3 Basic block diagram of power line communication

Figure 10.3 depicts block diagram of power line used as a media for communication, the power line coupling network interface the transmitter/receiver and the power line.

10.3 | 'NO NEW WIRE' COMMUNICATION

In present digital age of information technology there is an increasing need for sending digital voice, video, and Internet data to and around home, office, or other buildings. Installation of wires to support this is expensive, disruptive, and time consuming process. In the context of a home networking environment, 'no new wires' is the term applied to a suit technologies that utilize the existing wiring systems to distribute high-speed data and video throughout the home (or small office). Phone line and power line systems are two dominant 'no new wires' technologies. The functions of electric utility providers can be classified into three major types: power generation, transmission, and local distribution. The utilities can take advantage of the existing wiring infrastructure for provision of certain services. Telecommunication carriers are interested in a reliable way to move their content and services to the various devices in the home.

As present broadband networks establishing new benchmarks in terms of speed and reliability, there is a rapidly growing small office home office (SOHO) networking scenario, where a consumer has two or more PCs, printers, scanners, or digital home entertainment devices. The need for enabling all these devices to communicate to each other as well as the Internet, along with the control of home appliances by the consumer are some of the driving factors demanding a home networking solution. Home networking has long been characterized as a solution in search of a problem.

Commercial networks are being designed specifically to carry data between computers. They typically use fibre optic, twisted pair, or coaxial cables to minimize noise and interference over the communication medium on the network. Most homes today do not have dedicated high-speed network cabling installed and the labour costs requirement for the installation of such dedicated cabling is very high for the home owners to fund on their own. Home presents some novel and unusual challenges that have not been primary concerns in network deployments until now. For home networking to be successful, solutions must exist that utilize the existing wiring infrastructures. Home networking techno-logies are based on the following criteria:

- Existing wiring infrastructure should be utilized by the technology
- Ease of installation and maintenance
- Ease of use and simplicity (use of existing standards and software platforms)
- Quality of service (QoS) mechanism should be included providing low latency for telephony and other voice application
- Data rates of 10 Mbps or higher should be supported to allow consumers distribute multimedia in real time
- Extensibility
- Data type versatility (audio, video, etc.)
- Should provide automatic security to protect intrusion and leakage of data
- Technology needs to be relatively cheap to other existing solutions

10.3.1 Home Networking Technologies

Many types of broadband home networking technologies are now available. With each type addressing certain user needs and application requirements, none has been comprehensive enough to satisfy the need for all applications and thus new technologies are being built constantly to better address the needs. Practically the ideal solution is a combination of technologies that would be used in many homes. Broadband home networks (BHNs) can operate over various physical media. BHN fall into three major categories: structured wiring, existing wiring, and wireless.

Structured wiring   It requires the installation of new cabling in the walls. Structured wiring provides high bandwidth and excellent security. To handle the full range of current applications, a complete installation for today requires several cabling types, including UTP for telephone and data, and coax for video.

Fast Ethernet at 100 Mbps over UTP is widely used for data applications. While having sufficient bandwidth for video, it does not currently include the QoS support. With the introduction of HD video to home, it is believed that the home backbone network would be required based on the structured wiring to interconnect sections of the home. EIA and CEA are developing the R-7.4 VHN home network standard for this purpose.

Existing wiring It makes use of electrical, telephone, or coax wiring already installed in the walls. As the structured wiring is relatively expensive to install in the existing home, various companies now develop technologies based on the existing wiring in the walls of the home.

Phone line technologies use the existing telephone wiring. Home Phone Networking Alliance (HomePNA) has recently defined a 3.0 specification that reaches an unprecedented data rate of 128 Mbps with optional extensions reaching up to 240 Mbps. HomePNA technology complements wireless networking technologies providing the ideal high-speed backbone for a home multimedia network requiring a fast and reliable channel to distribute multiple, feature-rich digital audio and video applications throughout a home. The International Telecommunication Union (ITU) has already adopted global phoneline networking standards G.989.1, G989.2, and G989.3 based on the HomePNA 2.0 specification. The HomePNA 3.0 physical interface is based on version 2.0 physical layer technology and is fully backwards compatible and interoperable with HomePNA version 2.0 network components.

Wireless It avoids the use of wires by transmitting through the air. Wireless local area networking (WLAN) does not belong to the evolution path of the mobile networks, but started as a wireless extension to the enterprise LAN networks. Wireless networking avoids the cost of pulling new wires and the challenges of using the existing wiring. There are many competing technologies and associated standards and advocacy groups in this area.

IEEE 802.11 is a family of evolving standards, originally designed for enterprise networking and now moving into home. 802.11b at 2.4 GHz is the current version. 802.11a at 5 GHz is the future, although the proposed 802.11 g at 2.4 GHz may be a 'bridge' technology.

Bluetooth short-range radio technology makes it possible to transmit signals over short distances between phones, computers, and other devices. Bluetooth was designed for short-range personal networking and is being extended for longer ranges.

10.3.2 Power Line Networking

In the power line carrier (PLC) communication systems, the power line is used not only for energy transmission, but also as a medium for data communication. Power line networking is an emerging home networking technology that allows the end-users or consumers to use their already existing electrical wiring systems to connect home appliances to each other and to the Internet. Home networks utilizing the high-speed power line networking technology are able to control anything which plugs into the AC outlet. This includes lights, television, thermostats, and alarms among others. Power line communications fall into two broad distinct categories: access and in-house.

In-house   (Fig. 10.4) In-house power line technologies communicate data exclusively within the consumer's premises and extend to all of the electrical outlets within the home. The same electrical outlets which provide AC power are acting as access points for the network devices. The access and in-house power line networking solutions both send data signals over the power line, however, the technologies differ fundamentally. The focus of access technologies is on delivering a long-distance solution, competing with xDSL, and broadband cable technologies. The in-house technologies focus on delivering a short-distance high-bandwidth solution ($\geq$10 Mbps) that competes with other existing in-home interconnection technologies like phone line and wireless.

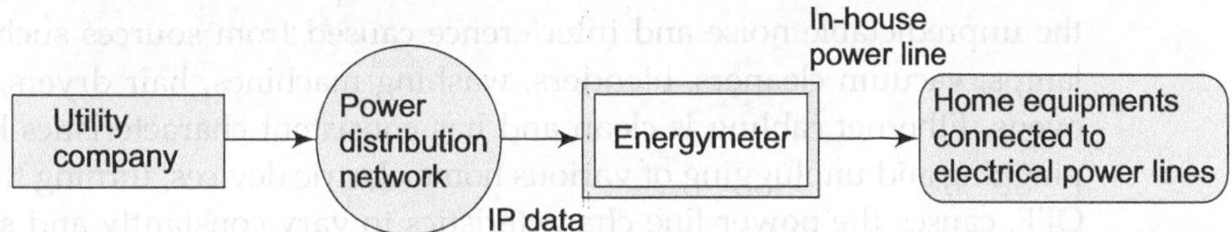

Fig. 10.4   In-house AC power line technology

In the in-house or low voltage (LV) power line technology, to support the data transmission over the power grid, a power line carrier controller (PLC controller) is installed, typically in the local transformer. The PLC controller is coupled upstream to a telecommunication network, either via traditional methods, or using some innovative/proprietary solution from the energy provider. Downstream, the PLC controller handles the data transmission (Internet and Voice) over the existing low voltage.

Access   Power line technologies are responsible for sending data over the low-voltage electric networks that connect the consumer's home to the electric utility provider. The power line access technologies enable a 'last mile' local loop solution which provides individual homes with the broadband connectivity to the Internet. The access or medium voltage (MV) power line technology is capable of providing broadband data transmission and provides that extra link where the telecommunication network does not reach without expensive infrastructure extensions. Broadband power line communication systems are commercially available. They provide data transmission over the low voltage power grid from the low voltage transformer station to the household power socket. Connection to the telecommunication network presently requires a direct connection via fibre, copper pairs, or wireless. Power line communication provides power grid owners with new, interesting business opportunities.

The medium voltage power line communication solution, combined with low voltage power line to the home allows the power utility to offer cost-effective, wide-coverage, broadband data services. The medium voltage power line solution closes the gap between the low-voltage network and the telecommunication network. Using power line modules at home, or in the office, provides network facilities for PCs and printers just by plugging into the in-house low-voltage network. Direct connection of a telephone to the power line modem provides voice connection. The low voltage in-house network is easily transformed into a local telephone network simply by using a number of power line modems plugged into power sockets as needed.

Communications over power line is different than communications over dedicated network wirings. Power line presents a difficult medium for transmission of information. Communication over the AC power line is difficult because of the unpredictable noise and interference caused from sources such as halogen lamps, vacuum cleaners, blenders, washing machines, hair dryers, microwave ovens. Ethernet cabling is clean and has consistent characteristics but constant plugging and unplugging of various home electric devices, turning them ON and OFF, causes the power line characteristics to vary constantly and significantly.

The residential power circuit (RPC) or the distribution line carrier (DLC), i.e., the low-voltage end of the power line communications, network generally consists of three major parts.

- *Multifunction node (MFN)*   It is a unit installed in each household either as an integrated or separate part of the electricity meter. It reads the meter value each hour and stores it in a memory. The memory can store up to many days of meter values.

- *Concentrator and communication node (CCN)*   It manages all MFNs in an area, e.g., a low-voltage grid, and it is responsible for collecting their meter values. It is installed in a sub station and it consists of a PC.

- *Operation and management system (OMS)*   It manages a set of CCN. The meter values collected by the CCN are stored in the OMS where they can be accessed and analysed.

10.3.3 Power Line Communication Technologies

The technologies and standards used presently in the power line communications include X-10, OFDM, and the HomePlug standard. The quality of service, data transmission rates, the limitations, the drawbacks and other important factors are taken into account.

X-10 It is a communications protocol that allows compatible home networking products to talk to each other via the existing home electrical wiring. The X-10 code format was first introduced in 1978 by X-10 Inc., for the Sears Home Control System and the Radio Shack Plug 'n' Power System. The 25 years old X-10 technology was initially developed to integrate with low cost lighting and appliance control devices.

X-10 enables the X-10 compatible devices, which are electrical components directly plugged into wall outlets, to communicate with each other. These devices are susceptible to damage by voltage spikes. Signal attenuation and line noises generated by household appliances or external sources can transiently interfere with X-10 communications. Due to interference of neighbourhood, reliability remains a major issue in X-10 power line networking. Complex and unanticipated faults are unavoidable in X-10 networking.

The X-10 modules are adapters connected to outlets and controlling simple devices. X-10 transmission rate is limited to only 60 bps, which makes it unsuitable for carrying Internet type traffic around the house. It is possible to control lights and virtually any other electrical device from anywhere in the house with no additional wiring.

Power packet technology Intellon Corporation's Power Packet™ technology which is the basis of the HomePlug Power line Alliance industry specification is a carefully crafted version of the orthogonal frequency division multiplexing (OFDM). Power packet is the brand name for Intellon's high-speed power line communications technology that now provides a 14 Mbps data rate over the existing power lines in the home.

Power packet is a complete solution that encompasses the physical (PHY) and media access (MAC) layers of the networking model. It supports advanced services such as voice over IP (VoIP), quality of service (QoS), and streaming media, which will provide new multimedia and telephony applications to the consumer. OFDM is a spectrum efficient modulation technique that enables transmission of very high data rates in frequency selective channels. Data rates in excess of 100 million bits per second (Mbps) are possible. The physical layer (PHY) uses the orthogonal frequency division multiplexing (OFDM) as the basic transmission technique. It is adopted for the IEEE's high rate wireless LAN standard (802.11a).

The high-speed data stream to be transmitted is processed as multiple parallel bit streams by OFDM, with each having low bit rate. Each bit stream then modulates one of a series of closely spaced carriers. The need for equalization is completely eliminated by using different phase modulation. OFDM waveforms are being typically generated using an inverse fast Fourier transform (IFFT), output is a time domain signal, called the OFDM signal. The data can be recovered using forward FFT, converting back to the frequency domain.

It is rate adaptive, the transmission time for a given packet size varies. Segmentation allows higher priority frames to jump in between the slower transmission's segments. Busting is uded to reduce the chance for collision among equal priority members.

Contention-free access improves QoS for certain types of multimedia traffic such as voice-over-IP (VoIP) or streaming media. Using the enhanced OFDM, the rate-adaptive design allows these frames to maintain an Ethernet-class connection throughout the power line network without losing any data. These devices connect via a USB or Ethernet cord to the computer or other devices and on the other end to the AC outlet. It uses a peer-to-peer network architecture and is rated at 14 Mbps, which is faster than existing phone line and wireless solutions. OFDM approach to power line networking is highly scalable allowing the technology to surpass 100 Mbps.

10.4 | INTEGRATED POWER LINE TRANSRECEIVER

Integrated power line transreceivers such as MAX2986 use CMOS design techniques which not only improves level of performance but imparts flexibility also. Technique is to combine the media access control layer (MAC) and the physical layer (PHY) on a single chip. The digital baseband and analog front-end (AFE) are used for high-speed power line communication. Readers may visit the Website http://www.maxim-ic.com for operational and interfacing details of these components.

10.4.1 USB Interface

Figure 10.5 shows the structure of a USB cable. The two pins USBD+ and USBD− are the data pins used in the USB interface, and correspond to D+ and D− in Fig. 10.5. V_{BUS} is nominally +5 V at the source. Figure 10.5 shows the upper layer interface pin setting to select USB. Refer to the universal serial bus specification, Revision 1.1 for more details on the USB interface. Figure 10.6 depicts application block diagram for USB interface.

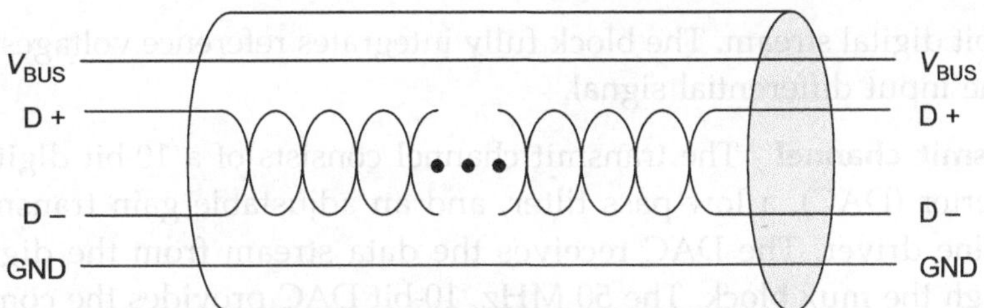

Fig. 10.5 USB cable

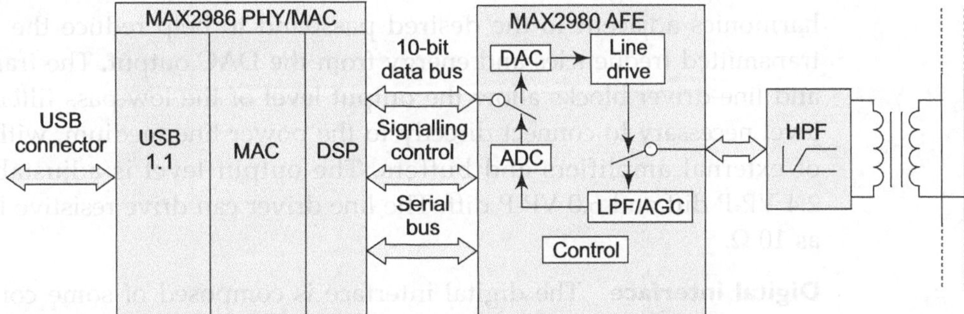

Fig. 10.6 Power line base band to USB application block diagram

10.4.2 Analog Front-end Transreceiver

The MAX2980 power line communication AFE integrated circuit is a state-of-the-art CMOS device that delivers high performance and low cost. This highly integrated design combines the ADC, DAC, signal conditioning, and line driver as shown in the functional diagram.

The MAX2980 substantially reduces previously required system components, while compatible to third-party HomePlug devices. This device interfaces with many companion digital PHY ICs to provide a complete power line communication solution. The advanced design of the MAX2980 allows operation without external control, enabling simplified connection to third-party digital PHY chips. Additional power resource management techniques can be employed in Rx and Tx modes through the use of various control signals.

Receive channel The receiver analog front-end consists of a low-noise amplifier (LNA), a low-pass filter (LPF), and an adaptive gain-control circuit (AGC). An ADC block samples the AGC output. The ADC communicates to the digital PHY chip through a mux block. The LNA reduces the receive channel input-referred noise by providing some signal gain to the AFE input. The AGC scales the signal for conversion from analog to digital. The scaling maintains the optimum signal level at the ADC input and keeps the AGC amplifiers out of saturation. The 50 MHz, 10-bit ADC samples the analog signal and converts it to a 10-bit digital stream. The block fully integrates reference voltages and biasing for the input differential signal.

Transmit channel The transmit channel consists of a 10-bit digital-to-analog converter (DAC), a low-pass filter, and an adjustable gain transmitter buffer and line driver. The DAC receives the data stream from the digital PHY IC through the mux block. The 50 MHz, 10-bit DAC provides the complementary function to the receive channel. The DAC converts the 10-bit digital stream to an analog voltage at a 50 MHz rate. The low-pass filter removes spurs and harmonics adjacent to the desired passband to help reduce the out-of-band transmitted frequencies and energy from the DAC output. The transmit buffer and line-driver blocks allow the output level of the low-pass filter to obtain a level necessary to connect directly to the power line medium, without the use of external amplifiers and buffers. The output level is adjustable between 2.4 VP-P diff and 6.0 VP-P diff. The line driver can drive resistive loads as low as 10 Ω.

Digital interface The digital interface is composed of some control signals and a 10-bit bi-directional data bus for the DAC and ADC. The control signals include a reset line, a transmit request, an I/O direction request, and a receiver shutdown control.

Readers may visit Website: http://www.maxim-ic.com for operational and interfacing details.

10.5 | POWER LINE CARRIER COMMUNICATION HARDWARE

Figure 10.7 depicts a schematic block diagram of a PLCC which may be implemented using PC and microcontroller.

Data is transmitted/received through power line. The basic function of the coupling network is to damp 230 V, 50 Hz frequency and provide antialiasing during receiving. Coupling network must behave like a band pass filter, with centre frequency equal to carrier frequency. The BP behaviour in transmission mode eliminates the unexpected harmonics of the digitized carrier and the aliasing component around the sampling frequency. The data communication takes place at zero crossing, which may be used to interrupt the microcontroller. Active band pass filter may be built with a low cost op-amp IC and a few external components. The input stage is directly connected to the output of the coupling network. Centre frequency is about 115.2 kHz. The output of the filter is connected to the RXIN input of the modem by means of the decoupling capacitor. Microcontroller receives the demodulated data from modem and transmits the data to power line, using external clock frequency equal to one-half of the modem clock via serial asynchronous communication controller.

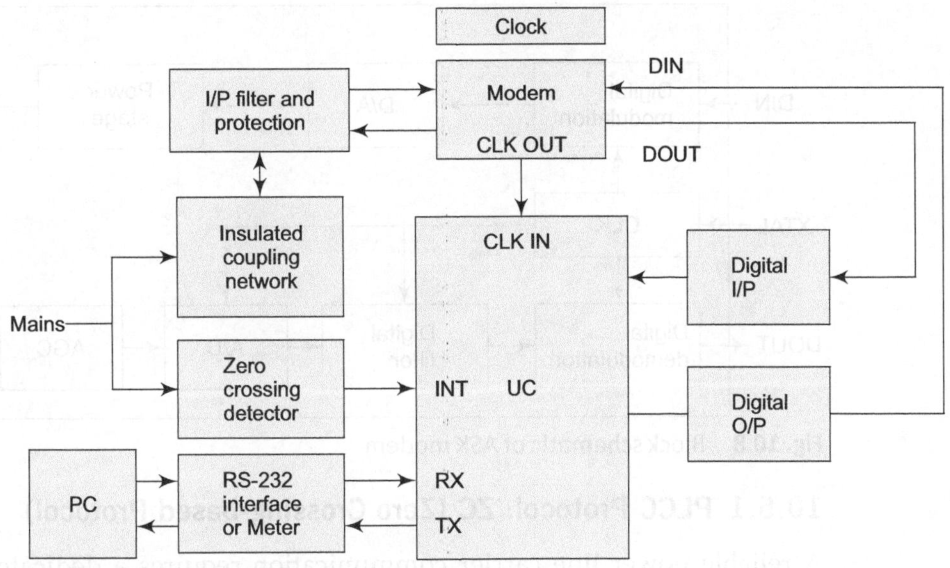

Fig. 10.7   Hardware system setup for PLCC

Amplitude shift key (ASK) modulation is the most commonly used techniques for data transmission on power line. Digital data is Logic 1 then the analog signal for that is 115.2 kHz carrier and when digital data is Logic 0 then analog signal is also 0. ASK is best suited for low cost design compared to traditional FSK.

Figure 10.8 depicts block diagram of modem which takes signal from controller to its DIN pin and modulate it, convert digital signal to equivalent analog then comes to power stage and transmit the data on power line. It receives signal from power line, amplifies and converts to digital and generates demodulated output at the DOUT pin. Typically, modem use external clock of 7.3728 MHz. Transfer the data at maximum 1200 baud rate. Carrier frequency is 115.2 kHz and a CLOCKOUT frequency of 3.6864 MHz (for the microcontroller).

The modem signals DIN/DOUT pin respectively connected to the P0.2 and P0.3 ports of the controller. Zener diode may be used to protect TXOUT/RXIN pin of the modem against transient over voltage which may occur during power up, for instance. Coupling network may be realized using capacitors. During the transmission on low impedance transmission line, the modem demands that power supply must be able to cop up with the high pick current (more than 150 mA peak), opt coupler is used for primary regulation in conventional SMPS.

The RS-232 interface needs two interfacing cards (master and slave card). One card is interfaced with energymeter card through I^2C and another one is placed at destination location and interfaces it with PC via RS-232 cable. Since power line cannot transmit the data at long distance, power amplifier are used or other techniques such as FSK or OFDM are used.

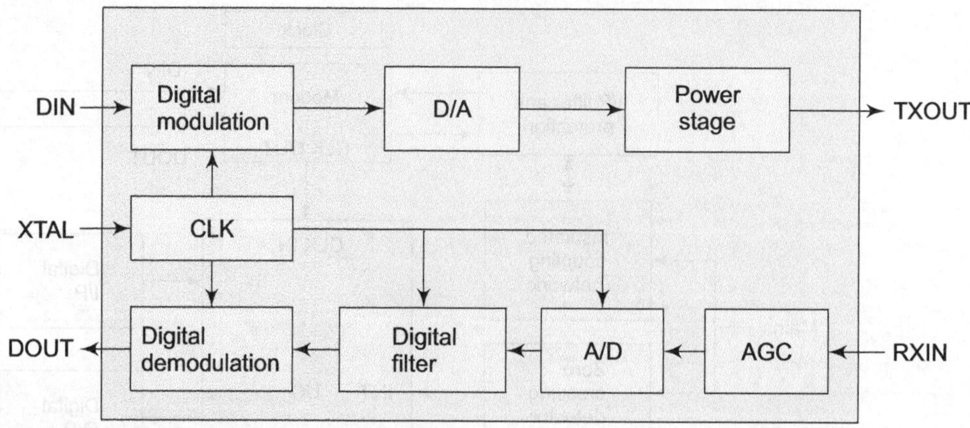

Fig. 10.8   Block schematic of ASK modem

10.5.1 PLCC Protocol: ZC (Zero Crossing-based Protocol)

A reliable power line carrier communication requires a dedicated protocol, acknowledgement, and several other techniques (described later). Two protocols may be used: user side protocol and power line side protocol for master and slaves network communication and management. The protocol should support features such as bit synchronization with the zero crossing point of the mains, error checking, bit multisampling, and acknowledgement.

The 'ZC' protocol uses the zero crossing point of the mains as clock synchronization for data exchange on the network. Since zero crossing is a quite reliable time reference for any device connected on a given phase, synchronous bit transfer can be achieved. When a slave receives a frame from the master, it checks all fields' integrity and checksum; then if the received address matches with its own address code, and only in that case, it replies to the master. After any transmission to a slave, the master waits for the slave's reply. If no reply is received, the master sends again the same frame 5 (max) times and then aborts transmission. Figure 10.9 depicts frame level format of ZC protocol.

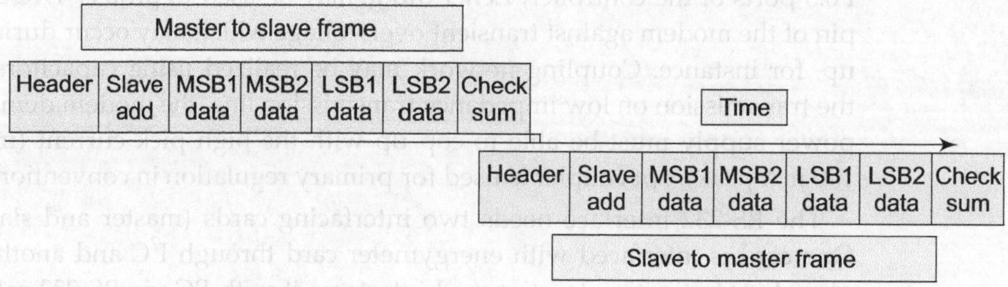

Fig. 10.9   Frame level ZC protocol

10.5.2 Transmission Mode

As shown in Fig. 10.10 each bit of a frame is sent (or received) just after the zero crossing point. A 'ONE' is represented by a carrier-modulated pulse of 1 ms; a 'ZERO' is an absence of modulation. Each bit must be sent 3 times, with the appropriate delay.

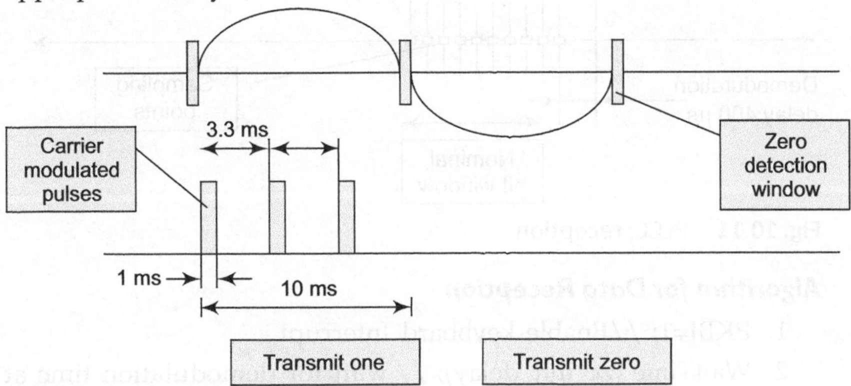

Fig. 10.10   PLCC: transmission

Algorithm for Sample Transmission

1. Wait for free power line // wait until get the consecutive 64-bits zero
2. Wait for zero crossing
3. If (transmitdata==1)

> {   Send 1 ms pulse //1st phase
> Wait for 2nd phase zero crossing
> Send 1 ms pulse //2nd phase
> Wait for 3rd phase zero crossing
> Send 1 ms pulse //3rd phase
> }

4. Else transmit zero to all 1st, 2nd, 3rd phase

10.5.3 Reception Mode

Figure 10.11 depicts reception process. To improve the communication reliability, each bit 'window' (a bit window is a time slot where a bit is expected) is sampled several time (10 times) at a 100 µs rate. This is done by sampling the output of the modem and a decision threshold is taken to determine if the sampled bit is zero or a one. Since the demodulation of any incoming bit takes several 100 µs. Sampling is done after an initial delay (demodulation delay), which varies with the signal amplitude, with an average value of about 400 µs seems to be a good compromise. This delay also depends on the crystal frequency; for higher frequency, demodulation delay is lower.

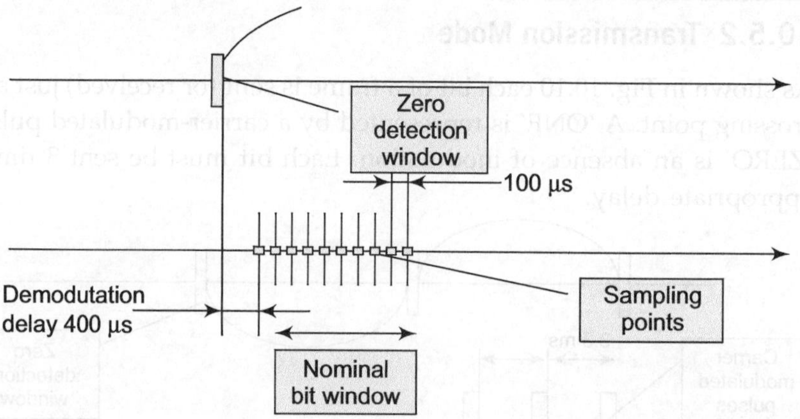

Fig. 10.11 PLCC: reception

Algorithm for Data Reception

1. EKBI=1; //Enable keyboard Interrupt
2. WaitTime (zc_init_delay); // wait for demodulation time at the time of receive data
3. While (! Wait Time (0)); // wait for free power line, check if power line free or not

   // Wait for consecutive 64 samples

   // Start sampling
4. For (sample_cntr = 0; sample_cntr< 10; sample_cntr++) // take the 10 samples

   {

   Wait for 100 us

   If (receivedata ==1) // check receive sample is high or low

   cnt_high++;   // if high then increment counter

   }
5. If (cnt_high> zc_treshold) // if high bit is greater than threshold value it

   // means receive bit is one

   recievedata =1;
6. Else

   recievedata=0;

Problem of synchronization arises when a certain phase relationship between the rising edge of DIN and clock of the modem causes a wrong behaviour of the modulator. This leads to an incomplete carrier shaping process. It is advisable to use same clock for both microcontroller and modem to provide the synchronization. This can be achieved by either using the modem clock to feed the microcontroller or by driving the modem with microcontroller so that the phase relation between DIN rising edge and the modem clock is same.

10.5.4 PLCC (Master, Slave) Card Software

PLCC master and slave card software code should include:

1. Initialization of various routine like timer, external interrupt, I2C, KBI, and watchdog.
2. Check for the hardware section of modem by loop back method (short the modem TX and RX).
3. Setting of ZC protocol for transmit to and receive from the power line.
4. I2C routine on master side to communicate with the other side of the power line.
5. Acknowledgement or the error message to master.

EXERCISES

10.1 Define PLCC.

10.2 What is the need for source encoding in digital communication?

10.3 State and explain important features of source encoder used in digital communication system.

10.4 What is the advantage of carrier modulation?

10.5 Enlist parameters of carrier modulation.

10.6 Explain the term: 'no new wire' communication.

10.7 Discuss broadband home networking technology.

10.8 How does the communication over power line differ from traditional wired communication?

10.9 Explain the standards used in power line communication.

10.10 With the help of schematic block diagram, explain interfacing and operation of MAX2986 and MAX2980.

10.11 With the help of schematic block diagram explain PLCC using microcontroller. Describe function of each block.

10.12 Define ASK.

10.13 Why ASK is commonly used in PLCC?

10.14 Explain dedicated protocols for PLCC.

10.15 Discuss zero crossing-based protocol and its format.

10.16 Explain process of transmitting data using PLCC. Write a program module for transmission over power line.

10.17 Explain process of receiving data using PLCC. Write a program module for receiving data over power line.

10.18 Write program modules for master-slave card communication used in PLCC.

Chapter 11

Application III: Projection Welding Machine Controller

This is the age of modernization, sophistication, and competition. Today, engineers are designing more and more sophisticated devices for automatic process. Most of the industrial processes need accurate and optimized control. Unoptimized setting of process parameters results in unnecessary utilization of electrical energy and poor output. Resistance welding is such a critical process. Controllers for resistance welding machines still use blind analog timers and it is difficult for an operator to set the exact timing of various timers, which results in wastage of electrical energy during weld cycle. Such a loss can be prevented by highly precise and accurate timing controller which allows optimum setting of weld time.

In projection welding machine system, control of timing and weld energy is highly important. It requires accurate and precise controller to supervise the actions of solenoids in required sequences for squeezing the material, perform welding operation on it, holding it, releasing it, removing it out, and welding energy control during weld time.

11.1 | RESISTANCE WELDING PROCESS

Two pieces of metal may be welded or fused together by passing large current (1000 to 100,000 A) through these pieces, while they are being forced together between the electrodes of the welding machine. But when pieces of metal are being joined or welded together by a bright electric arc, which melts a rod into a pool of metal, the process is called *arc welding*.

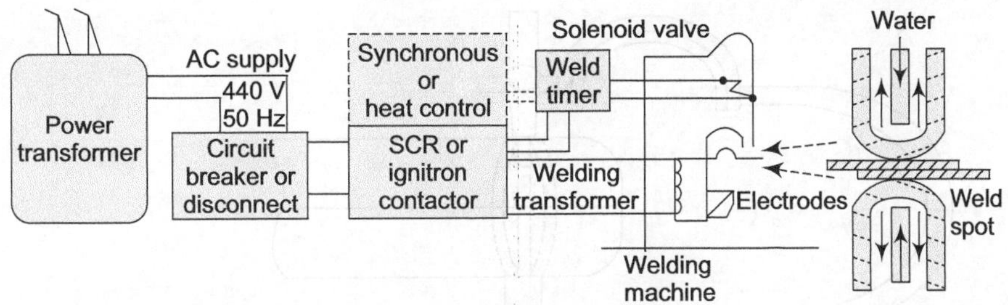

Fig. 11.1 Block schematic: setup for welding process

Figure 11.1 shows a block schematic of a setup for the welding process. The AC supply is received from the power line and transformer so that it passes, at 230 or 400 V, through a disconnecting device and then through a contactor before reaching the welding machine. Inside the machine, a welding transformer reduces the secondary voltage at the electrode tips to 1–60 V and supplies the large welding current (50–40000 A depending on the capacity of transformer) by drawing 5.0–200.0 A from the AC mains supply on primary side of the weld transformer. The electrode tips are water-cooled and must be kept clean. To make a weld, the required heat

$$H = I^2RT$$

where I is the weld current in A, R the resistance in Ω and T the time in s.

As shown in Fig. 11.2, there must be a resistance to current flow between the metal pieces, where most of the weld heat is produced, this process is called *resistance welding*. This resistance depends on the metal that is being weld. Steel has high resistance, and welding heat is easily produced but aluminum has low resistance, and the welding heat is harder to obtain. Resistance between the metal pieces decreases when they are forced together by the electrodes with greater pressure and it increases thereafter due to rise in temperature, as weld current flows through the work-piece. To make the weld, current needs to flow for only few seconds, so the contactor must close and open the circuit quickly and it must do this hundreds of times each hour. While magnetic contactors control many such welders, ignitron contactors and other electronic equipments are used where better welds must be made in shorter time with less contactor noise and maintenance.

To make a single spot weld, the pieces of metal are placed in the space between the two electrodes, one of which can move. When the operator presses the start button, a solenoid valve applies air pressures so that the electrodes come together and squeeze the metal pieces. Welding current then flows to heat the metal and make the weld. The metal is held under pressure for a moment until the weld hardens; then the electrodes separate so that the metal can move before the next weld is started. Control circuits control all these action of the welding machine. During the welding operation or cycle, AC controller controls the flow of primary current to weld transformer which ultimately produces the weld during weld time.

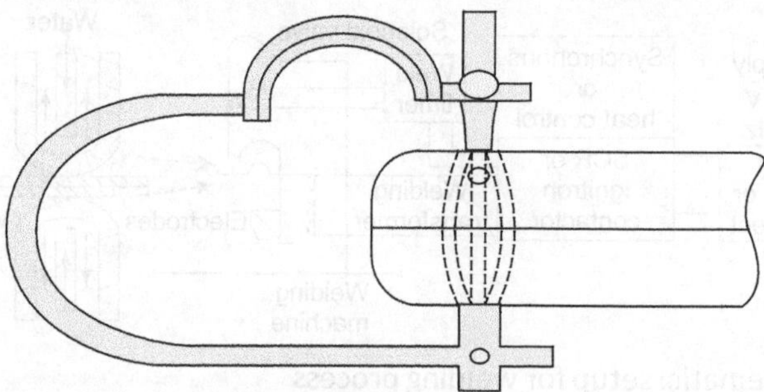

Fig. 11.2 Resistance welding

To control a projection welding machine, sequential timer is used. It measures four lengths of time (perhaps from 2 to 100 cycles or seconds each depending upon the machine and work-piece to be welded) in a certain order or sequence. After the start switch is closed, the sequential timer permits the electrodes to build the right pressure on the work. This time is called *squeeze time*. The *weld time* is the length of time during which the welding current flows. After the welding current stops, the electrodes continue to press against the metal pieces during the hold time, while the weld hardens. Then the electrodes separate. This time is called *release time*. If the operator still holds the foot switch closed, the electrodes will close again after a period called the *off time*, which gives time to move the work to a new position between the electrodes. The out timer is used to remove the work-place in automatic welding machines and it is useful in automatic machines only.

Metals like aluminum may weld better if the welding current starts at a small amount, then rises to full welding value. This rise of current may be obtained by 'slope control', a feature that may use with a welder that has phase-shift heat control. During each weld, slope control makes the single phase welding current increase at a gradual rate that may be adjusted for best welding results. The addition of slope control prevents the flash, which is often seen when heavy pieces are being welded; therefore, the electrode pressure may be decreased. Since this increases resistance, R, between the pieces, the line current or KVA needed for a heavy weld may be decreased up to 50 percent. Moreover, the small current during the first cycle softens or corrects the fit of the parts being welded, thus producing more consistent welds.

11.2 | AC CONTROLLER FOR WELD ENERGY CONTROL

AC controllers are widely used in industry for domestic and industrial heating, weld energy controller, induction motor speed control for fan and pump drives, and transformer tap changer in utility systems. In earlier days, transformer with taps or autotransformer were employed for these applications, which are replaced by thyristors or triac controllers due to high efficiency, compact size, and flexibility of control. TRIAC-based AC controllers are useful up to few kilowatts but generally SCR-based controllers are very popular for large AC power control. Table 11.1 gives specifications of power circuit for the AC controller.

Table 11.1 Specification: AC controller power circuit

Supply	440 V /230 V 50 Hz
No. of SCR	2
Max. current rating of SCR	40 A
SCR connection	Back to back
Display	LCD type
Cooling of thyristors	Yes, using natural convection
Max. loading capacity of AC controller	5 KVA weld transformer (for 440 V) 2.5 KVA weld transformer (for 230 V)
Voltage control	By potentiometer
Weld/no weld control	Yes, using SPDT switch

11.3 | HARDWARE SETUP

The system consists of three sub systems: microcontroller card, analog AC voltage controller card, and power circuit for AC voltage control. Table 11.2 depicts the specification of the controller card.

Table 11.2 System specifications: controller card

Supply	230 V AC, 50 Hz for microcontroller card
No. of relay outputs	5
Rating of contacts of outputs	Amp. at 230 V AC ,10 A at 24 V DC
No. of timers	5
Range of timers	0.1 to 99.9 s
Resolution of timers	0.1 s
Display	LCD type
Battery back-up	Not needed
No. of inputs	4 keys, 1 start process; input (optically isolated)
Memory retention time during switch-off condition	Unlimited

11.3.1 Microcontroller Card

Figure 11.3 depicts a microcontroller-based setup.

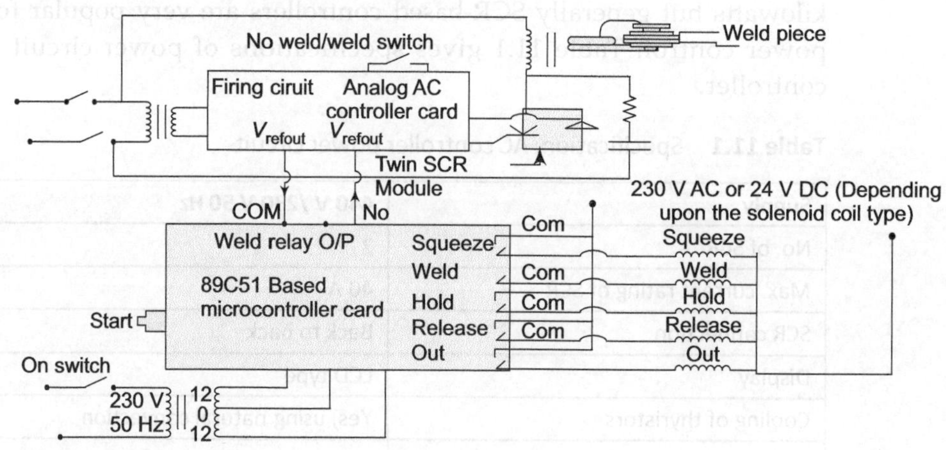

Fig. 11.3 Block diagram: microcontroller card

11.3.2 Power Supply

The function of power supply (Fig. 11.4) is to provide regulated +5 V DC supply to the microcontroller, except relay driver section. All the relays are PCB mounted type and needs +12 V DC supply. So +12 V DC unregulated supply is separately derived for this section. Power supply section contains step-down center-tap transformer for converting 230 V, 50 Hz AC to 12–0–12 AC. This 12 V AC is converted into DC by center tap full-wave rectifier, which after filtered by capacitor becomes +15 V DC at no load. This +15 V DC becomes 12–13 V when loaded. And it is unregulated. It is given to relay driver section for driving relay by TIP122 Darlington transistor. The unregulated supply of +12 V is given at Pin 1 input of 7805, and at the output Pin 3 +5 V regulated supply is available. Pin 2 of 7805 is connected to ground. The output voltage of +5 V DC available at Pin 3 of 7805 is used for microcontroller card.

The maximum current rating of 7805 is 500 mA, which is more than sufficient for this card due to low current demand by the LCD display and CMOS microcontroller and EEPROM (93C46).

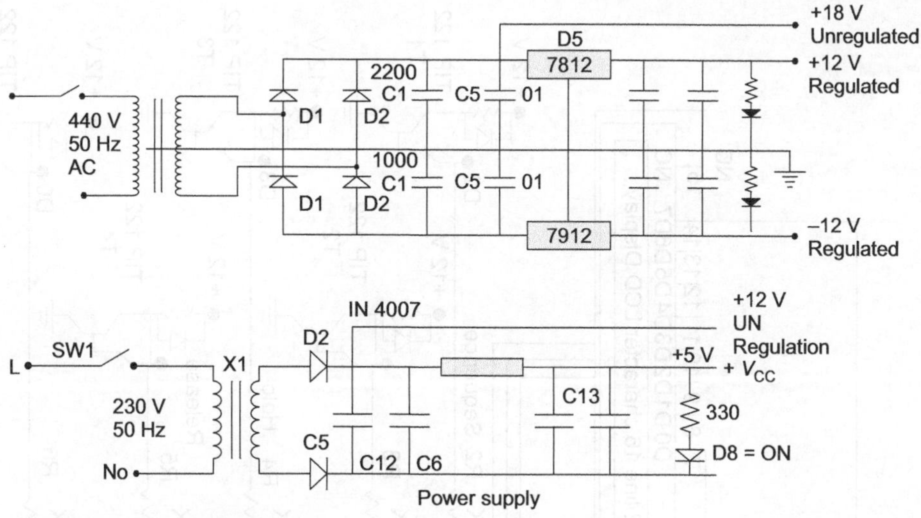

Fig. 11.4 Power supply

11.3.3 Microcontroller Section

Figure 11.5 shows microcontroller section using IC 89C51. No other extra static RAM or EPROM is interfaced with it because internal 4 K flash PROM and 128-byte is sufficient for the application. 89C51 is given a +5 V supply between Pin 40 and Pin 20, 0.1 µF capacitor is connected across the supply for power supply decoupling, and 12 MHz crystal is connected between Pins 19 and 18 for deriving clock. For the operation of microcontroller, two 22 pf capacitors are connected between crystal leads and ground for stability of clock, Pin 29 (PSEN) is left unused and Pin 31 (EA) is connected to $+V_{CC}$ as microcontroller does not have to access external memory, and Port 0 pins are connected to LCD data lines. Port 1 pins are used as control pins for LCD display, while Port 2 pins are used.

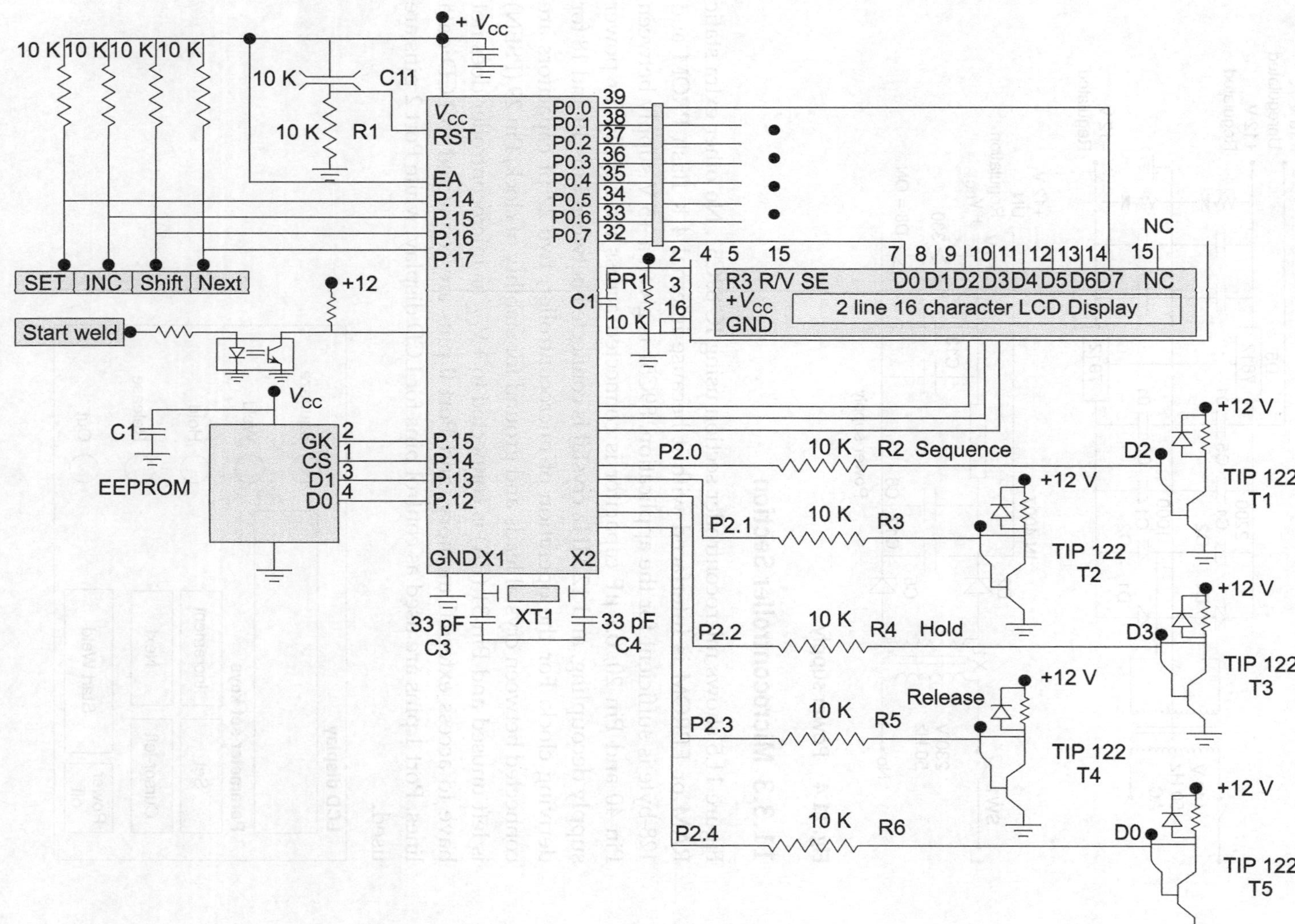

Fig. 11.5 Microcontroller card

11.3.4 Keyboard Section

Keyboard section contains four keys for data inputs of the timers. These keys are:

1. Enter and exit
2. Next timer
3. Up
4. Move cursor right

For keyboard, Port 1 is used. These port pins are internally pulled high and it is externally pulled low by pressing these keys. An external resistor is also connected in series to limit the current when they are pressed. The connection of various keys is shown in Table 11.3.

Table 11.3   Port pin assignment

Key	Port pin
Enter and Exit	P1.4
Next	P1.5
IP	P1.6
Move cursor right	P1.7

By using these keys, the user can vary the timer value of any timer, pointed in RAM location in BCD, of 16-bit size as maximum time can be 99.9 s. Internal software takes care of key-debouncing during the pressing of keys. These port pins are defined as input pins by storing 1's in their respective bit of Port 1, latch at address 90h.

11.3.5 LCD Display Section

A 2-line-16 characters display is used for user interface. Although, single line display is sufficient for this application, 2-line display is used because it is more user friendly. One can display more data as well as messages in details. Port Pin 1.0 and Port 1 is used to address 2-byte wide registers, one for command (RS = 0) and second for the character (RS = 1) to be displayed.

Port Pin 1.1 is connected to Pin 5 of LCD display, which is R/W. When P1.1 is '1', user can read from the internal RAM of the LCD display when it is '0', user can write to it. Pins 7 to 14 are the data pins of LCD display and connected to Port B of 8255. LCD display is highly useful and once interfaced properly, and appropriate software may be written.

11.3.6 Memory Section

In case of power failure, the data of timer values for various back up to 89C51 and put 89C51 in idle mode EEPROM 93C46 by Atmel is used. Since it is serial

EEPROM, uses 3-wire serial bus, and is interfaced as shown in Fig. 11.6; pin timer will get erased, if it is stored in RAM of 89C51. Alternate to this is a battery 8 and 5 are connected to supply (+5 V) and ground, respectively. Chip select (CS) that is an active high signal, is connected to port Pin 3.5. CLK signal (Pin 2 of 93C46) is connected to port Pin 3.3. Output Pin 4 of 93C46 is connected to port Pin 3.2 and it is defined as input by writing 1 into the Port 3 latches at address 0B0h. DI pin of 93C46 is connected to port Pin P3.1.

Reading and writing the EEPROM is not easy and fast as in static RAM. So definite minute delay (order of 10–20 ms) is needed each time one writes to it or reads from it. It also needs delay of approximately 100 ms after power ON. These delays needed are taken care by the software. Each time user changes the data of timer, same time it will be written into EEPROM as well as relevant RAM location. So in case of power failure, these data remains in memory and retrieved by the microcontroller whenever power resumes.

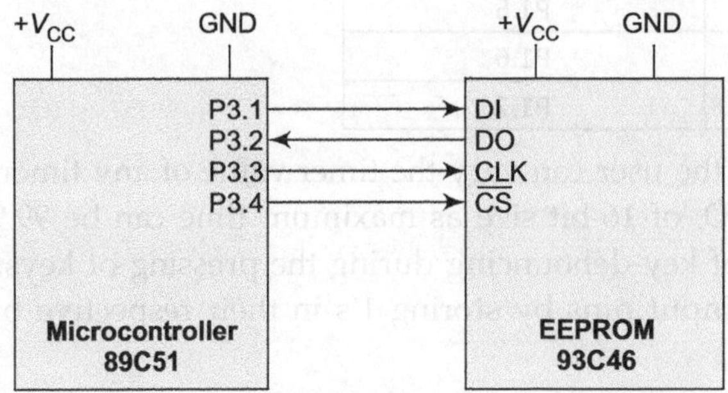

Fig. 11.6 Interfacing EEPROM with microcontroller

11.3.7 Input/Output Section

This is the most useful port of the system, as with the help of this section, it interfaces with the outer world. Projection welding machine needs 5-digital outputs for relay driver section and 8-pin output port for LCD display. These outputs are:

1. Squeeze output
2. Hold output
3. Weld output
4. Release output
5. Out output
6. 8-bit output port

A switch input (START WELD) from the operator is used to start the welding process. Table 11.4 depicts pin assignments and interfacing of relay drivers.

Table 11.4   Pin assignment for input section

Pin no.	Function
P2.0	SQUEEZE
P2.1	HOLD
P2.2	WELD
P2.3	RELEASE
P2.4	OUT
P0.0–P0.7	D0–D7 of LCD
P1.0	RS of LCD
P1.1	RW of LCD
P1.2	En of LCD

11.3.8 Relay Driver Section

Relay driver section consists of TIP 122 power Darlington transistor and relays. TIP 122 is a NPN Darlington power transistor having V_{BE} max = 100 V, V_{CE} max = 100 V, V_{BE} max = 5 V, IC max = 5 A, and Pd = 10 W with h_{fe} max = 1000.

Digital output available at Port C pins will be given to the base of TIP 122 transistor via 10 K resistor. As it is a high gain transistor, pins of 8155 can directly drive it and even 1 mA current is more than enough to saturate it.

When any of port pin is high, relevant transistor goes into saturation region. Hence, collector voltage drops to zero causing full 12 V across relay coil and it gets energized. A diode is connected in reverse parallel across relay coil to prevent damage to transistor during switching OFF of the relay due to high di/dt that time, which causes large spike voltage = $L \times (di/dt)$ across the relay coil (across collector and emitter of the transistor). Relay contacts are rated 5 A at 230 V AC and it drives the solenoids of the hydraulic or pneumatic cylinders of the projection welding machine for squeeze, hold, release, and out functions. The output of weld relay is connected to potentiometer of analog AC controller card, which controls the weld energy during weld time.

11.3.9 Firing Circuit of AC Controller

An analog AC controller section (Fig. 11.7) consists of blocks such as: regulated power supply; reference circuit; line synchronizing transformer; synchronized ramp-generator; comparator; a stable multivibrator circuit; gating circuit; pulse amplifier, and isolation circuit.

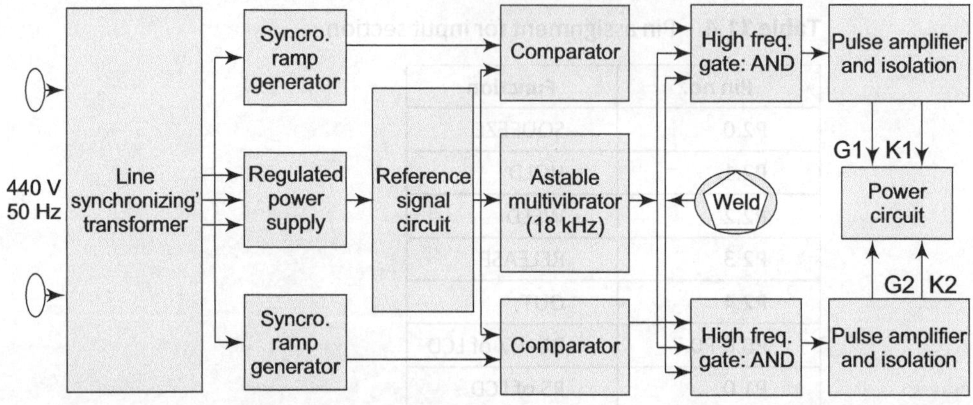

Fig. 11.7 Firing circuit: analog controller

Regulated power supply section Regulated power supply is obtained from 440 V, 50 Hz supply. First of all, 440 V supply is stepped down to 15-0-15 V AC by a step-down transformer, which is converted to DC by a bridge rectifier and further filtered by shunt capacitor filter and smooth +/− 15 to +/−16 V DC is available at the output. +15 V unregulated DC is regulated by 7812 IC to obtain +12 V DC output and −12 V DC is available at the output of 7912 as its input is converted −15 V DC. Capacitor of 0.1 μF is connected across input and output of both these controller IC for improving the transient response of controller.

Reference signal section Reference circuit provides reference voltage to the AC controller. This reference voltage decides the firing angle of the AC controller. +12 V is connected to common of the weld relay while NO contact of relay is connected to potentiometer. When weld relay sets energize, +12 V reaches to resistor connected in series with the potentiometer. So at the midpoint, V_{ref} will be available which is given to comparators of comparator section.

Line synchronizing transformer section Line synchronizing transformer is a step-down transformer used in the power supply section, which has 440 V AC primary and 15-0-15 V AC center tapped secondary. It gives two low voltage AC synchronized sine waves of 21 V max. amplitude, which are 180° phase shifted to each other, in same time relationship to line voltage. Thus, line-synchronized voltage is taken from the power supply section itself, and given to synchronized ramp-generator.

Line-synchronized ramp-generator section Line synchronized ramp-generator section generates two ramp waves which are synchronized to line voltage waveforms and both are 180° phase shifted to each other. Synchronized and ramp-generator consists of transistorized circuit. Synchronized AC waveforms from synchronizing transformer are given at the input to the base of transistor which gets saturated during positive cycle of AC waveform and it goes in cut-off region when negative cycle arrives, ring negative cycle, transistor remains in cut-off region, and voltage across capacitor rises exponentially with time constant RC and line synchronized ramp is available at the collector. The two ramp waveforms available at this section are given to comparator section.

Comparator section   Comparator section is wired around 741, and used as comparator. In this section, following two inputs are given at the inputs of 741.

1. V_{ref} from reference circuit
2. Ramp input

V_{ref} is connected to inverting pin (–) of the 741 while synchronized ramp is connected at the non-inverting pin (+) of 741. When instantaneous ramp waveform voltage becomes more than V_{ref}, output will become high and remain high until the cycle reverses. Thus, its output remains high from the instant of firing angle to the end of that half cycle. Comparator 2 is wired same way and works in same manner.

Astable-multivibrator section   This section is a 555-based astable-multivibrator section, which gives 18 kHz square waves at its output. Frequency of 18 kHz is selected so that at least 180 pulses comes during each half cycle, i.e., 10 ms period for 50 Hz line voltage waveform which gives a maximum firing angle error of 1°.

The value of R and C for 18 kHz are derived from equations:

$$T_{on} = 0.69 \times (R_A + R_B) \times C$$
$$T_{off} = 0.69 \times R_B \times C$$

and values of timing components chosen from this equation are

$$R_A = 3.3 \text{ K}, R_B = 1.5 \text{ K}, \text{ and } C = 0.01 \text{ }\mu\text{F}$$

Gating circuit   The purpose of gating circuit is to modulate the firing angle pulses with the frequency of 18 kHz. Sustained pulses are necessary to prevent half-firing of SCR's (firing of only one SCR of single phase full wave AC controller, i.e., half wave rectification) of the AC controller, which causes saturation of weld-transformer when single pulse firing scheme is employed for R-L load. A CMOS AND gate chip 4081 is used for this task. TTL-type AND gate chip can be used here but it will need additional +5 V regulated supply, hence it is avoided. The one input of AND gate is connected to output of 555 and second input is connected to output of comparator. Two such AND gates are used here having same wiring logic and works in same manner. Modulated pulses available at the output of the first gate stage is given to the input of second gating stage of AND gate. Here it is logically ANDed with the weld/no-weld switch input from operating panel and inhibits the firing pulses to the gates of SCRs when no-weld switch is pressed for trial checking of microcontroller.

Pulse amplifier and isolator section   The function of this stage is to raise the power level of the signals available at AND gate output and to provide electrical isolation by means of pulse transformer between power circuit and control circuit.

Output of AND gate is connected to the base of transistor T1 and collector is connected to primary of pulse transformer. A signal diode 1N4148 is connected across the primary in reverse parallel to prevent damage to the transistor during switching OFF from ON condition due to inductive peaks. Pulse transformer provides electrical isolation between control circuit and power circuit. Secondary of pulse transformer is half wave rectified by 1N4148 signal diode preventing any negative voltage, reaching the gate of SCR. Only positive pulses are successful to reach across gate cathode of relevant SCR. When positive pulses reach the gate of the SCR, it becomes ON and supply voltage appears across the weld transformer and weld current is induced in secondary. So welding occurs due to energy loss (I^2Rt) in the work piece.

11.3.10 Power Circuit

Power circuits control the AC voltage input to the weld transformer using phase angle control and connected in series with the weld transformer.

Power circuit is made up of two SCR connected back to back (Fig. 11.8). Since welding transformer is R-L load having high inductance, di/dt protection is not necessary. But dv/dt protection is needed. So R-C snubber network is connected across them, which prevents the damage of SCR due to excessive dv/dt at the time of switching ON. Thus, whole system works and controls the weld voltage to weld transformer which allows control of primary current in the weld transformer directly and hence the secondary weld current indirectly. This ultimately allows the open loop control of weld energy which is given by

$$\text{Weld energy (during weld time)} = I^2_{\text{sec}} \times R_{\text{m}} \times T_{\text{w}}$$

where $\quad I_{\text{sec}}$ = secondary current of weld transformer (A)

$\qquad\qquad R_{\text{m}}$ = resistance of material (Ω)

$\qquad\qquad T_{\text{w}}$ = weld time (s)

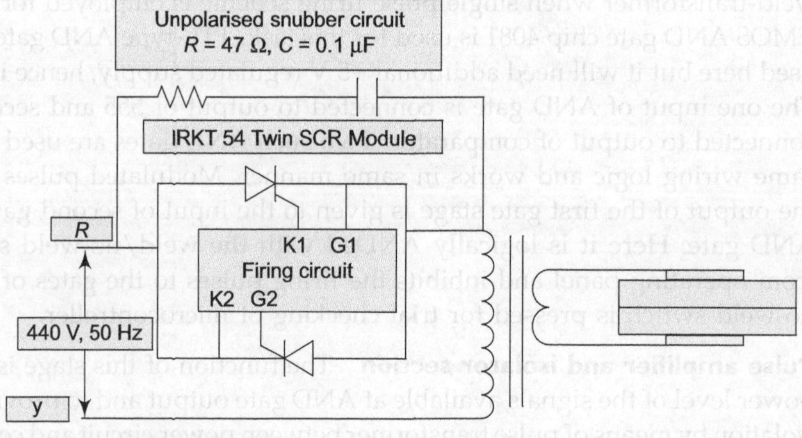

Fig. 11.8 Power circuit

11.4 | SYSTEM SOFTWARE

The software can be developed in a modular form. The algorithms for the few important modules will be described in this section.

11.4.1 Main Module

The function of this module is to carry out initialization of the system for use. The default values of the various parameters are loaded from memory to make the controller ready for use. When power is switched ON, following steps are carried out by the microcontroller:

1. Initialize on chip components
   - Load timer count (EF08h)
   - Clear accumulator
   - Clear RAM locations
   - Display initial message on LCD
2. Read and store default timer setting from EEPROM.
3. Display OK message.
4. Call timer setting module if SET key is pressed which indicates set mode allowing user to change default timer data, otherwise wait for START key.
5. Load count in the Timer 0 and enable Timer 0 interrupt.
6. Enable squeeze timer
   - Display remaining squeeze time
   - Decrement time and wait until the squeeze timer is zero
   - Disable squeeze timer
7. Enable hold timer
   - Display remaining hold time
   - Decrement time and wait this step untill hold timer is zero
   - Disable hold timer
8. Enable release timer
   - Display remaining release time
   - Decrement time and wait till release timer is zero
   - Disable release timer
9. Enable out timer
   - Display remaining out time
   - Decrement time and wait till out time is zero
   - Disable release timer
10. Goto Step 4

11.4.2 intO Module

It is necessary to check at a regular interval (100 ms) whether any key is closed or not? If any key is closed, it is necessary to identify the key pressed and take appropriate action. This can be done using interrupt. The intO is used for this purpose. The desired services from ISR module for intO are:

- Check for the key closure and identify the key pressed.
- Decrement relevant count stored in 2-bytes at location/register.

Following steps must be carried out by the microcontroller to provide these services:

1. Load Timer T0 count and enable timer.
2. If **weld-enable** is set

       Goto Step 3

       Else if **weld-start key** is pressed

           Goto Step 5

           Else

               return

3. If **not enable squeeze timer**

       Goto Step 7
4. Decrement
5. Display OK message
6. Call timer **setting module** if **SET** is pressed which indicates set mode allowing user to change default timer data, otherwise wait for **START** key.
7. Load count in the Timer 0 and enable Timer 0 interrupt.
8. Enable squeeze timer
   - Display remaining squeeze time
   - Decrement time and wait till it is zero
   - Disable squeeze timer
9. Enable weld timer
   - Display remaining squeeze time
   - Decrement time and wait till it is zero
   - Disable squeeze timer
10. Enable hold timer
    - Display remaining hold time
    - Decrement time and wait till it is zero
    - Disable hold timer
11. Enable release timer
    - Display remaining release time
    - Decrement time and wait it is zero
    - Disable release timer

12. Enable out timer
    - Display remaining out time
    - Decrement time and wait till it is zero
    - Disable release timer
13. Goto Step 4.

11.4.3 Accessing EEPROM Data

The data is stored in the EEPROM. Unlike RAM, operations such as: read, write, or delete/erase with this type of memory are not simple and cannot be implemented using data transfer instructions such as mov. The section describes steps required for storing or retrieving data to/from EEPROM.

Read Data$_{16}$ from EEPROM

The steps to be carried out to read data byte from any location in EEPROM are:

1. Issue READ command to IC 93C46 (the address in bits b5–b0 in an accumulator)
2. Read upper byte in Register B
    - Load count (09h)
    - Read data bit from DI pin
    - Rotate accumulator left, save in Register B
    - Decrement count
      repeat the process till the count is zero
3. Read lower byte in Register B
    - Load count (09h)
    - Read data bit from DI pin
    - Rotate accumulator left
    - Decrement count
      repeat the process till the count is zero
4. Ignore upper 3-bits (as the bit b5–b0 contains the address). Rotating accumulator to the left three times can do this
5. Load count (06) for address bits
6. Rotate accumulator to the left (address bit is now in carry)
7. Send carry to DI pin
8. Wait for 10 ms
9. Make clock **LOW**
10. Wait for 10 ms
11. Decrement count
12. Goto Step 5 if count is non-zero
13. Disable 93C46
14. Reset DI pin
15. Reset CLK pin

Write Data$_{16}$ in EEPROM

The steps to be carried out to write 16-bit data in the data-pointer (dptr) at the location in EEPROM specified by bits b5–b0 in an accumulator are:

1. Select the chip: 93C46
2. Give start bit
3. Issue write command to IC 93C46 (The address in bits b5-b0 in an accumulator) by sending op-code (01)
4. Send address through DI pin of 93C46.
5. Load lower byte of dptr (dpl) in accumulator
6. Write lower byte in memory:
   - Load count (00h)
   - Rotate accumulator left without carry
   - Send carry to DI pin
   - Wait for 20 ms
   - Decrement count and repeat the step till the count is zero
7. Load upper byte of dptr (dph) in accumulator
8. Write upper byte in memory:
   - Load count (00h)
   - Rotate accumulator left without carry
   - Send carry to DI pin
   - Wait for 20 ms
   - Decrement count and repeat the step till the count is zero
9. Disable 93C46
10. Reset DI pin
11. Reset CLK pin

Erase Data EEPROM

The steps to be carried out to erase data byte from any location in EEPROM are

1. Select the chip: 93C46
2. Wait for 20 ms
3. Ignore upper 3-bits (as the bit b5–b0 contains the address). Rotating accumulator to the left three times can do this
4. Load count (06) for address bits
5. Rotate accumulator to the left (address bit is now in carry)
6. Send carry to DI pin
7. Wait for 10 ms
8. Make clock LOW
9. Wait for 10 ms

10. Decrement count
11. Goto Step 5 if count is non-zero
12. Disable 93C46
13. Reset DI pin
14. Reset CLK pin

The other modules required to be developed are: defining variables, displaying time on LCD, getting default data from EEPROM.

Variable Definitions

Constants defined for timer parameter (EEPROM memory constant parameters)		Outputs for relays of the unit	
c start	05h	squeeze	01h
c squeeze	06h	hold	02h
c hold	06h	weld	04h
c_weld equ	07h	release	08h
C_release	08h	out	10h
C_out	09h	No-output	00h
totalcount	0ah		
Delay-cnt	092h		
Internal RAM memory allocation			
result1	30h	sec-time	44h
result2	31h	min-time	45h
result3	32h	chg-disp	46h
result4	33h	string-length	47h
squeezetimer low	34h	pointer	48h
squeezetimer high	35h	squeezecount high	49h
holdtimer low	36h	squeezecount low	4ah
holdtimer high	37h	holdcount high	4bh
weldtimer-low	38h	holdcount low	4ch
weldtimer-high	39h	weldcount high	4dh
releasetimer-low	3ah	weldcount_low	4eh
releasetimer-high	3bh	releasecount high	4fh
outtimer-low	3ch	releasecount low	50h
outtimer-high	3dh	outcount high	51h
alarm-time	3fh	outcount low	52h
Timeoureg	40h	Step-cnt	53h
key-reg	41h	final-op	54h
disp-dly	43h		

Bit locations		
Flags	**Enable/disable different timers**	**EEPROM (port-pins)**
key-flag bit 00h	en-squeeze-timer bit 10h	
set-flag bit 01h	en-hold-timer bit 11h	
timer-start bit 02h	en-weld-timer bit 12h	do-9346 bit p3.2
key-1 bit 03h	en-release-timer bit 13h	di-9346 bit p3.3
key-2 bit 04h	en-out-timer bit 14h	clk-9346 bit p3.4
key-3 bit 05h	en-input bit 15h	cs-9346 bit p3.5
key-4 bit 06h	ce-lcd equ 10h	
units bit 07h		
tens bit 08h		
hundred bit 09h		

Default Data

This code loads data from EEPROM

```
mov  result1, #00h
clr  select9346             ; cs=0(chip disable)
clr  clock9346              ; clock = 0
clr  input9346             ; data i/p bit =0
mov  a,#c_squeeze          ; squeeze timer count
lcall  read-16bit          ; module to be written by the reader based
                           ; on algorithm described in Section 11.4.

mov  releasecount_high,  result1
mov  releasecount_low,  result2
mov  a, #c_hold            ; hold-timer count
lcall  read-16bit          ; module to be written by the reader based
                           ; on algorithm described in Section 11.4.

mov  holdcount_high,  result1
mov  holdcount_low,  result2
mov  a,#c_weld            ; weld-timer count
lcall  read-16bit          ; module to be written by the reader based
                           ; on algorithm described in Section 11.4.

mov  weldcount_high,  result1
mov  weldcount_low,  result2
mov  a,#c_release          ; release
lcall  read-16bit          ; module to be written by the reader based
                           ; on algorithm described in Section 11.4.
```

```
        mov releasecount_high, result1
        mov releasecount_low, result2
        mov a, #c_out                      ; out-timer count
        lcall rd9346                       ; module to be written by the reader based
                                             on algorithm described in Section 11.4.

        mov outcount_high, result1
        mov outcount_low, result2
        ret
```

Display: Remaining Time for Ongoing Operation

display-time MACRO addr, dataH, dataL

```
            mov dptr, #squeeze-msg ; pointer to the current timer
            lcall displayline_1    ; string display
            lcall display2
            mov a, #' '
            lcall data-lcd
            mov r2, dataH
            mov r3, dataL
            lcall display-bcd
            ret
dis_squeeze:                          ; ' squeeze timer '
display-time squeeze-msg,
squeezetimer_high, squeezetimer_low
dis_hold:                             ; ' hold timer '
display-time hold-msg,
hold-timer-high, hold-timer-low
dis_weld:                             ; ' weld timer '
display-time weld-msg,
weld-timer-high, weld-timer-low
dis_release:                          ; ' release timer '
display-time release-msg,
release-timer-high, release-timer-low
dis_out:                              ; ' out timer '
display-time out-msg,
out-timer-high, out-timer-low
```

LCD: Display Module

```
lcd-display:
jnb set-flag, continue                ;setting mode: display settings
ret
```

```
continue:
jb en-input, key-input
mov dptr,#startmsg1
lcall displayline_1            ;string display routine
mov dptr,#startmsg2
lcall displayline_2            ;start string
sjmp over
key-input:
jnb en-squeezetimer, skip-squeeze
lcall dis_squeeze
sjmp end
skip-squeeze:
jnb en-holdtimer, skip-hold
lcall dis_hold
sjmp end
skip-hold:
jnb en-weldtimer, skip-weld
lcall dis_weld
sjmp end
skip-weld:
jnb enreleasetimer, skip-release
lcall dis_release
sjmp end
skip-release:
jnb enouttimer, end
lcall dis_out
end:
mov dptr, #sec-msg             ; " * 0.1 s " start string
lcall char-display            ; using look-up table
over:
ret
```

The topmost comment for the code reads: `;if not in setting mode then normal display`

EXERCISES

11.1 With the help of a schematic block diagram, explain resistance welding process.

11.2 Explain: slope control.

11.3 Draw a circuit diagram for microcontroller section based on description given in Section 11.3.4.

11.4 Write a program module to display status of machine at any time.

11.5 State advantage of using 93C46 in the welding machine controller of Fig. 11.5.

11.6 Specify significance of synchronizing transformer.

11.7 Why is isolator needed in Fig. 11.7?

11.8 Write a program module to initialize the microcontroller card.

11.9 Write a program module to change the default setting of the machine timers.

11.10 Write an interrupt routine for intO to check status of key closure and identification.

11.11 How does the operation for EEPROM differ from conventional RAM?

11.12 Write a program module to write 16-bit data in the EEPROM at specified address.

11.13 Write a program module to erase data from the EEPROM at specified address.

11.14 Write a program module to display remaining time on LCD.

Chapter 12

Application IV: Add-on Features for Telephones

Telephones lines are popular for communication. The subscribers are demanding more facilities such as caller ID, metering, and conferencing. There are many offices or homes where more than one telephone lines are installed. It will be a desirable facility to have a single meter to keep records of telephones. The weekly usage of each line can be locally monitored. By studying the intensity of calls on each line, arrangements can be made to ensure that no excessive billing occurs on one line, i.e., all the telephones are equally utilized. In residential complexes where there are many subscribers, one such meter can be centrally located to bill all the lines. The concept of teleconferencing and implementation using microcontroller is described in this chapter. Design of a multiphone metering system which will be capable of user detection and carry out accounting and billing for every user separately is also discussed in this chapter.

12.1 | INTRODUCTION

Graham Bell demonstrated a point-to-point telephone connection. In such a network, a calling subscriber chooses the appropriate link to establish connection with the called subscriber. In order to draw the attention of the called subscriber before information exchange can begin, some form of signalling is required with each link. For example, if the called subscriber is engaged, suitable indication should be given to the calling subscriber by the means of signalling.

In a general case with n entities, there are $n(n-1)/2$ links. Networks with point-to-point links among all the entities are known as fully connected networks. The number of links required in fully connected network becomes very large even with moderate values of n. Consequently, practical use of Bell's invention on a large scale or even on a moderate scale demanded not only the telephone sets and the pairs of wires but also the so-called switching system or the switching office or the exchange.

With the introduction of the switching system, the subscribers are not connected directly to one another, instead, they are connected to the switching system. When a subscriber wants to communicate with another, a connection is established between the two at the switching system. In this configuration, only one link per subscriber is required between the subscriber and the switching system and the total number of such links is equal to the number of subscribers connected to the system. The signalling is now required to draw the attention of the switching system to establish or release a connection. It should also enable the switching system to detect whether a called subscriber is busy and indicate the same to the calling subscriber. The functions performed by a switching system in establishing and releasing connections are known as *control functions*.

Early switching systems were manual and operator oriented. Automatic switching systems can be classified as electromechanical and electronic. Electro-mechanical switching systems include step-by-step and crossbar systems. Step-by-step systems increment automatically by steps while the crossbar system has hardware control subsystems which use relays and latches. These sub-systems have limited capability and it is virtually impossible to modify them to provide additional functionalities. In the electronic system, computer or a processor performs the control functions. Hence, these systems are called *store program control (SPC) systems*. The switching scheme used by the electronic switching systems may be either space division switching or time division switching. In space division switching, a dedicated path is established between the calling and the called subscribers for the entire duration of the call.

In time division switching, sample values of speech signals are transferred at fixed intervals. Time division switching may be analog or digital. In analog switching, the sampled voltage levels are transmitted as they are, whereas in digital switching, they are binary coded and transmitted. If the coded values are transferred during the same time interval from input to output, the technique is called *space switching*. If the values are stored and transferred to the output at a later time interval, the technique is called *time switching*.

Subscribers all over the world cannot be connected to a single switching system unless we have a gigantic switching system in the sky and every subscriber has direct access to the same. Technical and engineering constraints of signal transfer on a pair of wires necessitate that the subscriber be located within a few kilometers from the switching system. For subscribers in different localities to communicate, it is necessary that the switching systems are interconnected in the form of a network. The links that run between the switching systems are called *trunks*, and those that run to the subscriber premises are known as *subscriber lines*.

A number of signalling functions are involved in establishing, maintaining, and releasing a telephone conversation. An operator in a manual exchange performs these functions. In the automatic switching system, the verbal signalling of the operator is replaced by a series of distinctive tones. Five functions are performed by the operator in the subscribers related signalling.

1. Respond to the calling subscriber to obtain the identification of the called party.
2. Inform the calling subscriber that the call is being established.
3. Ring the bell of the called party.
4. Inform the calling subscriber, if the called party is busy.
5. Inform the calling subscriber, if the called party line is unobtainable for some reason.

Distinctive signalling tones are provided in all automatic switching systems for functions 1, 3, 4, and 5. Most of the modern exchanges provide a call in-progress or routing tone for function 2.

The above signalling function 1 is fulfilled by sending a dial tone to the calling subscriber. This tone indicates that the exchange is ready to accept dialed digits from the subscriber. The subscriber should start dialing only after hearing the dial tone. Otherwise, initial dial pulses may be missed by the exchange, which may result in the call landing on a wrong number. The dial tone is a 33 or 50 or 40 Hz continuous tone. The 400 Hz signal is usually modulated with 25 or 50 Hz.

When the called party line is obtained, the exchange control equipment sends out the ringing current to the telephone set of the called party. This control equipment sends out a ringing tone to the calling subscriber.

The two rings in a double-ring pattern are separated by a time gap of 0.2 s and two double-ring patterns by a gap of 2 s. The ring burst has a duration of 0.4 s. The frequency of the ringing tone is 133 or 400 Hz, sometimes modulated with 25 or 33 Hz. Busy tone pattern is a bursty 400 Hz signal with silence period in between. The burst and silence duration has the same value of 0.75 or 375 s. A busy tone is sent to the calling subscriber whenever the switching equipment or junction line is not available to put through the call or the called subscriber line is engaged. The number unobtainable tone is a continuous 400 Hz signal.

This tone may be sent to the calling subscriber due to number of reasons such as the called party line is out-of-order or disconnected, and an error in dialing.

The routing tone or call in-progress tone is a 400 or 800 Hz intermittent pattern. In electromechanical systems, it is usually 800 with 50 percent duty ratio and 0.5 s ON/OFF period. In digital exchanges it is a 400 Hz pattern with 0.5 s ON period and 2.5 s OFF period.

12.2 | TOUCH TONE DIAL TELEPHONE

Pulse dialing or rotary dialing is limited to signalling between the exchange and the subscriber and no signalling is possible end-to-end, i.e., between two subscribers. End-to-signalling is a desirable feature and is possible only if the signalling is in the voice frequency band so that the signalling information can be transmitted to any point in the telephone network to which voice can be transmitted. Rotary dial signalling is limited to 10 distinct signals, whereas a higher number would enhance signalling capability significantly. Finally, a more convenient method of signalling in rotary dialing is preferred from the point of view of human factors. These considerations led to the development of touch tone dial telephones.

In the touch tone dialing scheme, the rotary dial is replaced by a push button keyboard. Touching button generates a tone, which is a combination of two frequencies, one from the lower band and the other from the upper band. For example, pressing push button 9 transmits 852 and 1477 Hz.

The need for touch tone signalling frequencies to be in the voice band brings with it the problem of vulnerability to talk-off, which means that the speech signals may be mistaken for touch tone signals and unwanted control actions such as terminating a call may occur. Another aspect of talk-off is that the speech signal may interfere with the touch tone signalling if the subscriber happens to talk while signalling is being attempted. The main design consideration for touch tone signalling system from the need for protection against talk-off include the following factors:

1. Choice of code
2. Band separation
3. Choice of frequencies
4. Choice of power levels
5. Signalling duration

The choice of code for touch tone signalling should be such that imitation of code signals by speech and music should be difficult. Simple single frequency structures are prone to easy imitation as they occur frequently in speech or music. Hence, some form of multifrequency code is required. Such code are easily derived by selecting a set of N frequencies and restricting them in a binary fashion to either present or absent in a code combination. However, some of the 2eN combinations are not useful as they contain only one frequency. Transmitting simultaneously N frequencies involves N-fold sharing of a restricted amplitude range, and valid code word. These factors lead to the consideration of P-out-of-N code. Here a combination of P frequencies out of N frequencies constitutes a code word. The code yields $[N!/P! (N–P)!]$ code words.

Prior to touch tone, P-out-of-N multifrequency signalling known as multifrequency key pulsing (MFKP) was used between telephone exchanges by the operators. When 2-out-of-6 code were used, it is known to give a talk-off performance of less than 1 in 5000. However, this degree of talk-off performance is inadequate for subscriber level signalling. In order to improve the performance, two measures are adopted:

- Retaining pass two, N is chosen to be seven or eight, depending upon the number of code words desired.
- Chosen frequencies are placed in two separate bands, and restriction is applied such that one frequency from each band is chosen to form a code word.

When multiple frequencies are present in a speech signal, they are closely spaced. Band separation of touch tone frequencies reduced the probability of speech being able to produce touch tone combinations. The number of valid combinations is now limited to $N_1 \times N_2$, where N_1 and N_2 represent the number of frequencies in each band.

12.3 | INTEGRATED SERVICES DIGITAL NETWORK (ISDN)

The integrated services digital network (ISDN) can be defined as a network, in general evolving from a telephony, that provides end-to-end digital connectivity to support a wide range of services, including voice and non-voice services, to which user have access by limited set of standard multipurpose user–network interfaces.

The first set of ISDN standards G 705 laid down six conceptual principles on which ISDN should be based:

1. ISDN is evolve from the telephony IDN by progressively incorporating additional functions and network features including those of any other dedicated network so as to provide for the existing and new services.
2. New services introduced into the ISDN should be so arranged so as to be compatible with 64 kbps switched digital connections.
3. The transmission from the existing networks to a comprehensive ISDN may require period of time extending over one or two decades.

4. During the transition period, arrangement must be made for the inter-working of services on ISDN and services on other networks.

5. The ISDN will contain intelligence for the purpose of providing service features maintenance and network management functions. This intelligence may not be sufficient for new service and may have to be supplemented by either additional intelligence within the network or possibly compatible intelligence within the customers terminals.

6. A layered functional set of protocol is desirable for the various access arrangement to the ISDN. Access from the customer to ISDN resources may vary, depending upon the services required and on the status of evolution of national ISDN.

There are three types of fundamental channels in ISDN around which the entire information transmission is organized.

1. Basic information channel   B-channel, 64 kbps
2. Signalling channel   D-channel, 16 or 64 kbps
3. High-speed channel   H-channels

B-channel and D-channel are basically adopted from telephone digital networks with common channel signalling. Digital networks have evolved around A-law and Mu-law encoding of speech signal at 64 kbps.

ISDN has the same rate for the basic information channel although the technology has advanced to a stage where quality voice can be transmitted at 32 kbps or even less. ISDN permits lower rate signals to be transmitted by using rate adaptation. Alternatively a number of low rate information channels may be multiplexed and sent on the B-channel. Information stream at the rate of 8, 16, and 32 kbps are rate adapted by placing the information bits in the first 1, 2, and 4-bits, respectively of an octet (8-bit quantity) in the B-channel and filling the remaining bits by binary ones. Stream at the rate other than 8, 16, or 32 kbps and lower than 32 kbps are rate adapted in two stages. First they are rate adapted to one of 8, 16, or 32 kbps rates and then to 64 kbps.

D-channel is primarily used for carrying signalling information to control and monitor ISDN services. It acts as the common signalling channel for all services. For example, call establishment for a voice call using a B-channel is entirely handled over by the D-channel without the involvement of the B-channel. Information transfers on the D-channel is governed by a layered protocol along the lines of ISO-OSI reference model. Spare capacity of D-channel may be used to carry other packet switched data conforming to the layered model. Main application envisaged on the D-channel is user-to-user signalling, telemetry signals for energy management, etc., and low-speed videotext and computer terminal data.

H-channels are used for high-speed applications like high-speed facsimile, video, high fidelity audio, and high-speed data. A user may also use an H-channel as a high-speed trunk on to which a number of low-rate information streams are multiplexed.

12.4 | CONFERENCING

In rural areas or remote places, distant communication becomes a problem. The solution to such a problem is to provide an STD booth in such areas. Now again consider a situation in which a person has to walk down a kilometer away from his house just to find an STD booth and that too possibly a locked one. Considering such a situation and again from privacy point of view one would prefer to have access to the STD facility while sitting at home. This facility can well be provided to the customer via the conference system. In such a system, the procedure is as follows:

1. The customer dials into the STD booth, through the local telephone line, wherein he finds the operator.
2. The customer then gives his STD number to the operator who then dials the number through another telephone line with the STD facility.
3. The operator then connects the STD line with the local line of the customer via the conference circuit, for the customer to have far distant communication.

This way customers, without having an STD facility in their telephone line, can have access to the STD facility sitting at their house via the STD booth. Now again consider a pathetic condition in which a customer dials into the STD booth and is left unanswered due to the unavailability of the operator or if the STD booth is locked up. Such a situation is undesirable in case of emergencies. Again consider the condition of the operator who has to unnecessarily waste his time waiting for the incoming calls and connecting the STD line with the local line manually. The solution to such a situation is *auto-conference system*.

In such a system, the procedure is as follows:

1. The customer dials in the STD booth where he is answered with a short music.
2. He dials his password.
3. After that he dials his account number.
4. The call gets incremented in his account automatically and then the customer gets the dial tone of the STD line.
5. The customer then dials in and can have far distant communication.

The system reduces the tedious job of the operator and the user can have an access to the STD facility at any time. The advantages are as follows:

1. In a manual system, the subscriber needs to communicate with the operator and a common language becomes an important factor. In multilingual areas this aspect may pose problems, the operation of an automatic system is language independent.

2. A greater degree of privacy is obtained in automatic system as no operator is normally involved in setting up and monitoring calls.

3. Establishment and release of calls are faster in automatic systems.

4. In an automatic system, the time required to establish and release a call remains more or less of the same order irrespective of the load on the system or the time of the day. In a manual system, this may not be true.

5. Access to the STD facility is available 24 hours a day.

12.4.1 System Hardware

The system operation is as follows:

- The customer dials in through the local line in the STD booth.
- It is being detected by the ring detector IC, the output of which is given to the microcontroller (89C51).
- The controller fires the local relay, i.e., relay no. 1. Thus, the local line gets connected to the conference circuit.
- The controller then fires the music relay, i.e., relay no. 2, for a period of few seconds.
- The customer then dials his password.
- This password is being detected by the DTMF (dual tone multiple frequency) IC and then sent on to the controller where it is being verified.
- If the password is correct, then
  - Controller again fires the music relay, i.e., relay no. 2 for few seconds, which is an indication for the customer to dial his account number.
  - After verification of the account number a call is automatically incremented for that account number by the controller.
  - Simultaneously, the controller also fires the STD relay, i.e., relay no. 3 and this way the local line gets connected to the STD line via the conference circuit.

Figure 12.1 depicts the hardware setup for conferencing. In order to connect two independent telephone lines together without disturbing the DC conditions on either line, a transformer is used. This is the heart of the circuit which allows audio information on one line to be impressed on another line with no DC connection between them. T1 is a telephone line coupling transformer which has been designed for telephone line use. Each telephone line is connected to separate windings of T1 through coupling capacitors.

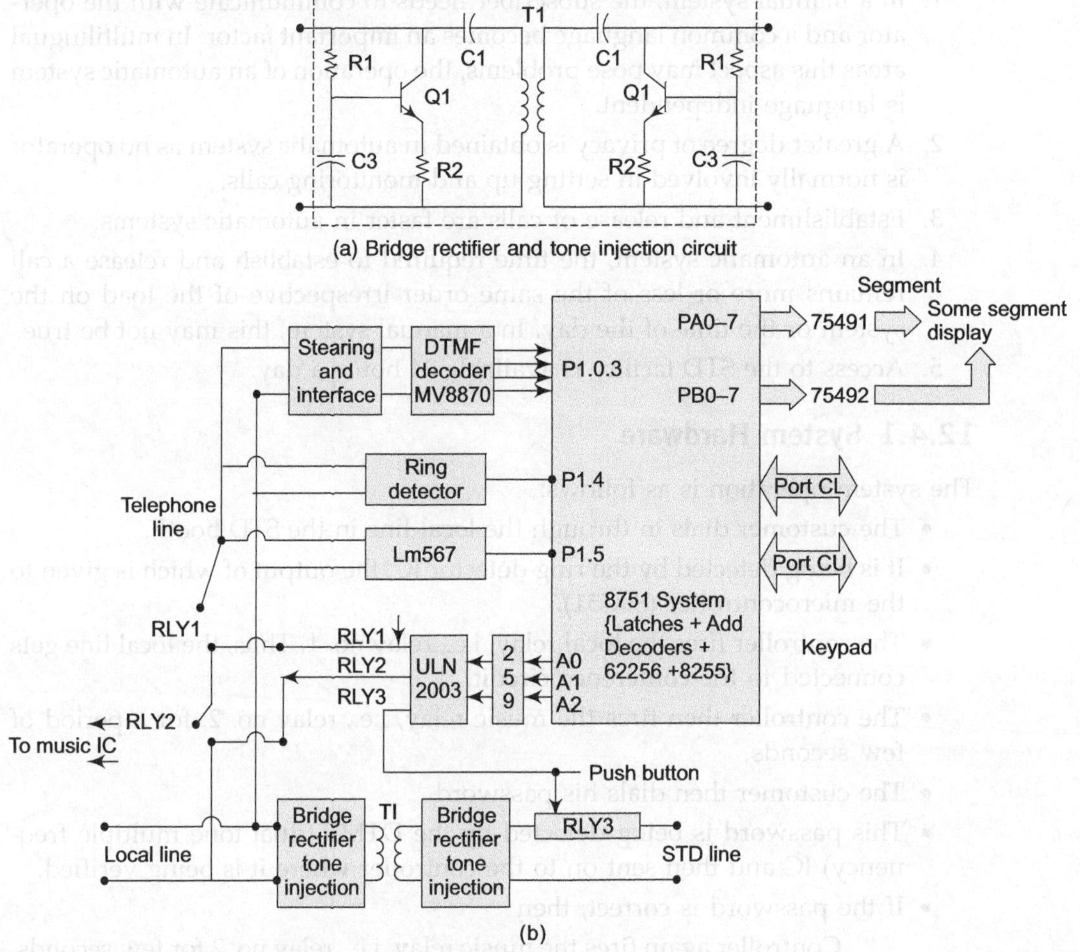

Fig. 12.1 Hardware setup for conferencing

Telephone line no. 1 may be connected across one winding of the transformer through the diode bridge rectifier and electrolytic capacitor C1. The diode bridge rectifier has been provided to maintain the polarity of the telephone line. The capacitor prevents any DC from flowing through the transformer winding, while coupling the audio frequency information appearing across the telephone line to the transformer winding. The DC isolation provided by C1 ensures that the normal DC conditions of the telephone lines are not disturbed. In a similar manner telephone line number 2 is connected to the second winding of the transformer. Each telephone line can be independently controlled by the action of the relays.

An additional circuit has been included in both the telephone line circuits. This is a DC holding circuit which provides the proper DC characteristic to hold telephone line no. 2 in operation even though no telephone on that line is present. The circuit composed of R1, R2, and Q1 acts like a resistance to DC, and as high impedance to audio signals. The high impedance of the circuit is provided by C3, which prevents any audio signals from appearing at the base of Q1. Thus, any audio voltage appearing across telephone line no. 2. will not cause a corresponding current in Q1.

The components used are DTMF decoder, ring detector, 89C51 microcontroller, 8255A, 74LS138 decoder, 74LS244 and 741s245-buffer, 74LS373 octal latch, 75491 and 75492-segment and digit driver, ULN2803-relay driver and 62256 external RAM

DTMF decoder (MV8870)   The input arrangement of the MV8870 provides a differential input operational amplifier as well as bias sources. V_{ref} is used to bias the inputs at mid-rail. Provision is made for connection of a feedback register to the op-amp output (GS) for adjustment of gain. In a single ended configuration the input pins are connected with the op-amp connected for unity gain and V_{ref} biasing the input at ½ V_{dd}.

The MV8870 monolithic DTMF receiver offers small size, low power consumption and high performance. Its architecture consists of a bandsplit filter section, which separates the high and low tones of the received pair, followed by a digital counting section which verifies the frequency and duration of the received tones before passing the corresponding code to the output bus.

Filter section   Separation of the low-group and high-group tones is achieved by applying the dual-tone signal to the inputs of two-sixth order switched-capacitor bandpass filters, the bandwidths of which correspond to the bands enclosing the low-group and high-group tones. The filter section also incorporates notches at 350 Hz and 440 Hz for exceptional dial-tone rejection. Each filter output is followed by a single-order switched capacitor section to smooth signals prior to limiting.

Limiting is performed by high gain comparators which are provided with hysteresis to prevent detection of unwanted low-level signals and noise; the outputs of the comparators provide full-rail logic swings at the frequencies of incoming tones.

Steering circuit   DTMF decoder chip MV8870 may be used for registration of decoded tone pair. Before registration of a decoded tone-pair, the receiver checks for a valid signal duration. This check is performed by an external *RC* time constant driven by Tst pin on the MV8870 chip. Logic HIGH on Est causes V_c to rise as the capacitor discharges.

V_c reaches threshold V_{tst} of the steering logic to register the tone pair, latching its corresponding 4-bit code into the output latch. At this point, the GT output is activated and drives V_c to V_{dd}. GT continues to drive high as long as Est remains high. Finally, after a short delay to allow the output latch to settle, the delayed steering output flag, STD, goes high, signalling that a received tone pair has been registered. The contents of the output latch are made available on the 4-bit output bus by raising the 3-state control input (TOE) to a logic HIGH. The steering circuit works in reverse to validate the interdigit pause between signals.

Table 12.1 Code generation (DTMF)

LOW	HIGH	KEY	TOE	SEL	Q4	Q3	Q2	Q1
697	1209	1	H	L	0	0	0	1
697	1336	2	H	L	0	0	1	0
697	1477	3	H	L	0	0	1	1
770	1209	4	H	L	0	1	0	0
770	1336	5	H	L	0	1	0	1
770	1477	6	H	L	0	1	1	0
852	1209	7	H	L	1	1	1	1
852	1336	8	H	L	0	0	0	0
852	1477	9	H	L	1	0	0	1
941	1209	0	H	L	1	0	1	0
941	1336	.	H	L	1	0	1	1
941	1477	#	H	L	1	1	0	0
687	1633	A	H	L	1	1	0	1
770	1633	B	H	L	1	1	1	0
852	1633	C	H	L	1	1	1	1
941	1633	D	H	L	0	0	0	0
697	1209	1	H	H	0	0	0	1
697	1336	2	H	H	0	0	1	0
697	1477	3	H	H	0	0	1	1
770	1209	4	H	H	0	1	0	0
770	1336	5	H	H	0	1	0	1
770	1477	6	H	H	0	1	1	0
852	1209	7	H	H	0	1	1	1
852	1336	8	H	H	1	0	0	0
852	1477	9	H	H	1	0	0	1
941	1209	0	H	H	0	0	0	0
941	1336	.	H	H	1	0	1	0
941	1477	#	H	H	1	0	1	1
697	1633	A	H	H	1	1	0	0
770	1633	B	H	H	1	1	0	1
852	1633	C	H	H	1	1	1	0
941	1633	D	H	H	1	1	1	1
		ANY	L	ANY	Z	Z	Z	Z

L : Logic low   H : Logic high   Z : High impedance

The decoder uses digital counting techniques to determine the frequencies of the limited tones and to verify that they correspond to standard DTMF frequencies (Table 12.1). When the detector recognizes the simultaneous of two valid tones, it raise the early steering flag Est Any subsequent loss of signal condition will cause Est to fall.

12.4.2 Ring Detector

ZN480E is a ring detector 8-pin IC in DIL package by Plessey Semiconductors which may be used when a special function is required in response to an incoming ringing signal. It has built-in lightning protection. A logic output is provided for direct interface to answering machines, modems, lamp indicators, and optoisolators. A standard 560 kHz ceramic resonator controls the clock oscillator which provides a reference for the dial pulse rejection circuitry giving an active high only in the presence of a ringing signal.

The incoming ringing voltage from the line, V_r typically 75 V, 25 Hz is connected to pins 1 and 8 via a DC blocking capacitor C1, and a current surge limiting resistor R. This applies the ringing voltage to the built-in full wave rectifier ringing voltage to the built-in full wave rectifier bridge and the output is filtered by the reservoir capacitor C2. This smoothed DC supply is then used power up the remaining circuits. An AC voltage appearing on the line causes the voltage across C2 to rise but output is inhibited until this voltage exceeds the threshold voltage, V_t, and the dial pulse discrimination has detected the presence of the correct ringing frequency. ICS1240 which is used by TATA TELECOM as a ring detector is used. This IC though not pin compatible with ZN480E performs quite similar functions. An optoisolator MCT2E is used at the output of LS1240 in order to provide a logic level output which is feed to the controller. The optoisolator also protects the controller from the high voltage surges coming at times from the exchange.

12.4.3 Microcontroller (Intel 89C51)

The controller forms the heart of the entire auto-teleconferencing machine as it is used to control all the other chips used, through the program feeded into the EEROM, which has a memory of 4 K. For current applications, the port connections of the microcontroller 89C51 are as follows:

Port 0:   Lower order add/data port demultiplexed by latch 74LS375

          Current capacity is increased by buffer 74LS245

Port 2:   Higher order add port

          Current capacity increased by unidirectional buffer 74LS244

          Data lines :   Connected to RAM and 8255

Address lines : Connected to RAM; three address lines can be decoded using 74LS138; A15-To select RAM

Port 1:　P1-0, P1-1, P1-2, P1-3 → output of 8870 (DTMF decoder)

P1-4 → Connected to the ringer circuit, i.e., LS1240

Port 3:　P3-2 → Connected to the strobe pin of 8870

8255 is used to interface a keypad and a six digit, 7-segment LED display. Port A, B of 8255 are used in output mode and C as input mode. Latch 74LS373 provides demultiplexing of address and data bus. Decoder 74LS138 provides chip select signals, while buffers 74LS244 (unidirectional); 74LS245 (bidirectional) buffers generate address bus and data buses. 75491, 75492 are used as segment and digit drivers to increase the current capacity of the input and output data lines, used for designed scanned displays. 75491 is a quad device that has four Darlington pair transistors in a package and can source a total current of about 50 mA, i.e., approximately 12.5 mA/pair. 75492 has six Darlington pairs in a package and can sink a total current of 250 mA.

The ULN 2803: RELAY DRIVER is used to convert the 5 V (TTL logic level) into 12 V to connect/disconnect the required relay according to the data input to the 74LS259. The input to 2803 is the output from 74LS259. External RAM (62256) is used to store the data during the execution of the program such as the number of calls at a particular account no. (which in this case is an RAM memory address). A battery backup is provided for power failures.

Keypad–LED interface is provided for the operator to display the number of calls in any particular account number. This enables easy billing and a long term record maintenance for the operator. Each time the operator wishes to see the number of calls in a particular account, he should press the keys of that particular account number. It should also be noted that while the operator is having a look at the number of calls in a particular account, the controller is in polling mode and hence cannot perform the auto-conferencing functioning. Thus, it is necessary that during that time the conferencing is done manually.

12.5　USERS MANUAL

1. To make an STD call, first dial your STD operator's phone number.
2. The call is acknowledge by music of 2 s which is an indication for the caller to dial his/her password.
3. A correct password is acknowledge by a music of 5 s which is an indication to dial the account number. An incorrect password breaks the connections preventing any wrong access to the facility.
4. Before dialling the account number it is necessary to dial (*) key, otherwise it results in breaking of connection.

5. After the account number is dialed, there is an increment in the number of calls made on that particular account, simultaneously an STD dial tone is also obtained on the dialers telephone lines.

6. After completion of the STD call it is necessary to press (#) key for deliberately breaking the connections.

12.6 | MULTI-PHONEMETER

This section describes the development of a microcontroller-based multi-phonemeter having the following characteristics:

1. The meter should monitor multiple calls from maximum eight subscribers simultaneously.

2. It should display the number dialed, duration of an STD call in terms of number of local calls, and STD rates. Selector switches may be provided for the selection of the function to be displayed.

3. Records of the previous telephone numbers and the calls should be stored in the memory.

4. A facility to connect a printer for the hardcopy of the record.

5. 16-digit LCD display are provided to all the subscriber should display.

    - Dialled digits even when the dialling is going on, to provide a visual check of the number being dialled and prevent dialling of a wrong number.

    - The rate at which the STD calls are being charged, i.e., full rate (F), half rate (H), quarter rate (Q) or one-third rate (T).

6. Once the call is initialized, STD bill in terms of the local calls should be continuously updated and displayed.

12.6.1 Block Diagram

Figure 12.2 depicts the multi-phonemeter system. It continuously scans input from the telephone lines. Scanning means checking the status of all subscribers. All subscribers are assigned one port and each line is checked sequentially. A hardware interrupt is generated at every 10 ms for scanning each line. As this time is in terms of 10 ms, whenever one telephone is being handled, the chances of losing information from other telephones do not arise.

Once it detects off-hook, it starts to detect the dialled number and display as-well-as store in memory record. Once dialling is completed, it waits for battery reversal as indication of start of conversation. The process of finding the STD rate is performed only after the call is initialized (i.e., after getting reversal).

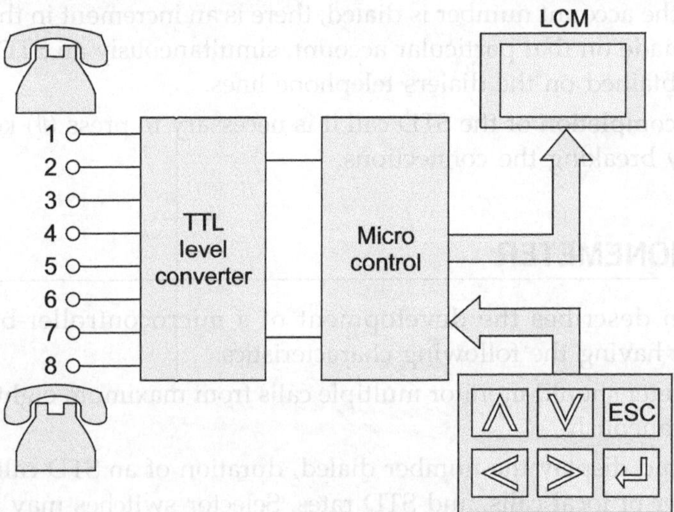

Fig. 12.2 Block diagram of a multi-phonemeter

Since the metering pulses are not available at the subscriber junction, these pulses are to be locally generated or may be simulated using software.

Once the conversation starts, system starts accumulating the duration of the call in terms of local calls. It is displayed when display function is selected. The process of accumulating time consumed will be continued until the subscriber is on-hook. The process is applied to all the subscribers in parallel.

The calculation of local calls will be based on current time and specific rate to be applied according to the time-table. The time-table is made according to the rates at which the STD calls are being charged, i.e., full-rate, half-rate, one-third rate, and quarter rate. A 16-digit LCD is used for displaying the required information like telephone number dialled, billing, etc.

The successful calls will be stored in memory (the storage of records will be done in FIFO). The selection of lines (i.e., 1 to 8) and records of successful calls will be available on the command, which can be given by keyboard and can be controlled by user-friendly menu driven software.

12.6.2 Terminology

Scanning Scanning means checking the status of all subscribers one-by-one. There is a scanning routine in the software, which performs the functions such as: checking the status of the telephones (on-hook or off-hook) collecting and storing the dialled digit, sensing line reversals. All the subscribers are assigned one port and each line is checked sequentially. A timer interrupt is generated at fix rate for scanning each line. Since this time is in terms of microseconds, whenever one telephone is being handled, the chances of losing information from other telephones do not arise. Hence, it is a type of parallel processing. Each user has a virtual control of the meter.

Actual metering   Since the metering pulses are not available at the subscriber junction, these pulses are to be locally generated or they have to be simulated using software. Our meter follows the first approach. The B/R signal is simulated using extra key on test keyboard. The units on display will be incremented as per number of key pressed by user for simple simulation. Actually these pulses will be expected from the line.

Call storage   Once the call has been completed, the telephone number of the called party and the bill in terms of total number of calls are stored. All the subscribers are allocated separate memory page in which the corresponding number and calls are stored. Since 8 K of RAM has been made available to each user, a total 8 telephone calls can be stored at a time. Once the memory is used up, the records should be taken out using any printer or by any other means. This process is necessary due to the arrangement that a new number will be stored at the first location and hence old data will be over written and lost. The storage is done on FIFO basis and amount of storage can be made to move by simply updating software, as the RAM is available 8 K.

12.6.3 System Hardware

The microcontroller 80C31 is used. It generates 11.059 MHz clock using Pin 18 and Pin 19 from the crystal input.

Port 1 is used as input/output line for getting data from telephone interface and user key is scanned by reading the latch IC74541. These are the inputs for the control software. In actual, we can use eight optocouplers for eight telephone lines for isolation and to get TTL level output. For that optocouplers are interfaced with Port 1 of microcontroller.

A 6-key push-button type keyboard is directly interfaced with the data bus. Port 0 is used as address/data bus and Port 2 is used as address bus. The ALE signal is fed to 74245 chip for providing driving capacity to the data bus. EA pin of 80C31 is put at logic level low to use external program codes, and execute the program from external EPROM using PSEN signals. IC27256 is a 32 K EPROM chip which is connected with A0–A14 address lines. IC6264 is 8 K RAM. A0–A12 address lines are connected with this RAM. It is used as general purpose RAM. Its main job is to store the data regarding all the telephones.

74LS138 generates chip selections for RAM, LCM, and the keyboard input latch. It uses A13–A15 address lines as inputs. 7410 is used to generate logic for selection of display write. The output is LCM display, whereas 74245 is used as driver to the LCM interfaced with the microcontroller data lines. The complete details of the system hardware and the circuit description for the power supply port and the multi-telephone unit is described here.

Power supply Regulated power supply is required for all active components/ICs used in the system. The power supply of multi-telephone meter is conventional and based on fixed voltage regulators (i.e., linear type). The power supply generated +5 V regulated output for all active components/ICs used in our module, derived from 230 V, 50 Hz AC mains.

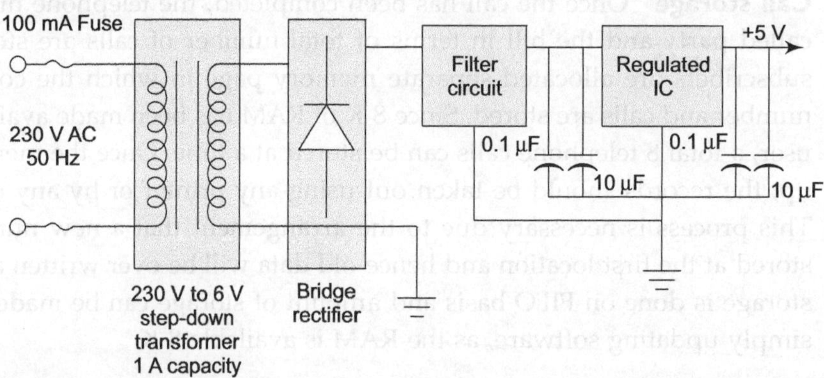

Fig. 12.3 Power supply

Regulated power supply is required for all active components/ICs used in the system. The power supply of multi-phonemeter is conventional and based on fixed voltage regulators (i.e., linear type). The power supply generated +5 V regulated output for all active components/ICs used in our module, derived from 230 V, 50 Hz AC mains.

Figure 12.3 shows power supply circuit. A 100 mA fuse is connected at the primary of the transformer for protection. The transformer is a 230 to 6 V step-down having current capacity of 1 A. The 6 V output of the transformer secondary is fed to bridge rectifier (using four IN4007 diodes).

The output of the bridge rectifier is unregulated DC voltage having unidirectional pulses is smoothened by a capacitor of 1000 µF. The regulation at the output (constant voltage +5 V output) is done using three terminal regulated IC7805. The capacitors of 0.1 and 10 µF between input voltage and ground are used to avoid spikes at input as well as output.

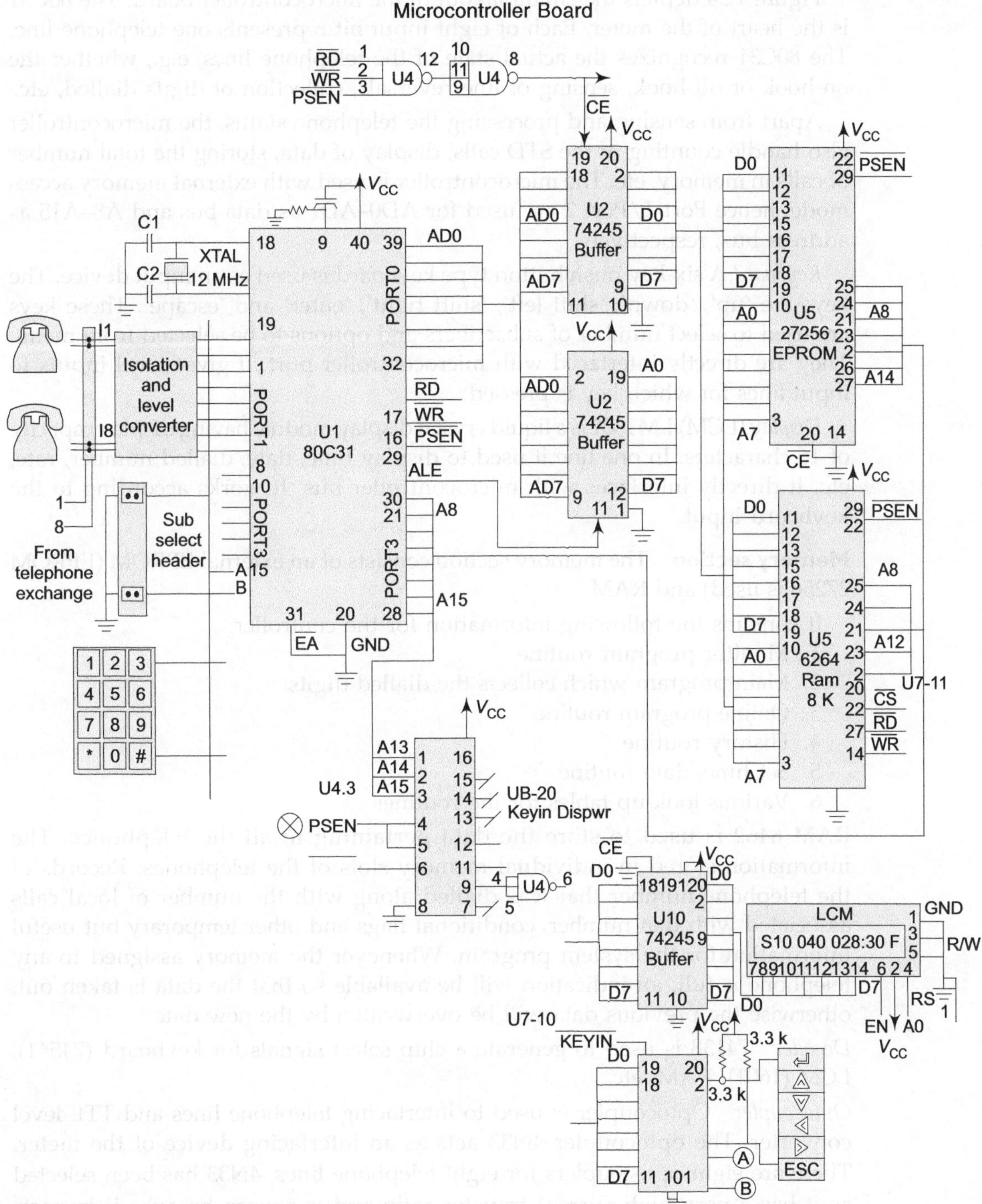

Fig. 12.4 Hardware layout of multi-phonemeter

Figure 12.4 depicts the circuit layout of the microcontroller board. The 80C31 is the heart of the meter. Each of eight input bit represents one telephone line. The 80C31 recognizes the actual state of the telephone lines, e.g., whether the on-hook or off-hook, sensing of line reversals, collection of digits dialled, etc.

Apart from sensing and processing the telephone status, the microcontroller also handle counting of the STD calls, display of data, storing the total number of calls in memory, etc. The microcontroller is used with external memory access mode, hence Port 1/Port 2 are used for AD0–AD7 as data bus and A8–A15 as address bus, respectively.

Keyboard A six-key push button type keyboard is used as an input device. The keys are 'up', 'down', 'shift left', 'shift right', 'enter' and 'escape'. These keys are used to select number of subscribers and options to be selected from menu. They are directly interfaced with microcontroller port. It gives level inputs to input lines for which key is pressed.

Display (LCM) LM1601 is a liquid crystal display module having display capacity of 16 characters. In one line it used to display time, date, dialled number, rate, etc. It directly interfaces with microcontroller bus. It works according to the keyboard input.

Memory section The memory section consists of an external EPROM (EPROM 27256 is used) and RAM.

It contains the following information for the controller:
1. Monitor program routine
2. Main program which collects the dialled digits
3. Online program routine
4. History routine
5. Set time/date routine
6. Various look-up tables for the routines

RAM 6462 is used to store the data pertaining to all the telephones. The information stored in individual memory slots of the telephones: Records of the telephone number that was dialled along with the number of local calls associated with that number, conditional flags and other temporary but useful information for the system program. Whenever the memory assigned to any telephone is full, an indication will be available so that the data is taken out, otherwise the previous data will be overwritten by the new date.

Decoder 74138 is used to generate a chip select signals for keyboard (74541), LCM (1601), RAM, etc.

Optocoupler Optocoupler is used to interfacing telephone lines and TTL level converter. The optocoupler 4N33 acts as an interfacing device of the meter. There are eight optocouplers for eight telephone lines. 4N33 has been selected as it has a very high current transfer ratio and is chosen because it imparts excellent isolation of the meter from the C/O lines. The optocoupler receives various telephone voltages at its input and generates a corresponding TTL level output, which is processed by the microcontroller through the input/output ports. The levels information is shown in Table 12.2.

Table 12.2   TTL voltage level information generated by optocoupler

	Actual level	Optocoupler output	Logic level
On-hook	– 48 V	5 V	1
Off-hook	– 10 V	0 V	0
1st reversal	– 48 V	5 V	1
2nd reversal	– 10 V	0 V	0

12.6.4 Operating Procedure

The steps followed for the system operation are described below:

1. Power on the system
   "MICROCONTROLLER-BASED MULTI-PHONEMETER" is displayed for a moment: Then,   **DATE   Hr : MIN**
   will be displayed.

2. Set date and time using keys
   Press 'enter' (↵)
   DATE SET OK will be displayed

3. Press 'ESC' : to obtain main MENU, showing select options as:
   ON LINE
   HISTORY
   SET DATE/TIME
   Use [↑/↓] keys & [↵] for selection.

4. Select HISTORY : by pressing ENTER
   ENTER LINE NO:/REC NO : ___ / ___
   Will be displayed
   Pressing 'ENTER' again message ;
   'NOTHING STORED'
   is displayed

5. Select ON LINE by pressing 'ENTER'
   Select subscriber code (e.g., sub = 1)
   By connecting header
   SUB 1 OFF-HOOK
   Will be displayed
   Dial desired number
   Dialed number will be displayed
   Provide B/R by defined key

Display will show:

SUB.1 011 is displayed

[011 is no. of B/R provide]

Disconnect the header, following message will be displayed:

SUB 1 011 16:41   F

[F indicates full charge]

16:41- payable amount in Rs. ***

6. Similarly, we can enter many data on different subscribers numbers using Step 5.

7. Press 'ESC' for getting Main Menu again

8. Select HISTORY and enter line number and record number by the same procedure as in Step 4:

Then following information is obtained.

Date/Time of which call is done

Dialed No.

Units – Rate – (H/F/Q/T – change)

H - Half charge

F - Full charge

T - One-third charge

Q - Quarter charge

9. If querry for not existing record is made by pressing HISTORY 'DOES NOT EXISTS' is displayed.

10. SET TIME/DATE can be used to set any desired date/time during initialization.

12.6.5 System Software

The system is based on microcontroller 80C31 chip. The main peripherals include LCM, EPROM, RAM, ADDR decoder, buffers, and latch ICs. The control software inside the EPROM does the following things:

Initialize variables, devices microcontrollers and resistors during power ON. Give facility to date/time using keys like 'up', 'down', 'right', 'left', 'enter' and 'esc'. It gives control to user for selection of modes of operation: On line, History, and Date/time.

The 'On line' mode gives the line status of the subscriber on the display. It shows the dialled number, number of unit, etc.

(a) History option gives stored date records for the selected subscriber.

(b) Date/time option sets date/time and it can also be used as general clock.

(c) The main input/outputs to the software and keyboard and line selection are inputs to the controller and display (LCM) is an output of the controller unit.

After initialization, microcontroller waits for date/time setting, once it is settled, the system is not used as general clock unit. When Esc key is pressed, it gives Main Menu on display where On line/History/Date/Time options be available for user.

1. Once 'On Line' is selected from main menu, the system software searches for Off-hook condition, it waits for digits to be dialled. It collects it and display on LCM. After dialling is completed it waits for battery reversal (B/R) signal for successful call establishment. Once it gets B/R signal it starts collecting the same and display the number of units collected. The process will continue unit 'On-hook' condition on the same subscriber is detected.

2. Once 'On-hook' condition detected then the call will be considered as a matured call and the available data like subscriber number, dialled number, number of unit, time and date, etc. will be stored as a record for that subscriber. Now new call setup will be ready for that subscriber.

3. Once 'History' option is number to be shown from available records. The entry will be based on keyboard. Once correct entry has been done, it shows stored data one by one like dialled number, number of units, time, date, changes, etc. The history will be based on FIFO and only last eight records will be available.

EXERCISES

12.1 Define: fully connected networks, DTMF, ISDN.

12.2 Explain: time divison switching, space divison switching.

12.3 Enlist principles of ISDN standards.

12.4 What do you mean by conferencing?

12.5 Enlist advantages of auto-conferencing.

12.6 With the help of schematic block diagram explain operation of auto-conferencing system.

12.7 Why steering circuit is required in auto-conferencing system?

12.8 Develop a setup for auto-conferencing system employing LCD instead of LEDs.

12.9 What do you mean by term multi-phonemeter? In what situation are they useful?

12.10 Draw a block diagram showing essential components of a multi-phonemeter and explain their functions.

12.11 Explain address generation of the microcontroller system board shown in Fig. 12.4.

12.12 Explain memory interfacing of the setup shown in Fig 12.4 and specify the address mapping.

12.13 Describe the function of the microcontroller hardware for proper operation of multi-phonemeter. Describe features of the software requires for the same.

12.14 Explain operating procedure of the multi-phone-system.

12.15 Develop the program modules for the proto-type functions defined in the system software Section 12.6.5

* setflg (): To set the flag indication which decides the pusle rate depending on the time at which call is made.

* scankey(): To scan the key pressed.

* gethsno(): To obtain and identify the requesting subscriber

* history() : To obtain history information from multimeter

* set-dtm(): To set or initialize the date information for the operation of multi-meter

Application V: IR-Based Wireless Communication

In process control industries, it is necessary to transmit data using control signals between central unit and remote units. The communication done using telephone lines or power lines is termed as *wired communication*. In cases where remote units are located at long distances, wired communication becomes costly, complex, and impractical.

13.1 | WIRELESS COMMUNICATION

Various techniques used in wireless communication are:

1. **Infrared communication** It uses frequencies in infrared region. The wavelengths are below 400 mm. It is the cost-effective solution for remote controller but has to compromise with the line-of-sight communication.
2. **Radio frequency (RF) communication** It is characterized by a small size transmitter and a long distance communication. The length depends on the height of the antenna.
3. **Ultrasonic communication** It uses sound waves having frequency of 40 kHz. The length of communication depends on the transmitted power.

13.2 | IR COMMUNICATION SYSTEM

Figure 13.1 depicts the block schematic of a general IR communication system. To transmit the frame, it is required to drive the light source such that sufficient power is launched at the desired frequency to produce adequate reception. The primary concern competing with the feeble IR transmitted signal is the ambient light environment created by light sources of relatively high power, such as local incandescent source, flourescent light, and sunlight. These produce large amount of infrared energy, especially sunlight.

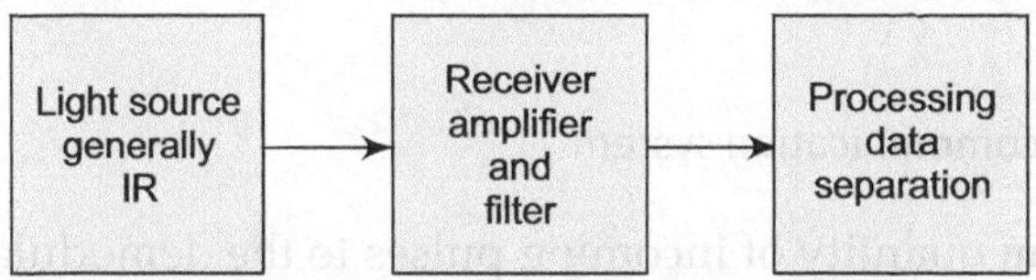

Fig. 13.1 General IR communication system

There are three basic detection schemes using the photo-transistor, the photo diode and the Darlington pair. In almost all remote control applications the diode is the detector of choice. It is free from saturation even in most sunlit environment. The sensitivity is a disadvantage, but it is balanced by the response speed. The available transmission frequency is 50–100 kHz. The range of IR remote control varies from 5 cm to 100 m and makes use of SL207 (IR LED).

As shown in Fig. 13.1, the purpose of filter is to attenuate the visible portion of the spectrum while leaving the IR intact. It may be Kodak filter series, or acrylic plastic (red non-opaque plastic). This filter does not alter IR signal heavily. This filter is placed in front of the detector. The detector diode behind the filter is usually constructed as a large geometric device specifically designed for IR remote control. The diode has large area, simply for more IR energy reception. In addition to visible light filtering mentioned above, electrical filtering must also be applied to attenuate the low frequency. The interference is present in visible and IR region.

The signal is amplified by an amplifier. After the signal is brought up to a level sufficient for detection, the data is extracted from this signal. The peak detector or PLL can be used for data extraction. The process may be controlled by the microcontroller. A typical setup for the IR communication system is shown in Fig. 13.2.

The environment has various IR sources such as our body, lamps or hot bodies; hence the remote controls pulsate their infrared in a certain frequency (e.g., 30 kHz or 60 kHz). The IR receiver module in the device 'tunes' to this certain frequency and ignores all other IR received. Commonly used frequencies are 36 kHz or 38 kHz. Infrared light emitted by IR diodes is pulsated at 36 kHz. It is difficult to receive and identify this frequency, hence the infrared receivers contain filters, decoding circuits, and the output shaper to generate a square wave for detection of existence of incoming pulsating infrared.

A square wave of approximately 27 µs injected at the base of a transistor drives an infrared LED to transmit this pulsating light wave (receiver) will switch its output to high level (+5 V).

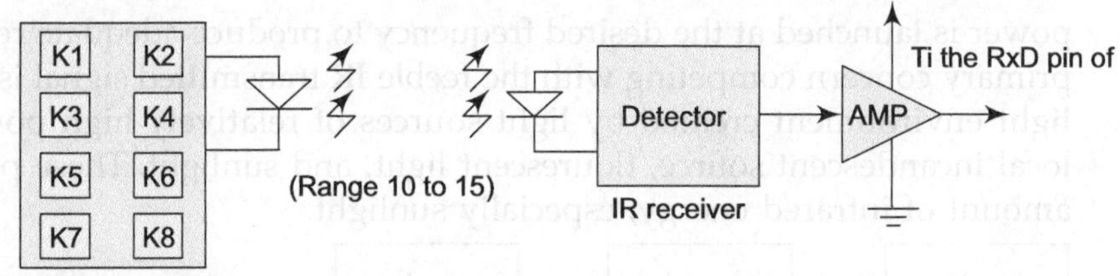

Fig. 13.2 Setup for IR communication system

There is a minimum quantity of incoming pulses to the demodulator, e.g., the number of pulses used at the Philips remotes are 32 pulses for each half-bit, 64 pulses per bit. So, a bit 0 to be transmitted means 32 square pulses of 27 µs each, then 32 × 27 µs of silence. Bit 1 is the opposite, 32 × 27 µs of silence followed by 32 square pulses of 27 µs. The microcontroller decode the received waveform at the demodulator normally by checking the transition state in the middle of the transmission pulse. The IR transmitter in Fig. 13.3 transmits the digital data stream after converting it into IR signal.

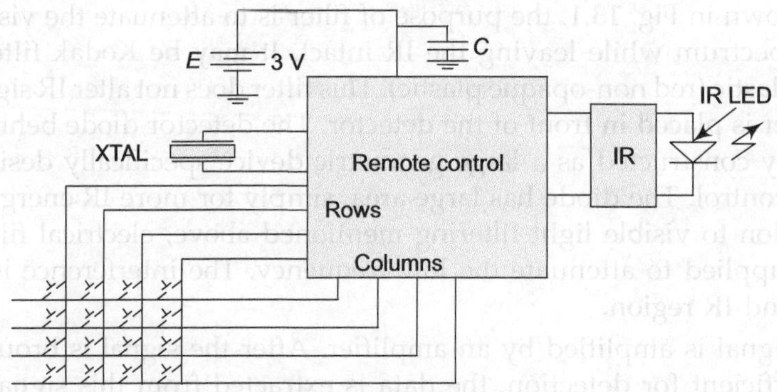

Fig. 13.3 Basic transmitter setup

The data stream is transmitted only after any one of the keys is pressed. The digital level TTL data stream before being converted into a IR signal is shown in Table 13.1.

Table 13.1 IR data stream

Frame ID	Data 1	Data 2	— — — — — — —	Data n	End frame

The frame ID is a unique signal which indicates the frame specification, i.e., start of the data. Data 1 to Data *n* varies depending upon the key pressed while the end of frame is fixed for all the types of data. Transmitter will transmit different frames for different keys when pressed, having same ID and END fields. IR transmitter uses IR LED to transmit the digital data stream generated by the unit according to the key pressed. The unit uses internal battery cell of (1.5 × 2). It consumes power during transmission only (i.e., when operation on the key is done). The IRT front is kept in front of the IRR device of the receiver to activate maximum distance coverage (normal distance is 10 to 15 ft).

Normally, a transmitter is isolated with a battery supply. To increase the life of a battery, the power is used only during transmission of the frame. Table 13.2 shows the code look-up table for keycode along with the data frame transmitted by a representative transmitter. The code changes with the make and model of the remote controller.

Table 13.2 Look-up table: keycode and data frame for remote keypad

Key no.	Key code	Data frame
1	912	Start pulse ... 33 ms ... 0.1 ms ... 4 ms ... IDP
2	732	
3	613	
4	514	
5	541	
6	414	
7	A2.1	
8	415	

13.3 | IR TRANSMITTER

IR remote transmitter works on the principle of pulse code modulation. Remote transmitter generates a fixed frequency to accomplish this. The generation of fixed frequency is accomplished with the help of a crystal. Carrier pulse equal to the reference frequency is generated by the sequence pulse generator. Information produced by pressing a key in the keyboard key matrix is encoded in the key-in encoder and comes out from this section as digital code waveform information.

This digital information is converted into the signal voltage with the help of carrier pulse in the code modulator section. The final signal before sending it to the LED driver is amplified in the output buffer/driver section. The output from the output buffer/driver is fed to the infrared range converter which converts this output into a range of infrared signals with the help of an IR LED.

The infrared signal or remote signal is then transmitted towards the remote-control receiver. The remote control receiver which is also known as the IR receiver, gives an appropriate command to the microcontroller on receiving these signals, after which the microcontroller performs the task specified by the user.

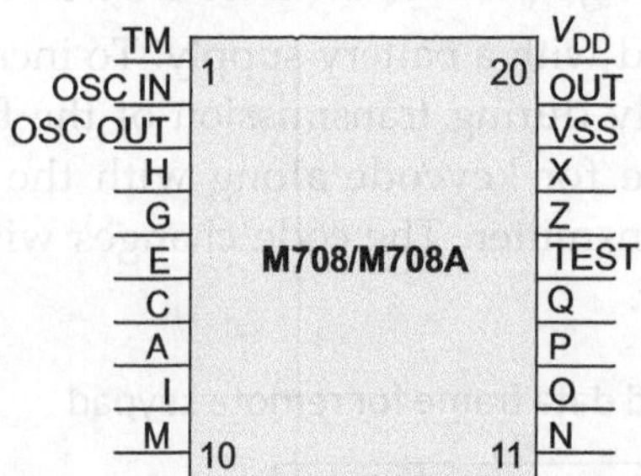

Fig. 13.4 PCM transmitter IC

Figure 13.4 depicts the pin configuration of PCM transmitter IC.

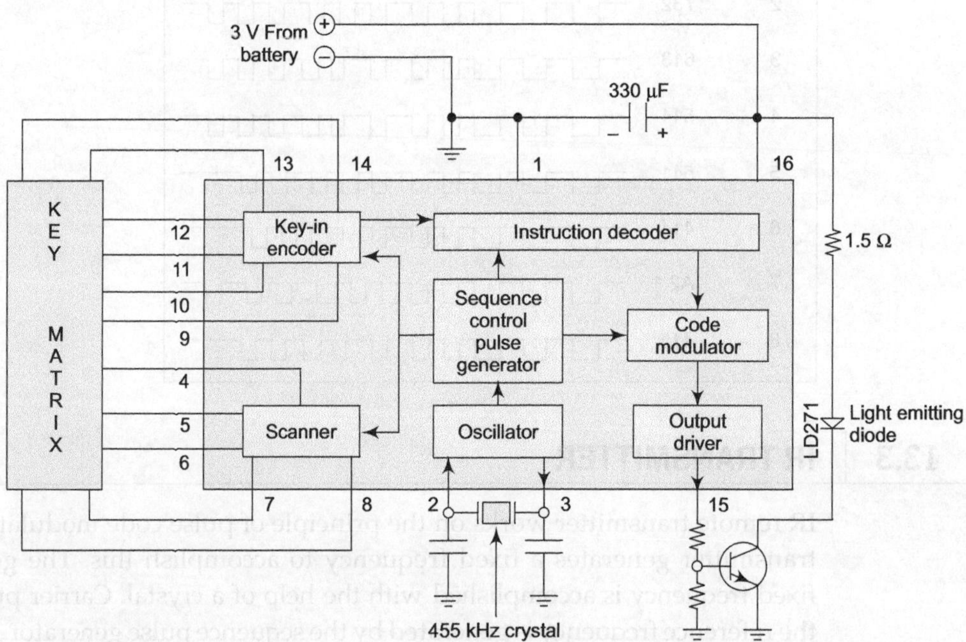

Fig. 13.5 Block schematic of IR transmitter

As shown in Fig. 13.5, the main sections of remote transmitter IC are as follows:

Oscillator It generates oscillation frequency required for the transmitter as reference frequency. This oscillation frequency is controlled by crystal and capacitors connected with the crystal. The frequency generated by the oscillator is used as the reference frequency. As the number of channels and other information for each transmitter is different, the value of crystal oscillator also differs in each transmitter. The frequency generated by the oscillator section is fed to the sequence control pulse generator section.

Sequence control pulse generator It is used to convert the frequency received from the oscillator section into a carrier pulse of fixed value. The reference frequency is converted in a pulse train by dividing it into various equal clock pulses. The pulse generated by this section is given to various other sections of the IC.

Scanner section This section together with the scanner section is used to find out the key being pressed on the remote control keyboard. The scanner section is used to find out the column number of the key being pressed. Together these two sections will give information about the exact key pressed on the key matrix. After recognizing the pressed key, this section converts this information into proper digital output which is sent to the instruction decoder.

Instruction decoder It generates proper digital code based on the information received from the key-in encoder. Pressing of different keys such as volume up, down and brightness up, down. On the remote control, keyboard will make this section to generate different control words. This control word is understood by the remote control receiver in the TV and then proper action is taken.

Code modulator The control word generated by the instruction decoder is send to the code modulator before transmitting. At this section the control word is converted into proper format for transmission. It is converted into proper format using the pulse code modulation (PCB) method where the 1s and 0s based on their position [odd or even] will be converted into pulses having 1T, 2T, or 3T pulse width, where T is the time base. For this job, the carrier pulse generated from the sequence pulse generator section is also used. Technically, it can be said that this section modulates the code waveforms received from instruction decoder into the format required for transmission output buffer/ driver. The modulated code waveform is send from the code modulator section to the output buffer/driver section before sending it to the infrared LED for transmission. This section stores this signal from code modulator section and amplifies it into a proper level.

Infrared range converter This section is not a part of the transmitter IC, it is outside the IC, but as it is a part of the transmitter, we are explaining this section here. This section amplifies the decoded signal coming out of the buffer/ driver section to their park level. This section also converts this amplified signal into a infrared signal with the help of the IR LED. This infrared signal from LED is then transmitted toward the remote-control receiver.

The transmitter may be designed around the PCM transmitter IC M708/ M708A (Fig. 13.6).

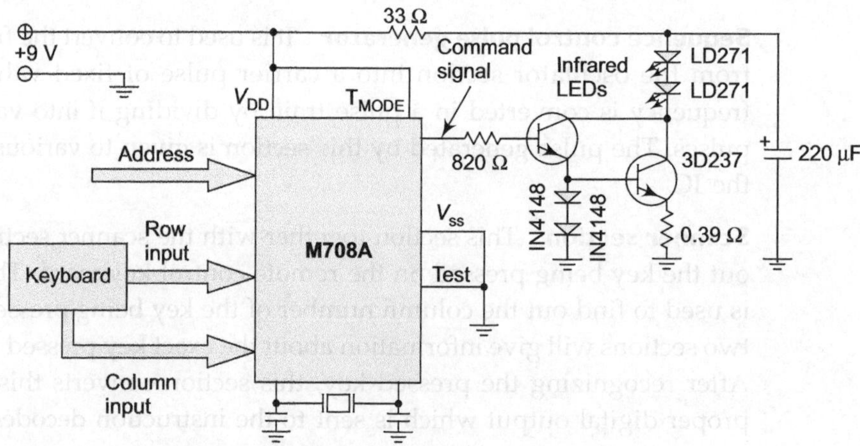

Fig. 13.6 IR transmitter: block schematic

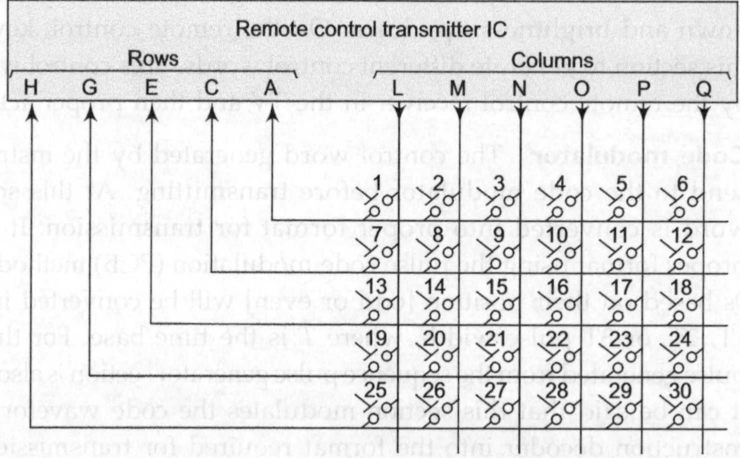

Fig. 13.7 Interfacing matrix keyboard with IR transmitter IC

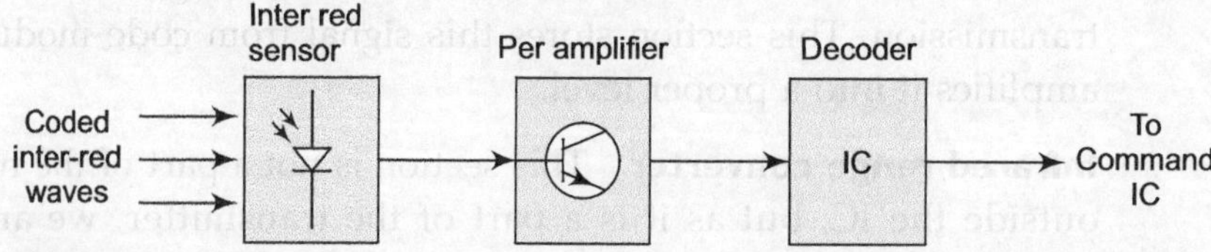

Fig. 13.8 IR receiver section

Figure 13.7 depicts the interfacing of a matrix keyboard with the transmitter IC. Table 13.3 gives function codes for the key on the panel. This may be used for development of a software routine for the key closure, key and function code identification to provide the desired service by the controller.

Table 13.3 Key code command signal generator

Key	Row					Column						Command signals					
0	A	C	E	G	H	L	M	N	O	P	Q	CI	C2	C3	C4	C5	C6
1	1	0	0	0	0	1	0	0	0	0	0	1	0	0	0	0	0
2	1	0	0	0	0	0	1	0	0	0	0	1	1	0	0	0	0
3	1	0	0	0	0	0	0	1	0	0	0	0	0	1	0	0	0
4	1	0	0	0	0	0	0	0	1	0	0	1	0	1	0	0	0
5	1	0	0	0	0	0	0	0	0	1	0	0	1	1	0	0	0
6	1	0	0	0	0	0	0	0	0	0	1	1	1	1	0	0	0
7	0	1	0	0	0	1	0	0	0	0	0	1	0	0	0	1	0
8	0	1	0	0	0	0	1	0	0	0	0	1	1	0	0	1	0
9	0	1	0	0	0	0	0	1	0	0	0	0	0	1	0	1	0
10	0	1	0	0	0	0	0	0	1	0	0	1	0	1	0	1	0
11	0	1	0	0	0	0	0	0	0	1	0	0	1	1	0	1	0
12	0	1	0	0	0	0	0	0	0	0	1	1	1	1	0	1	0
13	0	0	1	0	0	1	0	0	0	0	0	1	0	0	0	0	1
14	0	0	1	0	0	0	1	0	0	0	0	1	1	0	0	0	1
15	0	0	1	0	0	0	0	1	0	0	0	0	0	1	0	0	1
16	0	0	1	0	0	0	0	0	1	0	0	1	0	1	0	0	1
17	0	0	1	0	0	0	0	0	0	1	0	0	1	1	0	0	1
18	0	0	1	0	0	0	0	0	0	0	1	1	1	1	0	0	1
19	0	0	0	1	0	1	0	0	0	0	0	1	0	0	0	1	1
20	0	0	0	1	0	0	1	0	0	0	0	1	1	0	0	1	1
21	0	0	0	1	0	0	0	1	0	0	0	0	0	1	0	1	1
22	0	0	0	1	0	0	0	0	1	0	0	1	0	1	0	1	1
23	0	0	0	1	0	0	0	0	0	1	0	0	1	1	0	1	1
24	0	0	0	1	0	0	0	0	0	0	1	1	1	1	0	1	1
25	0	0	0	0	1	1	0	0	0	0	0	1	0	0	1	1	1
26	0	0	0	0	1	0	1	0	0	0	0	1	1	0	1	1	1
27	0	0	0	0	1	0	0	1	0	0	0	0	0	1	1	1	1
28	0	0	0	0	1	0	0	0	1	0	0	1	0	1	1	1	1
29	0	0	0	0	1	0	0	0	0	1	0	0	1	1	1	1	1

13.4 | PROTOCOLS FOR IR REMOTE CONTROLLER

The key code transmitted indicates the services desired by the user. IR transmitter uses standard communication protocols for transmission manufactures of the remote controllers for TV/VCD or other electronic equipment use different protocols.

The section provides a brief summary of most commonly used protocols in IR remote control.

13.4.1 ITT

ITT is a very old one and does not use a modulated carrier frequency to send the IR messages. A single command is transmitted by a total of 14 pulses with a width of 10 µs each as shown in Fig. 13.9. The command is encoded by varying the distance between the pulses. It consumes very little power hence has long battery life. Three different time intervals between the pulses are used to get the message across: 100 µs for a Logic 0, 200 µs for a Logic 1, and 300 µs for the lead-in and lead-out.

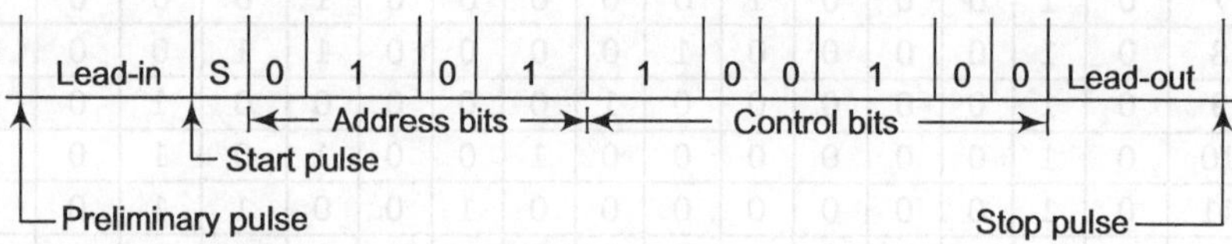

Fig. 13.9   Protocol frame format: ITT

13.4.2 JVC

This format is used by JVC equipment. It has 8-bit address and 8-bit command length. It uses pulse distance modulation with a carrier frequency of 38 kHz (Fig. 13.10). It uses pulse distance encoding of the bits. Each pulse is a 526 µs long 38 kHz carrier burst. A Logic 1 takes 2.10 ms to transmit, while a Logic 0 takes 1.05 ms with a carrier duty cycle is 1/4 or 1/3.

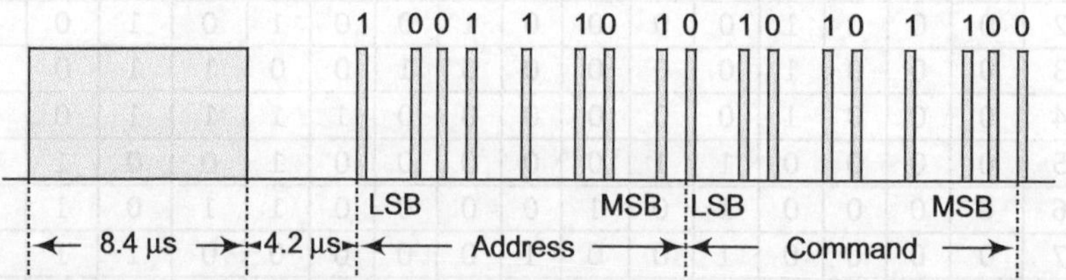

Fig. 13.10   Protocol frame format: JVC

A message is started by a 8.4 ms AGC burst which sets the gain of the earlier IR receivers. This AGC burst is then followed by a 4.2 ms space, the address and command. An IR command is transmitted every 50–60 ms for as long as the key on the remote is held down. First command is preceded by the 8.4 ms pre-pulse and 2.4 ms space, which is used by the receiver to determine multiple key closure.

13.4.3 NEC

This protocol also uses pulse distance encoding of the bits. Each pulse is a 560 µs long 38 kHz carrier burst. A Logic 1 takes 2.25 ms to transmit, while a Logic 0 takes 1.12 ms with carrier duty-cycle is 1/4 or 1/3.

13.4.4 NRC

The Nokia Remote Control protocol uses 17-bits to transmit the IR commands (Fig. 13.11). The protocol was designed for Nokia consumer electronics. Currently, it is mainly used in Nokia satellite receivers and set-top boxes. It has 8-bit command, 4-bit address, and 4-bit sub-code length and uses bi-phase coding with a carrier frequency of 38 kHz. The protocol uses bi-phase (NRZ) modulation of a 38 kHz IR carrier frequency. All bits are of equal length of 1 ms, with half of the bit time filled with a burst of the 38 kHz carrier and the other half being idle. A Logic 1 is represented by a burst in the first half while a Logic 0 in the second half of the bit time. The pulse/pause ratio of the carrier frequency is 1/4.

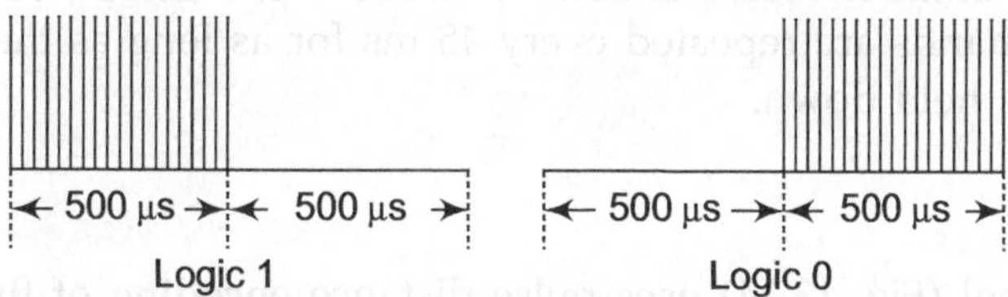

Fig. 13.11   Protocol frame format: NRC

13.4.5 Philips RC-5

The RC-5 code from Philips is possibly the most used protocol due to the wide availability of cheap remote controls. Philips has started using a new protocol called RC-6 which has more features. It has 6-bit command and uses bi-phase coding with a carrier frequency of 38 kHz. A message consists of a total of 14 bits, which adds up to a total duration of 25.2 ms. A toggle bit is used by the receiver to detect multiple key closure The protocol uses bi-phase (aka Manchester coding) modulation of a 36 kHz. The bit time is 1.8 ms. The pulse/ pause ratio of the 36 kHz carrier frequency is 1/3 or 1/4 which reduces power consumption. Frame timings are shown in Fig. 13.12.

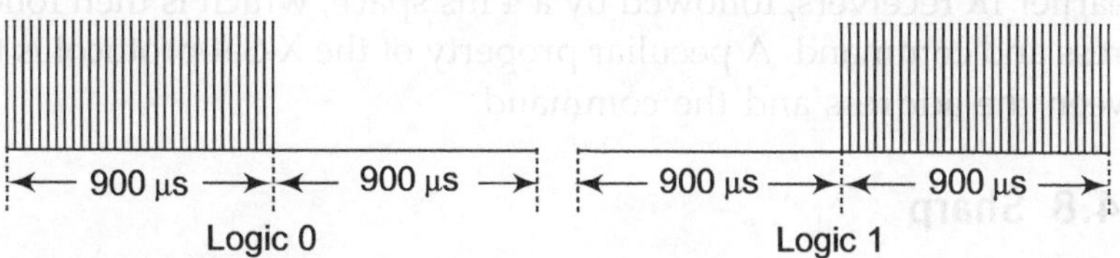

Fig. 13.12   Protocol frame format: Phillips RC-5

13.4.6 Sony SIRC

The Sony SIRC protocol (Fig. 13.13) is 12-bit and uses a pulse width encoding of the bits. The pulse representing a Logic 1 is a 1.2 ms long burst of the 40 kHz carrier, for a Logic 0 it is 0.6 ms long. All bursts are separated by a 0.6 ms long space interval with a carrier duty-cycle of 1/4 or 1/3.

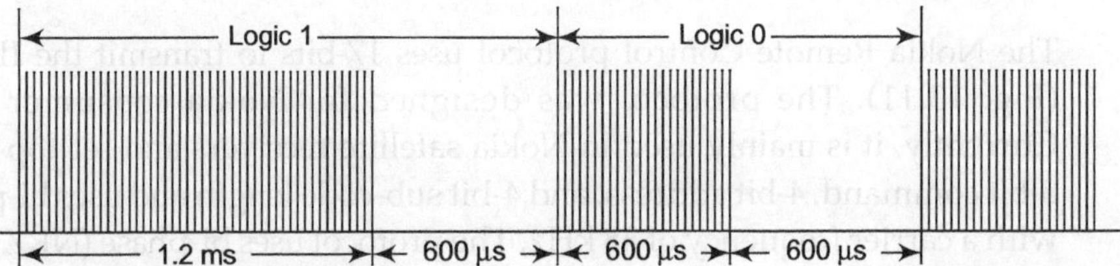

Fig. 13.13 Protocol frame format: Sony SIRC

The LSB is transmitted first. The start burst is always 2.4 ms wide, followed by a 0.6 ms space. In addition to signalling, the start of a SIRC message is used to adjust the gain of the IR receiver. Command uses 7-bits and the device address has 5-bits. Commands are repeated every 45 ms for as long as the key on the remote control is held down.

13.4.7 X-Sat

The X-Sat protocol (Fig. 13.14) uses pulse distance encoding of the bits. Each pulse is a 526 µs long 38 kHz carrier burst. A Logic 1 takes 2.0 ms to transmit, while a Logic 0 takes 1.0 ms with the carrier duty cycle of 1/4 or 1/3.

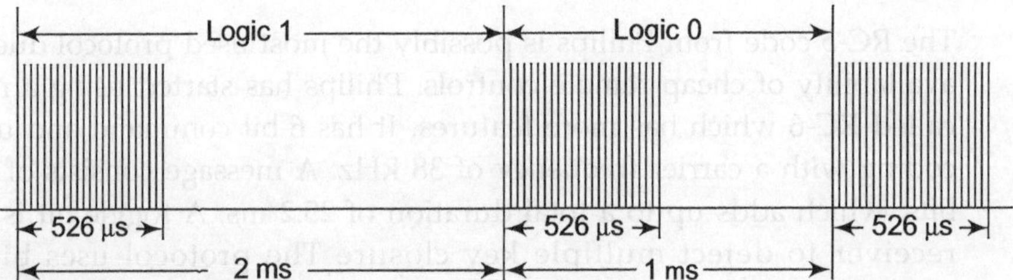

Fig. 13.14 Protocol frame format: X-Sat

A message is started by a 8 ms AGC burst, which was used to set the gain of the earlier IR receivers, followed by a 4 ms space, which is then followed by the address and command. A peculiar property of the X-Sat protocol is the 4 ms gap between the address and the command.

13.4.8 Sharp

The Sharp protocol uses a pulse distance encoding of the bits. Each pulse is a 320 µs long 38 kHz carrier burst. A Logic 1 takes 2 ms to transmit, while a Logic 0 takes 1 ms with the carrier duty-cycle is 1/4 or 1/3, as shown in Fig. 13.15.

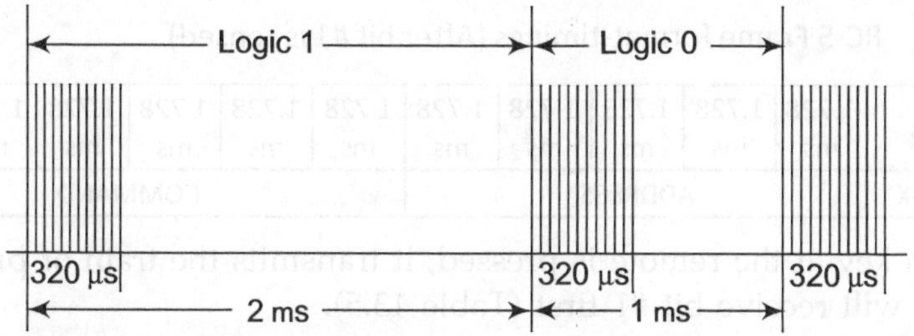

Fig. 13.15 Protocol frame format: Sharp

13.5 | DIGITAL CONTROLLER FOR WIRELESS COMMUNICATION

Our goal in this section is to design a controller that has the following features:

1. It can identify the remote transmitter used.

2. A facility to display function codes.

3. An interface for most of the I/O devices used in the industrial environment.

The purpose is to create a conceptual platform for understanding the use of microcontrollers in wireless communication using IR. The procedure for code detection and identification is simple: Select appropriate transmitter protocol or generate same protocol using existing microcontroller. Use bit bagging polling a bit or interrupt for frame grabbing.

Table 13.4 RC-5 frame format

1	2	3	4	5	6	7	8	9	10	11	12	13	14
AGC		CHK			ADDRESS						COMMAND		

Table 13.4 shows 14-bits of the RC-5 system. The first 2-bits, #1 and #2, are called *AGC calibration*. They are ON level , used to calibrate the IR receivers auto-gain control.

The bit #3 is the check bit, every time you press a key at the remote, even pressing repeatedly the same key, this bit flips state. To receive this signal using a microcontroller, follow the timing from Table 13.4. Note that the infrared receiver invert the bit signal, low level means bit ON. During inactivity (no infrared present) the output of the infrared receiver is UP (bit zero). Connect the IR receiver output to any input port pin or interrupt pin of the microcontroller for polling or prepare the interrupt routine to trigger after the first low level sensed.

Table 13.5 RC-5 Frame format-timings (After bit #1 is sensed)

4.762 ms		1.728 ms	1.728 ms	1.728 ms	1.728 ms	1.728 ms	1.728 ms	1.728 ms	1.728 ms	1.728 ms	1.728 ms	1.728 ms
AGC	CHK	ADDRESS					COMMAND					

When a key at the remote is pressed, it transmits the train of pulses, microcontroller will receive bit #1 first (Table 13.5).

It will be sensed right after the middle of the bit, when it changes from HIGH to LOW level to detect start. Wait for 3-bits (2.75-bits time) receive the address bits.

On sensing the first low level the software waits for 4.752 ms, read the next 11-bits spaced 1.728 ms each (first 5-bits: address and next 6-bits: command). Use always key number ZERO for this software calibration.

The operation/function to be carried out by the digital controller is specified by pressing a code on the IR remote keypad. The transmitter will send the data frame of the key code (as shown in Fig. 13.2 of the key pressed). The digital controller grabs frame from remote transmitter in a 10-bit UART format via serial port.

The transmitted code is detected. by the micorcontroller and the operation specified by the function key at the remote end is carried out. The IR detector detects IR signal and converts after pre-amplification and amplification into TTL level output. The IR detector is operated by a +5 V power supply and generates TTL level data output. The signal TTL is given to the microcontroller. The microcontroller identifies the key code-code and activates or deactivates the relay drive circuit.

The keypad connected to the controller board is termed as manual keypad having functions similar to those on the remote transmitter. When manual key is pressed, corresponding relay will be activated or deactivated through relay driver.

The system operates on sensing the identification mark. While receiving, it collects serial data until the frame ends. It compares with the key code sent and takes appropriate action, i.e., it toggles that output line and outputs the data. The action is made only when valid frame is received, otherwise it rejects the frame. To avoid noise pick up, three consecutive frames are checked before making decision regarding validity of the frame.

The output lines are fed to the buffers to avoid the data loading and misconnection between output connections and the microcontroller. Each output of the buffer is fed to the transistor which in turn activates/deactivates the relay coil. The contact of the relay will be used as contact point corresponding to either manual or the remote controller key.

When HIGH output is fed from the buffer. Transistor goes into saturation and operates relay by collector current following through the transistor and the contact gets through. When LOW output is given through the buffer, transistor is switched OFF, current become zero and relay contact opens. LEDs glow when output line controlled by the microcontroller becomes HIGH. While it is OFF when the output is low.

13.5.1 Hardware Setup

Figure 13.16 shows block diagram of the digital controller. It has following sections:

Power supply section Two power supplies one for on-board hardware and another for the auxiliary hardware such as stepper motor and DC motor. This section requires bridge rectifier, filter and regulator to rectify, filter and regulate voltages as per requirement. This section supplies +5 V for all other section and raw DC for EM relays, solenoid valves, stepper motor, etc. Total reliability of the system based on reliability of this section. For driving stepper motor, separate power supply is required.

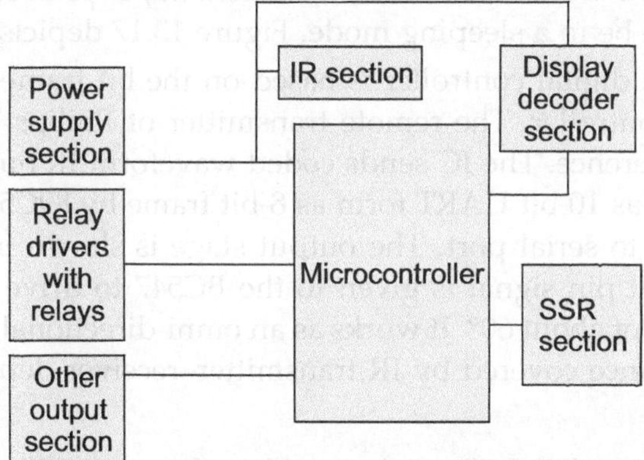

Fig. 13.16 Block diagram of remote operated digital controller

Relay driver section Relay driver section drives EM relays. It has clamp diodes and connectors for each relay for external connection. To drive relays, either individual transistors or npn Darlington arrays may be used.

Display decoder section It multiplexes eight displays via display decoder and has capability to drive segment bus as well as digit selection. To drive display segment, bus driver has to be used.

Solid state relay driver It has a ZCD for reference. It is capable to drive large current triac for application.

IR sensor IR sensor senses IR codes from remote transmitter and gives them to the microcontroller programmed unit via serial port.

Microcontroller unit It consists of the 89C51/89C52 controller along with the reset and the clock circuit.

Other outputs:

(a) PWM for DC motor

(b) Pulse output for stepper motor

(c) Relays for 230 V operation

(d) Variable AC output

13.5.2 Hardware Implementation

The system hardware implementation consists of design of two main units: remote transmitter and the microcontroller board with IR sensor and auxiliary interfaces for connecting input or output devices.

Remote Transmitter

Different ICs are available as one-time programmable (OTP) for transmitter design. Using 89C51, we can design remote transmitter. We use a transmitter that uses IC SAA3010. IC is such that until and unless key is pressed there is no consumption as IC will be in a sleeping mode. Figure 13.17 depicts the setup.

The operation of the digital controller is based on the bit frame sent by the domestic TV remote controller. The remote transmitter of Philips TV of 21 in. TV set is taken as a reference. The IC sends coded waveform IR out pin of that IC and that is detected as 10-bit UART form as 8-bit frame by 89C51 serial port with sensor connected to serial port. The output stage is shown in Fig. 13.17. The output from IR out pin signal is given to the BC547 to drive the IR LED having the beam width of about 60°. It works as an omni-directional for distance of about 10–12 ft. Distance covered by IR transmitter–receiver depends on the following factors:

Supply voltage of anode of IR LED   It is fixed here because it is a 3 V battery handy tool.

Supply voltage of IR sensor   It is also fixed and it is +5 V due to direct coupling with microcontroller serial port.

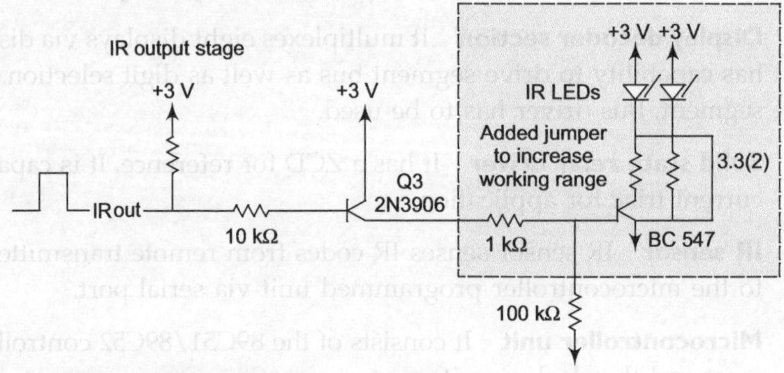

Fig. 13.17   Setup for IR transmitter

Peak current from IR LED It depends on supply; series resistor that is nearly zero as well as duty cycle of waveform. For a system with 1 kHz continuous square wave output having duty cycle of 50%, average current through IR LED is half of the peak current. As the duty cycle decreases with the frequency of 1 kHz and on time 1 ms. The average current is 1/10th of peak current.

Pre-amplifier sensitivity of IR sensor The sensors available in market are 3-pin sensors. If matched pair is established then the distance of about 20 ft can also be OTP. ICS are also available from Philips for remote transmitter and IR OUT pin is active low and able to sink current of about 27 mA. So only one pnp transistor is sufficient to drive an IR LED. Generally, LD271 is used. It is having plastic package, transparent, and emitting cup is exactly at focal point of the cover lens. Lens is such that it gives beam width of about 60°.

Figure 13.18 depicts a front layout of a representative IR transmitter unit with key codes.

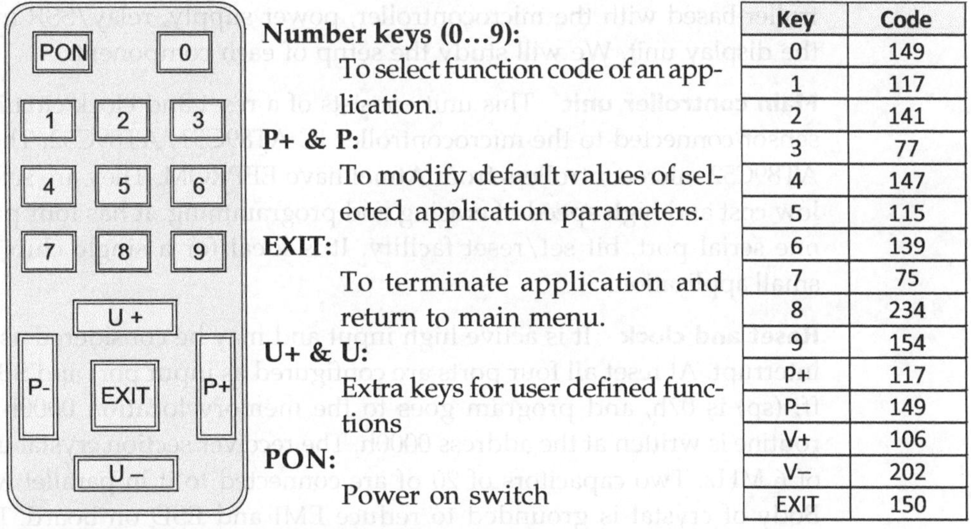

Number keys (0...9):
: To select function code of an application.

P+ & P:
: To modify default values of selected application parameters.

EXIT:
: To terminate application and return to main menu.

U+ & U:
: Extra keys for user defined functions

PON:
: Power on switch

Key	Code
0	149
1	117
2	141
3	77
4	147
5	115
6	139
7	75
8	234
9	154
P+	117
P–	149
V+	106
V–	202
EXIT	150

Fig. 13.18 IR remote transmitter panel

Remote Receiver

Three steps of the receiver design are

1. Detection of any effect on serial RxD pin while any remote transmitter or any switch is pressed. This step is implemented by connecting IR sensor output pin to the RxD pin. Load the baud rate deciding register with F3h (1200-bits/sec at 6 MHz). In the software at vector address of serial interrupt 0023h provide a jump instruction to the desired location. The

program corresponding to the interrupt complements a bit connected to the ULN 2003 IC, so a relay is activated or deactivated at the reception of each serial interrupt occurrences.

2. The second step is receiving the code byte in the form of 8-bit. Detect the byte received on the RxD pin when any remote transmitter key is pressed. TH1 is used in Mode 1 of Timer 1 in auto-reload mode to generate the desired baud rate. The value in the TH1 can be selected for different standard baud rates helps user to select rate for the IR transmitter in use. Apply the detected code theory for the functional implementation using controller.

3. This step converts hex data received in SBUF to decimal and display it on the 7-segment display unit. This reflects the effect of each switch is displayed.

Remote Operated Controller Board

Consider the block diagram of Fig. 13.16. The controller unit is a microcontroller-based with the microcontroller, power supply, relay/SSR interface, and the display unit. We will study the setup of each component.

Main controller unit This unit consists of a reset and clock circuit and the IR sensor connected to the microcontroller IC AT89C51/AT89C52. The AT89C51/AT89C52 microcontrollers from Atmel have EEPROM. They are selected due to low cost and high speed of erasing and programming. It has four parallel ports, one serial port, bit set/reset facility. It is ideal for a single chip solution for small application.

Reset and clock It is active high input and may be considered as an ultimate interrupt. At reset all four ports are configured as input port and SBUF contains ff, (sp) is 07h, and program goes to the memory location 0000h. A boot up routine is written at the address 0000h. The receiver section crystal uses a crystal of 6 MHz. Two capacitors of 20 pf are connected to it in parallel with ground. Body of crystal is grounded to reduce EMI and ESD on board. Figure 13.19 shows the clock and the reset circuit. Figure 13.19 shows interfacing of clock generator and reset key. Crystal generates clock pulses by which all internal operations are synchronized. Crystal connected between XTAL1 and XTAL2. Typically quartz crystal and capacitors are used as resonant network. Frequency range of 1–16 MHz is fixed by manufacturers of the AT89C51.

Oscillator generates pulse train of crystal frequency and it establishes the smallest interval time within microcontroller. One-pulse time is phase (P) and two-pulse time is state (S). Machine cycle is made up of six states as shown in Fig. 13.20. We will use a quartz crystal of 6 MHz.

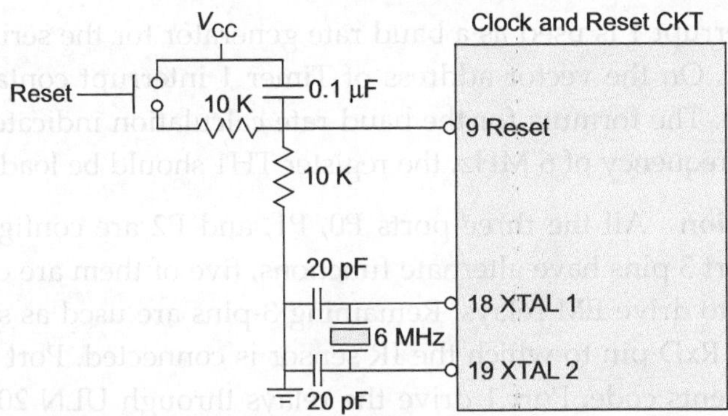

Fig. 13.19 Microcontroller clock

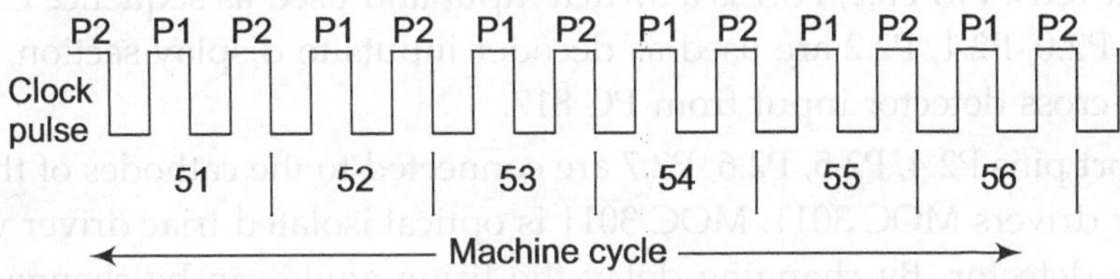

Fig. 13.20 Machine cycle of the 8-bit microcontrollers

IR sensor Remote sensor is the most important component. They have a preamplifier, post amplifier in a plastic package of 3-pin configuration. For example, SFH506 from Sharp semiconductor. It produces TTL compatible output signal. Specification of IR sensor used are as follows:

1. Voltage rating: Max up to 12 V via resistor
2. Maximum frequency range: up to 250 kHz
3. Distance covered: 20 ft with a perfect matched pair of remote transmitter IR LED
4. Beam width: about 60°
5. Colour: Black
6. Pins are as 5-pin sip 1-GND, 2-VCC, 3-NC
7. 4-O/P,5-NC

Interrupt allocation

1. On board switches are connected to the microcontroller interrupt pins to use the external interrupts 0 and 1.
2. Timer 0 interrupt is used as a refreshing mechanism for 7-segment display units used in multiplexed mode. On vector address of Timer 0, a program for the periodical refreshing is to be developed. The values in the th0, t l0 registers decide the refreshing time value. The Timer 0 is used in Mode 0.

3. Timer interrupt 1 is used as a baud rate generator for the serial port used in Mode 1. On the vector address of Timer 1 interrupt contains the (rti) instruction. The formula for the baud rate calculation indicate that for an oscillator frequency of 6 MHz, the register TH1 should be loaded with f3h.

Port configuration All the three ports P0, P1, and P2 are configured as the output ports. Port 3 pins have alternate functions, five of them are connected to transistor array to drive EM relays. Remaining 3-pins are used as switch input and serial input RxD pin to which the IR sensor is connected. Port 0 is used to generate 7-segments code. Port 1 drive the relays through ULN 2003.

Port 3 pins P3.3, P3.4, P3.5, P3.6, P3.7 are used to drive relay. P3.0 is sensor input RxD. Pin P3.2, P3.1 is a switch input and used as sequence control. Port 2 pins P2.0, P2.1, P2.2 are used as decoder inputs to display section, P2.3 has the zero cross detector input from PC 817.

Port pins P2.4, P2.5, P2.6, P2.7 are connected to the cathodes of the solid state relay drivers MOC 3011. MOC 3011 is optical isolated triac driver without zero cross detector. By changing delay the firing angle can be changed and hence rms voltage across load changes which results in variation in speed of motor.

Power supply Figure 13.21 depicts circuit diagram of a power supply. A 12 V, 60 VA transformer is used. Bridge rectifier is configured using diodes 1N4007 Filter capacitor 1000 μF/35 V and decoupling capacitor of 0.1 is used. Voltage regulator IC 7805 is used. Pin 1 is input rectified and filtered voltage 16.42 V. Pin 2 is common ground for the circuit board. Pin 3 is output + 5 V that is filtered by 470 μF/35 V to remove the ripple.

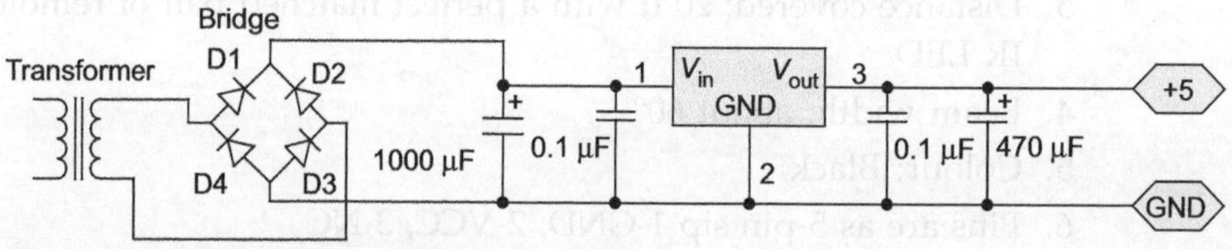

Fig. 13.21 Circuit diagram of power supply

Relay driver ULN2003 It is peripheral driver array. Seven npn Darlington transistors, shown in Fig. 13.22 are well suited for driving lamps or relays in variety of industrial applications.

The internal suppression diodes insure freedom for driving inductive loads and peak in rush currents of 600 mA permits to drive incandescent lamps. Specification of the IC ULN2003 are as follows:

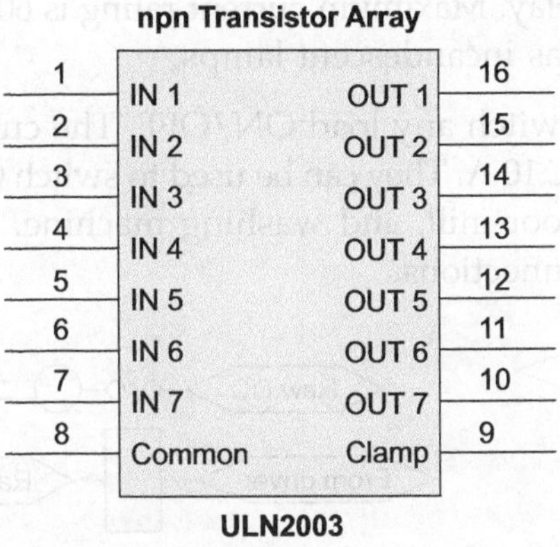

npn Transistor Array

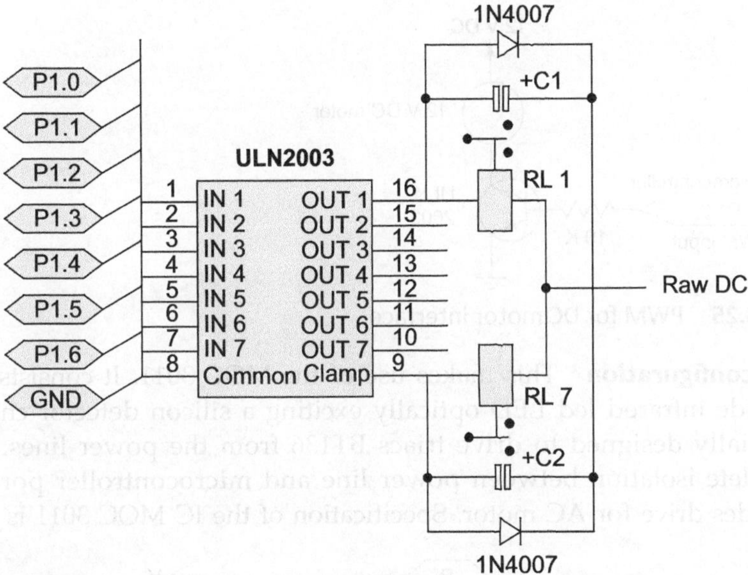

ULN2003

Fig. 13.22   Relay driver

1. Max output voltage: 50 V
2. Max input voltage: 30 V
3. Max I_c: 500 mA continuous
4. Max I_b: 25 mA continuous
5. Temperature range: 0 to 85° C
6. 16-pin dip package
7. Internal series resistor: 2.7 K

Fig. 13.23   Interfacing relay drivers

Figure 13.23 shows interfacing of the relay driver. The transistor array is used to drive relays and 7-segment codes generated at the microcontroller port. An npn Darlington configured array of seven transistors so that a logic HIGH '1' from the port can switch ON the relay. Maximum current rating is 600 mA. It can drive the solenoid valves as well as incandescent lamps.

Relays Relays are used here to switch any load ON/OFF. The current ratings are at 28 V DC 20 A and at 240 V AC 10 A. They can be used to switch ON domestic appliances such as freezer, TV, floor-mill, and washing machine. Figure 13.24 depicts the relay interface and connections.

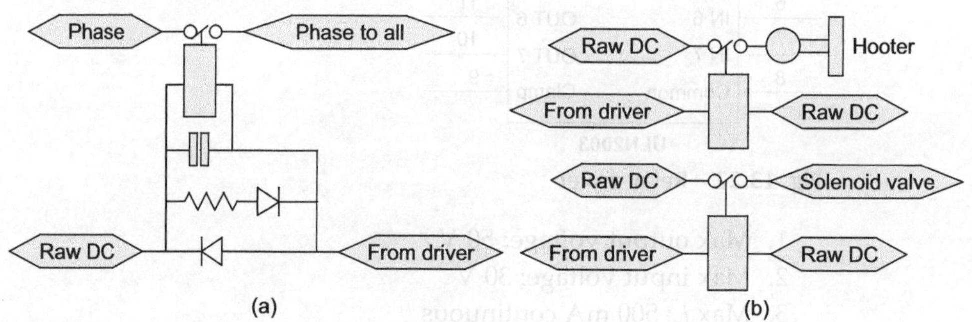

Fig. 13.24 Relay interface and connections

PWM interface The PWM waveform is generated by the microcontroller at the port pin P1.0. To drive a DC motor of 12 V, the npn Darlington array. ULN 2003 is used. Figure 13.25 shows the connection.

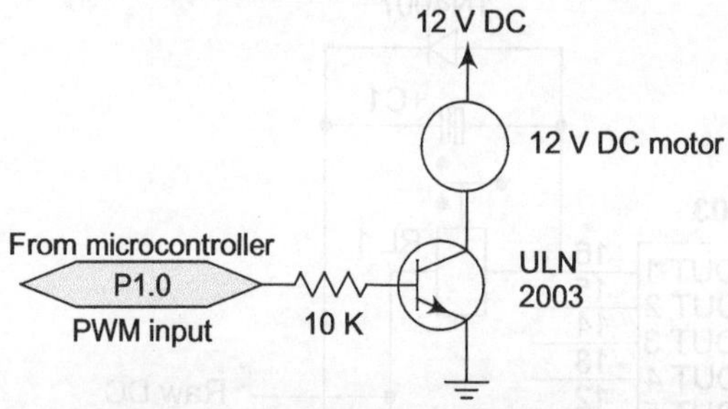

Fig. 13.25 PWM for DC motor interface

SSR configuration This makes use of the MOC 3011. It consists of gallium-arsenide infrared led LED optically exciting a silicon detector chip, which is especially designed to drive triacs BT136 from the power lines. It provides complete isolation between power line and microcontroller port as well as provides drive for AC motor. Specification of the IC MOC 3011 is as follows:

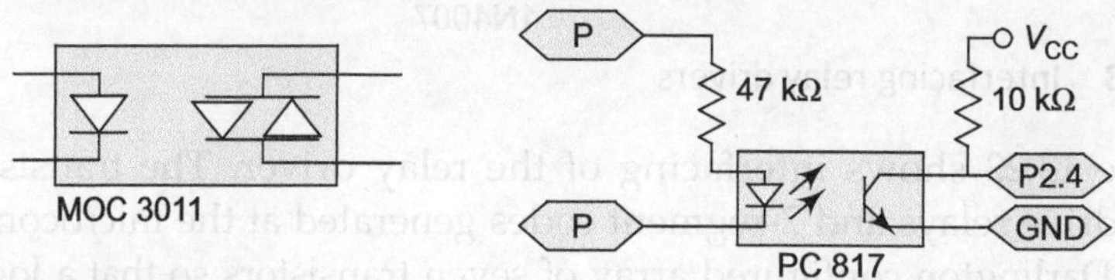

1. Max reverse voltage for emitter (LED): 3 V
2. Max forward current for emmiter: 60 mA
3. Off-state output voltage: 250 V
4. Peak repetitive surge current: 1 A
5. Isolation surge voltage: 7500 V
6. Temperature range: $-40°C$ to $85°C$

Figure 13.26 shows the setup for SSR.

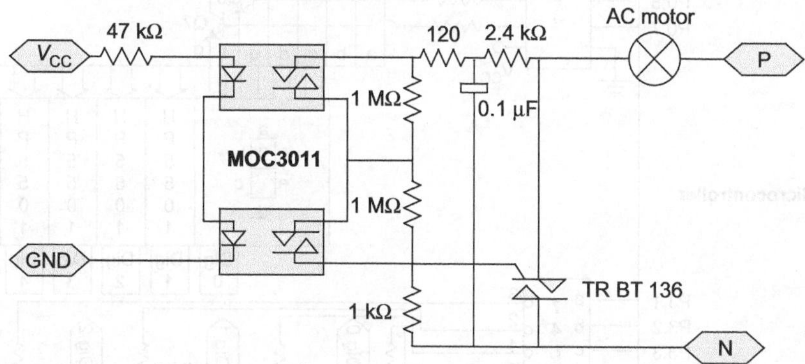

Fig. 13.26 SSR interfacing setup

Display module Display module consists of eight common anode display HP 5501 or LT 542R. Digits in the display are driven by the transistor BC557. Segments of all the LED form segment bus which is driven by the ULN2003. Output is TTL compatible and is directly connected to the Port 0 of microcontroller with external pull up resistors. The hardware is refreshed repeatedly as it uses multiplexing technique. Figure 13.27 depicts the display module. Layout  and circuit diagram of complete microcontroller board is in Online Resource Centre of the book.

13.5.3 Testing and Troubleshooting

Testing is an important step for proper operation of entire system. Testing is carried out in stages starting for power supply unit to the final controller board.

Testing of power supply The AC voltage input to the bridge rectifier and its output is to be checked. For example, in our setup input is 13 V and the output of the rectifier is 16 V. Next check whether regulator output is +5 V or not because the output of the IC 7805 is +5.01 V. Now checking is required at all points for the +5 V supply. In case there is any problem, check the tracks on the board.

Transformer should be capable to drive all relays without loading of it. This can be tested by directly testing of all relays ON/OFF operation.

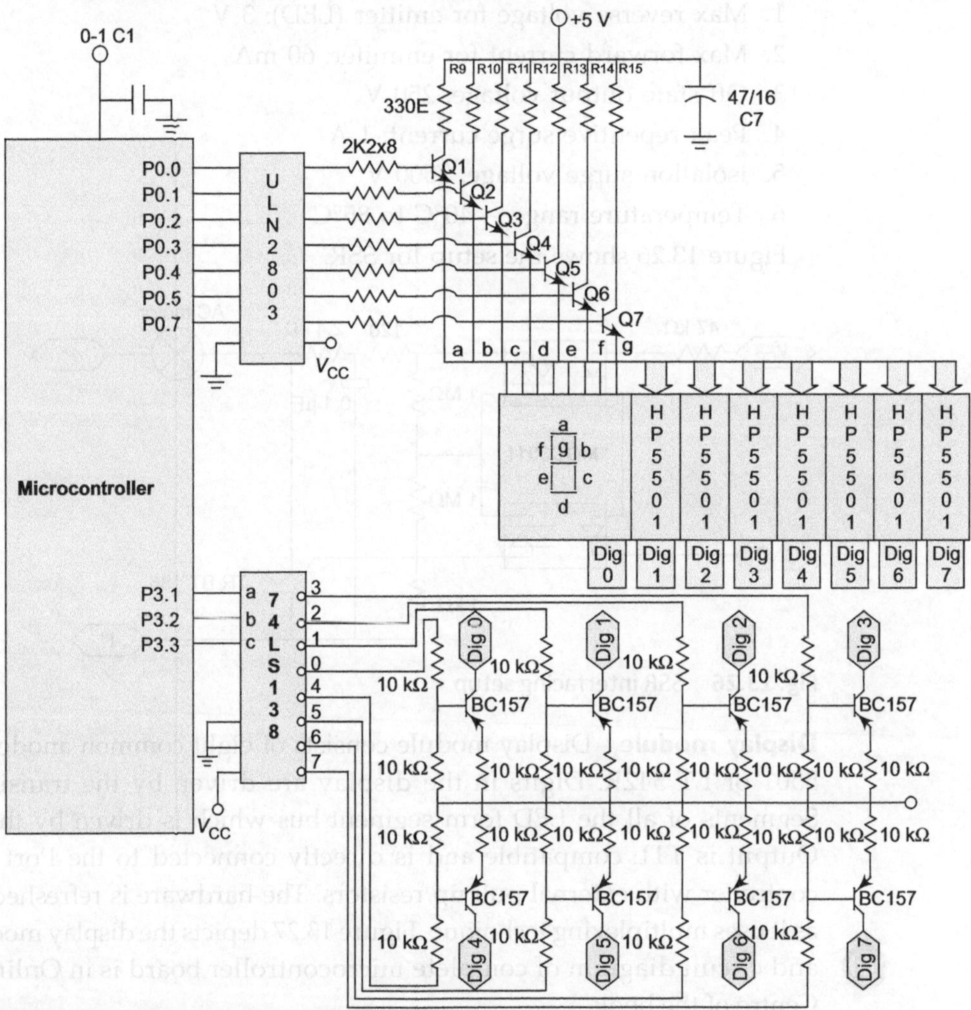

Fig. 13.27 Display module using 7-segment module

Testing of relay driver section NPN Darlington array and the invertors are used as relay driver. Apply logic HIGH (+5 V) to input of the driver, output is driven low and the relay connected to that output becomes ON. This way the relay drivers are tested. For improper operation first check track-on board, otherwise check IC of driver by interchanging both the IC.

Testing of display decoder section Displays used here are common anode type and connected in multiplexed configuration. PNP transistors are used as digit driver and they are driven by output of the 74LS138 decoder. Testing of this section has to be done in two modes: static and dynamic.

Static testing   Apply logic LOW (0 V) to one-by-one digit drivers and digit 8 on the 7-segment display one-by-one on each of the units.

Dynamic testing   It is carried out by the display routine written to display a sequence of digits on the units. For example: 1, 2, 3, 4, 5, 6, 7, 8 with necessary hardware.

Testing of SSR section   First observe a 50 Hz square wave on the output pin of PC817. In case of failure, check the connection, resistor of 47 k, and the IC. Value of the series resistor selection depend upon input voltage and maximum LED current, 47 k is used in the setup. 10 k pull-up resistor is used.

Next check the MOC 3011 logic LOW is applied to the cathode of LED of the IC and check the internal SBS switching. Resistor connected to SBS or pin 6 of the MOC 3011 depends upon supply 390 Ω is used. It is also connected to the pin 4 of IC, which fires TRIAC gate. Firmware testing can be done by running routine and connecting a lamp as load. For driving inductive load greater attention is given, because it generates high electrical noise and EMI. Designing any system in electrically noisy environment requires special techniques. Running routines for the application can do auxiliary hardware connection testing.

13.5.4 System Software

At reset, the hardware components such as: timers and counters are initialized in appropriate modes and display message for example: OK to acknowledge the reset operation and system initialization. Since the interrupt pins are used, the interrupt service programs must be written at the address specified as per Table 13.6.

Table 13.6   Interrupt code addresses and operations

Pin	IBA	Interrupt type	Operation
RESET	00h	Power up or reset	Initialization
EXTI0	03h	External interrupt 0	
TIMER0	0Bh	Timer 0 interrupt	Refresh display
EXTI 1	13h	External interrupt 1	
TIMER 1	1Bh	Timer 1 interrupt	
SINT	23h	Serial port interrupt	Receive data from RxD pin

A program segment for power On initialization is given below:

```
Prog-post:                    ;symbol and identifier definitions
      digit equ 40            ;digit no. to be display
      msg equ 41              ;message buffer
```

```
            htodmsb equ 50          ;hex to dec msb
            dl0 equ 51              ;delay count lsb
            dl1 equ 52              ;delay count msb
            rcode equ 53            ;code address
            num equ 0f4h            ;baud number to be loaded in th1
            stack equ 90            ;stack starts
                                    ;interrupt pointer table initialization
            org 0000h              ;branch addresses for the interrupts
            ajmp init              ;initialization program at reset
            org 0003h
            reti
            org 000bh
            ajmp display           ;program to refresh display at regular interval
            org 0013h
            reti
            org 001bh
            reti
            org 0023h
            ajmp function-code
```

It is clear that it is necessary to develop three modules

1. Initialization
2. Display
3. Function-code

Initialization Program

This is a program segment to initialize registers and SFR at reset.

```
init:       clr p3.3
            clr p3.4
            clr p3.5
            clr p3.6
            clr p3.7               ;clear all Port 1 pins and p3.3 to p3.7
                                   ;to disable relays
            mov p1, #0
            mov r7, #250           ;delay count for serial interrupt
            mov sp, #stack         ;initialize stack
            setb ea                ;enable all interrupts
            setb et1               ;enable Timer 1 interrupt
            setb es                ;enable serial interrupt
            setb et0               ;enable Timer 0 interrupt
```

```
                setb  pt0              ; set priority of Timer 0 interrupt
                mov   tmod, #21h       ; Timer 0; Mode 0 and Timer 1; Mode 2
                mov   th0,  #0fch
                mov   tl0,  #17h
                setb  tr0              ; start Timer 0
                anl   pcon, #7fh       ; reset SMOD
                mov   th1,  #num        ; load baud number here
                mov   tl1,  #num
                setb  tr1              ; start Timer 1 in auto reload mode
                mov   scon, #40h       ; clear RI, set Mode 1 for serialport
main:           setb  ren              ; set receive enable bit
                mov   digit, #00h
                mov   r0,   #msg
                mov   @r0,  #1dh       ; display message "****" to confirm
                                       ; init.and return to main menu
                inc   r0
                mov   @r0,  #0eh
                inc   r0
                mov   @r0,  #1ch
                inc   r0
                mov   @r0,  #1dh
                acall delay            ; delay to stabilize
                acall delay
wait:           sjmp  wait             ; wait for the interrupts
delay:          mov   dl0,  #0ffh
loop2:          mov   dl1,  #0ffh
loop1:          nop
                djnz  dl1,  loop1
                djnz  dl0,  loop2
                ret
```

Display Program

Port 0 is used as segment bus to generate 7-segment code. The code is applied as the input to the ULN2003 transistor array. The pin allocation is shown in Table 13.7. Logic HIGH on the port pin will be inverted by ULN2003 and the corresponding segment will be turned ON.

Table 13.7 Allocation on Port 0 pins

```
P0.0-dp
P0.1-seg a
P0.2-seg b
P0.3-seg c
P0.4-seg d
P0.5-seg e
P0.6-seg f
P0.7-seg g
```

```
display: push acc
         push psw
         mov th0, #0fch
         mov tl0, #17h              ;refresh time may be programmed
         setb psw.3
         setb psw.4
         mov a, digit
         mov c, acc.0              ;p2.0, p2.1, p2.2 are connected to A,B,C
                                   ;inputs of the 74LS138

         mov p2.0, c
         mov c, acc.1
         mov p2.1, c
         mov c, acc.2
         mov p2.2, c
         mov a, digit
         add a, #msg
         mov r0, a
         mov a, @r0
         acall con
         mov p0, a
         mov a, digit
         inc a
         mov digit, a
         cjne a, #08h,cal          ;calculate no. of digits scanned
         mov digit, #00h
         pop psw
         pop acc
         reti
```

```
con:                                    ;converts data to 7-segment codes
        inc a
        movc a, @a+pc
        ret
```

Look-up table for 7-segment code converter

db	7eh	; 0	db	0e0h	; k 14
db	0ch	; 1	db	70h	; l 15
db	0b6h	; 2	db	2ah	; m 16
db	9eh	; 3	db	0a8h	; n 17
db	0cch	; 4	db	7eh	; o 18
db	0dah	; 5	db	0e6h	; p 19
db	0fah	; 6	db	0ceh	; q 1a
db	0eh	; 7	db	0a0h	; r 1b
db	0feh	; 8	db	0dah	; s 1c
db	0deh	; 9	db	0f0h	; t 1d
db	0eeh	; 0a	db	7eh	; u 1e
db	0f8h	; 0b	db	38h	; v 1f
db	72h	; 0c	db	0cah	; w 20
db	0bch	; 0d	db	0ech	; x 21
db	0f2h	; 0e	db	0d6h	; y 22
db	0e2h	; 0f	db	0b6h	; z 23
db	0deh	; g 10	db	00h	; space 24
db	0ech	; h 11	db	01h	; '.'25
db	0ch	; i 12	db	80h	; '-' 26
db	0eh	; j 13	end		

Function Code

The algorithm to detect a function code on the serial interrupt pin can be described by the pseudocode as follows:

```
While (code !=150)
{
display_it(code);
}
   serial_interrupt()
{
code=sbuf;
}
```

A program can be developed to implement the above algorithm. The purpose of this program module is to accept the function code from the remote transmitter keypad and display it on the upper four LED of the display unit. The function code is stored in the memory location rcode.

```
        function code:
                djnz r7, ext            ; this routine returns data code
                mov r7, #250            ; detected as byte from SBUF, stored in the
                                        ; location rcode and displayed at the upper four
                                        ; LED, RI is cleared

                mov rcode, sbuf
                mov r6, rcode
                mov r5, #0
                acall h2dstore
                clr ri
        ext:    reti
        h2dstoreup:
                mov a, r5
                anl a, #0fh
                mov htodmsb, a
                mov a, r6
                acall htod
                push psw
                setb psw.4
                clr psw.3
                mov a, r4
                anl a, #0fh
                mov r1, #msg+4
                mov @r1, a
                inc r1
                mov a, r5
                swap a
                anl a, #0fh
                mov @r1, a
                inc r1
                mov a, r5
                anl a, #0fh
                mov @r1, a
                pop psw
                ret
        htod:                           ; module to convert hex to decimal
                push psw
                setb psw.4
```

```
                    clr  psw.3
                    mov  r4, #00h
                    mov  r5, #00h
                    mov  r2, #00h
                    mov  r2, a
                    mov  r3, #16
        ocnt:       clr  c
                    mov  a, r2
                    rlc  a
                    mov  r2, a
                    mov  a, htodmsb
                    rlc  a
                    mov  htodmsb, a
                    mov  a, r5
                    addc a, r5
                    da   a
                    mov  r5, a
                    mov  a, r4
                    addc a, r4
                    da   a
                    mov  r4, a
                    djnz r3, ocnt
                    pop  psw
                    ret
```

The program should be modified if we wish to allocate operation to the key on the transmitter keypad. This depends upon the application on hand.

13.6 | APPLICATION SOFTWARE

This section describes the most common applications of the digital controller. Assuming the hardware setup developed in Section 10.5.1, the application software will be developed for these applications.

Example 13.1   Develop a program to detect the remote key code and display it on the display module.

Solution

Let us assume that Philips TV remote transmitter is used. The codes for a representative remote transmitter are shown in Fig. 13.28.

For detecting codes, baud number loaded in TH 1 register is F3h and crystal frequency is 6 MHz to generate baud rate of 1200-bits/s. For detecting other remote codes, trial and error method can be use by varying baud rate. The pseudo code of detection algorithm is as follows:

```
serial_ interrupt()
      {delay();
      code=sbuf;
      hextodec();
      display()
      clear RI;}
      }
```

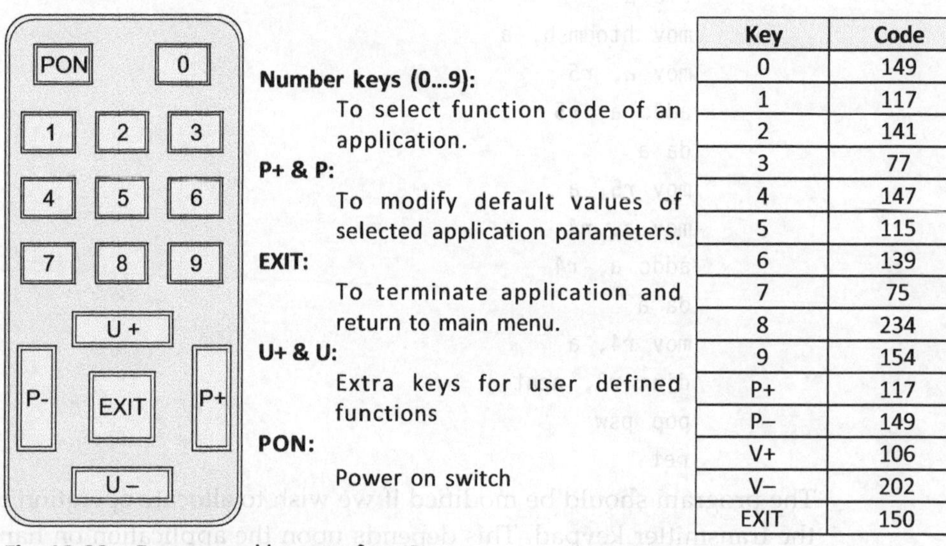

Number keys (0...9):
  To select function code of an application.

P+ & P:
  To modify default values of selected application parameters.

EXIT:
  To terminate application and return to main menu.

U+ & U:
  Extra keys for user defined functions

PON:
  Power on switch

Key	Code
0	149
1	117
2	141
3	77
4	147
5	115
6	139
7	75
8	234
9	154
P+	117
P–	149
V+	106
V–	202
EXIT	150

Fig. 13.28 Front panel layout of an IR

The program module is similar to one develop in previous section. It is left as an exercise to write program.

Example 13.2   Write a program segment to make the appliances ON/OFF connected to the digital controller through the relay 1 to relay 4 connected at Port 1 pins.

Solution

Port 1 is used as an actuator bus to drive eight relays through ULN2003 transistor array. Assuming the hardware setup of Fig. 13.28 (circuit configuration on ORC) logic HIGH on the port pin will be inverted by ULN2003, will turn the relay ON. Table 13.8 gives pin connection of Port 1 pins.

Table 13.8   Pin connection of Port 1 pins

```
P1.0-relay1
P1.1-relay2
P1.2-relay3
P1.3-relay4
P1.4-relay5
P1.5-relay6
P1.6-relay7
P1.7-relay8
```

The operations required are detection of codes and carry out the operation. The code is detected. It can be compared with the look-up table to identify the operation. It is possible to switch appropriate relay and display name of the device and status on the 7-segment display unit.

Let us assume that the key 1, 2, 3, 4 are allocated the ON/OFF function for the relay 1, relay 2, relay 3, and relay 4, respectively. The algorithm is as follows:

```
serial_interrupt()
{
code=sbuf;
}
wlile (code !=150)
{
switch(code)
{
case code1:     display_it(code);
                toggle relay1;
                delay(1 sec);
                break;
case code2:     display_it(code);
                toggle relay2;
                delay(1 sec);
                break;
case code3:     display_it(code);
                toggle relay3;
                delay(1 sec);
                break;
case code4:     display_it(code);
                toggle relay4;
                delay(1 sec);
```

```
                                  break;
                          }
                          }
```

The process can be extended for the operation of more than one or all the 13 relays.

The program listing is as follows:

```
init:
                  mov p1, #0              ; initialize Port P0 pins
                  mov r7, #250           ; delay count for serial interrupt
                  mov sp, #stack         ; initialize stack here
                  setb ea                ; enable all interrupts
                  setb et1               ; enable Timer 1 interrupt
                  setb es                ; enable serial interrupt
                  setb et0               ; enable Timer 0 interrupt
                  setb pt0               ; set priority of Timer 0 interrupt
                  mov tmod, #21h         ; Timer 0-Mode 0, Timer 1-Mode 2
                  mov th0, #0fch
                  mov tl0, #17h
                  setb tr0               ; start Timer 0
                  anl pcon, #7fh         ; reset SMOD register
                  mov th1, #num          ; load number for the baud rate
                  mov tl1, #num
                  setb tr1               ; start Timer 1 auto-reload mode
                  mov scon, #40h         ; clear RI, serial port-Mode 1
start:    setb ren                       ; set receive enable bit
                  mov p1, #00h           ; initialize Port P1
                  mov digit, #00h
                  mov r0, #msg
                  mov @r0, #data1        ; display "****" to confirm hardware setup
                  inc r0
                  mov @r0, #data2
                  inc r0
                  mov @r0, #data3
                  inc r0
                  mov @r0, #data4
                                         ; wait for the function
main:     mov a, rcode                   ; rcode is the location where the serial
                                         ; interrupt returns the byte from SBUF register
```

```
            cjne  a, #117, skip-1     ;relay-1 operation
            cpl  p1.1
            acall  display
            acall  delay
            sjmp  start
skip-1:
            cjne  a, #141,skip-2      ;relay-2 operation
            cpl  p1.2
            acall  display
            acall  delay
            sjmp  start
skip-2:
            cjne  a, #77,skip-3       ;relay-3 operation
            cpl  p1.3
            acall  display
            acall  delay
            sjmp  start
skip-3:
            cjne  a, #147,error
            cpl  p1.4                 ;relay-4 operation
            acall  display
            acall  delay
            sjmp  start
error:
            acall  display           ;display error message
            acall  delay
            sjmp  main               ;wait for valid relay number
```

Example 13.3   Write a program segment to vary speed of an AC universal motor using the remote transmitter keypad.

Solution

Solid state relay is an electronic system consisting of zero cross detector (ZCD), optical isolator with monolithic IR LED and silicon bi-lateral switch (SBS) capable to drive SCR, and the triacs with higher current rating. The types of optical isolated triacs and SCR available with in-built ZCD and without ZCD. The formers are suitable for ON/OFF operation of the AC appliances.

Port 2 pins in the hardware setup of Section 13.5.2 are used as decoder input as well as an SSR driver along with ZCD input as shown in Table 13.9.

Table 13.9   Pin connection of Port 2 pins

P2.0-input A	; these three inputs A, B, and C are decoded by the 74LS138
P2.1-input B	; they are incremented at each Timer 0 interrupt to refresh
P2.2-input C	; multiplexed 7-seg. display
P2.3-ZCD input	; reference time for zero crossing of sinusoidal 230 V/ 50 Hz
P2.4-SSR1	
P2.5-SSR2	
P2.6-SSR3	
P2.7-SSR4	

Logic LOW on port pin turns ON the LED inside MOC 3011 and the internal SBS. The signal results in firing of external high rated triac at desired angle to change the RMS voltage across load using phase control. To drive resistive load through MOC 3011 at 240 V AC/50 Hz two MOC 3011 are required in series connection because MOC 3011 is designed for 115 V AC line. MOC 3011 is designed for 115 V AC line, two MOC 3011 are used in series to drive resistive load through MOC 3011 at 230 V AC/50 Hz.

Incandescent lamps and resistive heating elements are the two main classes of resistive load. Triac must be protected by the fuse or rated high enough to sustain proper inrush loads. Incandescent lamps can sometimes draw pick current known as *flash over* which can be transient voltage disturbance on AC line may exceed static dv/dt rating of MOC 3011 . This creates a problem in the noisy environment as the MOC 3011 and triac commute OFF at the next ZCD input of line voltage.

Inductive loads such as motor, solenoid, and magnet present a problem, as the voltage and current are not in phase with each other. Since triac turns OFF when applied current is zero but the applied voltage is high. The use of snubber networks reduces the rate of voltage rise seen by the device. Figure 13.29 shows the circuit diagram of the interface.

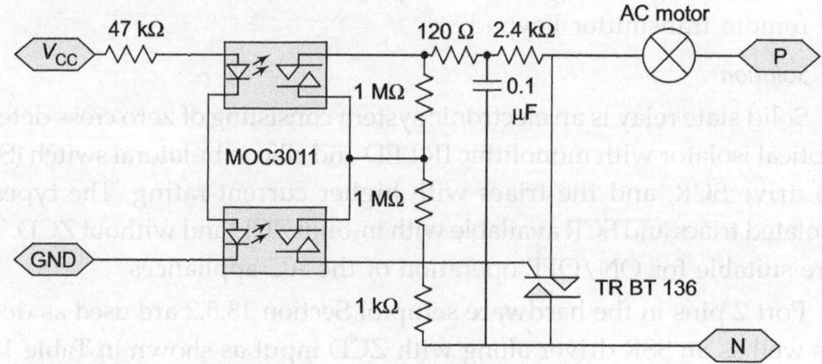

Fig. 13.29   Interface to drive AC motor through SSR

Function key 0 defines the operation of speed control of AC motor. (v+) and (v–) keys on the transmitter keypad is used to increase or decrease the speed, respectively. The program requires initialization of the system and wait for the serial interrupt. Detect the received code for operation, i.e., to increase or to decrease the speed and take appropriate action. The program listing is given below.

Speed-control:

```
                                        ; symbol and identifier definitions
        digit   equ   40                ; digit number to be display
        msg     equ   41                ; message buffer
        htodmsb equ   50                ; hex to dec msb
        dl0     equ   51                ; delay count lsb
        dl1     equ   52                ; delay count msb
        rcode   equ   53                ; code address
        num     equ   0f4h              ; baud number to be loaded in th1
        stack   equ   90                ; stack starts
                                        ; interrupt pointer table initialization
        org     0000h                   ; branch addresses for interrupts
        ajmp    init                    ; initialization program at reset
        org     0003h
        reti
        org     000bh
        ajmp    display
        org     0013h
        reti
        org     001bh
        reti
        org     0023h
        ajmp    wait-0
init:   mov     p1, #00h                ; initialize Port P1 pins
        mov     p2, #00h                ; initialize Port P2 pins
        mov     r7, #250                ; delay count for serial interrupt
        mov     sp, #stack              ; initialize the stack
        setb    ea                      ; enable all interrupts
        setb    et1                     ; enable Timer 1 interrupt
        setb    es                      ; enable serial interrupt
```

```
                setb et0                ; enable Timer 0 interrupt
                setb pt0                ; set priority of Timer 0 interrupt
                mov tmod, #21h          ; Timer 0-Mode 0 and Timer 1-Mode 2
                mov th0, #0fch
                mov tl0, #17h
                setb tr0                ; start Timer 0
                anl pcon, #7fh          ; reset SMOD register
                mov th1, #num           ; load number for the baud rate
                mov tl1, #num
                setb tr1                ; start Timer 1 in auto reload mode
                mov scon, #40h          ; clear RI and serial port-Mode 1
start:
                setb ren                ; set receive enable bit
                mov p1, #00h            ; initialize Port P1
                mov digit, #00h
                mov r0, #msg
                mov @r0, #data1         ; confirm hardware setup
                inc r0
                mov @r0, #data2
                inc r0
                mov @r0, #data3
                inc r0
                mov @r0, #data4

                                        ; wait for the function 0
wait0:          mov a, rcode
                cjne a, #149, wait0     ; speed control operation
start:
                setb p1.7               ; 230 V supply thro'.relay connected to p1.7
                setb p2.3               ; set ZCD as input
                mov b, #10              ; initialize angle for firing for initial speed
continue:
                mov a, rcode
                cjne a, #106, chk-dec   ; check code of (v+)
                inc b
                ajmp end
chk-dec:        cjne a, #202, chk-exit  ; check code of (v-)
```

```
                dec b
                ajmp end
chk-exit:
                cjne a, #150, ok1              ;check exit switch (menu)
                ajmp start
end:            mov rcode, #255
                mov r6, b
                mov r5, #0
                acall h2dstore
                jnb p2.3, $                    ;loop while ZCD = 0
                acall delay-on                 ;when ZCD = 1 start delay Td which can be
                                               ;varied by Vol switches
                clr p2.4                       ;fire Triac BT136(LED-MOC 3011ON)
                nop                            ;small delay
                setb p2.4                      ;make LED of MOC 3011 OFF
                jb p2.3, $                     ;loop while ZCD = 1
                acall delay-on                 ;when ZCD = 0 Start delay Td
                clr p2.4
                nop
                setb p2.4
                sjmp continue
delay-on:
                mov r2,b
loop:           mov d10, #10
                djnz d10, $
                djnz r2, loop
                ret
```

Example 13.4   Write a program segment to vary speed of a DC motor using the remote transmitter keypad.

Solution

For the speed control of a DC motor the pulse width modulation technique will be used. PWM technique uses a square wave of 1 kHz with variable duty cycle. Switching the transistor SL100 via port pin of the microcontroller generates PWM of 1 kHz square wave and varying duty cycle from 1% to 99%. As the time-on increases voltage applied to the DC motor increases, which increases the speed,

$$\text{time-on} = \text{total-time} - \text{time-off}$$

On the port pin p1.0, PWM output is generated. Pin is connected to the 12 V, 1500 rpm tape recorder (DC motor) via transistor array ULN2003 or SL100 as shown in Fig. 13.30.

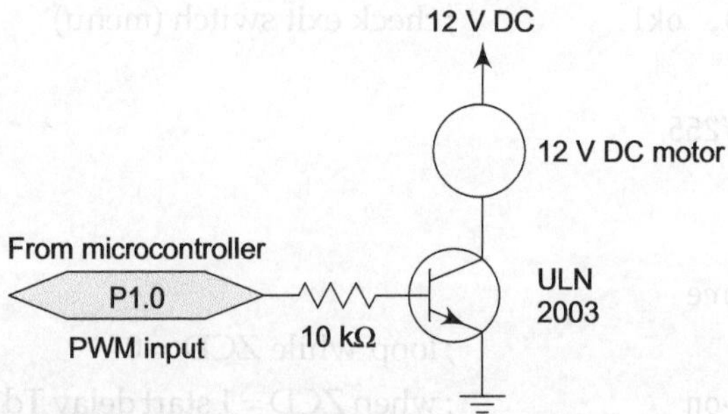

Fig. 13.30 Interfacing DC motor

The speed control process can be described by the pseudo-code as follows:

```
Main()
{While (1)
{
Set port pin ;
delay-time-on  ();
Reset port pin ;
Delay-time-off () ;
  }
}
Wlile (code !=150)
{
set_pwmoutput();
delay(ton);
reset_pwmoutput();
delay(toff);
if (code=inc)
{ton++;
toff-;}
else if (code=dec)
{toff++;
ton-;}
}
serial_interrupt()
```

```
{
code=sbuf;
}
```

Program is similar to one developed in the previous example and is left as an exercise to the reader.

Example 13.5   Write a program segment to drive stepper motor using the remote transmitter keypad. Assume that the stepper motor is driven through relays connected to Port 1 pins.

Solution

Stepper motor requires electrical pulses in pre-defined sequence. Reversing the sequence reverses the direction. High power switch like power transistor, MOSFET or relays may be used to generate pulse with enough strength to drive the motor. Relays are used to generate the required pulses for stepper motor as shown in Fig. 13.31.

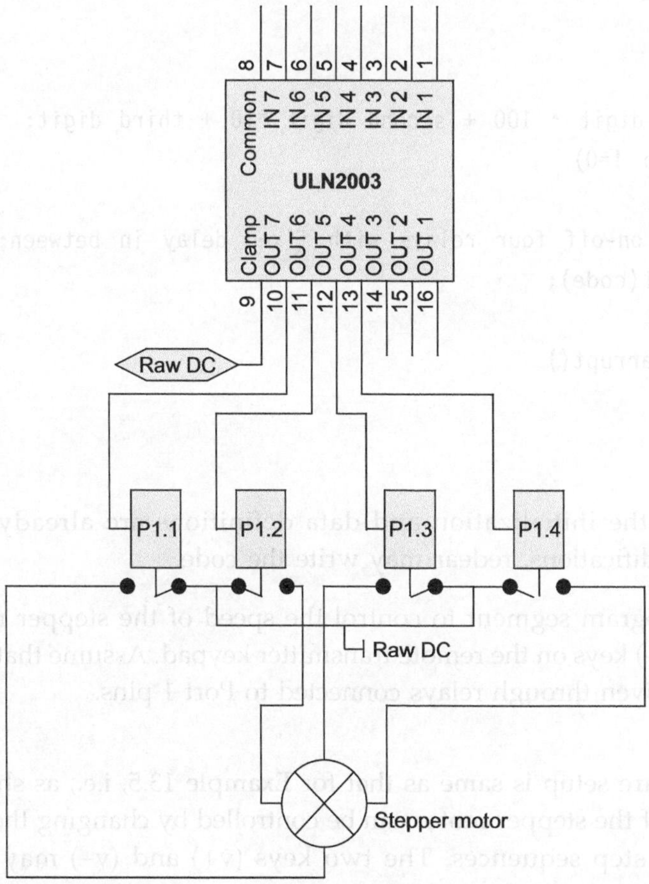

Fig. 13.31   Stepper motor interfaced via relays

Port pin p1.1, p1.2, p1.3, p1.4 are used to supply four phases of the stepper motor. They are connected to stepper motor via ULN2003 and EM relays. Remote transmitter is used to enter the number of steps for rotation. Stepper motor starts rotating after a 3-digit number is entered.

This program starts when power ON/OFF switch (PON) on the remote transmitter is pressed. The message ANG is displayed on the display unit which informs user to enter angle of rotation. Stepper motor rotates by value of the angle entered. The procedure steps are as follows:

```
while (code !=150)
{
if (digit != 3)
{
display digit;
save it in memory;
}
else
{
Step=first digit * 100 + second digit *10 + third digit;
while (step !=0)
{
alternaely on-off four relays with fixed delay in between;
}display_it(code);
}
serial_interrupt()
{
code=sbuf;
}
```

Assume that the initialization and data definitions are already done with necessary modifications, redear may write the code.

Example 13.6 Write a program segment to control the speed of the stepper motor using (v+) and (v–) keys on the remote transmitter keypad. Assume that the stepper motor is driven through relays connected to Port 1 pins.

Solution

The hardware setup is same as that for Example 13.5, i.e., as shown in Fig. 13.19. Speed of the stepper motor can be controlled by changing the time delay between two step sequences. The two keys (v+) and (v–) may be used to increment and decrement time delay between two pulses. Upper speed limited by the relay switching. To remove the upper speed limit transistor switching may be used. The speed control process is described by the pseudo-code given below:

```
wlile (code !=150)
{
if (digit != 3)
{
display digit;
save it in memory;
}
else
{
Step=first digit * 100 + second digit *10 + third digit;
while (step !=0)
{
alternaely on-off four relays with programmable delay;
if (code = incr)
delay_count++;
else if (code=decr)
delay_count-;
}
display_it(code);
}
serial_interrupt()
{
code=sbuf;
}
```

Assume that the initialization and data definitions are already done with necessary modifications, reader may write the code.

Example 13.7   Write a program segment to design an electronic lock for the security purpose.

Solution

The system accepts a 4-digit password entered via remote keypad and compares with the standard (programmed password).

The lock is opened via solenoid valve switched through relay only if a match is found, otherwise hooter is switched ON to indicate the unauthorized entry from the remote. A password is also programmed to switch OFF the hooter known by the *authorized person*. The lock remains open for 40 s and is closed automatically.

On reset there is a 4-digit password (maximum 8-digits) called a *master password*. Entering the authorized old password can change it. The changed password is live until system reset due to battery voltage problem or any other problem. Provision is made for the four trials for the wrong entry.

Correct entry of the password will energize solenoid plunger via relay. The pseudo-code for the process is as follows:

```
while(code!=150)
{if(code=change)
display 'OLD'
set flag OLD;
else
{save serially digits entered in FIFO;
compare master password with entered word
inFIFO;
if comparison fails make hooter ON;
else if(OLD=0)
{operate lock for 40 secs;}
else{
display 'NEW';
after digit entry screen becomes blank;
save in pass word location;}
}
serial_interrupt()
{
code=sbuf;
}
```

EXERCISES

13.1 Explain types of remote communication.

13.2 With the help of a schematic block diagram explain IR communication system.

13.3 With the help of a schematic block diagram explain operation of a remote transmitter.

13.4 With the help of a schematic block diagram explain operation of a IR transmitter.

13.5 Explain use of TSET, TM pins of M708/708A.

13.6 Explain interfacing of a matrix keyboard with IR transreceiver A708/M708A.

13.7 What are IR protocols? With the help of frame level format explain all popular protocols used in IR communication.

13.8 What do you mean by a pulse coding? What are its advantages?

13.9 Enlist the factors which affect distance of operating IR transreceiver.

13.10 Modify Fig. 13.7 to interface LCD instead of LED.

13.11 Write a display module for LCD output.

13.12 Explain the algorithm for code detection by IR receiver.

13.13 Write a program module to detect key closure and display identified key on LCD.

13.14 Explain significance of the function module serial_interrupt().

13.15 Write a program module to drive stepper motor if function code is 3.

Application VI: RF-Based Intercontroller Wireless Communication

RF generally refers to radio frequency data communications. In the past, it has also been referred to as RFDC, to differentiate it from radio frequency identification (RFID), which is a different and potentially complementary technology. Wireless local area networks (WLANs) are one of the popular forms of RFDC.

RF is the wireless transmission of data by digital radio signals at a particular frequency. It maintains a two-way online radio connection between the mobile terminal and the host computer. The mobile terminal, which can be portable, collects and displays data at the point of activity. The host computer can be a PC, a minicomputer, or a much larger mainframe. The end result is a seamless flow of information to and from the host, allowing users to go wherever they need to go in order to get their job done without fear of being out of touch with the data they need. RFDC improves the timeliness of information, and therefore the value of information, especially in time-sensitive operating environments.

In the past decade, a dramatic increase in demand on wireless communication services has been observed. For example, the number of cellular users has increased by a factor of ten in the past 10 years. This trend is expected to continue for the next decade. Moreover, it is forecasted that in the next few years, users' thirst for non-voice applications will drive the demand for faster and more reliable wireless communication services. Specifically, the increasing demand on voice and multimedia communications mandates future generation wireless systems to support transmission rates up to several Mbps and to accommodate a large number of users. In order to meet this goal, innovations in technology are imperative.

Communication channels can be a physical connection between the correspondents. Wires, coaxial cables, and optical fibers are used for this purpose. Roughly speaking, the bandwidth and expense increases along this list. When distances or the cost of cable laying become very large, it can be advantageous to use radio waves as the channel. Microwave links are high bandwidth line-of-sight links. Satellite links (usually) use geostationary satellites to provide inter-continental high-bandwidth communications. Optical links (i.e., the transmission through free-space of light waves) can be used for short and high bandwidth transmission. The different types of communication channels are

- Transmission lines
- Optical fiber wave guide
- Propagation in free-space and the atmosphere
- Microwave link communication
- Satellite communication
- Optical fiber cables
- Mobile communications

Radio and television broadcasting is the natural method for communicating unselectively with large numbers of correspondents, or when one or both of the correspondents are moving. Mobile communications employ networks of radio transmitter/receiver stations.

Radio frequency and wireless originated nearly a century ago with Alexander Popov and Sir Oliver Lodge laying the groundwork for Guglielmo Marconi's wireless radio developments in the early 20th century. The most prominent of his experiments was in December 1901, where he successfully transmitted Morse code from Cornwall, England to St-John's, Canada.

The advantages of an RF communication system are many. Cable is expensive, less flexible than RF coverage, and is prone to damage. For new facilities, implementing a wireless infrastructure may be more cost effective than running cable through industrial environments, especially if the space configuration may change to support different storage space allocation or flexible manufacturing stations.

Accessibility is a key benefit. If workers are within the range of the system (and they always should be if a proper site survey is performed) they are always in touch with their data. This advantage cannot be overstated. To always have one's data literally at one's fingertips whenever needed means there is no break in productivity and no empty or 'deadhead' trips to a stationary terminal, docking station, or dispatch location to receive pick or put away instructions. Critical decisions can be made and action can be taken immediately at the point of activity. Less wasted time means one can work significantly faster, without adding additional employees.

Other general advantages of real-time RF communication include a significant improvement in order of accuracy (>99%), the elimination of paperwork, replacement of time-consuming batch processing by rapid real-time data processing, prompt response times, and improved service levels. Complementing a real-time data collection system with automated data entry by bar code scanning or another automatic data collection technology improves the accuracy of information and eliminates the need for redundant data entry, which provides another set of time- and cost-saving advantages.

This chapter describes the establishment of an RF link between two microcontrollers using the CYWUSB6935 IC from Cypress Inc.

14.1 | RF COMMUNICATION

The introduction of RF technology and its working along with theoretical details are covered in this section. Design specifications and trends in RF system design, and antenna specifications are also discussed along with an introduction to the spread spectrum.

RF communication works by creating electromagnetic waves at a source and being able to pick up these electromagnetic waves at a particular destination. These electromagnetic waves travel through air nearly at the speed of light. The wavelength of an electromagnetic signal is inversely proportional to the frequency; the higher the frequency, the shorter the wavelength:

$$C = f\lambda$$

Frequency is measured in hertz (cycles per second) and radio frequencies are measured in kilohertz (kHz or thousands of cycles per second), megahertz (MHz or millions of cycles per second), and gigahertz (GHz or billions of cycles per second). Higher frequencies result in shorter wavelengths. The wavelength for a 900 MHz device is longer than that of a 2.4 GHz device.

In general, signals with longer wavelengths travel a greater distance and penetrate through and around objects better than signals with shorter wavelengths.

Table 14.1 Frequency band designations

f	λ	Band	Description
30–300 Hz	10^4–10^3 km	ELF	Extremely low frequency
300–3000 Hz	10^3–10^2 km	VF	Voice frequency
3–30 kHz	100–10 km	VLF	Very low frequency
30–300 kHz	10–1 km	LF	Low frequency
0.3–3 MHz	1–0.1 km	MF	Medium frequency
3–30 MHz	100–10 m	HF	High frequency
30–300 MHz	10–1 m	VHF	Very high frequency
300–3000 MHz	100–10 cm	UHF	Ultra-high frequency
3–30 GHz	10–1 cm	SHF	Super high frequency
30–300 GHz	10–1 mm	EHF	Extremely high frequency
			(millimeter waves)

Table 14.2 Microwave letter band designations

f (GHz)	Letter band designation
1–2	L band
2–4	S band
4–8	C band
8–12.4	X band
12.4–18	Ku band
18–26.5	K band
26.5–40	Ka band

The frequency spectrum is quite packed and the various RF bands are allocated for different purposes. This is one of the reasons that we are constantly pushing applications into higher and higher frequencies. However, some of the other reasons accounting for this push into higher frequencies include the following:

1. Efficiency in propagation,
2. Immunity to some forms of noise and impairments, and
3. The size of the antenna required. The antenna size is typically related to the wavelength of the signal and in practice is usually 1/4 of the wavelength.

14.2 | SYSTEM OPERATION

Imagine an RF transmitter wiggling an electron in one location. This wiggling electron causes a ripple effect, somewhat akin to dropping a pebble in a pond.

The effect is an electromagnetic (EM) wave that travels out from the initial location resulting in electrons wiggling in remote locations. An RF receiver can detect this remote electron wiggling.

The RF communication system then utilizes this phenomenon by wiggling electrons in a specific pattern to represent information. The receiver can make this same information available at a remote location, communicating with no wires. In most wireless systems, a designer has two overriding constraints:

1. It must operate over a certain distance (range); and
2. It must transfer a certain amount of information within a time frame (data rate).

Then, the economics of the system must work out (price) along with acquiring government agency approvals (regulations and licensing).

14.2.1 Data Rates

Data rates are usually dictated by the system—how much data must be transferred and how often does the transfer need to take place. Lower data rates allow the radio module to have better sensitivity and thus more range. Higher data rates allow the communication to take place in less time, potentially using more power to transmit.

14.2.2 Regulations and Licensing

The Federal Communications Commission (FCC) and other regulatory bodies around the world have set up a series of regulations defining the emission levels and usage for all the different frequencies. Most radios operate within the Industrial, Scientific and Medical (ISM) bands that offer licence for free operation within certain frequencies. Within the United States, the most popular ISM band are at 902–928 MHz and 2.4–2.4835 GHz. Portions of the 902–928 MHz band are also available in Canada, Mexico, Australia, and Israel. The 2.4 GHz band is generally more accepted worldwide.

At certain power levels, some regulatory agencies require some form of spread spectrum. Spread spectrum can either be done by frequency hopping or by direct sequence. Frequency hopping consists of rapidly moving from one channel to the next while maintaining synchronization with the receiver. Direct sequence is more complex, but works by slicing the carrier up with a code that can be decoded at the other end.

14.3 | TERMINOLOGY: RF COMMUNICATION

This section describes the terms used with reference to RF communication.

Voltage standing wave ratio   Voltage standing wave ratio (VSWR) is the ratio of reflected to transmitted waves. Any impedance mismatches along a transmission line causes partial reflection of the propagating signals. The impedance difference determines the magnitude of the reflection. The length of a mismatched section determines the lowest signal frequencies that reflect from the section. VSWR is a measure of that signal reflection.

With an incident sine wave into the switch module, some of the signals reflect down the line. This reflected wave interferes with the incident wave. VSWR is the ratio of maximum to minimum amplitude in the resulting interference wave, as shown in the following formula:

$$\text{VSWR} = \frac{1 + [p]}{1 - [p]}$$

where p is the reflection coefficient.

Reflections can also be represented as a logarithmic ratio of the reflected signal to the input signal. This ratio is called *return loss*.

$$RL = 10 \log (P_{IN}/P_{REFLECTED})$$
$$(dB) = 20 \log (V_{IN}/V_{REFLECTED})$$
$$= -20 \log (V_{REFLECTED}/V_{IN})$$
$$= -20 \log |p|$$

Frequency response (Flatness)   Frequency response is the gain and phase response of a circuit or other unit under test (UUT) at all frequencies of interest. Although the formal definition of frequency response includes both the gain and the phase, in common usage, the frequency response often implies only the magnitude (gain). The frequency response $H(f)$ is defined as the inverse Fourier transform of the impulse response $h(\tau)$ of a system,

$$H(f) = \int_{-\infty}^{\infty} h(\tau)e^{-j2\pi f\tau}d\tau$$

Frequency response measurements require the excitation of the UUT with the energy at all relevant frequencies. The fastest way to perform the measurement is to use a broadband excitation signal that excites all frequencies simultaneously, and use FFT techniques to measure all of these frequencies at the same time. Noise and non-linearity is best minimized by using random noise excitation, but short impulses or rapid sweeps (chirps) may also be used.

When the desired resolution bandwidth of interest is less than about 100 kHz, the fastest way to measure the frequency response functions is to use FFT-based techniques.

14.4 | RF SYSTEM DESIGN

RF system development can be somewhat complex, requiring a keen understanding of RF design, circuit design, signal processing, and microcontroller technologies. Some of the most critical RF parameters that should be verified in a wireless product design are: transmit output power, receiver sensitivity, antenna impedance matching and antenna design, occupied bandwidth, RF spurious and harmonic signals, crystal oscillator operation and accuracy, power supply voltage and noise, current consumption, and critical component selection. The common problems encountered in RF system designs are:

- **Range problem** Low transmit power, poor receive sensitivity, poor antenna impedance matching, poor antenna design, crystal oscillator offset exceeds 50 ppm, power supply noise, and interference from test environment.

- **High data error rate** Crystal oscillator offset exceeds 50 ppm, improper correlator threshold settings, and power supply noise.

14.5 | ANTENNA DESIGN

An antenna essentially provides a means of converting the electrical energy into electromagnetic waves for transmission and reconverting the electromagnetic waves into electrical energy for reception. A properly designed antenna facilitates the evaluation, characterization, and production test correlation of the wireless system. In designing short-range radio data communication system, one of the most important task is the antenna design. The key parameters in antenna design are the antenna size, cost of implementation, radiation effectiveness, ease of manufacturability and range performance constraints, and product packaging.

There are many different types of antennas. Antennas most relevant to 2.4 to 2.5 GHz designs are: dipole antennas, sleeve dipole antennas, loop antennas, helical antennas, whip (monopole) antennas, ceramic chip antennas, slotted antennas, printed/planar inverted–F antennas, printed trace wiggle antennas, and microstrip patch antennas.

Each type of antenna has its own advantages and disadvantages depending upon its applications. The optimal antenna solution is one that is part of the mechanical structure or physical housing for achieving low cost and consistency in performance. The rules of thumb for antenna choice, selection, and implementation are as follows:

- Antenna design significantly affects wireless performance. The antenna should be designed to maximize the coupling of signals in the ISM frequency band (2.400–2.4835 GHz). Ideally, the antenna should have poor coupling for signals outside of this range, especially in frequency bands where common sources of radiated energy may exist (e.g., 1.9-GHz PCS).

- The performance of the antenna is dependent on its immediate surroundings, packaging, and proximity to the ground plane. The placement of antenna position should be identified early in the design process.

- The orientation of the device and the product usage model during the operation should be considered in mounting the antenna inside the device.

- If one is using an external antenna connecting the antenna with a coaxial cable assembly, the cable routing needs to be designed in such a manner so as to keep it away from motors and battery packs.

- Large and continuous ground plane surfaces will provide better radiation performance than small surfaces.

- Product applications using keypads, LCD, or other types of displays, battery packs, and other metallic surfaces will affect and degrade the symmetry of the radiation pattern, reflections, and multipath. Therefore, the location of the antenna placement is critical. The antenna should be placed for the best balance of the distribution of these objects.

- The effects of human body and the operator's hand should be examined and validated away during the product operation. By keeping the antenna away, the specific absorption rate (SAR) will be reduced and pattern symmetry will be improved.

- It is better to eliminate connectors and interconnect transmission lines to avoid insertion losses on transmit power and receive sensitivity on the receiver.

- It is disadvantageous to use any form of EMI/RFI shield coatings on plastic housing to solve EMI problems without considering the effect of shielding on antenna placement and location.

14.6 | SPREAD SPECTRUM

Spread spectrum uses wide band and noise-like signals. Since spread spectrum signals are noise-like, they are hard to detect. Spread spectrum signals are also hard to intercept or demodulate. Further, these signals are harder to jam (interfere with) than narrowband signals. These low probability of intercept (LPI) and anti-jam (AJ) features are useful for military. Spread signals are intentionally made to be much wider band than the information they are carrying to make them more noise-like.

Spread spectrum signals use fast codes that run many times the information bandwidth or data rate. These special 'Spreading' codes are called 'Pseudo

Random' or 'Pseudo Noise' codes. They are called 'Pseudo' because they are not real Gaussian noise.

Spread spectrum transmitters use similar transmit power levels to narrow band transmitters. Since spread spectrum signals are so wide, they transmit at a much lower spectral power density, measured in watts per hertz, than narrow-band transmitters. This lower transmitted power density characteristic gives spread signals a big plus. Spread and narrow band signals can occupy the same band, with little or no interference. This capability is the main reason for all the interest in spread spectrum today.

Over the last 50 years, a class of modulation techniques usually called '*spread spectrum*', has been developed. This group of modulation techniques is characterized by its wide frequency spectra. The modulated output signals occupy a much greater bandwidth than the signal's baseband information bandwidth. To qualify as a spread spectrum signal, two criteria should be met.

1. The transmitted signal bandwidth is much greater than the information bandwidth.
2. Some function other than the information being transmitted is employed to determine the resultant transmitted bandwidth.

Spread spectrum systems transmit an RF signal bandwidth as wide as 20 to 254 times the bandwidth of the information being sent. Types of spread spectrum systems are: direct sequence, time/frequency hopping type, or hybrid (combination of these two types). Time hopped spread spectrum systems have found no commercial application to date. However, the arrival of cheap random access memory (RAM) and fast microcontroller chips make time hopping a viable alternative spread spectrum technique for the future. 'Chirp' signals are often employed in radar systems.

14.6.1 Direct Sequence Systems

They employ a high-speed code sequence, along with the basic information being sent, to modulate their RF carrier. The high-speed code sequence is used directly to modulate the carrier, for setting the transmitted RF bandwidth. Binary code sequences as short as 11-bits or as long as $[2^{89} - 1]$ have been employed for this purpose, at code rates from under a bit per second to several hundred megabits per second.

The result of modulating an RF carrier with such a code sequence is to produce a signal centred at the carrier frequency, direct sequence modulated spread spectrum with a $(\sin x/x)2$ frequency spectrum. The main lobe of this spectrum has a bandwidth twice the clock rate of the modulating code, from null to null. The sidelobes have a null-to-null bandwidth equal to the code's clock rate. Figure 14.1 illustrates the most common type of direct sequence modulated spread spectrum signal. Direct sequence spectra vary somewhat in spectral shape depending upon the actual carrier and data modulation used. The signal illustrated is that for a binary phase shift key (BPSK), which is the most common modulation signal type used in direct sequence systems.

14.6.2 Frequency Hopping Systems

The desired wideband frequency spectrum is generated in a different manner in a frequency hopping system. It 'hops' from frequency to frequency over a wide band. The specific order in which frequencies are occupied is a function of a code sequence, and the rate of hopping from one frequency to another is a function of the information rate. The transmitted spectrum of a frequency hopping signal is quite different from that of a direct sequence system. Instead of a $[(\sin x)/x]^2$-shaped envelope, the frequency hopper's output is flat over the band of frequencies used. The bandwidth of a frequency hopping signal is simply w times the number of frequency slots available, where w is the bandwidth of each hop channel.

14.7 │ SYSTEM SETUP: INTERCONTROLLER COMMUNICATION

Figure 14.1 depicts the schematic block diagram of an RF-based communication system having one base station and multiple user remote stations. The system operation is controlled by PC through serial link provided by UART. The base station and remote stations are connected through RF links.

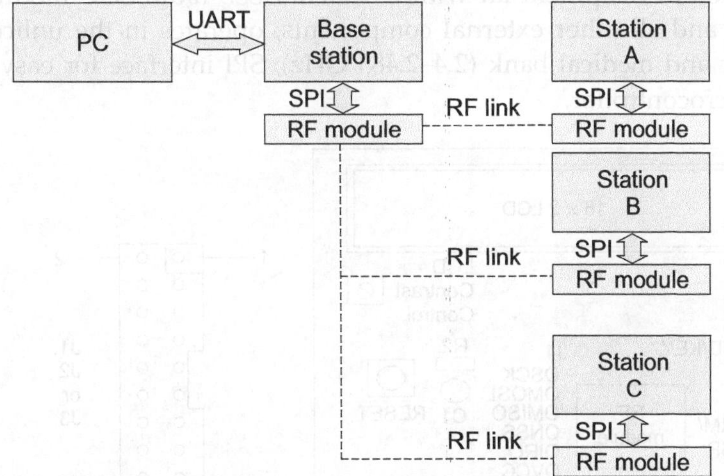

Fig. 14.1   Block diagram of RF-based communication system

The system under consideration consists of a base station, many other modules, and a PC. The base station is connected to the PC via a UART (RS-232) link. It also forms a wireless link with the other modules using an RF module. The base station consists of the LPC 2148 as the processor. It is connected to the RF module, consisting chiefly of the CYWUSB6935 transreceiver from Cypress Inc. using the serial peripheral interface (SPI).

14.8 | SYSTEM HARDWARE

The hardware section consists of two cards: microcontroller board and RF board with the following specifications:

Microcontroller board having the X51 or variant microcontroller with memory : 32 K SRAM, 512 K flash memory; 2 RS-232 interface 20-pin connectors may be designed as discussed in Chapter 5. Design is left as an exercise for the readers. EVMs or kits may be used for testing purpose.

RF board consists of RF transciever IC: CYWUSB6935 from Cypress Inc.; operating frequency: 13 MHz; transmission/reception frequency: 2.4–2.479 GHz; antenna type: dual trace wriggle antenna; 16 × 2 LCD display with contrast control; 20-pin connectors; reset switch; 4 user switches and LED indicator.

The RF application development kit offers wireless radio module working at 2.4 GHz. It is tested for functional operation and comes with dual integrated PCB trace antennas. The development kit works on 3.2 V supply and can be directly interfaced to educational practice board for ARM. Features are: RF application development kit with 2–4 GHz DSSS radio SOC module; includes antenna and all other external components; operates in the unlicensed band scientific and medical bank (2.4–2.483 GHz); SPI interface for easy connection with microcontroller.

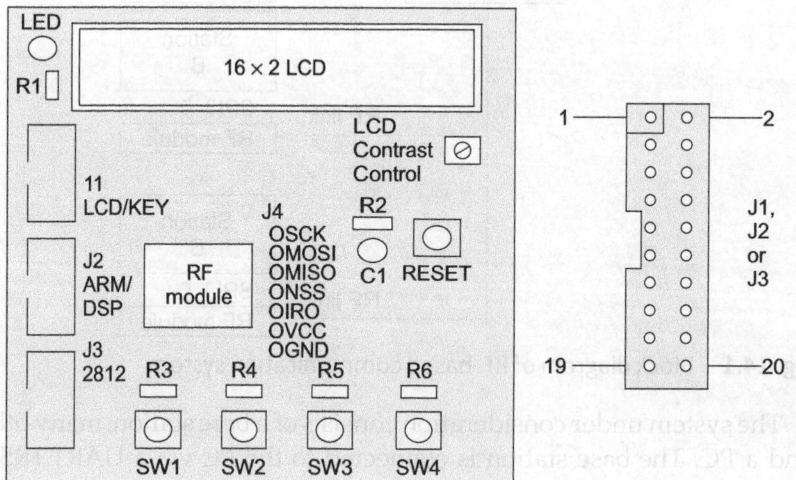

Fig. 14.2 RF board layout

Figure 14.2 depicts the RF board layout of RF board. 20-pins connector J1 is used to interface LCD and different switches to the CPU board, while J2 and J3 may be used for interfacing microcontroller board (Table 14.3).

- All the required lines are brought out to the 20-pins connector and as a test points on the board.
- 'RESET' switch is used to reset the RF module.

Table 14.3   Connector details

Pin number	Connector: J1	Connector: J2	Connector: J2
1	SW1	N.C.	MOSI
2	SW2	SCK	MISO
3	SW3	MISO	SCK
4	SW4	MOSI	nSS
5	N.C.	nSS	N.C.
6-9	N.C.	N.C.	N.C.
10	LCD Data Line – D4	N.C.	N.C.
11	LCD Data Line – D5	N.C.	N.C.
12	LCD Data Line – D6	N.C.	N.C.
13	LCD Data Line – D7	N.C.	N.C.
14	LCD RS Line	N.C.	N.C.
15	LCD EN line	N.C.	N.C.
16-17	N.C.	N.C.	N.C.
18	VCC(3.3V)	VCC(3.3V)	VCC(3.3V)
19	VCC(5V)	VCC(5V)	VCC(5V)
20	Ground	Ground	Ground

14.9   SYSTEM SOFTWARE

The software written for the project can be explained in a simpler way through the use of flowcharts. Explanation of the flow diagram is also given, though it is self explanatory. The flowcharts portray a step-by-step implementation of the proposed software. Also, note that all the hex codes generated using the GNU compiler are stored in the flash memory of the LPC 2148 and are then executed.

14.9.1 Implementation Algorithm

Flow diagram of Fig. 14.3 provides a general idea of the data flow in the proposed project. The software is developed in the modular form. Separate subroutines are written for base station and remote stations. Implementation algorithm: as per the flowchart of Fig. 14.3.

- Initially, all the stations except the base station are in receiver mode and waiting for the command from the base station. The base station initializes and waits for the response from the user to start the communication.

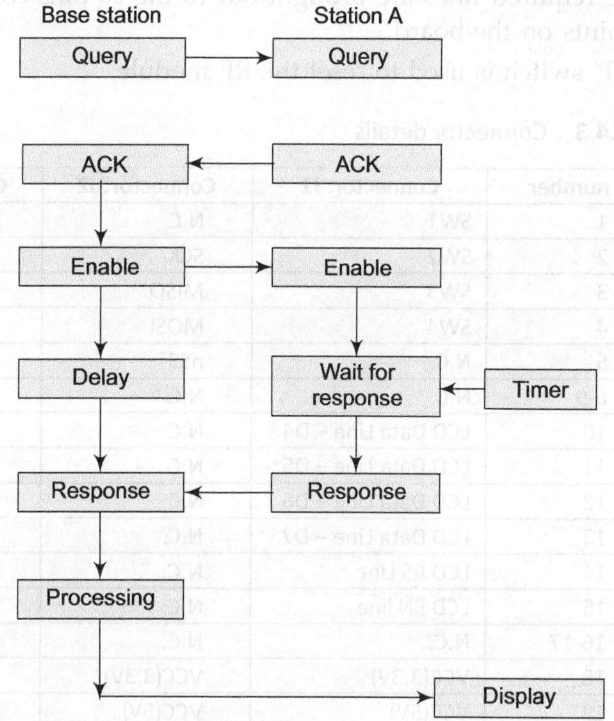

Fig. 14.3   System flow diagram

- On pressing a key, the base station first of all checks which of the stations are in ready state and which are switched OFF. It does this by sending a query signal to each station and waits for its response. If an acknowledgement is received from the station, the base station recognizes that station as ready and polls the next station. If the query times out, it considers that the station is switched OFF or unable to respond.

- After all the stations have been polled, the base station enables all the stations simultaneously and sets a timer to wait for the response of the other stations.

- The base stations are now in the transmit mode. First of all, they wait for a response from the user wishing to communicate with the base station. If the user responds (presses a key) within a certain time limit, the response is stored in the flash memory of the corresponding LPC 2148. If the timer set in the station runs out before the user responds, a default response is stored.

- After this, all the stations again go into reception mode and the base station now polls each of the stations to obtain their stored responses.
- It then compares the responses of each of the stations with a predefined response and displays the results on the LCD. It also sends the data to the PC connected to the base station via the UART (RS-232) interface where it is displayed.

After this process is over, the base station again waits for the user to start the communication and the other stations are in reception mode. Hence, the process is repeated.

14.9.2 Base Station Module

It is the main controlling unit of the project. It is a module similar to the other stations but is programmed in a different manner to communicate with the PC using the UART port and with the other stations using a wireless RF link at 2.402 GHz frequency.

Routines written for the base station include polling, enabling the other stations, waiting for response, performing specific operation on the obtained response, and finally displaying the results. Figure 14.4 depicts the flowchart for the base station.

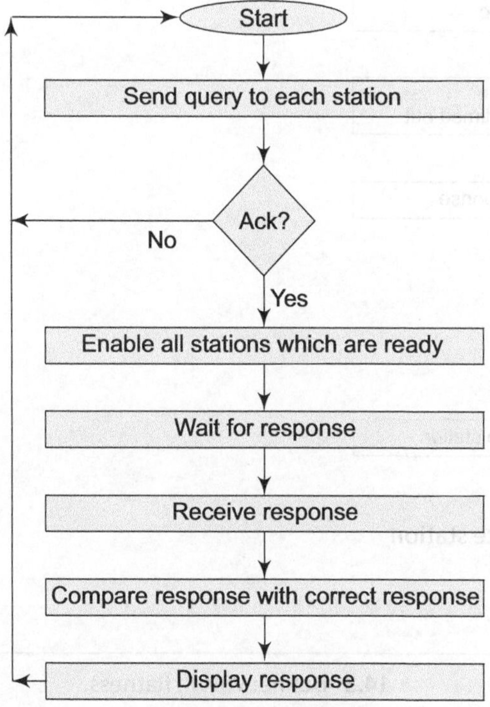

Fig. 14.4   Operation of base station

14.9.3 Remote Station Module

These stations are programmed to communicate with the base station using only a wireless RF link at 2.402 GHz frequency.

Routines written for these stations include waiting for command of base station, enabling the keys, obtaining the response from the user, displaying the response and sending it to the base station. Figure 14.5 depicts the flowchart for the remote stations.

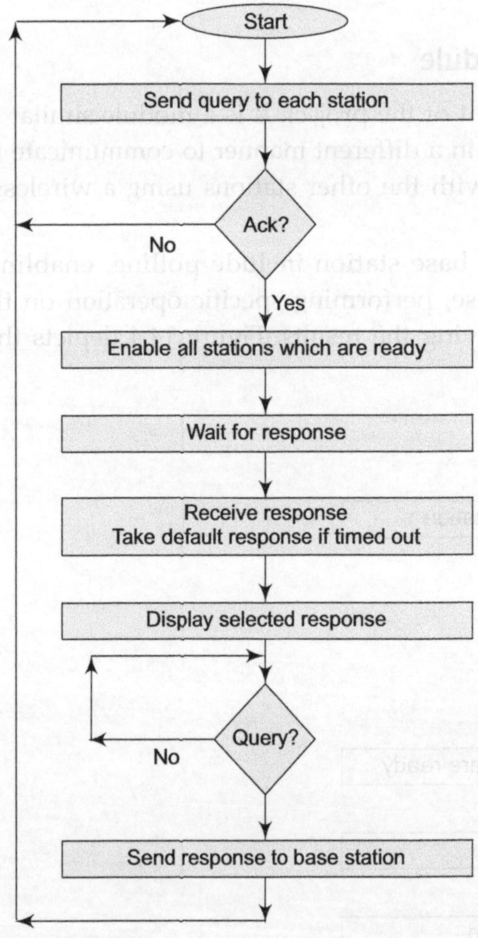

Fig. 14.5   Operation of remote station

EXERCISES

14.1 What is a communication channel?

14.2 State and explain the advantages of RF communication system.

14.3 Define: VSWR, flatness.

14.4 State and explain problems encountered in RF system design.

14.5 What is the function of antenna? Enlist parameters to be considered in design of an antenna.

14.6 Explain the effect of antenna design on the performance of wireless communication.

14.7 What is spread spectrum?

14.8 With the help of schematic block diagram, explain the operation of RF-based communication system with one source and multiple remote stations.

14.9 Describe the features required for an RF board to set up wireless channel.

14.10 Explain the operation of a base station and write software for operation.

14.11 Explain the operation of a remote station and write software for operation.

14.12 Explain how base and remote stations communicate with each other.

Chapter 15

Application VII: Mobile Communication

The vision of the emerging mobile and personal communication services for the new century is to enable communication with a person, at any time, at any place, and in any form, with a paradigm shift from the current focus on voice and low speed data services to that on high speed data and multimedia services. The current second generation digital mobile and personal communication systems are based on national or regional standards that are optimized for region or country-specific regulatory and operating environments. They are incompatible with each other and can provide mobility only within their radio environments.

The second generation (2G) of the wireless mobile network was based on low-band digital data signalling. The most popular 2G wireless technology is known as global system for mobile communications (GSM). GSM technology is a combination of frequency division multiple access (FDMA) and time division multiple access (TDMA). The first GSM systems used a 25 MHz frequency spectrum in the 900 MHz band. FDMA is used to divide the available 25 MHz of bandwidth into 124 carrier frequencies of 200 kHz each. Each frequency is then divided using a TDMA scheme into eight time slots. The use of separate time slots for transmission and reception simplifies the electronics in the mobile units. Today, GSM systems operate in the 900 MHz and 1.8 GHz bands throughout the world with the exception of America where they operate in the 1.9 GHz band.

3G wireless technology represents the convergence of 2G wireless telecommunications systems into a single global system that includes both terrestrial and satellite components. 3G wireless has ability to unify existing cellular standards, such as CDMA, GSM, and TDMA.

Wideband CDMA (W-CDMA) is compatible with 2G-GSM networks. It can be overlaid onto existing GSM, TDMA (IS-36), and IS-95 networks. Subscribers can access 3G wireless services, initially via dual band terminal devices. 3G wireless networks consist of a radio access network (RAN) and a core network. The core network consists of a packet-switched domain, which includes 3G-SGSNs and GGSNs, which provide the same functionality that they provide in a GPRS system, and a circuit-switched domain, which includes 3G-MSC for switching of voice calls. Charging for services and access is done through the Charging Gateway Function (CGF), which is also a part of the core network. RAN functionality is independent from the core network functionality. The access network provides a core network technology, independent access for mobile terminals, different types of core networks, and network services. Core network domain can access any appropriate RAN service; e.g., it should be possible to access a 'speech' radio access bearer from the packet-switched domain.

Communications on shared media are no longer private. Privacy and authentication are lost unless some method is established to regain it. Cryptography provides the solution to regain control over privacy and authentication. All digital mobile systems provide security through some kind of encryption system. Data can be encrypted in many ways, but algorithms used for secure data transfer fall into one of the two broad categories: symmetric and asymmetric. The security algorithms can be embedded in the software, but these are not considered in this chapter.

15.1 | SHORT MESSAGE SERVICE (SMS)

The features of the short message service (SMS) as defined within the GSM digital mobile phone standard are as follows:

- A single short message can be up to 160 characters of text in length comprising words or numbers or an alphanumeric combination. Non-text based short messages (e.g., in binary format) are also supported.
- The SMS is a store and forward service. In other words, short messages are not send directly from sender to recipient, but via an SMS centre instead. Each mobile telephone network that supports SMS has one or more messaging centres to handle and manage the short messages.
- Short message can be sent and received simultaneously with GSM voice, data, and fax calls. This is possible because short message travels over and above the radio channel using the signalling path. As such, users of SMS can get a busy or engaged signal during peak network usage times.

Short message service is a mechanism of delivery of short messages over the mobile networks. It is a store and forward way of transmitting messages to and from mobiles. The message (text only) from the sending mobile is stored in a central short message centre, which then forwards it to the destination mobile. If the recipient is not available, the short message is stored and can be sent later. Each short message cannot longer than 160 characters. These characters can be text (alphanumeric) or binary non-text short messages.

An interesting feature of SMS is return receipts. The sender can get a small message notifying whether the short message has been delivered to the intended recipient. Since SMS uses signalling channel as opposed to dedicated channels, these messages can be sent/received simultaneously with the voice/data/fax service over a GSM network. SMS supports national and international roaming. This means that one can send short messages to any other GSM mobile user around the world. With the PCS networks based on all the three technologies, GSM, CDMA, and TDMA supporting SMS, SMS is more or less a universal mobile data service.

15.2 | AT COMMANDS

They are also called *Hayes AT commands* as they are based on the Hayes *ATTENTION Commands* for modems. They are used to communicate with the modem. These commands modify the modem's behaviour or instruct the modem to do something specific, such as dialing a telephone number. The 'AT' refers to getting the *ATTENTION Commands* of the modem.

AT commands allow a mobile phone to be steered by an external device— for example, it is possible to connect a mobile to a PC and execute the dialing process by sending the appropriate AT commands from the PC to mobile phone. Similarly, the GSM modem steers mobile phones using a 'set' of AT commands. These set of commands activate the necessary 'basic' functionality to use a mobile phone with the modem, such as the initialization of the device, dialing, or hanging-up. These commands are stored as a property of the device being used. The following list describes the AT command set used for sending and receiving messages:

- AT—check if serial interface and GSM modem is working
- ATE0—turn echo OFF, less traffic on serial line
- AT + CNMI—display of new incoming SMS
- AT + CPMS—selection of SMS memory
- AT + CMGF—SMS string format, now they are compressed
- AT + CMGR—read new message from a given memory location
- AT + CMGS—send message to a given recipient
- AT + CMGD—delete message

15.3 | SMS CONFIGURATION COMMANDS

Since the DTE is the microcontroller and DCE is the mobile phone, data is transmitted in SMS format. EEPROM has to be programmed with certain SMS commands which are required to make it a compatible phone. The commands to be used are in the following order:

'AT + CESP' (enter SMS block mode protocol) Execution of this command sets the TA (terminal adaptation) in SMS block mode.

If it is set in SMS block mode,

$$\text{Response} = \text{'OK'} \qquad \text{(success)}$$
$$= \text{'ERROR'} \qquad \text{(failure)}$$

'AT + CMGF' (message format) This command tells the TA, which input and output format of messages to use. <mode> : Format of message used with send, list, and write commands. The mode can be either

PDU mode (entire TP data units used) or

TEXT mode (header and body of the message given as separate parameters)

Command	Possible response
+ CMGF = [<mode>]	
+ CMGF ?	+CMGF: <mode>
+ CMGF =?	+CMGF: (list of supported <mode>s)
<mode>	

O : PDU mode (default when implemented)

 I : text mode

Since we are in the default mode, i.e., PDU mode → 'AT + CMGF = O'; will set TA in PDU mode.

'AT + CSMS' (selects message service) This command selects messaging service. It returns the type of messages supported by ME:

<mt> for mobile terminated message

<mo> for mobile originated message

<bm> for broadcast type message

Command	Possible response
+ CSMS = <service>	+CSMS: <mt>,<mo>,<bm>
	+CMS ERROR: <err>
+CSMS?	+ CSMS: <service>,<mt>,<mo>,<bm>
+CSMS =?	+CSMS: <list of supported services>
<service>	

O   : GSM 03.40 and 03.41

1… 127 : reserved.

123.. : Manufacture specific.

<mt>,<mo>,<bm>

O : type not supported.

 I : type supported.

For example: To set in GSM mode → "AT + CSMS = O"

'AT + CMGS' (sending message) Execution of this command sends a message from TE to the network. Message reference value

<mr> is returned to the TE on successful message delivery

AT + CMGS =<length> where <length> indicates the number of octets in the message.

For example:

If the message to be sent is "hello world," the ASCII code is as follows:

$$h\ (104) - 1101000$$
$$e\ (101) - 1100101$$
$$l\ (108) - 1101100$$
$$l\ (108) - 1101100$$
$$o\ (111) - 1101111$$
$$w\ (119) - 1110111$$
$$o\ (111) - 1101111$$
$$r\ (114) - 1110010$$
$$l\ (108) - 1101100$$
$$d\ (99) - 1100011$$

ASCII code is 7-bit but it has to be converted in 8-bit to send in SMS format. It can be as follows:

$$h \rightarrow 1101000 \qquad 11101000\ (E8)$$
$$c \rightarrow 1100101 \qquad 00111010\ (32)$$
$$l \rightarrow 1101100 \qquad 10011101\ (9B)$$
$$l \rightarrow 1101100 \qquad 10011101\ (9B)$$

Similarly, all characters are converted. Now the message is sent as,

DLE	STX	MESSAGE	DLE	ETX	BCS	BCS

In order to send the message through the microcontroller to a mobile, it is necessary to convert the 7-bit ASCII code to 8-bit GSM code.

15.4 | REMOTE TEMPERATURE MONITORING USING SMS

We will develop a system which will sense temperature at remote places with the help of microcontroller and this data could be received in the form of a SMS message on a mobile phone. In this case, the temperature is taken at any one port of the microcontroller by means of a digital transducer such as DS1820. The microcontroller is programmed in to convert data to a mobile compatible SMS format. This data is obtained at a regular interval of 5 min and after a predetermined time (e.g., 1 h) a graph may be plotted using the values obtained during the time interval. This can further be transmitted to a PC and can be obtained when the user logs on. The data obtained over a given duration can be stored and then the graph indicating the temperature variations with respect to time can be plotted.

Max 232   It is used for serial data communication between 89C52 and PC. It converts TTL levels of 89C52 to RS-232 levels and converts RS-232 levels from PC to TTL levels. The device consists of three sections.

Dual charge pump voltage converter   The MAX 232 chip is internally connected with four capacitors C1, C2, C3, and C4 as shown in Fig. 15.1. These along with the internal circuitry form the dual charge pump voltage converter. Two charge pump voltage converters perform the conversion. The first uses capacitor C1 to double the $+5$ V to 10 V, storing $+10$ V on the V+ output converter filter capacitor C3, second charge pump converter uses capacitor C2 to invert $+10$ V to -10 V, storing -10 V on V+ output filter capacitor. C4 the MAX232 supplies current, ½ mA of excess output load current.

Driver (transmitter) section   The transmitter on the line drivers are inverting level translator, which convert TTL input levels to RS-232 voltage levels with $+5$ V V_{CC}, the typical output voltage swing is $+/- 9$ V. When loaded with the normal 5 kΩ input resistance of an RS-232 receiver the output swing is guaranteed to meet the RS-232 specifications of 15 V minimum output swing under the worst case conditions of all transmitters. Driving the 3 kΩ minimum allowable load impedance, $V_{CC} = 4.5$ V. The input thresholds of TTL compatible drivers are about 25% of V_{CC}. The inputs of unused drivers can be left unconnected and internal 400 kΩ input pull up the inputs high. The source about 12 μA of current. The slew rate is limited to less than 30 V/us. This limits the maximum usable band rate to 19,200 baud.

Receiver section   Receivers convert the ±5 V to ±15 V RS-232C/g's to 5 V TTL outputs. Since the RS-232C specifications define a voltage level greater than $+3$ V as a 0, the receivers are inverting. Maxim has set the input thresholds of receivers to 0.8 V minimum and 2.4 V maximum. This allows the receivers to respond both to RS-232 levels and TTL level inputs. The receivers are protected against input over voltage up to 30 V. The receivers have a hysterisis of approx 0.5 V, with a minimum guaranteed hysterisis. The propagation delays of the receiver are 350 ns for negative going input signals and 650 ns for positive going input signals.

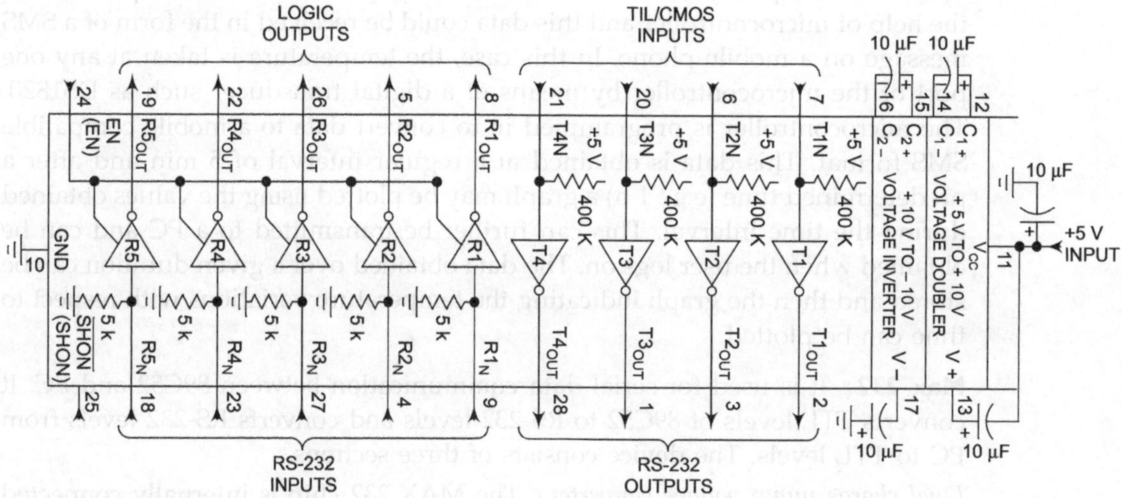

Fig. 15.1   Datasheet of MAXIM 232

RS-232C   As mentioned earlier, serial data communication is used to establish connection between microcontroller and PC/mobile phone. RS-232 standard cable is used for this purpose. RS-232 cable with 25-pin and 9-pin connectors may be used to connect the PC to the serial port RxD pin of the microcontroller.

DS180   The DS18S20 digital thermometer provides 9-bit centigrade temperature measurements and has alarm function with nonvolatile user-programmable upper and lower trigger points. The DS18S20 communicates over a 1-wire bus that by definition requires only one data line (and ground) for communication with a central processor. It has an operating temperature range of –55°C to +125°C and is accurate to = 0.5°C over the range of –10°C to +85°C. In addition, the DS18S20 can derive power directly from the data line (parasite power), eliminating the need for an external power supply. The pin description is as follows:

GND – Ground
DQ – Data In/Out
VDD – Power Supply Voltage
NC – No Connect

The core functionality of the DS18S20 is its direct-to-digital temperature sensor. The temperature sensor output has 9-bit resolution, which corresponds to 0.5°C steps.

SMS interface   A typical system is made up of two types of devices data communication equipment (DCE) and data terminal equipment (DTE).

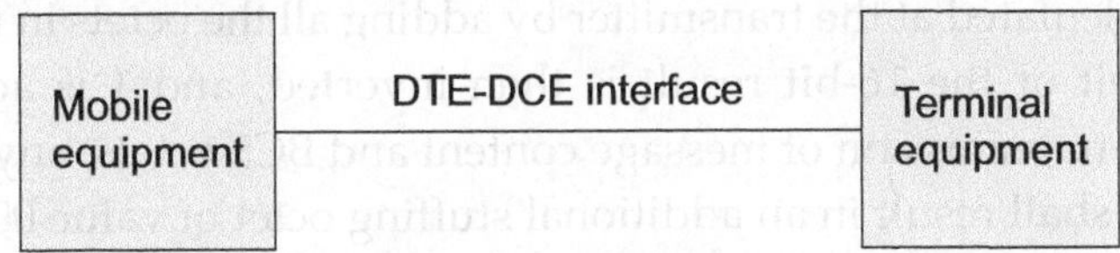

Fig. 15.2 Setup: mobile communication

In our case DCE consist of the mobile equipments. The null modern cable is used to connect the two DTE devices. The operation of the interface circuits for SMS is as given below:

TxD (transmit data): All commands from the TE (microcontroller) to the MT (mobile phone) are transferred across this circuit.

RTS (ready to send): This signal allows the TE to flow control the MT when in the block mode and at other times if hardware flow control is enabled.

CTS (clear to send): This signal allows the MT to flow control the TE in the block mode and is used in response to RTS.

DSR (data set ready): This signal shall be set to the ON condition before entry into the block mode, and shall remain in the ON condition during block mode. If the TE detects that this circuit has returned to the OFF condition during the block mode then TE returns OTR to the OFF condition and exits the block mode.

DTR (data terminal ready): This signal shall be set in the ON condition before the SMS block mode command is send from the TE to begin the block mode. It is returned to the OFF condition after the 'END SMS MODE' command has been accepted and acknowledged by the MT.

DCD (data carrier detect): It is used by DCE to indicate to the DTE that it has received a carrier signal and data is being transmitted. DCD is set to the ON condition before entry into the block mode and remains ON the mode.

The communication path between the MT and the TE across DCE/DTE interface should be quite reliable if it uses a short wire link. Each message sent from the mobile terminal (MT) to the terminal equipment (TE) or vice versa consist of a data block (Data) and block check sum (BCS).

The notation DLE, STX, NUL, and ETX refer to the control characters having the values as follows:

DLE	STX	Message content	DLE	ETX	BCS	BCS
10H	02 H		10H	03H	MSB	LSB

The data block consist of a start transmission sequence set to 0001 0000 0000 0010 (10 02h), the message content as defined and an end of transmission sequence set to 00001 0000 0000 0011 (10 03 hex). The LSB of each octet is always transmitted first.

The BCS is calculated at the transmitter by adding all the octets in the message content. Each bit of the 16-bit result is then inverted, and 1 is added to the answer. During transmission of message content and BCS octets, any occurrence of 10 hex (DLE) shall result in an additional stuffing octet of value 00 hex (NUL) being transmitted immediately following the octet containing 10 hex. This is to ensure that the start and end markers are unambiguous. The receiver shall remove stuffing octets by discarding any octet value 00 hex (NUL), which immediately follows an octet of value 10 hex (DLE). After removal of any stuffing octets the receiver can check the BCS by adding all of the octets in the message content. The transmitter shall only send DLE when it is followed by STX, NUL or ETX. Therefore, if the receiver sees a DLE followed by anything else then the receiver shall assume that some data has been lost, and shall start to search for the start marker. A start marker shall always be treated as the start of a new block, regardless of which state the receiver is in.

15.5 SOFTWARE DESIGN

Here the mobile phone is used as a link between the microcontroller and remote mobile. The data collected by the microcontroller from the temperature sensor are in ASCII or HEX format. The messages are to be converted into mobile compatible format, i.e., GSM format.

Since we are using the mobile phone only as a link between the microcontroller and the computer/another mobile phone, we are not using its display or the keypad. Hence it is necessary to initialize the mobile phone in the SMS mode using configuration commands such as: 'AT + CMGF', 'AT + CSMS', and 'AT + CESP'.

The microcontroller must be programmed to send these commands to the mobile phone so as to initialize it in the SMS mode. All these commands are fed to the mobile phone from the microcontroller in the required GSM format. The mobile phone is ready to transmit the data obtained.

For testing purpose data sending can be simulated by monitored opening and closing of the switch. Connect a switch to one of the pins of the microcontroller. When the switch is turned ON, it was fed as logic HIGH to the microcontroller. Once the logic HIGH is received at a particular pin of the microcontroller, it should indicate this on the terminating equipment (i.e., either the computer or a mobile phone) as 'SWITCH IS ON'. If Logic '0' is received, 'SWITCH IS OFF' is displayed. Thus, the microcontroller is programmed to send either 1 or 0 indicating the status of the switch.

The data will be the temperature values taken after regular intervals of time. To sense the temperature, we are using the digital temperature sensor like DS 1820 so that it is not required to perform analog to digital conversion. The sensor senses the temperature after a regular interval of time and these values are fed to the mobile phone using the SMS facility. These values are then sent to the computer and are taken as input for the program to plot the temperature variation with time. Thus, temperature at a remote place can be measured and monitored using a mobile communication.

System operation is as follows:

1. START: When start button is clicked, the user will be asked for the values at regular intervals (say ten minutes).

2. STOP: When all the values are entered, then stop button is clicked. This will stop asking the values.

3. PLOT: When PLOT button is clicked, the graph of temperature v/s time is plotted using the values entered.

15.5.1 Initialization and Testing of the Microcontroller

Before sending any message on any one port of the microcontroller, the timers, flags, ram, stack pointer, serial ports, and circumference of the microcontroller need to be initialized.

1. Disable all interrupts, enable global interrupt.

2. Set high priority for T2, low for serial and T0, priory of T1 can be high/low, set all bit flags to '0'.

3. Set the serial port for communication as: clear receive/transmit interrupts, enable receiver and serial interrupt.

4. Initialize Timer 0 in 16-bit timer mode, initialize count for 1 ms, start Timer 0, enable Timer 0 interrupt.

5. Initialize Timer 1 in autoreload mode, initialize TH1 with the count for 9600 bauds (11.069 MHz), start Timer 1, to generate desired baud rate.

6. Initialize Timer 2 in 16-bit timer mode, initialize count for 100 μs, start Timer 2, enable Timer 2 interrupt.

7. Microcontroller is ready for the message on any selected port.

8. **To test the initialization of microcontroller:** The message given to any port of the microcontroller; the binary format will be displayed on the Vterm. The program listing depicts a segment for initialization

```
init:
        clr  et0              ; disable Timer 0 overflow int
        clr  et1              ; disable Timer 1 overflow int
        clr  et2              ; disable Timer 2 interrupt
```

```
                clr ex0             ; disable External int 0
                clr ex1             ; disable External int 1
                clr es              ; disable serial interrupt
                ;setb ea            ; enable global interrupt
                clr ea              ; global interrupt disable
                setb Ip. 5          ; high priority for Timer 2
                clr Ip. 4           ; low priority for serial
                clr Ip. 1           ; low priority for Timer 0 (display)
                clr Ip. 3           ; these do not matter (Timer 1)
                call init_bit_flags
        init_uart:
                mov scon, #01000000b ; init. serial control 8-bit UART mode (variable)
                clr ri              ; clear receive interrupt
                clr ti              ; clear transmit interrupt
                setb ren            ; enable receiver
                ;setb es            ; enable serial interru
        init_timer0:
                orl tmod, #01h      ; to in 16-bit timer mode
                mov t10, #1fh       ; init Timer 0 with count for 1 ms
                mov th0, #0fch      ; count = 0fclfh for 1 ms
                setb tr0            ; start Timer 0
                setb et0            ; enable Timer 0 interrupt
        init_timer1:
                orl tmod, #20h      ; Timer 1 in 8-bit auto-reload mode
                mov th1, #0fdh      ; init. TH1 with count for
                                    ; 9600 baud (11.059 MHz)
                setb tr1            ; start Timer 1 (baud rate)
        init_timer2:
                mov t2con, #00000001h ; Timer 2 in 16-bit timer mode
                mov t12, #0a3h      ; init Timer 2 with count for 100 µs
                mov th2, #0ffh      ; count = 0ffa3h for 100 µs
                setb tr2            ; start Timer 2
                setb et2            ; enable Timer 2 interrupt
                ret
        init_10sec_delay:
                mov to_counter1, #0ffh
                mov to_counter2, #40d
                clr to 12_flag
                ret
```

15.5.2 ASCII to GSM Conversion

This is basically conversion of a 7-bit code into 8-bit code. The steps to convert a ASCII code to GSM and transmit on the serial port are given as follows:

1. Input the message
2. Initialize the pointer to the starting of the message
3. Find length of the message
4. Goto Step 33 if the length of message = 0
5. If the length of message = 1
   - Clear bit A_7
   - Transmit the message
   - Goto Step 33
6. Move 2nd byte of message to TEMP
7. Move bit $TEMP_0 \rightarrow$ bit A_7; transmit contents of (A)
8. Move 2nd byte of message to A; rotate contents of A to the right by 1-bit
9. If the length of message = 2
   - Clear bits A_6, A_7
   - Transmit the message
   - Goto Step 33
10. Move 3rd byte of message to TEMP
11. Move bit $TEMP_0 \rightarrow$ bit A_6
    bit $TEMP_1 \rightarrow$ bit A_7 ; transmit contents of (A)
12. Move 3rd byte of message to A; rotate contents of A to the right by 2-bits
13. If the length of message = 3
    - Clear bits A_5, A_6, A_7
    - Transmit the message
    - Goto Step 33
14. Move 4th byte of message to TEMP
15. Move bit $TEMP_0 \rightarrow$ bit A_5
    bit $TEMP_1 \rightarrow$ bit A_6 and
    bit $TEMP_2 \rightarrow$ bit A_7; transmit contents of (A)
16. Move 4th byte of message to A; rotate contents of A to the right by 3-bits
17. If the length of message = 4
    - Clear bits A_4, A_5, A_6, A_7
    - Transmit the message
    - Goto Step 33
18. Move 5th byte of message to TEMP

19. Move bit $TEMP_0 \rightarrow$ bit A_4

     bit $TEMP_1 \rightarrow$ bit A_5

     bit $TEMP_2 \rightarrow$ bit A_6 and

     bit $TEMP_3 \rightarrow$ bit A_7; transmit contents of (A)

20. Move 5th byte of message to A; rotate contents of A to the right by 4-bits

21. If the length of message = 5

    - Clear bits A_4, A_5, A_6, A_7
    - Transmit the message
    - Goto Step 33

22. Move 6th byte of message to TEMP

23. Move bit $TEMP_0 \rightarrow$ bit A_3

     bit $TEMP_1 \rightarrow$ bit A_4

     bit $TEMP_2 \rightarrow$ bit A_5

     bit $TEMP_4 \rightarrow$ bit A_6 and

     bit $TEMP_5 \rightarrow$ bit A_7; transmit contents of (A)

24. Move 6th byte of message to A; rotate contents of A to the right by 5-bits

25. If the length of message = 6

    - Clear bits A_3, A_4, A_5, A_6, A_7
    - Transmit the message
    - Goto Step 33

26. Move 7th byte of message to TEMP

27. Move bit $TEMP_0 \rightarrow$ bit A_2

     bit $TEMP_1 \rightarrow$ bit A_3

     bit $TEMP_2 \rightarrow$ bit A_4

     bit $TEMP_3 \rightarrow$ bit A_5

     bit $TEMP_4 \rightarrow$ bit A_6 and

     bit $TEMP_5 \rightarrow$ bit A_7; transmit contents of (A)

28. Move 7th byte of message to A; rotate contents of A to the right by 6-bits

29. If the length of message = 7

    - Clear bits A_2, A_3, A_4, A_5, A_6, A_7
    - Transmit the message
    - Goto Step 33

30. Move 8th byte of message to TEMP

31. Move bit $TEMP_0 \rightarrow$ bit A_1

     bit $TEMP_1 \rightarrow$ bit A_2

     bit $TEMP_2 \rightarrow$ bit A_3

$$\text{bit TEMP}_3 \rightarrow \text{bit A}_4$$
$$\text{bit TEMP}_4 \rightarrow \text{bit A}_5$$
$$\text{bit TEMP}_5 \rightarrow \text{bit A}_6 \text{ and}$$
$$\text{bit TEMP}_6 \rightarrow \text{bit A}_7; \text{ transmit contents of (A)}$$

32. If the length of message > 8
    - Set the pointer to point at the 9th byte
    - Goto Step 2
33. return

15.5.3 Initialization of Mobile in SMS Mode

This will initialize MT in SMS block mode. The algorithm steps are

1. Set count = 3, msg = 'AT + CESP'
2. Transmit the message and wait for the response within 10 seconds
3. If 'there is a response'

    goto step 2

    else, decrement count

    if count $\neq$ 0

    goto step 2

    else,

    display message 'initialization failed'

    goto step 1
4. If reply $\neq$ 'ok' $\rightarrow$ goto step 1
5. Set msg = 'AT + CMFG = 0' ; transmit message
6. Set msg = 'AT + CSMS = 0' ; transmit message
7. End

A sample initialization code is given below

```
dd init:
init_sms_mode:
        mov dptr,#msg_ismsm
        call txm
        jnb ri,$
        mov a, sbuf
        cjne a, #'0', init_sms_mode
        jnb ri, $
        mov a, sbuf
        cjne a, #'K', init_sms_mode
        ret
```

```
set pdu mode:
                mov dptr, #msg_ismsm
                call tem
                ret
set gsm mode:
                mov dptr, #msg_sgsmm
                call tem
                ret
msg sgsmm: db "AT+CSMS=0", #03h
msg spdum: db "AT+CMGF=0", #03h
msg ismsm: db "AT+CESP", #03h
Msg_ismsm: db "AT+CESP", #03h
init_sms_mode:
                mov count1, #03d
loop_init: mov dptr, #msg_ismsm
                call txm
                call init_10sec_wait        ; wait for 10 s
loop_chk:   jnb ri, chk_wait
                jmp out_chk_wait
chk_wait:   jnb tol2_flag, loop_chk
                dec count1
                mov a, count1
                cjne a, #00d, loop_init
                jmp out_init_sms_fail
out_chk_wait:
                mov a, sbuf
                clr ri
                cjne a, #'0', init_sms_mode
                jnb ri, $
                mov a, sbuf
                clr ri
                cjne a, #'K', init_sms_mode
                ret
out_init_sms_fail:
                mov dptr, #init_sms_fail_m
                call txm
                ret
init_sms_fail_m: db "initialization failed", #03h
```

```
msg_spdum:           db  "AT+CMGF=0"#03h
set_pdu_mode:
             mov dptr,  #msg_ismsm
             call txm
             ret
msg_sgsmm: db "ATCSMS=0",  #03h
set_gsm_mode:
             mov dptr,  #msg_sgsmm
             call txm
             ret
```

EXERCISES

15.1 Describe evolution of wireless mobile network.

15.2 Define: CDMA, FDMA, TDMA, ODFM.

15.3 Discuss significance of security in wireless communication.

15.4 What is an SMS?

15.5 Enlist characteristic of SMS.

15.6 Illustrate significance of AT commands.

15.7 Enlist and discuss AT commands used for SMS.

15.8 Describe need for configuration commands.

15.9 Explain format of GSM code.

15.10 Define: CTS, RTS, DTR, DCD and their usage in establishing handshaking operation for asynchronous communication.

15.11 What is a block check sum (BCS)?

15.12 Explain the procedure to calculate BCS.

15.13 Write a program module to convert ASCII to GSM code based on algorithm discussed in Section 15.5.2.

15.14 Write a program module to initialize the controller board.

15.15 Write a C-code to initialize mobile in SMS mode.

```
msg_sndpdu:    db   "AT+CMGF=0",0Dh
               set_pdu_mode;
               mov dptr, #msg_sndpdu
               call txm
               ret
msg_sndsms:    db   "AT+CMGS=0;",#05h
               set_gsm_mode;
               mov dptr, #msg_sndsm
               call txm
               ret
```

EXERCISES

15.1 Describe evolution of wireless mobile network.

15.2 Define CDMA, FDMA, TDMA, ODFM.

15.3 Discuss significance of security in wireless communication.

15.4 What is an SMS?

15.5 Enlist characteristic of SMS.

15.6 What are significance of AT commands.

15.7 Enlist and discuss AT commands used for SMS.

15.8 Describe need for configuration commands.

15.9 Explain format of GSM code.

15.10 Define CTS, RTS, DTR, DCD and their usage in establishing handshaking operation for asynchronous communication.

15.11 What is a block check sum (BCS)?

15.12 Explain the procedure to calculate BCS.

15.13 Write a program module to convert ASCII to GSM code based on algorithm discussed in Section 15.5.2.

15.14 Write a program module to initialize the controller board.

15.15 Write a C code to initialize mobile in SMS mode.

APPENDIX A

Functional Grouping of Instructions

Notations used in the instruction set:

Rn   — Register R7–R0 of the currently selected register bank.

Direct   — 8-bit internal data location's address. This could be an internal data RAM location (0–127) or a SFR, i.e., I/O port, control register, status register, etc. (128–255).

@Ri   — 8-bit internal data RAM location (0–255) addressed indirectly through register R1 or R0.

#data   — 8-bit constant included in instruction.

#data 16   — 16-bit constant included in instruction.

Addr 16   — 16-bit destination address. Used by LCALL and LJMP. A branch can be anywhere within the 64 K byte program memory address space.

Addr 11   — 11-bit destination address. Used by ACALL and AJMP. The branch will be within the same 2 K byte page of program memory as the first byte of the following instruction.

Rel   — Signed (2's complement) 8-bit offset byte. Used by SJMP and all conditional jumps. Range is –128 to +127 bytes relative to first byte of the following instruction.

Bit   — Direct addressed bit in internal data RAM or special function register.

ARITHMETIC INSTRUCTIONS

ADD	A,Rx	Add the contents of register x to A, results in A, all flags affected.
ADD	A,direct	Add the contents of direct location to A, results in A, all flags affected.
ADD	A,@Rx	Add the contents of address from Rx to A, all flags affected.
ADD	A,#data	Add #data to A, result in A, all flags affected.
ADDC	A,Rx	Add the contents of Rx and C to A, result in A, all flags affected.
ADDC	A,direct	Add the contents of direct location and C to A, results in A, all flags affected.
ADDC	a,@Rx	Add the contents of address from Rx and C to A, results in A, all flags affected.
ADDC	A,#data	Add the data and C to A, results in A, all flags affected.
SUBB	A,Rx	Subtract the contents of Rx and C from A, results in A, all flags affected.
SUBB	A,direct	Subtract the contents of direct location and C from A, results in A, all flags affected.
SUBB	A,@Rx	Subtract contents of address given in Rx and C from A, results in A, all flags affected.
SUBB	A,#data	Subtract data and C from A, results in A, all flags affected.
INC	A	Increment A, results in A, no flags are affected.
INC	Rx	Increment contents of Rx, results in A, no flags are affected.
INC	direct	Increment contents of direct address, results in direct address, no flags are affected.
INC	@Rx	Increment contents of address in Rx, results in address, given in Rx, no flags are affected.
DEC	A	Decrement A, results in A, no flags are affected.
DEC	Rx	Decrement contents of Rx, results in Rx, no flags are affected.
DEC	direct	Decrement contents of direct address, results in direct address, no flags are affected.
DEC	@Rx	Decrement contents of address in Rx, results in address in Rx, no flags are affected.
INC	DPTR	Decrement 16-bit DPTR, results in DPTR, no flags are affected.
MUL	AB	Multiply A by B. Results: low byte in A. High byte in B. OV flag affected, C always cleared.
DIV	AB	Divide A by B, integer result in A remainder in B, OV, and C flags cleared.
DA	A	Adjust A for BCD addition, result in A, C flag affected.

LOGICAL OPERATIONS

ANL	A,Rx	AND A with contents of Rx, results in A, no flags are affected.
ANL	A,direct	AND A with contents of direct address, results in A, no flags are affected.
ANL	A,@Rx	AND A with contents of address in Rx, results in A, no flags are affected.
ANL	A,#data	AND A with data, results in A, no flags are affected.
ANL	direct,A	AND contents of direct address with A, results in direct address, no flags are affected.
ANL	direct,#data	AND contents of direct address with data, results in direct address, no flags are affected.
ORL	A,Rx	OR A with contents of Rx, results in A, no flags are affected.
ORL	A,direct	OR A with contents of direct address, results in A, no flags are affected.
ORL	A,@Rx	OR A with contents of address in Rx, results in A, no flags are affected.
ORL	A,#data	OR A with data, results in A, no flags are affected.
ORL	direct,A	OR contents of direct address with A, results in direct address, no flags are affected.
ORL	direct,#data	OR contents of direct address with data, results in direct address, no flags are affected.
XRL	A,Rx	Exclusive OR A with contents of Rx, results in A, no flags are affected.
XRL	A,direct	Exclusive OR A with contents of address in Rx, results in A, no flags are affected.
XRL	direct,A	Exclusive OR contents of direct address with A, results in direct address, no flags are affected.
XRL	direct,#data	Exclusive OR contents of direct address with data, results in direct address, no flags are affected.
CLR	A	Clear A to zero, A = 0, no flags are affected.
CPL	A	Complement A, bit-wise complement of A, no flags are affected.
RL	A	Rotate A left 1-bit, no flags are affected.
RR	A	Rotate A 1-bit right, no flags are affected.
RLC	A	Rotate A 1-bit left through C, C flags affected.
RRC	A	Rotate A 1-bit right through C, C flags affected.
SWAP	A	Exchange nibbles of A, no flags are affected.

BRANCHING INSTRUCTIONS (FLAGS ARE NOT AFFECTED)

JC	rel	Branch to code address PC + rel if c=1.
JNC	rel	Branch to code address PC + rel if c=0.
JB	bit,rel	Branch to code address PC + rel if contents of bit address is 1.
JNB	bit,rel	Branch to code address PC + rel if contents of bit address are 0.
JBC	bit,rel	Branch to code address PC + rel if contents of bit address = 1. Clear contents of bit address.
ACALL	11bit add	Subroutine call to absolute code address PC (15 – 11) + 11-bit address. Return address on stack.
LACLL	16bitadd	Subroutine call to code space at 16-bit address. Return address on stack.
RET		Return from subroutine. PC loaded from stack.
RET1		Return from interrupt service routine, PC loaded from stack.
AJMP	16bitadd	Unconditional branch to 16-bit address in code space.
LJMP	11bitadd	Unconditional branch to absolute code address PC(15-11) + 11-bit address.
SJMP	rel	Unconditional branch to code address PC + rel.
JMP	@A+DPTR	Unconditional branch to code address A + DPTR.
JZ	rel	Branch to code address PC + rel if A is not equal to 0.
CJNE	A,direct,rel	Branch to code address PC + rel if contents of A are not equal to contents of direct address.
CJNE	A,#data,rel	Branch to code address PC + rel if data are not equal to the contents of A.
CJNE	Rx,#data,rel	Branch to code address PC + rel if contents of Rx are not equal to data.
DJNZ	Rx.rel	Decrement contents of Rx, Branch to code address PC + rel if new value in Rx is not equal to 0.
DJNZ	direct,rel	Decrement contents of direct address. Branch to code address PC + rel if contents of direct address are not equal to 0.

DATA TRANSFER INSTRUCTIONS

MOV	A,Rx	Copy contents of Rn to A, no flags are affected.
MOV	A,direct	Copy contents of direct address to A, no flags are affected.
MOV	A,@Rx	Copy contents of address in Rx to A, no flags are affected.
MOV	A,#data	Load data into A, no flags are affected.

MOV	Rx,A	Copy contents of A to Rx, no flags are affected.
MOV	Rx,direct	Copy contents of direct address to Rx, no flags are affected.
MOV	Rx,#data	Load data into Rx, no flags are affected,
MOV	durect,A	Copy contents of A to direct address, no flags are affected.
MOV	direct,Rx	Copy contents of Rx to direct address, no flags are affected.
MOV	direct1,direct2	Copy contents of direct2 address to direct1 address, no flags are affected.
MOV	direct,@Rx	Copy contents of address in Rx to direct address, no flags are affected.
MOV	direct,#data	Load data into direct address, no flags are affected.
MOV	@Rx,A	Copy contents of A into address in Rx, no flags are affected.
MOV	@Rx,direct	Copy contents of direct address to address in Rx, no flags are affected.
MOV	@Rx,#data	Load data into address in Rx, no flags are affected.
MOV	DPTR,#16bitdata	Load 16-bit data into DPTR, no flags are affected.
MOVC	A,@A+DPTR	Copy contents of code address A + DPTR to A, no flags are affected.
MOVC	A,@A+PC	Copy contents of code address A + PC to A, no flags are affected.
MOVX	A,@Rx	Copy contents of external data address in Rx to A, no flags are affected.
MOVX	A,@DPTR	Copy contents of external data address in DPTR to A, no flags are affected.
MOVX	@Rx,A	Copy contents of A to external data address in Rx, no flags are affected.
MOVX	@DPTR,A	Copy contents of A to external data address in DPTR, no flags are affected.
PUSH	direct	Copy contents of direct address to next available stack location, no flags are affected.
POP	direct	Copy contents of current stack location to direct address, no flags are affected.
XCH	A,Rx	Exchange contents of A and Rx, no flags are affected.
XCH	A,direct	Exchange contents of A and direct address, no flags are affected.
XCH	A,@Rx	Exchange contents of A and address in Rx, no flags are affected.
XCHD	A,@Rx	Exchange low nibble of A and address in Rx, no flags are affected.

BOOLEAN INSTRUCTIONS

CLR	C	Clear C to zero. C flag is affected.
CLR	bit	Clear bit at bit address, no flags are affected.
SETB	C	Set C to one, C flag is affected.
SETB	bit	Set bit address to one, no flags are affected.
CPL	C	Complement C, C flag is affected.
CPL	bit	Complement bit at bit address, no flags are affected.
ANL	C,bit	AND C with contents of bit address, results in C, C flag affected.
ANL	C,/bit	AND C with complement of contents of bit address, results in C, C flags affected.
ORL	C,bit	OR C with contents of bit address, results in C, C flag is affected.
ORL	C,/bit	OR C with complement of contents of bit address, results in C, C flag is affected.
MOV	C,bit	Copy contents of bit address to C, C flag is affected.
MOV	bit,C	Copy contents of C to bit address, no flags are affected.

MISCELLANEOUS INSTRUCTIONS

NOP	Do absolutely nothing.

APPENDIX B

Instructions: Byte/Period

Mnemonic	Description	Byte	Oscillator period
ADD A,Rn	Add register to accumulator	1	12
ADD A,direct	Add direct byte to accumulator	2	12
ADD A,@Ri	Add indirect RAM to accumulator	1	12
ADD A,#data	Add immediate data to accumulator	2	12
ADDC A,Rn	Add register to accumulator with carry	1	12
ADDC A,direct	Add direct byte to accumulator with carry	2	12
ADDC A,@Ri	Add indirect RAM to accumulator with carry	1	12
ADDC A,#data	Add immediate data to Acc with carry	2	12
SUBB A,Rn	Subtract register from Acc with borrow	1	12
SUBB A,direct	Subtract direct byte from Acc with borrow	2	12

Mnemonic	Description	Byte	Oscillator period
SUBB A,@Ri	Subtract indirect RAM from ACC with borrow	1	12
SUBB A,#data	Subtract immediate data from Acc with borrow	2	12
INC A	Increment accumulator	1	12
INC Rn	Increment register	1	12
INC direct	Increment direct byte	2	12
INC @Ri	Increment direct RAM	1	12
DEC A	Decrement accumulator	1	12
DEC Rn	Decrement register	1	12
DEC direct	Decrement direct	2	12
DEC @Ri	Decrement indirect RAM	1	12
INC DPTR	Increment data pointer	1	24
MUL AB	Multiply A and B	1	48
DIV AB	Divide A by B	1	48
DA A	Decimal adjust accumulator	1	12
ANL A,Rn	AND register to accumulator	1	12
ANL A,direct	AND direct byte to accumulator	2	12
ANL A,@Ri	AND indirect RAM to accumulator	1	12
ANL A,#data	AND immediate data to accumulator	2	12
ANL direct,A	AND accumulator to direct byte	2	12
ANL direct,#data	AND immediate data to direct byte	3	24
ORL A,Rn	OR register to accumulator	1	12
ORL A,direct	OR direct byte to accumulator	2	12
ORL A,@Ri	OR indirect RAM to accumulator	1	12
ORL A,#data	OR immediate data to accumulator	2	12
ORL direct,A	OR accumulator to direct byte	2	12
ORL direct,#data	OR immediate data to direct byte	3	24
XRL A,Rn	Exclusive-OR register to accumulator	1	12
XRL A,direct	Exclusive-OR direct byte to accumulator	2	12
XRL A,@Ri	Exclusive-OR indirect RAM to accumulator	1	12
XRL A,#data	Exclusive-OR immediate data to accumulator	2	12
XRL direct,A	Exclusive-OR accumulator to direct byte	2	12
XRL direct,#data	Exclusive-OR immediate data to direct	3	24

Mnemonic	Description	Byte	Oscillator period
CLR A	Clear accumulator	1	12
CPL A	Complement accumulator	1	12
RL A	Rotate accumulator left	1	12
RLC A	Rotate accumulator left through the carry	1	12
RR A	Rotate accumulator right	1	12
RRC A	Rotate accumulator right through the carry	1	12
SWAP A	Swap nibbles within the accumulator	1	12
MOV A,Rn	Move register to accumulator	1	12
MOV A,direct	Move direct byte to accumulator	2	12
MOV A,@Ri	Move indirect RAM to accumulator	1	12
MOV A,#data	Move immediate data to accumulator	2	12
MOV Rn,A	Move accumulator to register	1	12
MOV Rn,direct	Move direct byte to register	2	24
MOV Rn,#data	Move immediate data to register	2	12
MOV direct,A	Move accumulator to direct byte	2	12
MOV direct,Rn	Move register to direct byte	2	24
MOV direct,direct	Move direct byte to direct	3	24
MOV direct,@Ri	Move indirect RAM to direct byte	2	24
MOV direct,#data	Move immediate data to direct byte	3	24
MOV @Ri,A	Move accumulator to indirect RAM	1	12
MOV @Ri,direct	Move direct byte to indirect RAM	2	24
MOV @Ri,#data	Move immediate data to indirect RAM	2	24
MOV DPTR, #data16	Load data pointer with a 16-bit constant	3	24
MOVC A, @A+DPTR	Move code byte relative to DPTR to Acc	1	24
MOVC A, @A+PC	Move code byte relative to PC to Acc	1	24
MOVX A,@Ri	Move Ext. RAM (8-bit Addr) to Acc	1	24
MOVX A,@DPTR	Move Ext. RAM (16-bit Addr) to Acc	1	24
MOVX @Ri,A	Move Acc to Ext. RAM (8-bit addr)	1	24
MOVX @DPTR,A	Move Acc to Ext. RAM (16-bit addr)	1	24

Mnemonic	Description	Byte	Oscillator period
PUSH direct	Push direct byte onto stack	2	24
POP direct	POP direct byte from stack	2	24
XCH A,Rn	Exchange register with accumulator	1	12
XCH A,direct	Exchange direct byte with accumulator	2	12
XCH A,@Ri	Exchange indirect RAM with accumulator	1	12
XCHD A,@Ri	Exchange low-order digit indirect RAM with accumulator	1	12
CLR C	Clear carry	1	12
CLR bit	Clear direct bit	2	12
SETB C	Set carry	1	12
SETB bit	Set direct bit	2	12
CPL C	Complement carry	1	12
CPL bit	Complement direct bit	2	12
ANL C,bit	AND direct bit to carry	2	24
ANL C,/bit	AND complement of direct bit to carry	2	24
ORL C,bit	OR direct bit to carry	2	24
ORL C,/bit	OR complement of direct bit to carry	2	24
MOV C,bit	Move direct bit to carry	2	12
MOV bit,C	Move carry to direct bit	2	24
JC rel	Jump if carry is set	2	24
JNC rel	Jump if carry not set	2	24
JB bit,rel	Jump if direct bit is set	3	24
JNB bit,rel	Jump if direct bit is not set	3	24
JBC bit,rel	Jump if direct bit is set and clear bit	3	24
ACALL addr11	Absolute subroutine call	2	24
LCALL addr16	Long	3	24
RET	Return from subroutine	1	24
RET1	Return from interrupt	1	24
AJMP addr11	Absolute jump	2	24
LJMP addr15	Long jump	3	24
SJMP rel	Short jump (relative addr)	2	24
JMP @A+DPTR	Jump indirect relative to the dptr	1	24
JZ rel	Jump if accumulator is zero	2	24
JNZ rel	Jump if accumulator is not zero	2	24

Mnemonic	Description	Byte	Oscillator period
CJNE A, direct,rel	Compare direct byte to acc and Jump if not equal	3	24
CJNE Rn, # data,rel	Compare immediate to register and Jump if not equal	3	24
CJNE @ Ri,#data,rel	Compare immediate to indirect and Jump if not equal	3	24
DJNZ Rn,rel	Dec. register and jump if not zero	2	24
DJNZ direct,rel	Dec. direct byte and jump if not zero	3	24
NOP	No operation	1	12

Select Bibliography

1. Amirshahi, P. and M. Kavehrad, *Medium Voltage Overhead Power Line Broadband Communication; Transmission Capacity and Electromagnetic Interference*, Dept. of Elec. Engg, The Pennsylvania State University.
2. Assimakopoulos, C. and F.N. Pavlidou, *Measurement and Modeling of In-House Power Lines Installation for Broadband Communications*, Dept. of Electrical Engg, Aristotle University of Thessaloniki.
3. Del Re, E., R. Fantacci, S. Morosi, and R. Seravalle, *Comparision of CDMA and OFDM Techniques for Downstream Power Line Communications on Low Voltage Grid*, IEEE Transaction on Power Delivery, Vol. 18, No. 4, Oct 2003.
4. Dostert, K., *EMC Aspects of High Speed Power Line Communications*, Proceedings of the 15th International Wroclaw Symposium and Exhibition on Electromagnetics Compatibility, Wroclaw, Poland, June 27-30.
5. Hanzo, L., B. Choi, and M. Mnster, *A Stroll Along Multicarrier Boulevard to Next-Generation Plaza (I) OFDM Background and History*, School of ECS, University of Southampton.
6. Hrasnica, H., A. Haidine, R. Lehnert, *Reservation MAC Protocol for Power Line Communications*, Dresden University of Technology, Germany.
7. Jung, S.Y., Tae_Hyun Kim*, Myung-Un Lee**, Wook-Hyun Kwon, *Modeling and Simulation of Power Line Channel*, *Dept. of Elec. Tech Myongji College, **Dept. of Power Equipment Automation Seoul College, School of Elect. Engg, Seoul Nat'l University.
8. Lee, F. King, *Space-Time and Space-Frequency Coded Orthogonal Frequency Division Multiplexing Transmitter Diversity Techniques*, Georgia Institute of Technology.
9. Meng, H. and S. Cheng, Modeling of Transfer Characteristics for the Broadband Power Line Communication Channel, *IEEE Transactions on Power Delivery*, Vol. 19, No. 3, July 2004.
10. Meng, H. and Y.L. Guan, Modeling and Analysis of Noise Effects on Broadband Power Line Communications, *IEEE Transactions on Power Delivery*, Vol. 20, No. 2, April 2005.
11. Pavlidou, N., Aristotle University of Thes-saloniki, A. J. Han Vinck, University of Essen, Javad Yazdani and Bahram Honary, University of Lancaster, *Power Line Communications: State of the Art and Future Trends*.
12. Power Line Modem Micro-Module, http://www.michat.com.
13. Purroy, A., A. Sanz, J.I. Garcia Nicolas, and I. Urriza, *Reaserch Area for Efficient Power Line Communication Modems*, Dept. of E & C Engg, University of Zaragoza, Spain.
14. ST7538 FSK Power Line Transceiver, Application Note AN1714, ST Microelectronics, 2003, www.st.com.
15. User Manual and Technical Brochures of NXP Semiconductors, www.maxim-ic.com.
16. Wang, H. and J. Lilleberg, *Novel OFDM Transceiver with Time-Domain Scrambling* Nokia Corporation, Oulu, Finland.
17. Yitran, Power Line Carrier (PLC) Communication Chips, www.Yitran.com.

18. Zhao, Z.W. and I-Ming Chen, *Moving Home Plug to Industrial Applications with Power Line Communication Network*, Singapore-MIT Alliance Program, Nanyang Technological University, Singapore.

19. A position paper prepared by the RSGB EMC Committee for the PLC Workshop in Brussels, *Compatibility between Radio Communication Services and Power Line Communication System*, 5 Mar 2001.

20. Cypress Application Notes: AN4008, AN5028, AN5032, AN5033.

21. Data Sheet: TDA5051 Home Automation Modem for ASK Modulation, Philips Semiconductors.

22. Telecommunication Switching Systems and Networks by *Thaigarajan Viswanathan*.

23. Telecom IC's, *PLESSEY Semiconductors*.

24. *Microcontroller Handbook*, INTEL Publication.

25. *Telephone Circuits*, BPB Publications.

Select Web References

(**Note:** All the below-mentioned websites were last accessed on or before 15th July, 2009)

- www.microchip.com
- www.pacificindia.com
- www.8052.com
- www. atmel. com
- www.analogdevices.com
- www.embedded.com
- www.phillips.com
- www.realtech.com
- www.wikipedia.com/eng
- www.modbus.org
- www.futurelec.com
- www.control.com
- www.allaboutcircuits.com
- www.artechengg.com
- www.hansonwelding.com
- www.spotco.com
- www.ptswelding.com
- www.weldtechnology.com
- www.chowel.com
- www.tjsnow.com
- www.wme-inc.com
- www.nsrw.com
- www.weldersnpresses.com
- www.olimex.com
- www.amd.com, ©2009 Advanced Micro Devices, Inc.
- www.arccores.com/, ©2009 ARC International.
- www.atmel.com/, ©2009 Atmel Corporation.
- www.maxim-ic.com/, ©2009 by Maxim Integrated Products, Dallas Semiconductor.
- www.sxlist.com/techref/hitachi/index.htm, ©2009 Hitachi Semiconductor.

- www-03.ibm.com/technology/index.html, ©Copyright IBM Corporation 1994, 2009.
- w1.siemens.com/entry/cc/en/, ©Siemens AG 1996-2009.
- www.idt.com/, ©1997-2009 Integrated Device Technology, Inc.
- www.intel.com/, ©Intel Corporation.
- www.issi.com/index.html, Integrated Silicon Solution, Inc.
- www.microchip.com/stellent/idcplg?IdcService=SS_GET_PAGE&nodeId=64, ©2008 Microchip Technology Inc.
- www.mitsubishi.com/e/index.html, mitsubishi.com committee.
- www.motorola.com, ©2009 MOTOROLA,INC.
- www.national.com/analog, National Semiconductor Corporation.
- www.nec.com/, ©NEC Corporation 1994-2009.
- www.nxp.com/#/homepage, ©2006-2009 NXP Semiconductors.
- www.samsung.com/global/business/semiconductor/index.html, ©1995-2009 SAMSUNG.
- www.ubicom.com/, Copyright ©2009 Ubicom, Inc.
- www.st.com/stonline/, ©2009 ST Microelectronics.
- www.ti.com/, Copyright ©1995-2009 Texas Instruments Incorporated.
- www.zilog.com/, Copyright ©2009 Zilog, Inc.
- www.cypress.com/, ©2009 Cypress Semiconductor Corporation.
- www.analog.com/en/index.html, ©1995-2009 Analog Devices, Inc.
- www.toshiba.com/tai/, ©2009 Toshiba America, Inc.

- www.ibm.com/technology/index.html ©Copyright IBM Corporation 1994, 2008.
- w1.siemens.com/entry/cc/en/ ©Siemens AG 1996-2009
- www.idt.com/ ©1997-2009 Integrated Device Technology, Inc.
- www.intel.com/ ©Intel Corporation.
- www.issi.com/index.html Integrated Silicon Solution, Inc.
- www.microchip.com/stellent/idcplg?IdcService=SS_GET_PAGE&nodeId=1 ©2008 Microchip Technology Inc.
- www.mitsubishi.com/e/index.html mitsubishi.com committee.
- www.motorola.com ©2009 MOTOROLA,INC.
- www.national.com/analog National Semiconductor Corporation.
- www.necel.com/ ©NEC Corporation 1994-2009
- www.nxp.com/#/homepage/ ©2006-2009 NXP Semiconductors.
- www.samsung.com/global/business/semiconductor/index.html ©1995-2009 SAMSUNG.
- www.ubicom.com/ Copyright ©2009 Ubicom, Inc.
- www.st.com/stonline/ ©2009 ST Microelectronics.
- www.ti.com/ Copyright ©1995-2009 Texas Instruments Incorporated.
- www.zilog.com/ Copyright ©2009 Zilog, Inc.
- www.cypress.com/ ©2009 Cypress Semiconductor Corporation.
- www.analog.com/en/index.html ©1995-2009 Analog Devices, Inc.
- www.toshiba.com/taa/ ©2009 Toshiba America, Inc.

Index